574.522
W61n   98984

DATE DUE			
May 16 '77			
Nov 29 79			
Nov 13 '80			
Mar 19 '81			
Apr 24 '81			
Apr 29 '81			
Dec 11 '81			

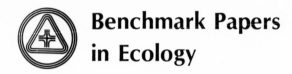

Benchmark Papers
in Ecology

Series Editor: Frank B. Golley
University of Georgia

PUBLISHED VOLUMES AND VOLUMES IN PREPARATION

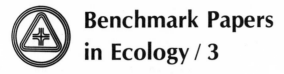

Benchmark Papers
in Ecology / 3

A BENCHMARK® Books Series

NICHE
Theory and Application

Edited by
ROBERT H. WHITTAKER
and SIMON A. LEVIN
Cornell University

Dowden, Hutchinson & Ross, Inc.

STROUDSBURG, PENNSYLVANIA

Distributed by

HALSTED
PRESS

A division of
John Wiley & Sons, Inc.

LIBRARY OF CONGRESS CATALOGING IN PUBLICATION DATA

Whittaker, Robert Harding, 1920-   comp.
   Niche ; theory and application.

   (Benchmark papers in ecology ; v. 3)
   1. Ecology--Addresses, essays, lectures. 2. Species
--Addresses, essays, lectures. 3. Biotic communities--
Addresses, essays, lectures. I. Levin, Simon A., joint
comp. II. Title.
QH541.145.W47   574.5'22   74-23328
ISBN 0-470-94117-0

Exclusive Distributor: **Halsted Press**
A Division of John Wiley & Sons, Inc.

ACKNOWLEDGMENTS AND PERMISSIONS

ACKNOWLEDGMENTS

NATIONAL ACADEMY OF SCIENCES—*Proceedings of the National Academy of Sciences USA*
  Niche Overlap and Diffuse Competition
  Niche Overlap as a Function of Environmental Variability

SIDGWICK & JACKSON LTD.—*Animal Ecology*
  Excerpt

SOCIETY FOR THE STUDY OF EVOLUTION—*Evolution*
  The Evolution of Bill Size Differences Among Sympatric Congeneric Species of Birds

SOCIETY OF SYSTEMATIC ZOOLOGY—*Systematic Zoology*
  Character Displacement

WILSON ORNITHOLOGICAL SOCIETY—*Wilson Bulletin*
  Ordinations of Habitat Relationships Among Breeding Birds

PERMISSIONS

The following papers have been reprinted with the permission of the authors and copyright holders.

AMERICAN ASSOCIATION FOR THE ADVANCEMENT OF SCIENCE—*Science*
  The Competitive Exclusion Principle
  Distributional Ecology of New Guinea Birds
  Genotype, Environment, and Population Numbers
  Population Regulation and Genetic Feedback

AMERICAN ORNITHOLOGISTS' UNION—*Auk*
  The Niche-Relationships of the California Thrasher
  The Origin and Distribution of the Chestnut-Backed Chickadee

THE BIOLOGICAL LABORATORY, LONG ISLAND BIOLOGICAL ASSOCIATION, INC.—
  Cold Spring Harbor Symposia on Quantitative Biology
  Concluding Remarks

BLACKWELL SCIENTIFIC PUBLICATIONS LTD.—*British Ecological Society Jubilee Symposium*
  Experimental Populations of Hydrida

CAMBRIDGE UNIVERSITY PRESS—*Biological Reviews*
  Gradient Analysis of Vegetation

Acknowledgments and Permissions

DUKE UNIVERSITY PRESS FOR THE ECOLOGICAL SOCIETY OF AMERICA
Ecological Monographs
Mesopsocus Populations on Larch in England—The Distribution and Dynamics of Two Closely-Related Coexisting Species of Psocoptera Sharing the Same Food Resource
The Niche Exploitation Pattern of the Blue-Gray Gnatcatcher
Ecology
The Competitive Structure of Communities; An Experimental Approach with Protozoa
Distribution on Environmental Gradients: Theory and a Preliminary Interpretation of Distributional Patterns in the Avifauna of the Cordillera Vilcabamba, Peru
On the Measurement of Niche Breadth and Overlap
Niche Breadth and Dominance of Parasitic Insects Sharing the Same Host Species
Some Aspects of the Ecology of Migrant Shorebirds

FÉDÉRATION FRANÇAISE DES SOCIÉTÉS DES SCIENCES NATURELLES—*Année Biologique*
Comparative Ecology of the Interstitial Fauna of Fresh-Water and Marine Beaches

DR. W. JUNK, NV, PUBLISHER—*Vegetatio*
Constellation of Frequent Herbage Plants, Based on Their Correlation in Occurrence

McGRAW-HILL BOOK COMPANY—*Animal Ecology*
Variations and Fluctuations of the Number of Individuals in Animal Species Living Together

PRINCETON UNIVERSITY PRESS
Evolution in Changing Environments: Some Theoretical Explorations
Excerpts
The Theory of Island Biogeography
Excerpt

SYRACUSE UNIVERSITY PRESS—*Population Biology and Evolution*
The Theory of the Niche

UNIVERSITY OF CHICAGO PRESS—*American Naturalist*
On Bird Species Diversity: II. Prediction of Bird Census from Habitat Measurements
Community Equilibria and Stability, and an Extension of the Competitive Exclusion Principle
Food Web Complexity and Species Diversity
Morphological Variation and Width of Ecological Niche
Niche, Habitat, and Ecotope
The Paradox of the Plankton

R. H. WHITTAKER—*Communities and Ecosystems*
Excerpt

JOHN WILEY & SONS, INC.—*A Treatise on Limnology: Vol. II. Introduction to Lake Biology and the Limnoplankton*
Excerpt

THE WILLIAMS AND WILKINS COMPANY—*The Struggle for Existence*
Excerpts

SERIES EDITOR'S PREFACE

Ecology—the study of interactions and relationships between living systems and their environment—is an extremely active and dynamic field of science. The great variety of possible interactions in even the simplest ecological system makes the study of ecology compelling but difficult to discuss in simple terms. Further, living systems include individual organisms, populations, communities, and ultimately the entire biosphere; there are thus numerous ecology subspecialties that focus on specific systems or interactions. Some ecologists are interested in wildlife and natural history; others are intrigued by the complex and apparently intractable problems of man and the environment. This means that a Benchmark series in ecology could be subdivided into innumerable subvolumes, each representing one of these diverse interests. However, rather than take this approach, I have tried to focus on general patterns or concepts that are applicable to two particularly important levels of ecological understanding: the population and the community. I have taken this dichotomy between these two as my major organizing concept in the series.

In a field that is rapidly changing and evolving, it is often difficult to chart the chain of ideas that lead to concepts and principles. Especially when the theoretical features of a field are relatively young, it is difficult to obtain the breadth of time to make judgments about the benchmarks of the subject. These twin problems, the interwoven ideas leading to the elucidation of theory and the youth of the subject, make difficult the development of a Benchmark series in the field of ecology. Each of the volume editors has recognized this problem, and each has acted to solve it in his or her own, unique way. Their collective efforts will, we anticipate, provide a survey of the most important concepts in the field. Thus we expect that the series will be useful not only to the student who seeks an authoritative selection of original literature but also to the professional who wants to quickly and efficiently expand his or her background in an area of ecology outside his special competence.

R. H. Whittaker and S. A. Levin have developed a thorough and comprehensive analysis of the niche for the present Benchmark volume. Niche has always been an exceedingly useful, yet difficult, concept. Older generations of ecologists were taught that the niche was the "profession" of a species, but close scrutiny of the ideas behind the idea of profession often resulted in confusion and ambiguity. Indeed, the ambiguity was such that some ecologists concluded that niche was a useless concept and should be discarded. Whittaker and Levin make clear that such a conclusion is superficial and that niche, although a complex concept, is central to an understanding of population and community organization. The two editors are well equipped to deal with these subjects. Robert Whittaker, who is Professor of Biology in the Section of Ecology and Systematics, Cornell University, is well known for his studies on a wide range of ecological topics, including community organization, gradient analysis, and primary production. Simon Levin is Chairman of the Section of Ecology and Systematics, but is also associated with the Center for Applied Mathematics and the Department of Theoretical and Applied Mechanics, Cornell University. Levin has utilized his mathematical background in the analysis of ecological problems, including community equilibria and stability.

FRANK B. GOLLEY

PREFACE

The niche has not had a long history as a scientific concept. First suggested in 1917 by Grinnell, it was identified as the species' position in a community by Elton (1927) and Gause (1934), and received from Hutchinson in 1957 the dimensional formulation with which much of this book is concerned. Since Hutchinson's article, research on niches has been expanding exponentially, and *niche* has become a central concept for the theory of natural communities and for the interpretation of species evolution.

We cannot hope, in a collection of readings, to represent the rapidly expanding research on niches adequately, but we can hope to represent the range of ideas about the niche. In our selections we have sought a balance among three categories: classic articles in which the concept developed; applied research, including some most recent work; and major theoretical formulations. We have sought also to make the book more than a collection; we should like it to be a kind of treatise on the niche that retains the flavor of various authors and differing viewpoints, but uses the general introduction and chapter commentaries to relate these into a reasonably coherent whole.

We are acutely aware of the subjectivity of our choices—the influence of selections that we have been struck by and have chosen to include in preference to articles that have at least equal claim. Our choice has been affected also—we hope not wrongly—by our view that the most essential role of the niche concept lies in the interpretation of the way species evolve and relate to one another in communities. The niche—the way a species relates to other species in the same environment and community—is thus distinguished from the species' habitat—its occurrence or distribution in a type or range of environments and communities. We believe that the clarity and usefulness of both these concepts suffer if they are not distinguished (such distinction does not mean that they are not closely related to one another). There are, however, others who prefer to use the term "niche" for species' relations to habitat, or in the sense of "ecotope," as developed in the last paper in the volume. In Part IV we seek to clarify some research approaches to both niche and habitat.

We cannot hope for universal approval of our choices, but we hope that we have done well by the reader. We have planned the book to offer (1) an introduction to and summary of niche theory and application, in the general introduction; (2) a survey of the history of the niche concept from its origin to current application; (3) a formulation of research approaches and ways of measuring niche characteristics; (4) some illustrations of research on niche relationships of different kinds of organisms and communities; and (5) a range of statements of niche theory. May the reader find in this selection interest not only in the niche and its application, but also in natural communities as living systems of interacting species, living systems for which the niche is one of our keys to understanding.

ROBERT H. WHITTAKER
SIMON A. LEVIN

CONTENTS

Contents

Contents

CONTENTS BY AUTHOR

INTRODUCTION

A familiar definition in ecology is that of the community as an assemblage of organisms living together in a particular environment. No one is very happy with this definition. The word "assemblage" in particular misses the idea of the community as a functional system of interacting populations and ignores the questions of how this system of species is organized, how the various species are interrelated and fit together into the whole, and the place of each species in the functional whole. This place or role of a species in this system is termed the *niche* and is the focus of this Benchmark volume. Our goal is to understand how niches are conceived, defined, and measured; how the niches of different species relate to one another; and how they are to be placed in an evolutionary context.

It is characteristic of this area of study that it links observation with theory by way of mathematics. The reader will find here a wide range of material of differing difficulty, from straightforward descriptive natural history to challenging theory. Consequently, the reader may not want simply to read, but rather to read *at*, the book—sampling the different selections according to their interest and accessibility. We have tried to increase the accessibility of some of the selections by our introductions to them, and our own views emerge in these introductions and the final article. There is little to be said in this general introduction that is not said somewhere else in the book. For such help as it may give a reader new to the area, however, we offer a kind of precis: an effort to state together some of the most essential ideas before they go their separate ways in later chapters.

The *niche* is the way a species population fits into a given com-

1

munity. A niche is therefore part of the whole set of relationships of the species to the environment. We can recognize three major aspects, or ways of approaching, these relationships. The niche concept focuses on ways the species relates to other species in the same community. Two bird species feeding on insects in the same forest occupy different niches if one feeds in the canopy and the other in the undergrowth, and in this way avoid using the same food resources. A hawk and an owl that feed on some of the same small mammals differ in niche along a time axis, because the hawk feeds by day and the owl by night. Two herbs of the forest floor differ in niche if one grows in spots where more light reaches the forest floor, and the other nearby in denser shade. Two insects in a forest differ in niche if one feeds on beech and the other maple, even though both insects are eating leaves in the same way at the same time. Niches thus differ according to different kinds of variables—among them place and time of activity and way of relating to other species—within a community.

The *habitat* of a species is the kind or range of environments in which it lives. Of two species of sparrows in a given area, one breeds in wet marshes and the other in dry fields. They differ in habitat along a gradient of soil moisture (and the kinds of communities that develop in response to soil moisture). A third sparrow differs in habitat from both; for it occurs in meadows at higher elevations in nearby mountains. The habitats of species are thus distinguished by intercommunity variables— differences such as those of topographic position, elevation, soil moisture and fertility, kind of rock, fertility or salinity of lake water, rock versus sand or mud bottom in a stream, depth of the ocean floor, and differences in kinds of communities. These variables we can term *extensive*, for they change from place to place on the earth's surface, and in some cases form well-defined spatial gradients (e.g., elevation or tidal gradients). Niche variables, in contrast, are *intensive* in the sense that they change within a given community and at a particular place on the earth's surface. Niche variables change with vertical height, or time, or as part of a small-scale and self-repeating horizontal pattern within the community, or as differences in kind and intensity of biological interactions within the community.

The third aspect of the species' environmental relationships is its *area*, its geographic range as this may be plotted on a map.

The environmental relationships of a species form a complex whole, and there is often no sharp distinction between niche and habitat, or between habitat and area. A given environmental factor can be involved in all three. A forest-floor snail, say, is sensitive to relative humidity of the air. Because of this, it becomes active only at night when the humidity is high; this time relation is part of its niche in a given forest. Also, because of this sensitivity, it occurs only in moist forests of valleys and lower slopes in a given area; humidity is thus restricting its. Furthermore, it is limited in area by humidity, for it occurs in t. valley forests only in the area of more humid climates in and near the southern Appalachian Mountains. The fact that niche, habitat, and area are not discontinuous with one another by no means reduces the importance of these as concepts. They differ in focus, in the kinds of questions to be asked, and in the kinds of evolutionary relationships to be emphasized. They are, to some extent, the concerns of different fields of research.

Areas are part of the concern of biogeography; study of species areas and other geographic relationships of organisms is sometimes termed *chorology*. Relationships to climate and geographic barriers, and long-term evolutionary histories of biotas and their past movements with changing climates, are among the bases of intepreting species areas. Much research deals with the kinds of habitats in which communities and species occur. The research approach termed "gradient analysis" (Part IV) studies the ways in which species populations are distributed in relation to one another and kinds of communities along habitat gradients. Physiological ecology and genecology seek the physiological, and the genetic, bases of species' habitat relationships. Studies of niche and of species diversity, finally, are part of an area of ecology that has no special name. This book, and a following Benchmark book, *Diversity* by Ruth Patrick, are complementary to one another in representing this area of ecology.

Four names are preeminent in the history of the niche concept—Grinnell, Elton, Gause, and Hutchinson. Selections from the writings of the first three are given in Part I. The linkage of the niche concept with the study of species diversity was recognized by Hutchinson, in an article given in our concluding section. This linkage involves ideas on the role of competition in communities. Simply stated: if two species are limited by the availability of the same single resource at the same time and place in the same stable community, one of these species will have the advantage and the other will become extinct. Experiments and mathematical reasoning (see Part II) are in accord in their support of this principle. This observation is variously termed the principle of competitive exclusion, or of Gause, or of Volterra and Gause, and it is the subject of our second part. It has two widely significant implications: (1) Many species survive together in a community because they differ in the kind of resource they use, or in place or time of activity in the community, or in their way of relating to other species, or in the manner in which their populations are controlled: they differ in niche. (2) Niche difference, because it makes possible the coexistence of species in communities, is the basis of the evolution of species diversity.

We have mentioned so far only a few of the ways species' niches may differ. Part III deals with some of the range of kinds of niche differences, or directions of niche differentiation among species. For birds, niche differences in vertical height of feeding, manner of food search, and size and kind of food are evident, and selections by MacArthur et al., Root, and Diamond illustrate some of these. For a group of species of closely related niches, Root (Paper 10) develops the concept of "guild."

Two species that differ in interaction with other species, particularly in the kinds of mechanisms that control their populations, are considered to differ in niche even though they may use the same resource. Paper 8 by Levin in Part II develops the idea of population controls as the key dimension of niches, and Paper 12 by Paine in Part III deals with the role of predation in defining niches and determining species diversity. Niche difference makes possible coexistence of species in stable communities. Many communities are unstable, however, and in these the adaptation to environmental instability becomes part of the species' niche, and part of the basis of species diversity as discussed by Hutchinson in Paper 18, "The Paradox of the Plankton."

If some niche relationships can be treated as axes or gradients, we can characterize niches by population responses along these axes. The study of population responses to habitat gradients may suggest some of the kinds of questions to ask; for research on habitat responses is easier, and has in some respects advanced further, than that on niche responses. As the first section of Part IV shows, we can ask (concerning species along a habitat gradient) the forms and the widths of their population distributions, the degree to which their populations overlap along the gradient, and the manner in which several species are spaced or arranged along the gradient. The same kinds of questions are asked about species responses to niche gradients in the second section of Part IV.

Next, the same questions can be asked about species relations to more than one axis. We might consider, for example, that the plant communities of an area are related by four habitat axes—elevation, topographic moisture, soil acidity, and intensity of disturbance by man. These four axes can be used as coordinates for an abstract, conceptual space, a habitat hyperspace. Each plant species has its limits of tolerance along these axes, and these limits confine a given species to some fraction of the hyperspace; the limits bound a hypervolume. One species may occur only in valley bottoms at elevations of 300–600 m on soils of pH 6.5 to 7.0 in communities undisturbed by man; another species may occur on dry to very dry slopes at elevations of 200–1000 m on soil of pH 4.0 to 6.0 but only after clearcutting of the forest by man. Mostly, however, the boundaries of species populations along habitat gradients are not sharp. The population has instead a center or optimum, from which its abundance tapers in all directions along gradients with increasing departure from that center. It is difficult to visualize such distributions in several dimensions in a hyperspace, but we can at least represent a population response to habitat gradients in two dimensions (Part VII, Paper 41, Fig. 2). For such distributions we can again ask: their form, breadth, extent of overlap with one another, and arrangement in the hyperspace. More broadly: how do the habitat relations of species evolve?

We may also consider several intracommunity variables as defining an abstract space; this is then a niche hyperspace. Along each variable a given species may have a range of values that it can tolerate, within which it is potentially able to occur. The limits of these ranges, for the several variables, bound a hypervolume, a fraction of the niche hyperspace; and this hypervolume is the "fundamental niche" of Hutchinson's formulation in Paper 39. Some part of the hypervolume in which a species potentially can occur may be preempted by a competing species. The more limited hypervolume in which the species actually occurs is then, in Hutchinson's formulation, its "realized niche." Competitors are normally part of the species' environment, and so most research concerns realized, rather than fundamental, niches. The niche hypervolume of our own article with our colleague R. B. Root, in Paper 41, is closer to Hutchinson's realized niche. We seek, however, to describe niches in terms of population measures, not just boundaries. The result is the niche response surface, as illustrated for two niche gradients from Root's work (Part VII, Paper 41, Fig. 4), and the conception of the species' niche, defined by a population measure, as a cloud-like response in niche hyperspace. It is easier to consider measurements applied to a two-dimensional response surface such as Root's. For such surfaces we can clearly ask the same questions of form, breadth, overlap, and arrangement as for habitat distributions.

Some interesting ideas have come from such approaches through measurements. First, what can be said about a set of species that occur in sequence along a particular niche (or habitat) gradient? Species population responses apparently form bell-shaped curves, as illustrated in Part IV, Paper 21, Fig. 8, and Part VII, Paper 41, Fig. 1. There is apparently a tendency, for species that are relatively close competitors, for the centers of these curves to be evenly spaced along the niche or habitat axis (see also Paper 26). As Hutchinson (1959) observed, when such a sequence relates species of different sizes, the linear size ratios of successive species in the sequence are rather consistently between 1.1 and 1.4, averaging around 1.26. A length ratio of about 1.26 for insects, or the bills of birds feeding on them, implies a mass ratio of about 2.0 for the insects, or the birds; it implies approximate doubling in mass from one species to another in such a sequence. Schoener in Paper 32, Root in Paper 10, Diamond in Paper 11, and Price in Paper 30 consider these ratios. May and MacArthur, in Paper 26, also offer a theoretical basis for a limit on the extent of overlap of species. In a sequence of species along a niche or habitat gradient, they predict the distance between species' centers to be about the same as the width of the species distributions, expressing the latter as standard deviations.

Ways of measuring niche and habitat width or breadth, and overlap, are considered by a number of the articles in Part IV. May and Mac-Arthur's hypothesis bears also on arrangement, for it suggests evolution toward relatively even spacing, with generally consistent widths and overlaps, for a set of species along a given axis. One asks next about arrangement of species in a multidimensional niche or habitat hyperspace. The techniques by which species, or community samples, can be arranged by relative positions along one or more habitat gradients (or axes of community difference) are termed "ordination." Ordinations of species along habitat gradients, or in habitat hyperspaces, are an essential basis of the research approach called "gradient analysis." We can with good reason speak also of niche ordination—arrangement of species by relative positions along niche axes, or in a niche hyperspace.

Species have often been arranged in relation to particular niche axes; the articles by Root (Paper 10), Diamond (Paper 11), and Recher (Paper 38) contain such one-dimensional ordinations, as does a paper by Price (1972) relating to his article in Part IV. Ordinations in relation to a niche space of more than one axis are difficult, few, and limited in effectiveness (see Paper 9, Fig. 4, and Paper 21, Fig. 2.7). The limited results are consistent with the expectation that species evolve toward difference in niche, so that no two species in the community have the same niche, and the centers of species niches are scattered in niche hyperspace. Habitat ordinations of species are easier, and there are numerous examples in an extensive literature. The results indicate that the species occurring together in a given community differ in their habitat distributions through other communities. Species appear to be scattered, no two alike, when they are ordinated in a habitat hyperspace. This state of affairs will be recognized as to some extent analogous (for habitats) to the principle of competitive exclusion (for niches). The observation that no two species are alike in their habitats was stated, quite independently, as the "principle of species individuality" of Ramensky (1924) and Gleason (1926). Species areas are not part of the concern of this book. We mention, however, that when the areas of the species that occur together in a community are mapped, these areas are different—diverse in their shapes and extents in different directions. As regards congeners, there is Jordan's rule: The species most closely related to a given species is likely to be found, not in the same district, but in a different, and often adjacent, area. Congeners being likely to be close competitors may thus differ in area, as other potential competitors may differ in habitat, or in niche.

Some other ideas on evolution and change of niche in time and space are treated in Part V. Suppose, despite Jordan's rule, that two congeners which are fairly close competitors do, in fact, overlap in an area and occur together in some of the same communities. It may then be found that these species are alike in some of their niche characteristics where they occur separately but diverge from one another (in bill size of birds, or some other indication of niche relationships) where they occur together. This phenomenon of "character displacement" is discussed by Brown and Wilson (Paper 31). Suppose a species that occurs on the continent, in communities rich in other species occupying similar niches, occurs also on an island in communities including fewer species. Because there are fewer competitors on the island, the species may have a broader niche (and also a broader habitat) on the island. This "character release" on islands is discussed by Van Valen (Paper 33).

Part VI gives four "case history" articles, examining niche relationships in particular communities. Such relationships can be most effectively studied for guilds, because studies of limited groups of species of closely related niches can identify the niche variables by which they are related and along one or more of which they differ. A guild can thus be interpreted in terms of species relations in a best-known fraction of the whole (and exceedingly complex) niche hyperspace of the community. The manner in which a set of apparent competitors—five species of warblers, all feeding on insects in a spruce forest—differ when the details of their niche relationships are examined is described by MacArthur (Paper 36). Broadhead and Wapshere (Paper 37) consider a guild of six bark lice and then examine in detail two apparently closely competing species, to find the critical difference that may permit their coexistence. Recher (Paper 38) deals with food relations and other adaptations of shorebirds in an estuary, and shows how they differ in both niche and habitat use. They differ, then, in *ecotope*— in overall environmental relations considering both niche and habitat relations (see Paper 41). Of the concluding section we mention only that it gives three essays on niche interpretation in general, by Hutchinson, by MacArthur, and by Whittaker, Levin, and Root.

There is, perhaps, a general perspective to be drawn from this introduction, this being the evolutionary meaning of the diversity of nature. Species evolve toward difference from one another. Evolution in directions that reduce competition is a major basis of this divergence;

7

but other aspects of evolution, including the accumulation of chance differences, are involved. Coexistence within a single community, without supportive migration from adjacent communities, is by difference in niches, and one aspect of community organization is implied by this: the community is a system of niche-differentiated, interacting species. Communities can accumulate species through evolutionary time by increasing complexity and subdivision of the niche hyperspace relating these species. Species that are too closely competitive to occur in the same community can evolve toward difference in habitat. Species that may be competitive in both niche and habitat requirements can survive in different areas. Because of difference in history and climate, and because of geographic barriers, different regions accumulate different biotas, or sets of species that differ in evolutionary history but in each region are differentiated in niche and habitat relations. The overall result is the existence of innumerable communities in diverse habitats in different regions, comprising all together a few millions of species that can survive because they differ in area or habitat, or because in particular communities they have evolved those differences in relationship to one another that we term *niche*.

REFERENCES

Gleason, H. A. 1926. The individualistic concept of the plant association. *Bull. Torrey Bot. Club* **53**: 7-26.

Hutchinson, G. E. 1959. Homage to Santa Rosalia, or why are there so many kinds of animals? *Amer. Naturalist* **93**: 145-159.

Price, P. W. 1972. Parasitoids utilizing the same host: adaptive nature of differences in size and form. *Ecology* **53**: 190-195.

Ramensky, L. G. 1924. Die Grundegesetsmässigkeiten im Aufbau der Vegetationsdecke (in Russian). *Vêstnik Opỹtnogo Dêla, Voronezh*, pp. 37-73. [Abstract in *Bot. Centralblatt* (n.f.) **7**: 453-455, 1926.]

Part I

ORIGIN OF THE NICHE CONCEPT

Editors' Comments
on Papers 1 Through 4

Joseph Grinnell was a prominent early student of vertebrate biology and ecology in California. From broad experience with birds and mammals in the complex landscape of that state, he sought understanding of species distributions. A result of that effort for one bird species is Paper 1 on the California thrasher, seeking to explain its distribution by close physiological and behavioral adjustment to critical environmental conditions, conditions that are to be recognized by examination of the thrasher's habitat (p. 428). In its dependence on cover for shelter and nesting, it is observed that the thrasher's limitation to, and position in, the chaparral seems defined by the shrub cover that only the chaparral offers among the plant communities in the area (p. 432). In its dependence on cover the thrasher occupies "one of the minor niches which with their occupants all together make up the chaparral association (p. 433)." This implies the nucleus of the niche concept: the species' requirements and position in relation to other species in a given community. But Grinnell did not really make clear the relation between habitat and niche, and his idea of niche was probably not so much the species' functional position in a community as those attributes of habitat that govern the species' distribution. We offer as first reading, however, the article in which he first used the term "niche" and the concept of niche began to take form.

In a most significant idea in the next-to-last sentence in this article, ". . . no two species regularly established in a single fauna have precisely the same niche relationships," Grinnell anticipates the principle of Gause, or of competitive exclusion, and we give a short passage from an earlier work in Paper 2, making clear his recognition of this principle.

Charles Elton's book *Animal Ecology* (1927) is one of the classics of ecology; in a small book he packed a wealth of observation and interpretations that have stimulated later students. The change of perspective from Grinnell's work was profound. Grinnell's interests were faunistic and autecological; he was most concerned with the distributions of species. Elton, in contrast, took the idea of community as the focus of interest and sought to characterize animal communities as distinctive systems with their own kinds of structure. It is in the context of the community that the concept of niche takes its effective meaning, and Elton uses the term *niche* "to describe the status of an animal in its community, to indicate what it is *doing* and not merely what it looks like (p. 63)." He observes animals of similar niche functions in different communities—sap-suckers, mice and other grazers, tick-feeders, earthworms, etc. We would not say now that "the same" mouse niche existed in different communities. We might say instead that there is often remarkable evolutionary convergence of species toward niche similarity, so that in different communities different species use similar resources and behave in similar ways. But in viewing the niche concept this way, Elton was initiating a most important modern function of the niche concept—as a basis for interpreting the organization of communities. Paper 3 gives Elton's statement of the niche concept, which in his book is part of his development of community organization, including also the ideas of food chains and sizes and the pyramid of numbers.

Paper 4, is a selection from another classic of ecology, also a small book embodying much observation and thought, by the Russian ecologist G. F. Gause. Gause's book differs from Elton's in many ways but overlaps with it in the way that is essential to our concern—the concept of niche. In the selected passage Gause adopts Elton's concept of niche and applies it to a group of terns that nest together but avoid competition by their different manners of feeding. Gause went beyond such observations to the experimental study of competition and recognition of its implication for the principle of competitive exclusion. That principle is treated in Part II, in a further selection from Gause.

Gaffney (1975) has pointed out that the earliest usage of the word "niche" in an ecological sense seems to have been by Johnson (1910). Johnson, however, did not develop niche as a concept.

REFERENCES

Gaffney, P. M. 1975. Roots of the niche concept. *Amer. Naturalist* **109**: 490.
Johnson, R. H. 1910. Determinate evolution in the color pattern of the ladybeetles. *Carnegie Inst. Washington Publ. 122.* 104 pp.

1

Reprinted from *Auk,* **34,** 427–433 (Oct. 1917)

THE NICHE-RELATIONSHIPS OF THE CALIFORNIA THRASHER.[1]

BY JOSEPH GRINNELL.

THE California Thrasher (*Toxostoma redivivum*) is one of the several distinct bird types which characterize the so-called "Californian Fauna." Its range is notably restricted, even more so than that of the Wren-Tit. Only at the south does the California Thrasher occur beyond the limits of the state of California, and in that direction only as far as the San Pedro Martir Mountains and

---

[1] Contribution from the Museum of Vertebrate Zoölogy of the University of California.

San Quintin, not more than one hundred and sixty miles below the Mexican line in Lower California.

An explanation of this restricted distribution is probably to be found in the close adjustment of the bird in various physiological and psychological respects to a narrow range of environmental conditions. The nature of these critical conditions is to be learned through an examination of the bird's habitat. It is desirable to make such examination at as many points in the general range of the species as possible with the object of determining the elements common to all these points, and of these the ones not in evidence beyond the limits of the bird's range. The following statements in this regard are summarized from the writer's personal experience combined with all the pertinent information afforded in literature.

The distribution of the California Thrasher as regards life-zone is unmistakable. Both as observed locally and over its entire range the species shows close adherence to the Upper Sonoran division of the Austral zone. Especially upwards, is it always sharply defined. For example, in approaching the sea-coast north of San Francisco Bay, in Sonoma County, where the vegetation is prevailingly Transition, thrashers are found only in the Sonoran "islands," namely southerly-facing hill slopes, where the maximum insolation manifests its effects in a distinctive chaparral containing such lower zone plants as Adenostoma. Again, around Monterey, to find thrashers one must seek the warm hill-slopes back from the coastal belt of conifers. Everywhere I have been, the thrashers seem to be very particular not to venture even a few rods into Transition, whether the latter consist of conifers or of high-zone species of manzanita and deer brush, though the latter growth resembles closely in density and general appearance the Upper Sonoran chaparral adjacent.

While sharply delimited, as an invariable rule, at the upper edge of Upper Sonoran, the California Thrasher is not so closely restricted at the *lower* edge of this zone. Locally, individuals occur, and numbers may do so where associational factors favor, down well into Lower Sonoran. Instances of this are particularly numerous in the San Diegan district; for example, in the Lower Sonoran "washes" at the mouths of the canyons along the south base of the San Gabriel Mountains, as near San Fernando, Pasadena, and

Azusa.   A noticeable thing in this connection, however, is that, on the desert slopes of the mountains, where *Toxostoma lecontei* occurs on the desert floor as an associational homologue of *T. redivivum* in the Lower Sonoran zone, the latter "stays put" far

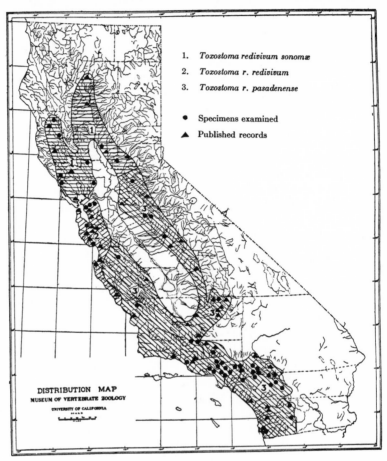

1.   *Toxostoma redivivum sonomæ*

2.   *Toxostoma r. redivivum*

3.   *Toxostoma r. pasadenense*

● Specimens examined

▲ Published records

DISTRIBUTION MAP
MUSEUM OF VERTEBRATE ZOOLOGY
UNIVERSITY OF CALIFORNIA

Figure 1.

more closely; that is, it strays but little or not at all below the typical confines of its own zone, namely *Upper* Sonoran.   The writer's field work in the vicinity of Walker Pass, Kern County,

2

Reprinted from *Auk,* 21, 375, 377 (July 1904)

THE ORIGIN AND DISTRIBUTION OF THE CHESTNUT-BACKED CHICKADEE

Joseph Grinnell

[*Editors' Note:* In the original, material precedes this excerpt.]

As to the distance to which a species may invade, we can surmise that, topography permitting, theoretically there is no limit so long as adaptive modifications continually take place. The geographic variation in *Melospiza* may be called to attention as an extreme illustration. But practically, in the case of *Parus rufescens barlowi*, much further invasion is improbable, because in adjoining areas are already firmly established members of the same family (*Bæolophus, Psaltriparus, Chamæa*) thoroughly adapted to prevailing food conditions. No one of these could probably be successfully competed against by a foreigner. Every animal tends to increase at a geometric ratio, and is checked only by limit of food supply. It is only by adaptations to different sorts of food, or modes of food getting, that more than one species can occupy the same locality. Two species of approximately the same food habits are not likely to remain long evenly balanced in numbers in the same region. One will crowd out the other; the one longest exposed to local conditions, and hence best fitted, though ever so slightly, will survive, to the exclusion of any less favored would-be invader. However, should some new contingency arise, placing the native species at a disadvantage, such as the introduction of new plants, then there might be a fair chance for a neighboring species to gain a foothold, even ultimately crowding out the native form. For example, several pairs of the Santa Cruz Chickadee have taken up their permanent abode in the coniferous portion of the Arboretum at Stanford University, while the Plain Titmouse prevails in the live oaks of the surrounding valley.

[*Editors' Note:* Material has been omitted at this point.]

3

Reprinted from *Animal Ecology*, Sidgwick & Jackson, London, 1927, pp. 63–68

ANIMAL ECOLOGY

Charles Elton

[*Editors' Note:* In the original, material precedes this excerpt.]

Niches

15. It should be pretty clear by now that although the actual species of animals are different in different habitats, the ground plan of every animal community is much the same. In every community we should find herbivorous and carnivorous and scavenging animals. We can go further than this, however : in every kind of wood in England we should find some species of aphid, preyed upon by some species of ladybird. Many of the latter live exclusively on aphids. That is why they make such good controllers of aphid plagues in orchards. When they have eaten all the pest insects they just die of starvation, instead of turning their attention to some other species of animal, as so many carnivores do under similar circumstances. There are many animals which have equally well-defined food habits. A fox carries on the very definite business of killing and eating rabbits and mice and some kinds of birds. The beetles of the genus *Stenus* pursue and catch springtails (*Collembola*) by means of their extensile tongues. Lions feed on large ungulates—in many places almost entirely zebras. Instances could be multiplied indefinitely. It is therefore convenient to have some term to describe the status of an animal in its community, to indicate what it is *doing* and not merely what it looks like, and the term used is " niche." Animals have all manner of external factors acting upon them—

chemical, physical, and biotic—and the " niche " of an animal means its place in the biotic environment, *its relations to food and enemies*. The ecologist should cultivate the habit of looking at animals from this point of view as well as from the ordinary standpoints of appearance, names, affinities, and past history. When an ecologist says " there goes a badger " he should include in his thoughts some definite idea of the animal's place in the community to which it belongs, just as if he had said " there goes the vicar."

16. The niche of an animal can be defined to a large extent by its size and food habits. We have already referred to the various key-industry animals which exist, and we have used the term to denote herbivorous animals which are sufficiently numerous to support a series of carnivores. There is in every typical community a series of herbivores ranging from small ones (*e.g.* aphids) to large ones (*e.g.* deer). Within the herbivores of any one size there may be further differentiation according to food habits. Special niches are more easily distinguished among carnivores, and some instances have already been given.

The importance of studying niches is partly that it enables us to see how very different animal communities may resemble each other in the essentials of organisation. For instance, there is the niche which is filled by birds of prey which eat small mammals such as shrews and mice. In an oak wood this niche is filled by tawny owls, while in the open grassland it is occupied by kestrels. The existence of this carnivore niche is dependent on the further fact that mice form a definite herbivore niche in many different associations, although the actual species of mice may be quite different. Or we might take as a niche all the carnivores which prey upon small mammals, and distinguish them from those which prey upon insects. When we do this it is immediately seen that the niches about which we have been speaking are only smaller subdivisions of the old conceptions of carnivore, herbivore, insectivore, etc., and that we are only attempting to give more accurate and detailed definitions of the food habits of animals.

17. There is often an extraordinarily close parallelism

between niches in widely separated communities. In the arctic regions we find the arctic fox which, among other things, subsists upon the eggs of guillemots, while in winter it relies partly on the remains of seals killed by polar bears. Turning to tropical Africa, we find that the spotted hyæna destroys large numbers of ostrich eggs, and also lives largely upon the remains of zebras killed by lions.[12a] The arctic fox and the hyæna thus occupy the same two niches—the former seasonally, and the latter all the time. Another instance is the similarity between the sand-martins, which one may see in early summer in a place like the Thames valley, hawking for insects over the river, and the bee-eaters in the upper part of the White Nile, which have precisely similar habits. Both have the same rather distinct food habits, and both, in addition, make their nests in the sides of sand cliffs forming the edge of the river valleys in which they live. (Abel Chapman [85c] says of the bee-eaters that " the whole cliff-face appeared aflame with the masses of these encarmined creatures.") These examples illustrate the tendency which exists for animals in widely separated parts of the world to drift into similar occupations, and it is seen also that it is convenient sometimes to include other factors than food alone when describing the niche of any animal. Of course, a great many animals do not have simple food habits and do not confine themselves religiously to one kind of food. But in even these animals there is usually some regular rhythm in their food habits, or some regularity in their diverse foods. As can be said of every other problem connected with animal communities, very little deliberate work has been done on the subject, although much information can be found in a scattered form, and only awaits careful coordination in order to yield a rich crop of ideas. The various books and journals of ornithology and entomology are like a row of beehives containing an immense amount of valuable honey, which has been stored up in separate cells by the bees that made it. The advantage, and at the same time the difficulty, of ecological work is that it attempts to provide conceptions which can link up into some complete scheme the colossal store of facts about natural history which has accumulated up to date in this rather haphazard manner. This applies with particular force to facts about the food habits of animals. Until more organised information about the subject is available, it is only possible to give a few instances of some of the more clear-cut niches which happen to have been worked out.

18. One of the biggest niches is that occupied by small sap-suckers, of which one of the biggest groups is that of the plant-lice or aphids. The animals preying upon aphids form a rather distinct niche also. Of these the most important are the coccinellid beetles known as ladybirds, together with the larvæ of syrphid flies (cf. Fig. 5) and of lacewings. The niche

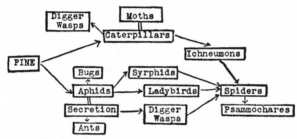

FIG. 5.—Food-cycle on young pine-trees on Oxshott Common.
(From Richards.[18])

in the sea and in fresh water which is analogous to that of aphids on land is filled by copepods, which are mainly diatomeaters. This niche occurs all over the world, and has a number of well-defined carnivore niches associated with it. If we take a group of animals like the herbivorous grass-eating mammals, we find that they can be divided into smaller niches according to the size of the animals. There is the mouse niche, filled by various species in different parts of the world; the rabbit niche, of larger size, filled by rabbits and hares in the palæarctic region and in North America, by the agouti and viscacha in South America, by wallabies in Australia, and by animals like the hyrax, the springbuck, and the mouse deer [56] in Africa. In the same way it can be shown that there is a special niche of carnivorous snakes which prey upon other snakes—a niche which is filled by different species in different countries. In

South America there is the mussarama, a large snake four or five feet in length, which is not itself poisonous, but preys exclusively upon other snakes, many of which are poisonous, being itself immune to the venoms of lachesis and rattlesnake, but not to colubrine poisons. In the United States the niche is filled by the king-snake which has similar habits, while in India there is a snake called the hamadryad which preys upon other (in this case non-poisonous) snakes.[86a]

19. Another widespread niche among animals is that occupied by species which pick ticks off other animals. For instance, the African tick-bird feeds entirely upon the ticks which live upon the skin of ungulates, and is so closely dependent upon its mammalian " host " that it makes its nest of the latter's hair (*e.g.* of the hartebeest).[12f] In England, starlings can often be seen performing the same office for sheep and deer. A similar niche is occupied on the Galapagos Islands by a species of scarlet land-crab, which has been observed picking ticks off the skin of the great aquatic lizards (*Amblyrhynchus*).[36c] Another niche, rather analogous to the last one, is that occupied by various species of birds, which follow herds of large mammals in order to catch the insects which are disturbed by the feet of the animals. Chapman [85d] saw elephants in the Sudan being followed by kites and grey herons ; Percival [12g] says that the buff-backed egret follows elephants and buffalo in Kenya for the same purpose ; in Paraguay [86c] there are the Aru blackbirds which feed upon insects disturbed by the feet of cattle ; while in England wagtails attend cattle and sheep in the same way.

20. There is a definite niche which is usually filled by earthworms in the soil, the species of worm differing in different parts of the world. But on coral islands their place may be largely taken by land-crabs. Wood-Jones [107a] states that on Cocos-Keeling Island, coconut husks are one of the most important sources of humus in the soil, and in the rotting husks land-crabs (chiefly of the genus *Cardiosoma*) make burrows and do the same work that earthworms do in our own country. (There are as a matter of fact earthworms as well on these islands.) On the coral reefs which cover such a large

part of the coast in tropical regions, there is a definite niche filled by animals which browse upon the corals, just as herbivorous mammals browse upon vegetation on land. There are enormous numbers of holothurians or sea-cucumbers which feed entirely in this way. Darwin [30] gives a very good description of this niche. Speaking also of Cocos-Keeling Island, he says :

" The number of species of Holothuria, and of the individuals which swarm on every part of these coral-reefs, is extraordinarily great ; and many ship-loads are annually freighted, as is well known, for China with the trepang, which is a species of this genus. The amount of coral yearly consumed, and ground down into the finest mud, by these several creatures, and probably by many other kinds, must be immense. These facts are, however, of more importance in another point of view, as showing us that there are living checks to the growth of coral-reefs, and that the almost universal law of ' consume and be consumed,' holds good even with the polypifers forming those massive bulwarks, which are able to withstand the force of the open ocean."

This passage, besides showing that the coral-eating niche has a geological significance, illustrates the wide grasp of ecological principles possessed by Darwin, a fact which continually strikes the reader of his works. We have now said enough to show what is meant by an ecological niche, and how the study of these niches helps us to see the fundamental similarity between many animal communities which may appear very different superficially. The niche of an animal may to some extent be defined by its numbers. This leads us on to the last subject of this chapter.

[*Editor's Note:* Material has been omitted at this point.]

REFERENCES

12a. Percival, A. B. 1924. A Game Ranger's Note-Book. London. pp. 158, 160.

12f. Percival, A. B. 1924. A Game Ranger's Note-Book. London. p. 344.

30. Darwin, C. 1874. On the Structure and Distribution of Coral Reefs, 2nd ed. Londen.

36c. Beebe, W. 1924. Galapagos: World's End. p. 122.

56. Christy, C. 1924. Big Game and Pygmies. London. p. 231.

85c. Chapman, A. 1921. Savage Sudan. London. p. 284.

85d. Chapman, A. 1921. Savage Sudan. London. p. 245.

References

86a. Roosevelt, T. 1914. Through the Brazilian Wilderness. London. pp. 16-18.

86c. Roosevelt, T. 1914. Through the Brazilian Wilderness. London. p. 88.

Reprinted from G. F. Gause, *The Struggle for Existence*, Williams & Wilkins, Baltimore, 1934, pp. 19–20 (reprinted by Hafner Publ. Co., New York, 1964)

THE STRUGGLE FOR EXISTENCE

G. F. Gause

[*Editors' Note:* In the original, material precedes this excerpt.]

One of these ideas is that of the "niche" (see Elton, '27, p. 63). A niche indicates what place the given species occupies in a community, i.e., what are its habits, food and mode of life. It is admitted that as a result of competition two similar species scarcely ever occupy similar niches, but displace each other in such a manner that each takes possession of certain peculiar kinds of food and modes of life in which it has an advantage over its competitor. Curious examples of the existence of different niches in nearly related species have recently been obtained by A. N. Formosov ('34). He investigated the ecology of nearly related species of terns, living together in a definite region, and it appeared that their interests do not clash at all, as each species hunts in perfectly determined conditions differing from those of another. This once more confirms the thought mentioned earlier, that the intensity of competition is determined not by the systematic likeness, but by the similarity of the demands of the competitors upon the environment. Further on we shall endeavor to express all these relations in a quantitative form.

(7) The above mentioned observations of A. N. Formosov on different niches in nearly related species of terns can be given here with more detail, as the author has kindly put at our disposal the following materials from his unpublished manuscript: According to

the observations in 1923, the island Jorilgatch (Black Sea) is inhabited by a nesting colony of terns, consisting of many hundreds of individuals. The nests of the terns are situated close to one another, and the colony presents a whole system. The entire mass of individuals in the colony belongs to four species (sandwich-tern, *Sterna cantiaca*; common-tern, *S. fluviatilis*; blackbeak-tern, *S. anglica*; and little-tern, *S. minuta*), and together they chase away predators (hen-harriers, etc.) from the colony. However, as regards the procuring of food, there is a sharp difference between them, for every species pursues a definite kind of animal in perfectly definite conditions. Thus the sandwich-tern flies out into the open sea to hunt certain species of fish. The blackbeak-tern feeds exclusively on land, and it can be met in the steppe at a great distance from the sea-shore, where it destroys locusts and lizards. The common-tern and the little-tern catch fish not far from the shore, sighting them while flying and then falling upon the water and plunging to a small depth. The light little-tern seizes the fish in shallow swampy places, whereas the common-tern hunts somewhat further from the shore. In this manner these four similar species of tern living side by side upon a single small island differ sharply in all their modes of feeding and procuring food.

[*Editors' Note:* Material has been omitted at this point.]

LITERATURE CITED

Elton, C. 1927. See Paper 3 this volume.
Formosov, A. N. 1934. Cited from manuscript.

Part II
THE COMPETITIVE EXCLUSION PRINCIPLE

Editors' Comments
on Papers 5 Through 8

The notions of niche and (competitive) exclusion are inseparably bound up with one another. Niche, as the complete functional role within a particular community, is strongly identified in ecological tradition with a particular species, its evolutionary *raison d'être*. Clearly, it would not then do for two species to "occupy" the same niche, for the characterization must be one to one, the species as unique to the niche as the niche is to the species.

However, that two species cannot occupy the same niche does not follow as a matter of logical consistency or via Hardin's "Axiom of Inequality" (Paper 5). The argument that it is "infinitely improbable" that two species occupy the same niche is not valid, for it views the niche of the species as a matter of chance. When the niche is understood as the end product of an evolutionary search for an optimal way of life, it is not unreasonable to expect that several species might be attracted toward the identical strategy. Moreover, there is good reason to expect such trends to convergent evolution when the several species do not confront one another in the same community. However, as will become clear in the pages to come, the combined weight of empirical and theoretical evidence does justify the unique identification of niche with species within a particular community. This is not as a matter of chance, but as the result of competitive exclusion, the dynamic process of natural selection at the species level. More precisely, when two or more species attempt to fill the same role in the same community, no stable balance can be struck between them. The principle does not rule out the possibility that in some situations, many species will be filling nearly identical niches with a highly unstable balance, as perhaps exists for the plankton (Paper 18).

The genesis of the present form of the competitive exclusion principle is often placed with Volterra (1926) and Gause (Paper 7), but it clearly dates back to at least Grinnell (Paper 2). Hutchinson finds elements of the principle in the work of Steere (1894); and as Crombie (1947) and others point out, its roots are surely with Darwin. Other discussions may be found in Cole (1960) and DeBach (1966). Hardin's summary article (Paper 5) provides a very clear introduction.

Following the lead of the naturalists, in particular Elton, a second line of investigation developed between 1926 and 1936, mathematical and experimental, designed to make precise the early qualitative notions. The fundamental mathematical treatment was that of Volterra (Paper 6), here reproduced in the translation by Mary Evelyn Wells reprinted by Chapman. Immediately following the appearance of Volterra's opus, the brilliant young Russian scientist Gause set out to give the theory experimental teeth. One critical test is summarized in Paper 7. Numerous field and laboratory examples of competitive displacement followed Gause's, and a comprehensive summary may be found in DeBach (1966).

Theoretical extensions of Volterra's work were carried out by MacArthur and Levins (1964), Rescigno and Richardson (1965), and others. However, because of the exclusive dependence of all these formulations upon competitive interactions, the principle came under criticism from those who cited apparent exceptions, based for example upon the ideas of Pimentel (1961) on genetic feedback, of Ayala (Paper 17) on interactions between age classes, and of Skellam (1951) on the importance of spatial distribution. Moreover, Paine (Paper 12) elegantly demonstrated the importance of predation in determining the outcome of competitive interactions, a point made also by Hutchinson (Paper 18), Brooks and Dodson (1965), Utida (1958), Slobodkin (Paper 13), and others. In short, population regulation can come about through other dimensions of the species' niche than competition. The requisite feedback can come in the form of predation or parasitism or other interspecific factors, or can come through response to changes in population structure (e.g., genetic structure, age structure, spatial distribution, social structure, etc.). This leads to the introduction of the more general notion of *limiting factor,* as developed in Levin (Paper 8), as any density-dependent response which feeds back to affect the dynamics of the population(s) in question. The limiting factor is, in short, a feedback loop in the overall system.

Several points are in order in relation to Levin's development:

1. When applied to oscillating populations, the minimal independent set of limiting factors refers to a set that will work throughout the ranges of the variables in question. Species that oscillate in numbers, or whose niches are different only at high or low densities, might still co-

exist although at times being affected by the same limiting factor, for example a limiting nutrient.

Moreover, when limiting factors interact nonlinearly, there are unresolved questions concerning how to count them when the populations vary over broad numerical ranges. If limitation of different species is keyed to different phases of temporal oscillations within the community, even those generated by limit cycle oscillations involving other species, then a large number of species can be packed along this periodic gradient. No problems arise if the species average over these fluctuations as they usually would for diurnal cycles, but there are difficulties if the affected populations fluctuate (e.g., seasonally) in response to the environmental fluctuations. In independent work, J. Yorke of the University of Maryland and R. Armstrong and J. McGehee of the University of Minnesota have recently examined this problem and provided answers for some of the questions posed in the Appendix to Levin (Paper 8). Yorke specifically negates the claim of Haigh and Maynard Smith (1972) that density-independent components of limiting factors are irrelevant.

2. Exclusion cannot be demonstrated through static arguments predicated upon the "theory of equations," as is sometimes invoked. Equilibria are possible for any number of species, even though they are dependent upon a single resource; and the evolutionary points made earlier indicate why such equilibria cannot be dismissed as "infinitely unlikely." The point is rather a dynamic one, based upon the notion that perfect competitors will not be able to stably coexist. An alternative way to convey the idea, at least in the case of a point equilibrium, is as follows.

Consider system (2) of Paper 8,

$$\frac{dx_1}{dt} = x_1 f_1(z_1, ..., z_p), ..., \frac{dx_n}{dt} = x_n f_n(z_1, ..., z_p),$$

where the functions $z_j(x_1, ..., x_n; y_1, ..., y_m)$ are the limiting factors. Here $y_1, ..., y_m$ are parameters and may effectively be regarded as constants. Independence of factors is always evaluated with regard to $x_1, ..., x_n$, the parameters $y_1, ..., y_n$ being fixed. We now make no linearity or other special assumption regarding either the f's or the z's.

Note (as discussed previously) that the general definition of a limiting factor includes density-dependent components (through dependence of z_j on $x_1, ..., x_n$) as well as density-independent components (dependence on $y_1, ..., y_m$). The density-dependent components are primarily responsible for the ultimate regulation of the populations, while the density-independent components determine the level of the regulation—for example, steady-state population densities. Population biologists have tended to focus on the density-dependent components, since interest is centered on the mechanisms of population regulation: competition, predation, etc. Ecosystem modelers tend to accept the fact of regulation and to focus on "sensitivity analysis," the dependence of the steady-state values on the density-independent components. Moreover, the parameters $y_1, ..., y_m$ define in essence the community in which the species exist. They vary between communities (in space and time) and provide the mathematical definition of habitat variables, as defined by Whittaker, Levin, and Root (Paper 41).

If an equilibrium set of values $z_1^0, ..., z_p^0$, exists, it is well known that the stability of the equilibrium is related to the eigenvalues of the Jacobian matrix

$$J = \left(\frac{\partial f_i}{\partial x_j} \right)$$

at equiiibrium, except that difficulty arises when $p < n$, since then J is always a *singular* matrix, which is to say that it has zero as one eigenvalue. Although this leaves the stability question up in the air for the deterministic theory, it guarantees instability when one allows for even the slightest amount of environmental fluctuation (May 1973).

3. Although regulation at fixed levels or through stable cycles requires essentially one limiting factor for each species, mechanisms such as predatory switching could result in the persistence of several species in a fluctuant balance. By concentrating only on species that exceed critical threshold levels, a predator could conceivably allow several such species to coexist in alternate boom and bust patterns. Further, the alternative possibilities for predator response to prey make it clear that a single predator can regulate any number of prey species at constant levels. Man effectively does this for managed populations, constructing separate feedback loops for each managed species.

4. Still unanswered are two major questions. Given that in effect the limiting factor is the principal distinguishing component of the niche, what forms do these factors take in nature, and how different must factors be to be counted as independent? The first question we address in Part III, wherein the roles of competition, predation, parasitism, genetics, dispersion, etc., are considered. The second is answered in part by the work of May and MacArthur (Paper 26) discussed in Part IV.

Finally, it behooves us to mention a controversy that has swirled around the experiments of Francisco Ayala (1969, 1971), who elegantly demonstrates that the Lotka-Volterra competition equations do not in general provide an adequate description of the interaction of competing species over the full range of their densities. In particular, to develop such approximate equations at equilibrium, one cannot use parameters that are applicable under very different competition regimes. Although this is a valid point, it has very little to do with the principle of "competitive" exclusion as developed in this chapter. The disputation of Ayala's interpretation was presented by Gilpin and Justice (1972). The epilogue to this particular debate is the formation of a very profitable collaboration (Ayala et al., 1973; Gilpin and Ayala, 1973), modifying the Lotka–Volterra models to provide ones that better explain Ayala's data.

REFERENCES

Ayala, F. J. 1969. Experimental invalidation of the principle of competitive exclusion. *Nature* **224**: 1076-1079.

Ayala, F. J. 1971. Competition between species: frequency dependence. *Science* **171**: 820-824.

Ayala, F. J., M. E. Gilpin, and J. G. Ehrenfeld. 1973. Competition between species: theoretical models and experimental tests. *Theoret. Pop. Biol.* **4**: 331-356.

Brooks, J. L., and S. I. Dodson. 1965. Predation, body size, and composition of plankton. *Science* **150**: 28-35.

Cole, L. C. 1960. Competitive exclusion. *Science* **132**: 348-349.

Crombie, A. C. 1947. Interspecific competition. *J. Animal Ecol.* **16**: 44-73.

DeBach, P. 1966. The competitive displacement and coexistence principles. *Ann. Rev. Entomol.* **11**: 183-212.

Gilpin, M. E., and F. J. Ayala. 1973. Global models of growth and competition. *Proc. Natl. Acad. Sci. USA* **70**: 3590-3593.

Gilpin, M. E., and K. E. Justice. 1972. Reinterpretation of the invalidation of the principle of competitive exclusion. *Nature* **236**: 273-274, 299, 301.

Haigh, J., and J. Maynard Smith. 1972. Can there be more predators than prey? *Theoret. Pop. Biol.* **3**: 290-299.

MacArthur, R. H., and R. Levins. 1964. Competition, habitat selection, and character displacement in a patchy environment. *Proc. Natl. Acad. Sci. USA* **51**: 1207-1210.

May, R. 1973. Stability and Complexity in Model Ecosystems. Princeton University Press, Princeton, N.J. 235 pp.

Pimentel, D. 1961. Animal population regulation by the genetic feed-back mechanism. *Amer. Naturalist* **95**: 65-79.

Rescigno, A., and I. W. Richardson. 1965. On the competitive exclusion principle. *Bull. Math. Biophys.* **27**: 85-89.

Skellam, J. G. 1951. Random dispersal in theoretical populations. *Biometrika* **38**: 196-218.

Steere, J. B. 1894. On the distribution of genera and species of non-migratory land-birds in the Philippines. *Ibis:* 411-420.

Utida, S. 1958. Population fluctuation, an experimental and theoretical approach. *Cold Spring Harbor Symp. Quant. Biol.* **22**: 139-151.

Volterra, V. 1926. Variazione e fluttuazione del numero d' invididui in specie animale conviventi. *Mem. Accad. Nazionale Lincei* (ser. 6) **2**: 31-113.

Reprinted from *Science*, **131**, 1292-1297 (Apr. 29, 1960)

The Competitive Exclusion Principle

An idea that took a century to be born has implications in ecology, economics, and genetics.

Garrett Hardin

On 21 March 1944 the British Ecological Society devoted a symposium to the ecology of closely allied species. There were about 60 members and guests present. In the words of an anonymous reporter (*1*), "a lively discussion . . . centred about Gause's contention (1934) that two species with similar ecology cannot live together in the same place. . . . A distinct cleavage of opinion revealed itself on the question of the validity of Gause's concept. Of the main speakers, Mr. Lack, Mr. Elton and Dr. Varley supported the postulate. . . . Capt. Diver made a vigorous attack on Gause's concept, on the grounds that the mathematical and experimental approaches had been dangerously over simplified. . . . Pointing out the difficulty of defining 'similar ecology' he gave examples of many congruent species of both plants and animals apparently living and feeding together."

Thus was born what has since been called "Gause's principle." I say "born" rather than "conceived" in order to draw an analogy with the process of mammalian reproduction, where the moment of birth, of exposure to the external world, of becoming a fully legal entity, takes place long after the moment of conception. With respect to the principle here discussed, the length of the gestation period is a matter of controversy: 10 years, 12 years, 18 years, 40 years, or about 100 years, depending on whom one takes to be the father of the child.

Statement of the Principle

For reasons given below, I here refer to the principle by a name already introduced (*2*)—namely, the "competitive exclusion principle," or more briefly, the "exclusion principle." It may be briefly stated thus: *Complete competitors cannot coexist.* Many published discussions of the principle revolve around the ambiguity of the words used in stating it. The statement given above has been very carefully constructed: every one of the four words is ambiguous. This formulation has

been chosen not out of perversity but because of a belief that it is best to use that wording which is least likely to hide the fact that we still do not comprehend the exact limits of the principle. For the present, I think the "threat of clarity" (*3*) is a serious one that is best minimized by using a formulation that is *admittedly* unclear; thus can we keep in the forefront of our minds the unfinished work before us. The wording given has, I think, another point of superiority in that it seems brutal and dogmatic. By emphasizing the very aspects that might result in our denial of them were they less plain we can keep the principle explicitly present in our minds until we see if its implications are, or are not, as unpleasant as our subconscious might suppose. The meaning of these somewhat cryptic remarks should become clear further on in the discussion.

What does the exclusion principle mean? Roughly this: that (i) if two noninterbreeding populations "do the same thing"—that is, occupy precisely the same ecological niche in Elton's sense (*4*)—and (ii) if they are "sympatric"—that is, if they occupy the same geographic territory—and (iii) if population *A* multiplies even the least bit faster than population *B*, then ultimately *A* will completely displace *B*, which will become extinct. This is the "weak form" of the principle. Always in practice a stronger form is used, based on the removal of the hypothetical character of condition (iii). We do this because we adhere to what may be called the axiom of inequality, which states that no two things or processes,

The author is professor of biology at the University of California (Santa Barbara), Goleta. This article is based on a Darwin centennial lecture given before the Society for the History of Medical Science of Los Angeles on 24 November 1959.

in a real world, are precisely equal. This basic idea is probably as old as philosophy itself but is usually ignored, for good reasons. With respect to the *things* of the world the axiom often leads to trivial conclusions. One postage stamp is as good as another. But with respect to competing *processes* (for example, the multiplication rates of competing species) the axiom is never trivial, as has been repeatedly shown (*5–7*). No difference in rates of multiplication can be so slight as to negate the exclusion principle.

Demonstrations of the formal truth of the principle have been given in terms of the calculus (*5, 7*) and set theory (*8*). Those to whom the mathematics does not appeal may prefer the following intuitive verbal argument (*2*, pp. 84–85), which is based on an economic analogy that is very strange economics but quite normal biology.

"Let us imagine a very odd savings bank which has only two depositors. For some obscure reason the bank pays one of the depositors 2 percent compound interest, while paying the other 2.01 percent. Let us suppose further (and here the analogy is really strained) that whenever the sum of the combined funds of the two depositors reaches two million dollars, the bank arbitrarily appropriates one million dollars of it, taking from each depositor in proportion to his holdings at that time. Then both accounts are allowed to grow until their sum again equals two million dollars, at which time the appropriation process is repeated. If this procedure is continued indefinitely, what will happen to the wealth of these two depositors? A little intuition shows us (and mathematics verifies) that the man who receives the greater rate of interest will, in time, have all the money, and the other man none (we assume a penny cannot be subdivided). No matter how small the difference between the two interest rates (so long as there is a difference) such will be the outcome.

"Translated into evolutionary terms, this is what competition in nature amounts to. The fluctuating limit of one million to two million represents the finite available wealth (food, shelter, etc.) of any natural environment, and the difference in interest rates represents the difference between the competing species in their efficiency in producing offspring. No matter how small this difference may be, one species will eventually replace the other. In the scale of geological time, even a small competitive difference will result in a rapid extermination of the less successful species. Competitive differences that are so small as to be unmeasurable by direct means will, by virtue of the compound-interest effect, ultimately result in the extinction of one competing species by another."

The Question of Evidence

So much for the theory. Is it true? This sounds like a straightforward question, but it hides subtleties that have, unfortunately, escaped a good many of the ecologists who have done their bit to make the exclusion principle a matter of dispute. There are many who have supposed that the principle is one that can be proved or disproved by empirical facts, among them (*9, 10*) Gause himself. Nothing could be farther from the truth. The "truth" of the principle is and can be established only by theory, not being subject to proof or disproof by facts, as ordinarily understood. Perhaps this statement shocks you. Let me explain.

Suppose you believe the principle is true and set out to prove it empirically. First you find two noninterbreeding species that seem to have the same ecological characteristics. You bring them together in the same geographic location and await developments. What happens? Either one species extinguishes the other, or they coexist. If the former, you say, "The principle is proved." But if the species continue to coexist indefinitely, do you conclude the principle is false? Not at all. You decide there must have been some subtle difference in the ecology of the species that escaped you at first, so you look at the species again to try to see how they differ ecologically, all the while retaining your belief in the exclusion principle. As Gilbert, Reynoldson, and Hobart (*10*) dryly remarked, "There is . . . a danger of a circular process here. . . ."

Indeed there is. Yet the procedure can be justified, both empirically and theoretically. First, empirically. On this point our argument is essentially an acknowledgement of ignorance. When we think of mixing two similar species that have previously lived apart, we realize that it is hardly possible to know enough about species to be able to say, in advance, which one will exclude the other in free competition. Or, as Darwin, at the close of chapter 4 of his *Origin of Species* (*11*) put it:

"It is good thus to try in imagination to give any one species an advantage over another. Probably in no single instance should we know what to do. This ought to convince us of our ignorance on the mutual relations of all organic beings: a conviction as necessary, as it is difficult to acquire."

How profound our ignorance of competitive situations is has been made painfully clear by the extended experiments of Thomas Park and his collaborators (*12*). For more than a decade Park has put two species of flour beetles (*Tribolium confusum* and *T. castaneum*) in closed universes under various conditions. In every experiment the competitive exclusion principle is obeyed—one of the species is completely eliminated, *but it is not always the same one.* With certain fixed values for the environmental parameters the experimenters have been unable to control conditions carefully enough to obtain an invariable result. Just how one is to interpret this is by no means clear, but in any case Park's extensive body of data makes patent our immense ignorance of the relations of organisms to each other and to the environment, even under the most carefully controlled conditions.

The theoretical defense for adhering come-hell-or-high-water to the competitive exclusion principle is best shown by apparently changing the subject. Consider Newton's first law: "Every body persists in a state of rest or of uniform motion in a straight line unless compelled by external force to change that state." How would one verify this law, by itself? An observer might (in principle) test Newton's first law by taking up a station out in space somewhere and then looking at all the bodies around him. Would any of the bodies be in a state of rest except (by definition) himself? Probably not. More important, would any of the bodies in motion be moving in a straight line? *Not one.* (We assume that the observer makes errorless measurements.) For the law says, ". . . in a straight line unless compelled by external force to change . . .," and in a world in which another law says that "every body attracts every other body with a force that is inversely proportional to the square of the distance between them . . .," the phrase in the first law that begins with the words *unless compelled* clearly indicates the hypothetical character of the law. So long as there are no sanctuaries from gravitation in space, every body is always "compelled." Our observer would claim that any body at rest or moving in a straight line verified the law; he would likewise claim that bodies moving in not-straight lines verified the law, too. In other words, any attempt to test Newton's first law *by itself* would lead to a circular argument of the sort encountered earlier in considering the exclusion principle.

The point is this: We do not test isolated laws, one by one. What we test is a whole conceptual model (*13*). From the model we make predictions; these we test against empirical data. When we find that a prediction is not verifiable we then set about modifying the model. There is no procedural rule to tell us which element of the model is best abandoned or changed. (The scientific response to the results of the Michelson-Morley experiment was not in any sense *determined*.) Esthetics plays a part in such decisions.

The competitive exclusion principle is one element in a system of ecological thought. We cannot test it directly, by itself. What the whole ecological system is, we do not yet know. One immediate task is to discover the system, to find its elements, to work out their interactions, and to make the system as explicit as possible. (*Complete* explicitness can never be achieved.) The works of Lotka (*14*), Nicholson (*15, 16*), and MacArthur (*17*) are encouraging starts toward the elaboration of such a theoretical system.

The Issue of Eponymy

That the competitive exclusion principle is often called "Gause's principle" is one of the more curious cases of eponymy in science (like calling human oviducts "Fallopian tubes," after a man who was not the first to see them and who misconstrued their significance). The practice was apparently originated by the English ecologists, among whom David Lack has been most influential. Lack made a careful study of *Geospiza* and other genera of finches in the Galápagos Islands, combining observational studies on location with museum work at the California Academy of Sciences. How his ideas of ecological principles matured during the process is evident from a passage in his little classic, *Darwin's Finches* (*18*).

"Snodgrass concluded that the beak differences between the species of *Geo-spiza* are not of adaptive significance in regard to food. The larger species tend to eat rather larger seeds, but this he considered to be an incidental result of the difference in the size of their beaks. This conclusion was accepted by Gifford (1919), Gulick (1932), Swarth (1934) and formerly by myself (Lack, 1945). Moreover, the discovery . . . that the beak differences serve as recognition marks, provided quite a different reason for their existence, and thus strengthened the view that any associated differences in diet are purely incidental and of no particular importance.

"My views have now completely changed, through appreciating the force of Gause's contention that two species with similar ecology cannot live in the same region (Gause, 1934). This is a simple consequence of natural selection. If two species of birds occur together in the same habitat in the same region, eat the same types of food and have the same other ecological requirements, then they should compete with each other, and since the chance of their being equally well adapted is negligible, one of them should eliminate the other completely. Nevertheless, three species of ground-finch live together in the same habitat on the same Galapagos islands, and this also applies to two species of insectivorous tree-finch. There must be some factor which prevents these species from effectively competing."

Implicit in this passage is a bit of warm and interesting autobiography. It is touching to see how intellectual gratitude led Lack to name the exclusion principle after Gause, calling it, in successive publications, "Gause's contention," "Gause's hypothesis," and "Gause's principle." But the eponymy is scarcely justified. As Gilbert, Reynoldson, and Hobart point out (*10*, p. 312): "Gause . . . draws no general conclusions from his experiments, and moreover, makes no statement which resembles any wording of the hypothesis which has arisen bearing his name." Moreover, in the very publication in which he discussed the principle, Gause acknowledged the priority of Lotka in 1932 (*5*) and Volterra in 1926 (*6*). Gause gave full credit to these men, viewing his own work merely as an empirical testing of their theory—a quite erroneous view, as we have seen. How curious it is that the principle should be named after a man who did not state it clearly, who mis-apprehended its relation to theory, and who acknowledged the priority of others!

Recently Udvardy (*19*), in an admirably compact note, has pointed out that Joseph Grinnell, in a number of publications, expressed the exclusion principle with considerable clarity. In the earliest passage that Udvardy found, Grinnell, in 1904 (*20*), said: "Every animal tends to increase at a geometric ratio, and is checked only by limit of food supply. It is only by adaptations to different sorts of food, or modes of food getting, that more than one species can occupy the same locality. Two species of approximately the same food habits are not likely to remain long enough evenly balanced in numbers in the same region. One will crowd out the other."

Udvardy quotes from several subsequent publications of Grinnell, from all of which it is quite clear that this well-known naturalist had a much better grasp of the exclusion principle than did Gause. Is this fact, however, a sufficiently good reason for now speaking (as Udvardy recommends) of "Grinnell's axiom?" On the basis of present evidence there seems to be justice in the proposal, but we must remember that the principle has already been referred to, in various publications, as "Gause's principle," the "Volterra-Gause principle," and the "Lotka-Volterra principle." What assurance have we that some diligent scholar will not tomorrow unearth a predecessor of Grinnell? And if this happens, should we then replace Grinnell's name with another's? Or should we, in a fine show of fairness, use all the names? (According to this system, the principle would, at present, be called the Grinnell-Volterra-Lotka-Gause-Lack principle—and, even so, injustice would be done to A. J. Nicholson, who, in his wonderful gold mine of unexploited aphorisms (*15*), wrote: "For the steady state [in the coexistence of two or more species] to exist, each species must possess some advantage over all other species with respect to some one, or group, of the control factors to which it is subject." This is surely a corollary of the exclusion principle.)

In sum, I think we may say that arguments for pinning an eponym on this idea are unsound. But it does need a name of some sort; its lack of one has been one of the reasons (though not the only one) why this basic principle has trickled out of the scientific consciousness after each mention during the last half century. Like Allee *et al.* (*21*) we should wish "to avoid further implementation of the facetious definition of ecology as being that phase of biology primarily abandoned to terminology." But, on the other side, it has been pointed out (*22*): "Not many recorded facts are lost; the bibliographic apparatus of science is fairly equal to the problem of recording melting points, indices of refraction, etc., in such a way that they can be recalled when needed. Ideas, more subtle and more diffusely expressed present a bibliographic problem to which there is no present solution." To solve the bibliographic problem some sort of handle is needed for the idea here discussed; the name "the competitive exclusion principle" is correctly descriptive and will not be made obsolete by future library research.

The Exclusion Principle and Darwin

In our search for early statements of the principle we must not pass by the writings of Charles Darwin, who had so keen an appreciation of the ecological relationships of organisms. I have been unable to find any unambiguous references to the exclusion principle in the "Essays" of 1842 and 1844 (*23*), but in the *Origin* itself there are several passages that deserve recording. All the following passages are quoted from the sixth edition (*11*).

"As the species of the same genus usually have, though by no means invariably, much similarity in habits and constitution, and always in structure, the struggle will generally be more severe between them, if they come into competition with each other, than between the species of distinct genera. We see this in the recent extension over parts of the United States of one species of swallow having caused the decrease of another species. The recent increase of the missel-thrush in parts of Scotland has caused the decrease of the song-thrush. How frequently we hear of one species of rat taking the place of another species under the most different climates! In Russia the small Asiatic cockroach has everywhere driven before it its great congener. In Australia the imported hive-bee is rapidly exterminating the small, stingless native bee. One species of charlock has been known to supplant another species; and so in other cases. We can dimly see why the competition

should be most severe between allied forms, which fill nearly the same place in the economy of nature; but probably in no one case could we precisely say why one species has been victorious over another in the great battle of life" (p. 71).

"Owing to the high geometrical rate of increase of all organic beings, each area is already fully stocked with inhabitants; and it follows from this, that as the favored forms increase in number, so, generally, will the less favored decrease and become rare. Rarity, as geology tells us, is the precursor to extinction. We can see that any form which is represented by few individuals will run a good chance of utter extinction, during great fluctuations in the nature or the seasons, or from a temporary increase in the number of its enemies. But we may go further than this; for, as new forms are produced, unless we admit that specific forms can go on indefinitely increasing in number, many old forms must become extinct" (p. 102).

"From these several considerations I think it inevitably follows, that as new species in the course of time are formed through natural selection, others will become rarer and rarer, and finally extinct. The forms which stand in closest competition with those undergoing modification and improvement, will naturally suffer most. And we have seen in the chapter on the Struggle for Existence that it is the most closely-allied forms—varieties of the same species, and species of the same genus or related genera—which, from having nearly the same structure, constitution and habits, generally come into the severest competition with each other consequently, each new variety or species, during the progress of its formation, will generally press hardest on its nearest kindred, and tend to exterminate them. We see the same process of extermination among our domesticated productions, through the selection of improved forms by man. Many curious instances could be given showing how quickly new breeds of cattle, sheep and other animals, and varieties of flowers, take the place of older and inferior kinds. In Yorkshire, it is historically known that the ancient black cattle were displaced by the longhorns, and that these 'were swept away by the short-horns' (I quote the words of an agricultural writer) 'as if by some murderous pestilence' " (p. 103).

"For it should be remembered that

the competition will generally be most severe between those forms which are most nearly related to each other in habits, constitution and structure. Hence all the intermediate forms between the earlier and later states, that is between the less and more improved states of the same species, as well as the original parent species itself, will generally tend to become extinct" (p. 114).

Those passages are, we must admit, typically Darwinian; by turn clear, obscure, explicit, cryptic, suggestive, they have in them all the characteristics that litterateurs seek in James Joyce. The complexity of Darwin's work, however, is unintended; it is the result partly of his limitations as an analytical thinker, but in part also it is the consequence of the magnitude, importance, and intrinsic difficulty of the ideas he grappled with. Darwin was not one to impose premature clarity on his writings.

Origins in Economic Theory?

In chapter 3 of *Nature and Man's Fate* I have argued for the correctness of John Maynard Keynes' view that the biological principle of natural selection is just a vast generalization of Ricardian economics. The argument is based on the isomorphism of theoretical systems in the two fields of human thought. Now that we have at last brought the competitive exclusion principle out of the periphery of our vision into focus on the *fovea centralis* it is natural to wonder if this principle, too, originated in economic thought. I think it is possible. At any rate, there is a passage by the French mathematician J. Bertrand (*24*), published in 1883, which shows an appreciation of the exclusion principle as it applies to economic matters. The passage occurs in a review of a book of Cournot, published much earlier, in which Cournot discussed the outcome of a struggle between two merchants engaged in selling identical products to the public. Bertrand says: "Their interest would be to unite or at least to agree on a common price so as to extract from the body of customers the greatest possible receipts. But this solution is avoided by Cournot who supposes that one of the competitors will lower his price in order to attract the buyers to himself, and that the other, trying to regain them, will set his price still lower. The two rivals will cease to pursue this path only when each has nothing more to gain by lowering his price.

"To this argument we make a peremptory objection. Given the hypothesis, no solution is possible: there is no limit to the lowering of the price. Whatever common price might be initially adopted, if one of the competitors were to lower the price unilaterally he would thereby attract the totality of the business to himself. . . ."

This passage clearly antedates Grinnell, Lack, *et al.*, but it comes long after the *Origin of Species*. Are there statements of the principle in the economic literature before Darwin? It would be nice to know. I have run across cryptic references to the work of Simonde de Sismondi (1773–1842) which imply that he had a glimpse of the exclusion principle, but I have not tracked them down. Perhaps some colleague in the history of economics will someday do so. If it is true that Sismondi understood the principle, this fact would add a nice touch to the interweaving of the history of ideas, for this famous Swiss economist is related to Emma Darwin by marriage; he plays a prominent role in the letters published under her name (*25*).

Utility of the Exclusion Principle

"The most important lesson to be learned from evolutionary theory," says Michael Scriven in a brilliant essay recently published (*26*), "is a negative one: the theory shows us what scientific explanations need not do. In particular it shows us that one cannot regard explanations as unsatisfactory when they are not such as to enable the event in question to have been predicted." The theory of evolution is not one with which we can predict exactly the future course of species formation and extinction; rather, the theory "explains" the past. Strangely enough, we take mental satisfaction in this ex post facto explanation. Scriven has done well in showing why we are satisfied.

Much of the theory of ecology fits Scriven's description of evolutionary theory. Told that two formerly separated species are to be introduced into the same environment and asked to predict exactly what will happen, we are generally unable to do so. We can only make certain predictions of this sort: either A will extinguish B, or B will extinguish A; or the two species

are (or must become) ecologically different—that is, they must come to occupy different ecological niches. The general rule may be stated in either of two different ways: *Complete competitors cannot coexist*—as was said earlier; or, *Ecological differentiation is the necessary condition for coexistence*.

It takes little imagination to see that the exclusion principle, to date stated explicitly only in ecological literature, has applications in many academic fields of study. I shall now point out some of these, showing how the principle has been used (mostly unconsciously) in the past, and predicting some of its applications in the future.

Economics. The principle unquestionably plays an indispensable role in almost all economic thinking, though it is seldom explicitly stated. Any competitor knows that unrestrained competition will ultimately result in but one victor. If he is confident that he is that one, he may plump for "rugged individualism." If, on the other hand, he has doubts, then he will seek to restrain or restrict competition. He can restrain it by forming a cartel with his competitors, or by maneuvering the passage of "fair trade" laws. (Laboring men achieve a similar end—though the problem is somewhat different—by the formation of unions and the passage of minimum wage laws.) Or he may restrict competition by "ecological differentiation," by putting out a slightly different product (aided by restrictive patent and copyright laws). All this may be regarded as individualistic action.

Society as a whole may take action. The end of unrestricted competition is a monopoly. It is well known that monopoly breeds power which acts to insure and extend the monopoly; the system has "positive feedback" and hence is always a threat to those aspects of society still "outside" the monopoly. For this reason, men may, in the interest of "society" (rather than of themselves as individual competitors), band together to insure continued competition; this they do by passing antimonopoly laws which prevent competition from proceeding to its "naturally" inevitable conclusion. Or "society" may permit monopolies but seek to remove the power element by the "socialization" of the monopoly (expropriation or regulation).

In their actions both as individuals and as groups, men show that they have an implicit understanding of the

exclusion principle. But the failure to bring this understanding to the level of consciousness has undoubtedly contributed to the accusations of bad faith ("exploiter of the masses," "profiteer," "nihilist," "communist") that have characterized many of the interchanges between competing groups of society during the last century. F. A. Lange (*27*), thinking only of laboring men, spoke in most fervent terms of the necessity of waging a "struggle against the struggle for existence"—that is, a struggle against the unimpeded working out of the exclusion principle. Groups with interests opposed to those of "labor" are equally passionate about the same cause, though the examples they have in mind are different.

At the present time, one of the great fields of economics in which the application of the exclusion principle is resisted is international competition (nonbellicose). For emotional reasons, most discussion of problems in this field is restricted by the assumption (largely implicit) that Cournot's solution of the *intra*national competition problem is correct and applicable to the *inter*national problem. On the less frequent occasions when it is recognized that Bertrand's, not Cournot's, reasoning is correct, it is assumed that the consequences of the exclusion principle can be indefinitely postponed by a rapid and endless multiplication of "ecological niches" (largely unprotected though they are by copyright and patent). If some of these assumptions prove to be unrealistic, the presently fashionable stance toward tariffs and other restrictions of international competition will have to be modified.

Genetics. The application of the exclusion principle to genetics is direct and undeniable. The system of discrete alleles at the same gene locus competing for existence within a single population of organisms is perfectly isomorphic with the system of different species of organisms competing for existence in the same habitat and ecological niche. The consequences of this have frequently been acknowledged, usually implicitly, at least since J. B. S. Haldane's work of 1924 (*28*). But in this field, also, the consequences have often been denied, explicitly or otherwise, and again for emotional reasons. The denial has most often been coupled with a "denial" (in the psychological sense) of the priority of the inequality axiom. As a result of recent findings in the fields of physiological genetics and population genetics, particularly as concerns blood groups, the applicability of both the inequality axiom and the exclusion principle is rapidly becoming accepted. William C. Boyd has recorded, in a dramatic way (*29*), his escape from the bondage of psychological denial. The emotional restrictions of rational discussion in this field are immense. How "the struggle against the struggle for existence" will be waged in the field of human genetics promises to make the next decade of study one of the most exciting of man's attempts to accept the implications of scientific knowledge.

Ecology. Once one has absorbed the competitive exclusion principle into one's thinking it is curious to note how one of the most popular problems of evolutionary speculation is turned upside down. Probably most people, when first taking in the picture of historical evolution, are astounded at the number of species of plants and animals that have become extinct. From Simpson's gallant "guesstimates" (*30*), it would appear that from 99 to 99.975 percent of all species evolved are now extinct, the larger percentage corresponding to 3999 million species. This seems like a lot. Yet it is even more remarkable that there should live at any one time (for example, the present) as many as a million species, more or less competing with each other. Competition is avoided between some of the species that coexist in time by separation in space. In addition, however, there are many ecologically more or less similar species that coexist. Their continued existence is a thing to wonder at and to study. As Darwin said (*11*, p. 363)—and this is one

more bit of evidence that he appreciated the exclusion principle—"We need not marvel at extinction; if we must marvel, let it be at our own presumption in imagining for a moment that we understand the many complex contingencies on which the existence of each species depends."

I think it is not too much to say that in the history of ecology—which in the broadest sense includes the science of economics and the study of population genetics—we stand at the threshold of a renaissance of understanding, a renaissance made possible by the explicit acceptance of the competitive exclusion principle. This principle, like much of the essential theory of evolution, has (I think) long been psychologically denied, as the penetrating study of Morse Peckham (*31*) indicates. The reason for the denial is the usual one: admission of the principle to consciousness is painful. [Evidence for such an assertion is, in the nature of the case, difficult to find, but for a single clear-cut example see the letter by Krogman (*32*).] It is not sadism or masochism that makes us urge that the denial be brought to an end. Rather, it is a love of the reality principle, and recognition that only those truths that are admitted to the conscious mind are available for use in making sense of the world. To assert the truth of the competitive exclusion principle is not to say that nature is and always must be, everywhere, "red in tooth and claw." Rather, it is to point out that *every* instance of apparent coexistence must be accounted for. Out of the study of all such instances will come a fuller knowledge of the many prosthetic devices of co-

existence, each with its own costs and its own benefits. On such a foundation we may set about the task of establishing a science of ecological engineering.

References

1. Anonymous, *J. Animal Ecol.* **13**, 176 (1944).
2. G. Hardin, *Nature and Man's Fate* (Rinehart, New York, 1959).
3. ———, *Am. J. Psychiat.* **114**, 392 (1957).
4. C. Elton, *Animal Ecology* (Macmillan, New York, 1927).
5. A. J. Lotka, *J. Wash. Acad. Sci.* **22**, 469 (1932).
6. V. Volterra, *Mem. reale accad. nazl. Lincei, Classe sci. fis. mat. e nat. ser. 6, No. 2* (1926).
7. ———, *Leçons sur la Théorie Mathématique de la Lutte pour la Vie* (Gauthier-Villars, Paris, 1931).
8. G. E. Hutchinson, *Cold Spring Harbor Symposia Quant. Biol.* **22**, 415 (1957).
9. G. F. Gause, *The Struggle for Existence* (Williams and Wilkins, Baltimore, 1934); H. H. Ross, *Evolution* **11**, 113 (1957).
10. O. Gilbert, T. B. Reynoldson, J. Hobart, *J. Animal Ecol.* **21**, 310–312 (1952).
11. C. Darwin, *On the Origin of Species by Means of Natural Selection* (Macmillan, New York, new ed. 6, 1927).
12. T. Park and M. Lloyd, *Am. Naturalist* **89**, 235 (1955).
13. R. M. Thrall, C. H. Coombs, R. L. Davis, *Decision Processes* (Wiley, New York, 1954), pp. 22–23.
14. A. J. Lotka, *Elements of Physical Biology* (Williams and Wilkins, Baltimore, 1925).
15. A. J. Nicholson, *J. Animal Ecol.* **2**, suppl., 132–178 (1933).
16. ———, *Australian J. Zool.* **2**, 9 (1954).
17. R. H. MacArthur, *Ecology* **39**, 599 (1958).
18. D. Lack, *Darwin's Finches* (University Press, Cambridge, 1947).
19. M. F. D. Udvardy, *Ecology* **40**, 725 (1959).
20. J. Grinnell, *Auk* **21**, 364 (1904).
21. W. C. Allee, A. E. Emerson, O. Park, T. Park, K. P. Schmidt, *Principles of Ecology* (Saunders, Philadelphia, 1949).
22. G. Hardin, *Sci. Monthly* **70**, 178 (1950).
23. F. Darwin, *The Foundations of the Origin of Species* (University Press, Cambridge, 1909).
24. J. Bertrand, *J. savants* (Sept. 1883), pp. 499–508.
25. H. Litchfield, *Emma Darwin. A Century of Family Letters, 1792–1896* (Murray, London, 1915).
26. M. Scriven, *Science* **130**, 477 (1959).
27. F. A. Lange, *History of Materialism* (Harcourt Brace, New York, ed. 3, 1925).
28. J. B. S. Haldane, *Trans. Cambridge Phil. Soc.* **23**, 19 (1924).
29. W. C. Boyd, *Am. J. Human Genet.* **11**, 397 (1959).
30. G. G. Simpson, *Evolution* **6**, 342 (1952).
31. M. Peckham, *Victorian Studies* **3**, 19 (1959).
32. W. M. Krogman, *Science* **111**, 43 (1950).

6

Reprinted from R. N. Chapman, *Animal Ecology*, McGraw-Hill, New York, 1931, pp. 412–414, 432–433 [originally published in *J. Conseil Intern. Exploration Mer*, III, 1, (1928)]

VARIATIONS AND FLUCTUATIONS OF THE NUMBER OF INDIVIDUALS IN ANIMAL SPECIES LIVING TOGETHER

V. Volterra

[*Editors' Note:* In the original, material precedes this excerpt.]

§2. BIOLOGICAL ASSOCIATION OF TWO SPECIES WHICH CONTEND FOR THE SAME FOOD

1. Let us suppose we have two species living in the same environment: let the numbers of the individuals be respectively N_1 and N_2 and let ϵ_1 and ϵ_2 be the values which their coefficients of increase would have if the quantity of the common food were always such as to amply satisfy their voracity. We shall have

$$\frac{dN_1}{dt} = \epsilon_1 N_1, \quad \frac{dN_2}{dt} = \epsilon_2 N_2 \qquad (\epsilon_1 > 0, \quad \epsilon_2 > 0).$$

Let it be admitted now that the individuals of the two species, continually increasing in number, diminish the quantity of food of which each individual can dispose. Let us suppose that the presence of the N_1 individuals of the first species diminishes this quantity by an amount $h_1 N_1$ and the presence of the N_2 individuals of the second species diminishes it by the amount $h_2 N_2$ and that therefore by the combination of the two, the diminution amounts to $h_1 N_1 + h_2 N_2$ and that by virtue of the unequal need of food of the two species, the two coefficients of increase are reduced to

$$\epsilon_1 - \gamma_1(h_1N_1 + h_2N_2), \qquad \epsilon_2 - \gamma_2(h_1N_1 + h_2N_2) \tag{1}$$

We shall then have the differential equations

$$\frac{dN_1}{dt} = [\epsilon_1 - \gamma_1(h_1N_1 + h_2N_2)]N_1, \tag{2_1}$$

$$\frac{dN_2}{dt} = [\epsilon_2 - \gamma_2(h_1N_1 + h_2N_2)]N_2, \tag{2_2}$$

in which we must suppose ϵ_1, ϵ_2, h_1, h_2, γ_1, γ_2 to be positive constants.

2. From the preceding equations it follows that

$$\frac{d \log N_1}{dt} = \epsilon_1 - \gamma_1(h_1N_1 + h_2N_2), \tag{3_1}$$

$$\frac{d \log N_2}{dt} = \epsilon_2 - \gamma_2(h_1N_1 + h_2N_2), \tag{3_2}$$

and hence

$$\gamma_2 \frac{d \log N_1}{dt} - \gamma_1 \frac{d \log N_2}{dt} = \epsilon_1\gamma_2 - \epsilon_2\gamma_1, \tag{4}$$

that is to say

$$\frac{d \log \dfrac{N_1{}^{\gamma_2}}{N_2{}^{\gamma_1}}}{dt} = \epsilon_1\gamma_2 - \epsilon_2\gamma_1 \tag{5}$$

and integrating and passing from logarithms to numbers,

$$\frac{N_1{}^{\gamma_2}}{N_2{}^{\gamma_1}} = Ce^{(\epsilon_1\gamma_2 - \epsilon_2\gamma_1)t}, \tag{6}$$

where C is a constant quantity

3. If the binomial $\epsilon_1\gamma_2 - \epsilon_2\gamma_1$ is not zero we can suppose it positive, for if it were not positive it would suffice to exchange species 1 with species 2 to make it positive.

In this case

$$\lim_{t = \infty} \frac{N_1{}^{\gamma_2}}{N_2{}^{\gamma_1}} = \infty .$$

For N_1 equal to or greater than ϵ_1/γ_1h_1, by virtue of (2_1), the differential coefficient dN_1/dt is negative, hence N_1 cannot exceed a certain limit.

N_2 then must approach zero.

It is easy to compute the expression asymptotic to N_1.

In fact when N_2 becomes small enough to remain negligible, equation (2_1) will become

$$\frac{dN_1}{dt} = (\epsilon_1 - \gamma_1 h_1 N_1)N_1$$

or, separating the variables

$$dt = \frac{dN_1}{N_1(\epsilon_1 - \gamma_1 h_1 N_1)}$$

and integrating and passing from logarithms to numbers,

$$\frac{N_1}{\epsilon_1 - \gamma_1 h_1 N_1} = C_o\, e^{\epsilon_1 t}$$

C_o being a constant.   Hence

$$N_1 = \frac{C_o \epsilon_1 e^{\epsilon_1 t}}{1 + \gamma_1 h_1 C_o e^{\epsilon_1 t}} = \frac{C_o\, \epsilon_1}{e^{-\epsilon_1 t} + \gamma_1 h_1 C_o}.$$

Therefore N_1 approaches asymptotically the value $\epsilon_1/\gamma_1 h_1$ for increasing or decreasing values according as C_o is positive or negative.

We can sum up the results we have obtained, in the following proposition: If $\epsilon_1/\gamma_1 > \epsilon_2/\gamma_2$ *the second species continually decreases and the number of individuals of the first species approaches* $\epsilon_1/\gamma_1 h_1$.

[*Editors' Note:* Material has been omitted at this point.]

§6. THE CASE OF ANY NUMBER WHATEVER OF SPECIES WHICH CONTEND FOR THE SAME FOOD

1. It is easy to extend what has been done in the case of two species living together, which contend for the same food, to the case of any number of species.

Let us take the number of species to be n and let us assume $\epsilon_1, \epsilon_2, \ldots \epsilon_n$, to be the coefficients of increase which each species would have if alone. Let us denote by $F(N_1, N_2, \ldots, N_n)dt$ the diminution in the quantity of food in the time dt, when the numbers of individuals of the different species are respectively $N_1, N_2, \ldots, N_n$. This function will be zero for $N_1 = N_2 = \cdots N_n = 0$; it will be positive and increasing and will increase indefinitely with an indefinite increase in each N_r. For simplicity we could take F linear, that is

$$F(N_1, N_2, \cdots, N_n) = \alpha_1 N_1 + \alpha_2 N_2 + \cdots + \alpha_n N_n$$

where the coefficients α_r are positive. But we shall leave F general.

The presence of N_1 individuals of the first species, of N_2 of the second, etc. will influence the coefficients of increase reducing ϵ_r to $\epsilon_r - \gamma_r F(N_1, \cdots, N_n)$ where the positive coefficient γ_r measures the influence which the diminution of the food has upon the increase of the species.

We shall have then the differential equations

$$\frac{dN_r}{dt} = N_r[\epsilon_r - \gamma_r F(N_1, \cdots, N_n)], \quad (r = 1, 2, \cdots, n) \quad (23)$$

from which follows

$$\frac{1}{\gamma_r N_r} \frac{dN_r}{dt} - \frac{1}{\gamma_s N_s} \frac{dN_s}{dt} = \frac{\epsilon_r}{\gamma_r} - \frac{\epsilon_s}{\gamma_s}$$

and integrating and passing from logarithms to numbers

$$\frac{N_r^{\frac{1}{\gamma_r}}}{N_s^{\frac{1}{\gamma_s}}} = Ce^{\left(\frac{\epsilon_r}{\gamma_r} - \frac{\epsilon_s}{\gamma_s}\right)t}$$

where C is a positive constant.

2. Let us arrange the ratios ϵ_r/γ_r in order of size, that is let us suppose[1]

$$\frac{\epsilon_1}{\gamma_1} > \frac{\epsilon_2}{\gamma_2} > \frac{\epsilon_3}{\gamma_3} > \cdots > \frac{\epsilon_n}{\gamma_n},$$

then we shall have, if $r < s$

---

[1] Let us exclude the cases of equality as of infinitesimally small probability.

$$\lim_{t\,=\,\infty} \frac{N_r^{\frac{1}{\gamma_r}}}{N_s^{\frac{1}{\gamma_s}}} = \infty\,.$$

As a consequence of this result either N_r, can, with increase of time, take values as large as we please, or

$$\lim_{t\,=\,\infty} N_s = 0.$$

But the first case is to be excluded, because F increases indefinitely with indefinite increase of N_r, therefore in (23) the second member becomes negative when N_r exceeds a certain limit; whence the upper limit of N_r is finite. Then the second case must hold. From this it follows that all the species tend to disappear except the first.

To have the asymptotic variation of N_1 it will suffice to repeat what has been done in the case of two species alone.

[*Editors' Note:* Material has been omitted at this point.]

7

Reprinted from G. F. Gause, *The Struggle for Existence*, Williams & Wilkins, Baltimore, 1934, pp. 103–111 (reprinted by Hafner Publ. Co., New York, 1964)

THE STRUGGLE FOR EXISTENCE

G. F. Gause

[*Editors' Note:* In the original, material precedes this excerpt.]

III

(1) Although the situation in our experiments with Osterhout's medium has been considerably simpler than in the case of the "oaten medium," it is still too complicated for a clear understanding of the mechanism of competition. In fact, why has one species been victorious over another? In the case of yeast cells we answered that the success of the species during the first stage of competition depends on definite relations between the coefficients of multiplication and the alcohol production, and that it can be exactly predicted with the aid of an equation of the struggle for existence. What will be our answer for the population of Paramecia?

To investigate this problem we made the conditions of the experimentation the next step in the simplification. We endeavored to make a medium with a very small concentration of nutritive bacteria and optimal in its physicochemical properties for Paramecia. Under such conditions the competition for common food between two species of Protozoa has been reduced to its simplest form.

(2) As Woodruff has shown ('11, '14), the waste products of Paramecia can depress the multiplication and be specific for a given species. In any case we are very far from an exact knowledge of their rôle and chemical composition. Therefore first of all we must eliminate the complicating influence of these substances. This problem is the reverse of the one we had to do with in the preceding chapter. There in the experiments with yeast we tried to set up conditions under which the food resources of the medium should be very considerable at the time when the concentration of the waste

products had already attained a critical value. Now with Paramecia our object is that the concentration of the waste products should still be very far from the critical threshold at the moment when the food is exhausted.

First of all we turned our attention to the hydrogen ion concentration (pH), which in the light of the researches of Darby ('29) can be of great importance for our species. When Paramecia are cultivated in Osterhout's medium, pH is near to 6.8 and unstable, where-

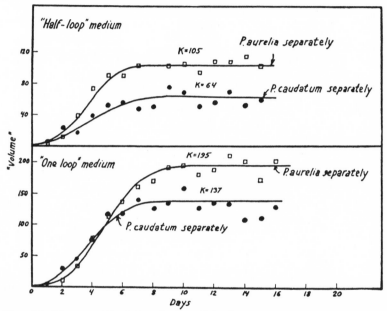

Fig. 23. The growth of the "volume" in *Paramecium caudatum* and *Paramecium aurelia* cultivated separately on the buffered medium ("half-loop" and "one-loop" concentrations of bacteria). From Gause ('34d).

as the reaction in our wild cultures is commonly near to 8.0. Therefore we, like Johnson, buffered Osterhout's medium by adding 1 cm³ of $\frac{m}{20}$ KH$_2$PO$_4$ to 30 cm³ of diluted salt solution, and bringing the reaction of the medium with the aid of $\frac{m}{20}$ KOH to pH = 8.0. At the same time we isolated new pure lines of Paramecia out of our wild culture, as the Paramecia which had been cultivated for a long time on Osterhout's medium could not stand a sudden transfer into a buffered medium.

In order to diminish the concentration of the bacteria we made a new smaller standard loop for preparing the "one-loop medium," and also arranged experiments in which the one loop medium was diluted twice ("half-loop medium"). The data obtained are given in Table 4 (Appendix) where every figure represents a mean value from the observations of two microcosms. This material is represented graphically in Figures 23, 24 and 25.

Let us examine Figure 23. The curves of growth of pure populations of *P. caudatum* and *P. aurelia* with different concentrations of the bacterial food show that the lack of food is actually a factor limiting growth in these experiments. With the double concentration of food the volumes of the populations of the separately growing species also increase about twice (from 64 up to 137 in *P. caudatum*; 64 × 2 = 128; from 105 up to 195 in *P. aurelia*; 105 × 2 = 210). Under these

TABLE XI

Parameters of the logistic curves for separate growth of Paramecium caudatum and Paramecium aurelia

Buffered medium with the "half-loop" concentration of bacteria

	P. AURELIA	P. CAUDATUM
Maximal volume (K).....................	$K_1 = 105$	$K_2 = 64$
Coefficient of geometric increase (b)........	$b_1 = 1.1244$	$b_2 = 0.7944$

conditions the differences in the growth of populations of *P. aurelia* and *P. caudatum* are quite distinctly pronounced: the growth of the biomass of the former species proceeds with *greater rapidity*, and it accumulates a *greater biomass than P. caudatum at the expense of the same level of food resources.*[1]   If we now express the curves of separate growth of both species under a half-loop concentration of bacteria with the aid of logistic equations we shall obtain the data presented in Table XI. This table shows clearly that *P. aurelia* has perfectly definite advantages over *P. caudatum* in respect to the basic characteristics of growth.

(3) We will now pass on to the growth of a mixed population of *P. caudatum* and *P. aurelia*. The general character of the curves on

[1] This is apparently connected with the resistance of *P. aurelia* to the waste products of the pathogenic bacterium, *Bacillus pyocyaneus* (see Gause, Nastukova and Alpatov, '35).

Figures 22, 24 and 25 is almost the same, but there are certain differences concerning secondary peculiarities. For a detailed acquaintance with the properties of a mixed population we will consider the growth with a half-loop concentration of bacteria (Fig. 24). First of all we see that as in the case examined before the competition between our species can be divided into two separate stages: up to the fifth

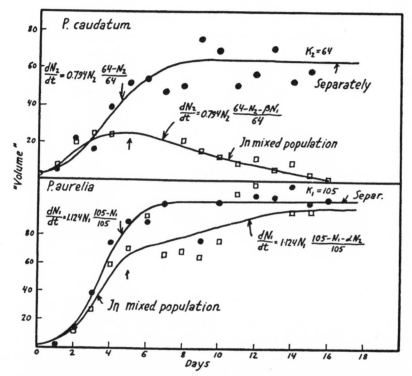

Fig. 24. The growth of the "volume" in *Paramecium caudatum* and *Paramecium aurelia* cultivated separately and in the mixed population on the buffered medium with the "half-loop" concentration of bacteria. From Gause ('34d).

day there is a competition between the species for seizing the so far unutilized food energy; then after the fifth day of growth begins the redistribution of the completely seized resources of energy between the two components, which leads to a complete displacement of one of them by another. The following simple calculations can convince one that on the fifth day all the energy is already seized upon. At the expense of a certain level of food resources which is a constant one in all "half-loop" experiments and may be taken as unity, *P. aurelia* growing separately produces a biomass equal to 105 volume units, and *P. caudatum* 64 such units.   Therefore, one unit of volume of *P. caudatum* consumes $\frac{1}{64} = 0.01562$ of food, and one unit of volume

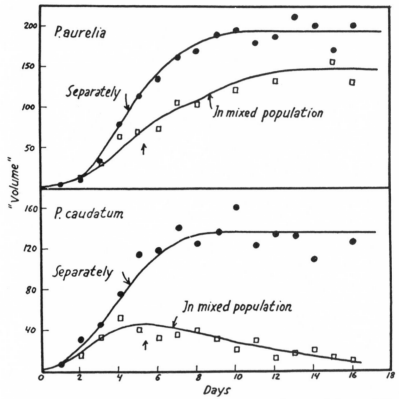

Fig. 25. The growth of the "volume" in *Paramecium caudatum* and *Paramecium aurelia* cultivated separately and in the mixed population on the buffered medium with the "one-loop" concentration of bacteria.   From Gause ('34d).

of *P. aurelia* $\frac{1}{105} = 0.00952$.   In other words, one unit of volume of *P. caudatum* consumes 1.64 times as much food as *P. aurelia*, and the food consumption of one unit of volume in the latter species constitutes but 0.61 of that of *P. caudatum*.   These coefficients enable us to recalculate the volume of one species into an equivalent in respect to the food consumption volume of another species.

On the fifth day of growth of a mixed population the biomass of
P. caudatum (in volume units) is equal to about 25, and of *P. aurelia*
to about 65.   If we calculate the total of these biomasses in equivalents of *P. aurelia*, we shall have: $(25 \times 1.64) + 65 = 106$ (maximal
free growth of *P. aurelia* is equal to 105).   The total of the biomasses
expressed in equivalents of *P. caudatum* will be $(65 \times 0.61) + 25 =$
65 (with the free growth 64).   This means that on the fifth day of
growth of the mixed population the food resources of the microcosm
are indeed completely taken hold of.

(4) The first period of competition up to the fifth day is not all
so simple as we considered it in the theoretical discussion of the third
chapter, or when examining the population of yeast cells.   The nature of the influence of one species on the growth of another does not
remain invariable in the course of the entire first stage of competition,
and in its turn may be divided into two periods.   At the very beginning *P. caudatum* grows even somewhat better in a mixed population
than separately (analogous to Fig. 22), apparently in connection
with more nearly optimal relations between the density of Paramecia
and that of the bacteria in accordance with the already mentioned
data of Johnson ('33).   At the same time *P. aurelia* is but very
slightly oppressed by *P. caudatum*.   As the food resources are used
up, the Johnson effect disappears, and the species begin to depress
each other as a result of competition for common food.

It is easy to see that all this does not alter in the least the essence
of the mathematical theory of the struggle for existence, but only
introduces into it a certain natural complication: the coefficients of
the struggle for existence, which characterize the influence of one
species on the growth of another, do not remain constant but in their
turn undergo regular alterations as the culture grows.   The curves
of growth of every species in a mixed population in Figure 24 up to
the fifth day of growth have been calculated according to the system
of differential equations of competition with such varying coefficients.
In the first days of growth the coefficient β is negative and near to
-1, i.e., instead of $-\beta N_1$ we obtain $+N_1$.   In other words, the
presence of *P. aurelia* does not diminish, but increases the possibility of growth of *P. caudatum*, which proceeds for a certain time
with a potential geometrical rate, outrunning the control culture
$\left(\dfrac{64 - N_2 + N_1}{64} \text{ remains near to unity} \right)$.   At this time the coefficient

α is equal to about $+0.5$; in other words, *P. aurelia* suffers from a slight depressing influence of *P. caudatum*. Later the inhibitory action of one species upon the growth of another begins to manifest itself more and more in proportion to the quantity of food consumed, because the larger is the part of the food resources already consumed the less is the unutilized opportunity for growth. In our calculations for *P. caudatum* from the second and for *P. aurelia* from the fourth days of growth we have identified the coefficients of competition with the coefficients of the relative food consumption, i.e., $\alpha = 1.64$, $\beta = 0.61$. It is obvious that this is but a first approximation to the actual state of things where the coefficients gradually pass from one value to another. The entire problem of the changes in the coefficients of the struggle for existence in the course of the growth of a mixed population (which apparently are in a great measure connected with the fact that the Paramecia feed upon living bacteria) needs further detailed investigations on more extensive experimental material than we possess at present.

(5) It remains to examine the second stage of the competition, i.e., the direct displacement of one species by another. An analysis of this phenomenon can no longer be reduced to the examination of the coefficients of multiplication and of the coefficients of the struggle for existence, and we have to do in the process of displacement with a quite new qualitative factor: the rate of the stream which is represented by population having completely seized the food resources. As we have already mentioned in Chapter III, after the cessation of growth a population does not remain motionless and in every unit of time a definite number of newly formed individuals fills the place of those which have disappeared during the same time. Among different animals this can take place in various ways, and a careful biological analysis of every separate case is here absolutely necessary. In our experiments the principal factor regulating the rapidity of this movement of the population that had ceased growing was the following technical measure: a sample equal to $\frac{1}{10}$ of the population was taken every day and then destroyed. In this way a regular decrease in the density of the population was produced and followed by the subsequent growth up to the saturating level to fill in the loss.

During these elementary movements of thinning the population and filling the loss, the displacement of one species by another took place. The biomass of every species was decreased by $\frac{1}{10}$ daily.

Were the species similar in their properties, each one of them would again increase by $\frac{1}{10}$, and there would not be any alteration in the relative quantities of the two species. However, as one species grows quicker than another, it succeeds not only in regaining what it has lost but also in seizing part of the food resources of the other species. Therefore, every elementary movement of the population leads to a diminution in the biomass of the slowly growing species, and produces its entire disappearance after a certain time.

(6) The recovery of the population loss in every elementary movement is subordinate to a system of the differential equations of competition. In the present stage of our researches we can make use of these equations for only a qualitative analysis of the process of displacement. They will show us exactly what particular species in the population will be displaced. However, the quantitative side of the problem, i.e., the rate of the displacement, still requires further experimental and mathematical researches and we will not consider it at present.

The qualitative analysis consists in the following. Let us assume that the biomass of each component of the saturating population is decreased by $\frac{1}{10}$. Then according to the system of differential equations, inserting the values of the coefficients of multiplication and of the coefficients of food consumption, we shall be able to say how each one of the components can utilize the now created possibility for growth. The result of the calculations shows that *P. aurelia*, primarily owing to its high coefficient of multiplication, has an advantage and increases every time comparatively more than *P. caudatum*.[2]

In summing up we can say that in spite of the complexity of the process of competition between two species of infusoria, and as one may think a complete change of conditions in passing from one period of growth to another, a certain law of the struggle for existence which may be expressed by a system of differential equations of competition remains invariable all the time. The law is that the species possess definite potential coefficients of multiplication, which are realized at

---

[2] It is obvious that in these calculations it is necessary to introduce varying coefficients of the struggle for existence. At the same time with our technique of cultivation corrections to the "elementary movements" must be also included in an analysis of the first stage of growth of a mixed population (an approximation to the asymptote). But at the present stage of our researches we have neglected them.

every moment of time according to the unutilized opportunity for growth. We have only had to change the interpretation of this unutilized opportunity.

(7) It seems reasonable at this point to coördinate our data with the ideas of the modern theory of natural selection. It is recognized that fluctuations in numbers resembling the dilutions we have artificially produced in our microcosms play in general a decisive rôle in the removal of the less fitted species and mutations (Ford, '30). An interesting mathematical expression of this process proposed by Haldane ('24, '32) can be formulated thus: how does the rate of increase of the favorable type in the population depend on the value of the coefficient of selection k? In its turn the coefficient of selection characterizes an elementary displacement in the relation between the two types per unit of time—one generation. Therefore the problem resolves itself into a determination of the functional relationship between the increase of concentration of the favorable type and the elementary displacement in its concentration. A recent theoretical paper by Ludwig ('33) clearly shows how the fluctuation in the population density alters the relation between the two types owing to the fact that one of them has a somewhat higher probability of multiplication than the other. It seems to us that there is a great future for the Volterra method here, because it enables us not to begin the theory by the coefficient of selection but to calculate theoretically the coefficient itself starting from the process of interaction between the two species or mutations.

[*Editors' Note:* Material has been omitted at this point.]

REFERENCES

Darby, H. H. 1929. The effect of the hydrogen ion cencentration on the sequence of protozoan forms. Arch. Protistenkunde 65, p. 1.

Ford, E. B. 1930. Fluctuation in numbers, and its influence on variation in *Militaea aurinia*. Trans. Ent. Soc. Lond. 78, p. 345.

Gause, G. F. 1934a. Experimental analysis of Vito Volterra's mathematical theory of the struggle for existence. Science 79, p. 16.

Gause, G. F., O. K. Nastukova, and W. W. Alpatov. 1935. The influence of biologically conditioned media on the growth of a mixed population of *Paramecium caudatum* and *Paramecium aurelia*. Journ. Anim. Ecol. (in press).

Haldane, J. B. S. 1924. Trans. Camb. Phil. Soc. 23, p. 19.

Haldane, J. B. S. 1932. The Causes of Evolution. London.

Johnson, W. H. 1933. Effects of population density on the rate of reproduction in *Oxytricha*. Physiol. Zool., 6:22.

Ludwig, W. 1933. Der Effekt der Selektion bei Mutationen geringen Selektion-swerts. Biol. Zbl. 53, p. 364.

Woodruff, L. L. 1911. The effect of excretion products of *Paramecium* on its rate of reproduction. Journ. Exp. Zool. 10, p. 551.

Woodruff, L. L. 1914. The effect of excretion products of Infusoria on the same and on different species, with special reference to the protozoan sequence in infusions. Journ. Exp. Zool. 14, p. 575.

8

Reprinted from *Amer. Naturalist,* **104**(938), 413-423 (1970)

COMMUNITY EQUILIBRIA AND STABILITY, AND AN EXTENSION OF THE COMPETITIVE EXCLUSION PRINCIPLE

SIMON A. LEVIN

Department of Mathematics, Center for Applied Mathematics, and Section of Ecology and Systematics, Cornell University, Ithaca, New York 14850

Beginning with the fundamental work of Volterra (1926), a large amount of the ecological literature has dealt with an elaboration of the concept of niche by means of one or another of the various forms of what is known alternatively as the "Gause hypothesis" (Slobodkin 1961*a*), "Gause's principle" (Odum 1959), or even "Gause's axiom" (Slobodkin 1961*a*), depending on how one feels about it. I shall take a middle course in this paper, referring to it as the "Gause principle" or the "competitive exclusion principle" and shall generalize the result to other than purely competitive situations, and ones involving an arbitrary number of species.

1. PRELUDE

Although the result is generally attributed to Gause because of his experimental evidence for it, the theory goes back at least to Grinnell (1904, 1917), and was formally developed by Volterra (1926, 1931), who set the stage for all that was to follow. Considering a situation in which all species are resource-limited, Volterra showed that only one species can survive on a single resource. Later theoretical work by MacArthur and Levins (1964; see also MacArthur and Wilson 1967; Levins 1968) extended this concept to show that, in general, there can be no more species than resources. The same theme was developed independently by Rescigno and Richardson (1965), still with the strong limitation that all species are resource-limited.

As stated earlier, an empirical substantiation of this principle—at least in the case of two species feeding on the same resource—came by way of Gause's experiments on competition between *Paramecium caudatum* and *P. aurelia* (Gause 1934).

In isolated cultures, each protozoan species reached a constant positive equilibrium concentration—64 per cubic centimeter in the case of *P. caudatum,* and 105 per cubic centimeter in the case of *P. aurelia.* However, after 16 days in a mixed culture containing both species, only *P. aurelia* survived, having won out in competition with *P. caudatum* for the common resource. Thus the principle became generally known as the ''Gause principle,'' rather than the ''Volterra principle,'' although some writers (e.g., Hutchinson 1953; MacArthur 1958) do refer to it as the ''Volterra-Gause'' principle.

Later work on competition (e.g., Utida 1953; Frank 1957; MacArthur 1958; Slobodkin 1961*b*; Park 1962) and on character displacement (see Brown and Wilson 1956) lent strong evidence for the robustness of the result, by somewhat different approaches to the problem. MacArthur studied five species of warbler which are congeneric and so similar in their ecological preferences as to constitute an apparent threat to the Gause principle. However, he succeeded in showing that the feeding habits of the five species were significantly different from one another, and that the species were thus occupying distinct ecological niches, as predicted by the Gause principle.

Character displacement evidence is somewhat more satisfying than the Gause experiments because it demonstrates that the result of such competition need not be the elimination of one species, but instead adaptive changes in the competing species.

In dealing with situations where resources are not the only factors limiting the populations, one must generalize the original statement of the result. If the resources are not in short supply and other factors become crucial, two species can coexist on the same resource, provided they are being limited by different factors. The rule is then modified to its most familiar form: (1) ''No two species can indefinitely continue to occupy the same ecological niche'' (Slobodkin 1961*a*). The danger in this statement is exemplified by the following further quotation from the same source: (2) ''Operationally, it seems most appropriate to define an ecological niche or ecological space as that space which no two species can continue to occupy for an indefinitely long period of time.''

If one accepts this view, one must decide whether one wishes to make statement (1) first, or statement (2); and in a sense it is not possible to make either until the other has been made. On the other hand, the advantage of this approach is that if one assumes that the set defined in (2) is well-defined, no one can any longer dispute the Gause principle, which becomes a tautology.

To get around this difficulty, one must attempt to form the result in terms of an independent definition of ''niche,'' as in the following discussion from Slobodkin (1961*a*) of the concept of the niche as developed by Hutchinson (1957; see also MacFadyen 1957).

Given a region of physical space in which two species do persist indefinitely at (or close to) a steady state, there exists one or more properties of the environment or species, or of both, that ensures an ecological distinction between the two species, and if one were able to construct the multi-dimensional, fundamental niche of these two species a region would be found in this multi-dimensional space that is part of the fundamental niche of one of the species but not of the other; and similarly, a region would be found that is part of the fundamental niche of the second species but not of the first. It would further be the case that the physical space in which the two species persist indefinitely at, or near, their steady state, represents a real-world projection of those portions of the fundamental niches of the two species that are not identical. If they seem identical the study is incomplete.

The point Slobodkin makes is that there is no extrinsic definition of niche possible that will make the Gause principle testable. The further and crucial objection to this form of the statement is that the conclusion is often drawn (although it certainly does not follow from the statement) that two species can coexist, provided their ecological niches—regarded as hypervolumes in multidimensional space—overlap in what amounts to a "sufficiently small" fraction of each. It is the purpose of this paper to show that this is not true, that there are instead certain dimensions of the hypervolumes of paramount importance. Which dimensions those are is determined by which factors are limiting those species, be those factors resources, predators, or others. Two species cannot coexist unless their limiting factors differ and are independent; this is the only criterion one need examine at a given time and place. That is, it is only necessary to determine whether the species differ in this aspect of the niche. Species which appear to fill different niches may be serious competitors. MacArthur (1958)—in the case where all species are resource-limited—comes close to saying this when he says that the proper statement of the Volterra-Gause principle is that "species divide up the resources of a community in such a way that each species is limited by a different factor." Root (1967) clearly recognizes this point when he refers to the "core of limiting factors that defines the fundamental niche of a particular species."

2. DEVELOPMENT OF THE MODEL

Let us consider an ecological community made up of n components in densities[1] $x_1, \ldots, x_n$. These components may or may not represent biological species. Depending on the situation, we might want to allow them to be various segments of species in order to study the effects of community interactions on natural selection, or they might be chosen to be collections of species which are acting similarly. Further, among the components, we might wish to include measures of various aspects of the community that do not represent biological populations: concentrations of chemical nutrients, of shelter, of available space. These present no special problems.

We similarly consider variables $y_1, \ldots, y_m$ which affect the densities of the components in the community, but are external to the community, in the sense that the values of the variables $y_1, \ldots, y_m$ are not significantly affected by the species $x_1, \ldots, x_n$. Examples of these are climatic variables and the densities of some invading species.

---

[1] We shall henceforth use the symbols here introduced both to refer to the components per se and to the densities of those components. No confusion is expected to result.

We therefore assume that the dynamics of the community are described by the following familiar modification of the equations of Lotka (1956):

$$\frac{dx_1}{dt} = x_1 f_1 (x_1, \ldots, x_n; y_1, \ldots, y_m), \ldots,$$

$$\frac{dx_n}{dt} = x_n f_n (x_1, \ldots, x_n; y_1, \ldots, y_m).$$

$$(1)$$

For technical purposes, the functions f_i are assumed to be smooth. Fluctuations in the environment are allowed for by means of the inclusion of the parameters $y_1, \ldots, y_m$, as noted above. Invasions can similarly be dealt with through the y variables.

The function $f_i(x_1, \ldots, x_n; y_1, \ldots, y_m)$ represents the growth rate per individual of the ith component. Anything which influences the function f_i is a factor regulating the ith component of the community, and we accordingly call it a *regulating factor* or a *limiting factor* for that component. That is, a limiting factor for component x_i is any x_j or y_k or combination (linear or nonlinear) of them which influences the function f_i, that is, on which f_i depends.

The first observation that must be made is that limiting factors are only locally defined in phase space (the space of variables $x_1, \ldots, x_n$). That is, it is clear that which factors will be so operating for a given component at a given time is determined by the densities of all components in the community and all parameters at that time. The intricacy of these effects is pointed up beautifully in the work of Paine (1966), discussed below in section 5.

A second point is that the definition specifically allows the limiting factors to be combination factors. If for example, a fine-grained species (MacArthur and Levins 1964) utilizes two resources, R_1 and R_2, with proportionate utilization efficiencies, α_1 and α_2, then the true limiting factor for the species is $\alpha_1 R_1 + \alpha_2 R_2$. Certainly R_1 and R_2 individually meet the qualifications listed above for limiting factors; but since there is clearly only one independent limiting factor, it is preferable to discuss limitation in terms of the single factor $\alpha_1 R_1 + \alpha_2 R_2$, rather than the two factors R_1, R_2.

More generally, for the entire community, we assume that there exists a minimal independent set of limiting factors $z_1 (x_1, \ldots, x_n; y_1, \ldots, y_m)$, $\ldots, z_p (x_1, \ldots, x_n; y_1, \ldots, y_m)$, where p will be $\leq n$.   By a "minimal independent set," we mean one with the property that, near the point in phase space in which we are interested at any given time, each f_i may be expressed as a function only of these factors, and that no smaller collection of factors would suffice for this purpose. In practice, of course, the exact determination of these functions may be extremely difficult.

As we said that $z_1, \ldots, z_p$ are the limiting factors for the components $x_1, \ldots, x_n$, we say in turn that $x_1, \ldots, x_n$ are limited by $z_1, \ldots, z_p$.

With the introduction of the factors $z_1, \ldots, z_p$, the system (1) may be rewritten

$$\frac{dx_1}{dt} = x_1 f_1 (z_1, \ldots, z_p), \ldots, \frac{dx_n}{dt} = x_n f_n (z_1, \ldots, z_p). \qquad (2)$$

As is natural and usually done (Volterra 1926; Kostitsyn 1937; Mac-Arthur and Levins 1964; Levins 1968; Rescigno and Richardson 1965), we make the assumption that the functions f_i are linear functions of $z_1, \ldots, z_p$. (No similar assumption is made however regarding the nature of the factors $z_1, \ldots, z_p$.) The linearity assumption makes graphic the results we are going to prove. However, such strong assumptions are unnecessary, as will be pointed out in the Appendix. The assumption is equivalent to the statement that each f_i can be written $f_i (z_1, \ldots, z_p) = \alpha_{i1}z_1 + \ldots + \alpha_{ip}z_p + \gamma_i$ for suitable constants $\alpha_{i1}, \ldots, \alpha_{ip}, \gamma_i$.

3. EQUILIBRIA AND STABILITY

For every choice of the parameters $y_1, \ldots, y_m$, the right-hand sides of the equations (2) define n functions of the variables $x_1, \ldots, x_n$. As the parameters $y_1, \ldots, y_m$ change, those functions $f_1, \ldots, f_n$ change, but presumably at much slower rates than the variables $x_1, \ldots, x_n$ themselves. Systems for which this is not valid obviously require a somewhat different analysis, as Hutchinson (1953) has suggested (see also Pimentel 1961).

We then are interested in whether the system can tend to some equilibrium, either constant or dynamic, and to what degree that equilibrium is resistant to slight perturbation. For the purposes of this paper, we require the equilibrium to be constant or cyclic (i.e., corresponding to a point orbit or periodic orbit), but for mathematical purposes, this notion could be generalized somewhat (see Appendix). We shall always be interested in the stability of equilibria for which all $x_i > 0$, since if some $x_i = 0$, we simply revise our original equations, removing x_i as a member of the community (and introducing it as a parameter if invasions by it are likely to affect the stability of the community). Thus the equilibria we are interested in are always bounded, and bounded away from the coordinate planes $x_i = 0$.

When one superimposes the random nature of the environment on the system, one expects to find the system oscillating about the purported equilibrium without ever attaining it, but this does not deny the existence of the equilibrium in the deterministic model.

There are several rather distinct measures of stability to apply to ecological situations, each yielding different information. I mention two, although it is only the second we shall apply, because the first deserves much more attention by ecologists than it has received in connection with the stability of ecological systems.

As pointed out at the beginning of this section, the functions f_i are changing, but on a time scale relative to which the x_i will in general appear to have already reached equilibrium. Thus, as we view the process on this time scale, we see the x_i in movement from one equilibrium to another, and generally by a smooth transition. Such an equilibrium—that is, one which changes smoothly (continuously) into a nearby equilibrium when the functions f_i are slightly altered—is called *structurally stable*. If, however, the community evolves in this smooth way only up to some point where a radical (discontinuous) change takes place, then the equilibrium when this occurs is structurally unstable. An example of this is when some physiographic factor (e.g., a glacial tongue) is gradually splitting a community. At some point, geographical isolation significant enough to allow speciation will result, and this is the point of structural instability. A similar situation occurs during centrifugal speciation (Brown 1957). For a further discussion of this topic and its relation to biological systems, one is referred to Thom (1969) and Lewontin (1969).

We shall be concerned here with what Lewontin (1969) refers to as "neighborhood stability," and what mathematicians know as "asymptotic orbital stability" (Coddington and Levinson 1955). Given that an equilibrium has been attained (and is varying with the changes in the parameters), is it stable with regard to fluctuations in the values of the x_i variables themselves? Any situation which does not satisfy this type of stability criterion cannot long prevail. In the case of the periodic orbit, this does not require that the perturbed state return to the point from which it was perturbed, but merely to some arbitrarily small neighborhood of the periodic orbit as a whole.

4. THE EXTENDED PRINCIPLE

We now state the major result of this paper:

No stable equilibrium can be attained in an ecological community in which some r components are limited by less than r limiting factors. In particular, no stable equilibrium is possible if some r species are limited by less than r factors.

It is not difficult to prove this result. Suppose, without loss of generality (since we may always relabel the components), that the first r components are limited by less than r factors. Then the functions $f_1, \ldots, f_r$, being linear functions of less than r variables, are linearly related; that is, there exist constants $\beta_1, \ldots, \beta_r$ and δ such that $\beta_1 f_1 + \ldots + \beta_r f_r = \delta$. If one now returns to system (2), one infers at once the relation

$$\beta_1 \frac{1}{x_1} \frac{dx_1}{dt} + \ldots + \beta_r \frac{1}{x_r} \frac{dx_r}{dt} = \delta,$$

which clearly may be integrated to yield $\beta_1 \ln x_1 + \ldots + \beta_r \ln x_r = \delta t +$ constant, i.e., $x_1^{\beta_1} \ldots x_r^{\beta_r} = K e^{\delta t}$.

Such a relationship must be valid for every solution to (2), with only the constant K allowed to vary from solution to solution. Thus, in particular the constant δ may be determined by examining any one solution, for example the proposed equilibrium. Since that equilibrium is constant or cyclic and has all $x_i > 0$ (see section 3 above), it is clear that $x_1^{\beta_1} \ldots x_r^{\beta_r}$ must stay bounded, and bounded away from zero. We may therefore eliminate the possibilities $\delta > 0$, for which $e^{\delta t}$ would grow without bound as t increased, and $\delta < 0$, for which $e^{\delta t}$ would tend to zero. Hence $\delta = 0$, and so the passage from any initial state to equilibrium (or nonequilibrium!) is confined to one of the surfaces $x_1^{\beta_1} \ldots x_r^{\beta_r} = K$. The proposed stable equilibrium must be confined to one such surface. A slight perturbation from it will in general carry the motion onto a different surface, to which it is thereafter constrained. Thus there can be no asymptotic return to the original equilibrium, which was, thus, not a stable equilibrium.

5. CONCLUDING REMARKS

The experiments of Paine (1966) in the intertidal waters at Mukkaw Bay, Washington showed that the removal of *Pisaster,* the top predator, resulted in the reduction of the number of species in the community from 15 to eight. Presumably, the presence of *Pisaster* had made possible the independent operation of a great many more limiting factors than was possible without *Pisaster*. In its absence, species which no longer had to contend with predation were now able to dominate several resources that could previously be divided among more species. Here then was a clear example of the operation of the extended limiting principle.

It can be shown that in certain situations the instability that occurs due to an insufficient number of limiting factors is related to the existence of numerous possible unstable equilibria in the neighborhood of every equilibrium point. When the number of limiting factors is only one less than the number of components, there will in general exist a one-parameter family of such equilibria; if it is two less, there will exist a two-parameter family. Allowing for a random environment being superimposed on the community, J. Dunn (personal communication) has remarked on the relationship between such a situation and a one- or two-dimensional random walk, particularly with regard to the ergodic properties of such walks. (The ergodic property is not valid for three-dimensional walks [Feller 1957].) The relationship of this phenomenon to apparent cyclicity in the size of animal populations bears further study.

Hutchinson (1964) has made this point in a slightly different way: "If the two species were almost equally efficient over a wide range of environmental variables, competitive exclusion would be a slow process. Both species then might oscillate in varying numbers, but persist almost indefinitely."

6. SUMMARY

It is shown in this paper that no stable equilibrium can be attained in an ecological community in which some r of the components are limited by less than r limiting factors. The limiting factors are thus put forward as those aspects of the niche crucial in the determination of whether species can coexist.

For example, consider the following simple food web:

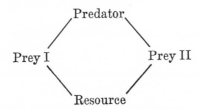

Despite the similar positions occupied by the two prey species in this web, it is possible for them to coexist if each is limited by an independent combination of predation and resource limitation, since then two independent factors are serving to limit two species.

On the other hand, if two species feed on distinct but superabundant food sources, but are limited by the same single predator, they cannot continue to coexist indefinitely. Thus these two species, although apparently filling distinct ecological niches, cannot survive together. In general, each species will increase if the predator becomes scarce, will decrease where it is abundant, and will have a characteristic threshold predator level at which it stabilizes. That species with the higher threshold level will be on the increase when the other is not, and will tend to replace the other in the community. If the two have comparable threshold values, which is certainly possible, any equilibrium reached between the two will be highly variable, and no stable equilibrium situation will result. This is not the same as dismissing this situation as "infinitely unlikely," which is not an acceptable argument in this case. Hutchinson's point of the preceding section vividly illustrates this.

The results of this paper improve on existing results in three ways. First, they eliminate the restriction that all species are resource-limited, a restriction persistent in the literature. Second, the results relate in general to periodic equilibria rather than to constant equilibria. Third, the nature of the proof relates to the crucial question of the behavior of trajectories near the proposed equilibrium, and provides insight into the behavior of the system when there is an insufficient number of limiting factors.

ACKNOWLEDGMENTS

The author's research has been supported in part by the Office of Naval Research, contract N00014-67-A-0077-0008.

The author wishes to express his deep appreciation to R. Root for his careful reading of this paper, and for his invaluable suggestions and general counsel. A further debt of gratitude is owed to all who listened and advised, including H. D. Block, W. L. Brown, Jr., D. Robson, J. Dunn, C. Dafermos, T. Chang, and especially, Carole Levin.

Finally, this paper cannot close without an exposition of certainly the most elegant form of the niche theorem (section 1) to appear anywhere (unfortunately without proof). The following is taken from Dr. Seuss's epic "On Beyond Zebra" (Geisel 1955).

> And NUH is the letter I use to spell Nutches
> Who live in small caves, known as Nitches, for hutches.
> These Nutches have troubles, the biggest of which is
> The fact there are many more Nutches than Nitches.
> Each Nutch in a Nitch knows that some other Nutch
> Would like to move into his Nitch very much.
> So each Nutch in a Nitch has to watch that small
>   Nitch
> Or Nutches who haven't got Nitches will snitch.

W. L. Brown, Jr., and R. Root were aware of this statement long before the author, but it should indeed have wider currency.

APPENDIX

The purpose of this section is to make clear the mathematical principles underlying the results of this paper.

We shall call a (finite) set A of functions dependent on a set B of functions, if every function in A is expressible in terms of the elements of B. In this paper, we shall simply call the set A dependent if it is dependent on a set B with fewer elements. It is a simple exercise to show that this property is inherited by any (finite) set C which includes A as a subset, since C is dependent on the set D which is formed by adding to B all the elements in C but not in A. Clearly, if B has fewer elements than A, then D has less than C.

Applied to the functions $f_1, \ldots, f_n$ in (2) when some r components are limited by less than r limiting factors, the definitions above simply say that the set $\{f_1, \ldots, f_n\}$ has a dependent subset containing r elements. By the above remarks, this means that the set $\{f_1, \ldots, f_n\}$ itself is dependent. The principal conclusion of the paper may thus be seen to depend only on the statement (subject to the linearity assumptions): if $p < n$, the system (2) has no point or periodic orbits which are strictly positive (that is, for which each x_i is always strictly positive) and asymptotically orbitally stable (section 3).

Because of the positivity assumption, one can take logarithms in system (2), and the theorem is equivalent to the theorem that if $p < n$ the system

$$\frac{dX_1}{dt} = F_1(Z_1, \ldots, Z_p), \ldots, \quad \frac{dX_n}{dt} = F_n(Z_1, \ldots, Z_p)$$

has no asymptotically orbitally stable periodic or point orbits. Here $Z_j(X_1, \ldots, X_n) = z_j(e^{X_1}, \ldots, e^{X_n})$, $j = 1, \ldots, n$. An analogous proof to that given previously works if the functions F_i are linear, and can be extended to a much wider class of functions F_i. On the other hand, one can alternatively obtain results for either system by making no assumptions on the F_i, but by taking the Z_j to be linear, for one can then simply introduce $Z_1, \ldots, Z_p$ as new dependent variables. The proof in the general case (that is, with no such assumptions on the F_i or the Z_j) is still not known (except in a slightly modified form for $n = 2$), and of course this means that the truth of the mathematical theorem itself is still an open question.

One other means of extension is possible, and that is by remarking that the mathematical results proved in this paper remain valid if the notion of a stable periodic orbit is replaced by the more general concept of a closed invariant set which is the union of finitely many orbits. This is a nontrivial mathematical point, but hardly worth discussing in detail here.

LITERATURE CITED

Brown, W. L., Jr. 1957. Centrifugal speciation. Quart. Rev. Biol. 32:247–277.

Brown, W. L., and E. O. Wilson. 1956. Character displacement. Syst. Zool. 5:49–64.

Coddington, E. A., and N. Levinson. 1955. Theory of ordinary differential equations. McGraw-Hill, New York. 443 p.

Feller, W. 1957. An introduction to probability theory and its applications. Vol. 1. 2d ed. Wiley, New York. 477 p.

Frank, P. W. 1957. Coactions in laboratory populations of two species of *Daphnia*. Ecology 38:510–519.

Gause, G. F. 1934. The struggle for existence. Williams & Wilkins, Baltimore. 163 p.

———. 1935. La théorie mathématique de la lutte pour la vie. Hermann & Cie., Paris. 61 p.

Geisel, Theodor Seuss (Dr. Seuss). 1955. On beyond zebra. Random House, New York. 70 p.

Grinnell, J. 1904. The origin and distribution of the chestnut-backed chickadee. Auk 21:364–382.

———. 1917. The niche relationships of the California thrasher. Auk 34:427–433.

Hutchinson, G. E. 1953. The concept of pattern in ecology. Amer. Acad. Natur. Sci., Proc. 105:1–12.

———. 1957. Concluding remarks. Cold Spring Harbor Symp. Quant. Biol. 22:415–427.

———. 1964. The lacustrine microcosm reconsidered. Amer. Sci. 52:334–341.

Kostitsyn, V. A. 1937. Biologie mathématique. Librarie Armand Colin, Paris. 223 p.

Levins, R. 1968. Evolution in changing environments. Monographs in Population Biology, No. 2. Princeton Univ. Press, Princeton, N.J. 130 p.

Lewontin, R. C. 1969. The meaning of stability. Pp. 13–24, *in* Diversity and stability in ecological systems. Brookhaven Symposium in Biology No. 22. Brookhaven National Laboratory. Available from Clearinghouse for Federal Scientific and Technical Information, Nat. Bur. Standards, U.S. Dept. Commerce, Springfield, Va.

Lotka, A. J. 1956. Elements of mathematical biology. Dover, New York. 460 p.

MacArthur, R. 1958. Population ecology of some warblers of northern coniferous forests. Ecology 39:599–619.

MacArthur, R., and R. Levins. 1964. Competition, habitat selection, and character displacement in a patchy environment. Nat. Acad. Sci., Proc. 51:1207–1210.

MacArthur, R., and E. O. Wilson. 1967. The theory of island biogeography. Monographs in Population Biology, No. 1. Princeton Univ. Press, Princeton, N.J. 215 p.

MacFadyen, A. 1957. Animal ecology, aims and methods. Pitman & Sons, London. 255 p.

Odum, E. P. 1959. Fundamentals of ecology. 2d ed. Saunders, Philadelphia. 546 p.

Paine, R. T. 1966. Food web complexity and species diversity. Amer. Natur. 100:65–76.

Park, T. 1962. Beetles, competition, and populations. Science 138:1369–1375.

Pimentel, D. 1961. Animal population regulation by the genetic feed-back mechanism. Amer. Natur. 95:65–79.

Rescigno, A., and I. W. Richardson. 1965. On the competitive exclusion principle. Bull. Math. Biophys. 27:85–89.

Root, R. B. 1967. The niche exploitation pattern of the blue-gray gnatcatcher. Ecol. Monogr. 37:317–350.

Slobodkin, L. B. 1961a. Growth and regulation of animal populations. Holt, Rinehart & Winston, New York. 184 p.

——. 1961b. Preliminary ideas for a predictive theory of ecology. Amer. Natur. 95: 147–153.

Thom, R. 1969. Topological models in biology. Topology 8:313–335.

Utida, S. 1953. Interspecific competition between two species of bean weevil. Ecology 34:301–307.

Volterra, V. 1926. Variazione e fluttuazione del numero d'individui in specie animali conviventi. Mem. Accad. Nazionale Lincei (ser. 6) 2:31–113.

——. 1931. Leçons sur la théorie mathématique de la lutte pour la vie. Gauthier-Villars, Paris. 214 p.

Part III
NICHE AXES

Editors' Comments
on Papers 9 Through 18

The principle of competitive exclusion states that in a stable community, no two species can be limited by the same factor. Species respond in differing ways to factors that vary within communities in space and time, and it is the differences in response patterns to these spatiotemporal differences that determine which species can coexist.

This chapter is directed toward the question of determining the nature of these intracommunity variables—identifying the critical variables that affect species composition. These include, most obviously, variations in resource availability, and competitor, predator, and parasite pressure, and these will be the first we shall discuss. One must go beyond this, however, to questions of seasonal time, genetic variation, and spatial variation, and these, too, will be given their due.

To begin, let us repeat the point of view developed in Levin (Paper 8). We assume the dynamics of the community to be described by equations

$$\frac{dx_1}{dt} = x_1 f_1 (x_1, ..., x_n; y_1, ..., y_m)$$

$$\cdot \qquad\qquad\qquad \cdot$$
$$\cdot \qquad\qquad\qquad \cdot$$
$$\cdot \qquad\qquad\qquad \cdot$$

$$\frac{dx_n}{dt} = x_n f_n (x_1, ..., x_n; y_1, ..., y_m);$$

and that these equations may be rewritten

$$\frac{dx_1}{dt} = x_1 f_1 (z_1, ..., z_p)$$
$$\cdot$$
$$\cdot$$
$$\cdot$$

$$\frac{dx_n}{dt} = x_n f_n (z_1, ..., z_p)$$

in terms of a minimal independent set of limiting factors $z_1 (x_1, ..., x_n; y_1, ..., y_m), ..., z_p (x_1, ..., x_n; y_1, ..., y_m)$. We do not consider at this point the complicating effects of spatial variables (Part V), time lags (differential-difference equations), or other forms of memory (e.g., integrodifferential equations).

Although limiting factors are locally defined in state space, the choice of a minimal independent set, the "core of limiting factors that defines the fundamental niche of a particular species" (Paper 10), must be so made as to be applicable to the entire community over the full range of the relevant variables. For the community, a minimal independent set of limiting factors provides the most compact set of niche axes. In practice, however, it may often be simpler to choose as axes a somewhat larger set of limiting factors which functionally combine to form the "core of factors." In this part we examine the most important such *niche axes.*

Several points are in order before we begin our discussion:

1. The notion of *niche* is intricately interwoven with those of *species* and *community*. If the species and community are well-defined and unitary entities, so is the niche, as the species' fundamental role within the commnuity. If the species and community are not well-defined units, neither is the niche. It is not our purpose to confront here the problems associated with the definition of "community"; for this purpose, the reader is referred to McIntosh (1967) or Whittaker (1967, 1975). Species, however, often comprise genetically different populations (ecotypes, subspecies, etc.) occurring in different biotopes and communities, interacting in these with different species. As a unitary concept a niche characterizes a particular species population in a particular community. A species possessing a range of different populations in different communities must possess a range of niche characteristics.

Even when the community is clearly defined, it may not provide the appropriate level at which to view the species' evolution. The opportunity for species to disperse among communities, in particular, argues that the appropriate level to view competitive exclusion may require consideration of many communities (Levin, 1974). In this case the notion of *niche* gives way to the broader notion of *ecotope* (Paper 41); and this solution applies as well to the problem posed in the preceding paragraph.

Just as the community may provide too narrow a context in which to view exclusion, a community-level approach may also prove too broad, neglecting the fine structure of intracommunity variation. We return to this point in Part IV.

2. The inclusion of parametric dependence in the definition of limiting factors is made necessary by the background of environmental variation to which the community is exposed. As the community evolves, it will follow a certain "integral path" in niche space (the space defined by the niche axes). The problem of how to treat the various parameters—e.g., which should be included as niche axes?—requires some delicate handling. We shall employ the following conventions.

The operational notion of the "rate of environmental change" is relative to individual species and the rate at which population characteristics are affected. Change which is so rapid that only time averages or new equilibria are observed by the populations is therefore to be viewed only in terms of those time averages or equilibria. For example, the logistic and Lotka–Volterra equations introduced earlier in this volume (Part II) by measuring competitive effects directly in terms of competitor densities implicitly assume that effects of changing competitor densities on resources are instantaneous, reaching new equilibria which perfectly track competitor densities.

Similarly, Gause (1934; Paper 4) modeled one of his laboratory systems by equations of the form

$$\frac{dN_1}{dt} = b_1 N_1 \frac{K_1 - N_1 - aN_2}{K_1}$$

$$\frac{dN_2}{dt} = b_2 f(C) N_2 \frac{K_2 - N_2 - \beta N_1}{K_2}$$

The notation differs slightly from Gause to enhance clarity. The equation for C is

$$\frac{dC}{dt} = g(N_1, N_2) - nC,$$

where N_1 and N_2 are the respective densities of *Paramecium bursaria* and *P. caudatum*; and where C is the concentration of waste products, considered inhibitory to *P. caudatum*. If waste levels adjust to species levels very rapidly, then a standard psuedo-steady-state approximation would set $dC/dt = 0$, so that $C = g(N_1, N_2)/n$. Substituting this into the system reduces it to

$$\frac{dN_1}{dt} = b_1 N_1 \frac{K_1 - N_1 - aN_2}{K_1}$$

$$\frac{dN_2}{dt} = b_2 f\left(\frac{g(N_1, N_2)}{n}\right) N_2 \frac{K_2 - N_2 - \beta N_1}{K_2},$$

in which the parameter C no longer appears explicitly.

When parametric change is slow compared to such interpopulation effects as exclusion, those parameters are treated first as constants, not as niche axes. The community then is viewed as reaching some equilibrium pattern in its niche space, a pattern that itself will evolve more slowly as the slow parameters change. Such parameters include those generally controlled in laboratory experiments, and most parameters of the natural environment of fast-breeding bacteria (Paper 18). They also include parameters that vary slowly in response to and on the same time scale as the community equilibria (e.g., late successional variables).

When parameters change on the same time scale as niche interactions, these parameters are usually appropriately included as niche axes. When nutrient levels change in response to consumer variation and on a comparable scale, the nutrient–consumer interaction is a classic prey–predator situation. The level of each "species" defines an axis in the niche space of the other. Hutchinson (Paper 18) suggests that the environment of the plankton and multivoltine insects probably occurs on this time scale and may actually prevent exclusion. If parameter change is coupled with population response in a mutual causal process, they must be treated together as "population variables." If, however, the parameter change is seasonal and unaffected by the community, it may be more appropriate to regard the niche axis as externally imposed on the community, and to treat packing along it as one would treat packing along a spatial gradient or a fixed intracommunity gradient (Part IV).

When parametric change is very rapid relative to life span, as perhaps with the fluctuation in the environments of birds and mammals, (Paper 18), these parameters are not treated as niche axes; but a very different viewpoint may be required. Differential equations may give way to stochastic differential equations; trajectories to evolving probability distributions.

In summary, niche axes relate to intracommunity gradients, including gradients over time when the time scale is comparable to that of exclusion. We turn now to a consideration of which niche axes are most crucial for an understanding of exclusion, hence of niche.

COMPETITION ALONG HEIGHT AND FOOD-SIZE AXES

Since the exclusion principle derives its notivation from the consideration of competitive processes, it makes sense to begin our enumeration of niche axes by discussing competitive axes. These include both resource characteristics and the densities of competitor species. Both are discussed in the laboratory work of Gause (Paper 7), Park (1962), and others. Resource axes are emphasized in the avian field work developed in the papers reprinted in this section, of MacArthur, MacArthur, and Preer (Paper 9), and of Root (Paper 10).

We have chosen to illustrate food-size and height relations with articles on birds. We should at least mention extensive work on niches of lizards, in which food size, height, perch size, and foraging behavior are among the directions of niche difference (Rand, 1964; Pianka, 1966; Schoener and Gorman, 1968; Schoener, 1970a, 1970b; Laerm, 1974; Huey et al., 1974; see reviews of Pianka, 1973, and Schoener, 1974). Schoener (1969) observes that males and females of island lizards may differ widely in size, with this niche difference between the sexes increasing the range of food sizes that can be used by the species; Storer (1966) has commented on the role of size difference between the sexes in North American hawks. Some further studies of food-size differences between bird species are by Kear (1962), Hespenheide (1966, 1971), Ashmole (1968), Ashmole and Ashmole (1967), Holmes and Pitelka (1968), and Pulliam and Enders (1971). We have not given a selection from, but should mention here, Lack's (1947) admirable study of the Galápagos finches. Food-size and other niche relationships in mammals are discussed for carnivores by Rosenweig (1966); rodents by Miller (1964, 1967), Rosenzweig and Winakur (1969), Brown and Lieberman (1973), and Smigel and Rosenzweig (1974); and bats by McNab (1971) and Fleming et al. (1972).

For MacArthur et al. (Paper 9), foliage density and height determine the most critical niche axes along which to study bird densities. The entire resultant "foliage profile" is the chief determinant of bird-species diversity. For Root (Paper 10) main axes of resource quality for foliage-gleaning birds were prey length and vertical distributions of the principal resource species. By these and other measurements Root was able to distinguish the niches of members of a *guild*—a group of species of closely related niches. In addition to Root's treatment we give a short selection on a guild of tropical pigeons by Diamond (Paper 11).

PREDATION AND PARASITISM

In the paper of Hutchinson (Paper 18) it is pointed out that the effects of competition can be affected to a large extent by other factors, especially densities of species other than the competitors themselves. Principal among these are predators and parasites. These need not be responsible for limiting species by themselves, but they could nevertheless operate as critical limiting factors, in the sense developed earlier, for one or for many competing species. This viewpoint is developed to some extent in the following studies by Paine (Paper 12), Slobodkin (Paper 13), and Utida (Paper 14). Our selections all concern predation effects on animals; some effects on plants are described by Harper (1969) and Janzen (1970). In Paine's study the role of the starfish *Pisaster ochraceus*, a top carnivore, is to reduce the level of the dominant competitor for intertidal space, the mussel *Mytilus californianus*, and thereby to permit coexistence with *Mytilus* of a much more diverse fauna and flora of space holders. Integral in this is the notion of spatiotemporal heterogeneity, to which we shall return. Further, the system is not closed. Competitively inferior species have a safe refuge from which new colonists may enter when opportunities appear.

The power of these examples is to point out that (because the competitive exclusion principle may be analogous to Liebig's Law of the Minimum), ecologists have often mistakenly sought to identify the limiting factors for species with variables that are clear biological entities (e.g. resource, competitor, parasite, or predator densities). It is misleading to ask whether a population is regulated by competition *or* predation, when in fact regulation may come about through some complex interaction of the two, as in Paine's intertidal experiments. The minimal choice of limiting factors is then some set of nonlinear functions of resource, competitor, and predator densities. It still is acceptable to embed these considerations in a higher-order hyperspace in which the integral entities are axes, and this may be much more convenient; but the complex nature of interaction and regulation should not be obscured.

SEASONAL AND SUCCESSIONAL TIME

Seasonal change involves parameter change on a time scale that populations can experience, hence an intracommunity gradient that must be treated as a niche axis. Seasonal relationships among bird species are discussed by Ricklefs (1966), Stallcup (1968), and Recher (Paper 38); Ricklefs tested, and rejected, the idea that difference in seasonal time relations was a basis for the high species diversity of tropical bird faunas. Broadhead and Wapshere (Paper 37) found seasonal time important as a direction of niche difference among species in a guild of bark lice. As a botanical example we give a short excerpt by Whittaker (Paper 15).

In many communities local population destructions form a mosaic of stages of successional recovery from disturbance; different species fit into different positions in this space and time mosaic. The effect is most familiar in land-plant communities affected by fire or windthrow, but, for example, the disturbance effects in littoral systems are similar in principle. Such effects are discussed in Levin and Paine (1974, 1975).

POPULATION STRUCTURE

The simplest equations of competition between pairs of species involve only two variables: the densities of the competing species. The parameters of these equations, however, may change through time as competition alters the structures of the populations (e.g., genetic structure, age composition, dispersion, sex ratio, or social structure). The dispersion problem is treated in Skellam (1951) and Levin (1974; Levin and Paine, 1974, 1975) and bridges the gap between intra- and intercommunity considerations. Complete competitors may coexist in an environment with spatial extent, even if one is an inferior competitor throughout, owing to the ability of that competitor to disperse. In a uniform environment in which either species can survive and resist invasion by the other once established, coexistence becomes possible due to a pattern of random colonization and consequent deversification (Levin, 1974; Colwell, 1973). Moreover, when extinctions of local populations are frequent, spatiotemporal diversification becomes possible (Levin, 1974; Levin and Paine, 1974, 1975).

Change in genetic structure, on a time scale to permit coexistence of complete competitors, requires inclusion of genetic variables as niche axes. Such an effect was suggested by Pimentel (1961) as being made possible by interspecific frequency-dependent selection favoring the rare species, and his arguments are summarized in Paper 16. Related mathematical considerations are treated by Levin (1972) for predator and prey.

The phenomenon of genetic change is fundamentally intertwined with change in age composition, and this connection is made in the work of Ayala (Paper 17), in which selection pressure varies with age. In his experiments, age composition and genetic structure equilibrate at levels that permit coexistence. Related mathematical considerations are treated in a series of papers by Charlesworth (1970, 1972, 1973) and by Charlesworth and Giesel (1972).

In place of our own summation for this section, we borrow from a master of exposition, G. E. Hutchinson for Paper 18. As developed by Hutchinson, the competitive exclusion principle becomes the concept about which further investigations are organized. Competitive exclusion is a fundamental force, as is gravity. That airplanes can fly does not negate the law of gravity. That species can coexist does not negate the principle of competitive exclusion. In an elegant article, "The Paradox of the Plankton," Hutchinson points out that in many natural situations, competitive exclusion may never have a chance to run its course because of the rapid rate at which the environment changes. Such environmental variation may swamp the effects of exclusion if it involves a variation in "most favored species"; and Hutchinson attributes the roots of the density-dependent versus density-independent modes of population regulation to the disparate laboratory and field experiences of the various proponents. The balance between exclusion and environmental variation will to some extent underlie the entries in Part IV; but Hutchinson's paper is reprinted here as a lesson in the care and use of the competitive exclusion principle.

REFERENCES

Ashmole, N. P. 1968. Body size, prey size, and ecological segregation in five sympatric tropical terns (Aves : Laridae). *Syst. Zool.* **17**: 292–304.

Ashmole, N. P., and M. J. Ashmole. 1967. Comparative feeding ecology of sea birds of a tropical oceanic island. *Peabody Mus. Nat. Hist., Yale Univ., Bull.* **24**: 1–131.

Brown, J. H., and G. A. Lieberman. 1973. Resource utilization and coexistence of seed-eating desert rodents in sand dune habitats. *Ecology* **54**: 788–797.

Charlesworth, B. 1970. Selection in populations with overlapping generations. I. The use of Malthusian parameters in population genetics. *Theoret. Pop. Biol.* **1**: 352–370.

Charlesworth, B. 1972. Selection in populations with overlapping generations. III. Conditions for genetic equilibrium. *Theoret. Pop. Biol.* **3**: 377–395.

Charlesworth, B. 1973. Selection in populations with overlapping generations. V. Natural selection and life histories. *Amer. Naturalist* **107**: 303–311.

Charlesworth, B, and J. T. Giesel. 1972. Selection in populations with overlapping generations. II. The relations between gene frequency and demographic variables. *Amer. Naturalist* **106**: 388–401.

Colwell, R. K. 1973. Competition and coexistence in a simple tropical community. *Amer. Naturalist* **107**: 737–760.

Fleming, T. H., E. T. Hooper, and D. E. Wilson. 1972. Three Central American bat communities: structure, reproductive cycles, and movement patterns. *Ecology* **53**: 555–569.

Harper, J. L. 1969. The role of predation in vegetational diversity. *Brookhaven Symp. Biol.* **22**: 48–62.

Hespenheide, H. 1966. The selection of food size by finches. *Wilson Bull.* **78**: 191–197.

Hespenheide, H. 1971. Food preference and the extent of overlap in some insectivorous birds, with special reference to the Tyrannidae. *Ibis* **113**: 59–72.

Holmes, R. T., and F. A. Pitelka. 1968. Food overlap among coexisting sandpipers on northern Alaska tundra. *Syst. Zool.* **17**: 305–318.

Huey, R. B., E. R. Pianka, M. E. Egan, and L. W. Coons. 1974. Ecological shifts in sympatry: Kalahari fossorial lizards *(Typhlosaurus). Ecology* **55**: 304–316.

Janzen, D. H. 1970. Herbivores and the number of tree species in tropical forests. *Amer. Naturalist* **104**: 501–528.

Kear, J. 1962. Food selection in finches with special reference to interspecific difference. *Proc. Zool. Soc. London* **138**: 163–204.

Lack, D. 1947. *Darwin's finches.* Cambridge University Press, Cambridge, England. 208 pp.

Laerm, J. 1974. A functional analysis of morphological variation and differential niche utilization in Basilisk lizards. *Ecology* **55**: 404–411.

Levin, S. A. 1972. A mathematical analysis of the genetic feedback mechanism. *Amer. Naturalist* **106**: 145–164.

Levin, S. A. 1974. Dispersion and population interactions. *Amer. Naturalist* **108**: 207–228.

Levin, S. A., and R. T. Paine. 1974. Disturbance, patch formation, and community structure. *Proc. Natl. Acad. Sci. USA* **71**: 2744–2747.

Levin, S. A., and R. T. Paine. 1975. The role of disturbance in models of community structure. *Ecosystem Analysis and Prediction: Proc. Conference Ecosystems, Alta, Utah, 1974,* ed. S. A. Levin, pp. 56–67. Soc. Indust. Appl. Math., Philadelphia.

McIntosh, R. P. 1967. The continuum concept of vegetation. *Bot. Rev.* **33**: 130–187.

McNab, B. K. 1971. The structure of tropical bat faunas. *Ecology* **52**: 352–358.

Miller, R. S. 1964. Ecology and distribution of pocket gophers (Geomyidae) in Colorado. *Ecology* **45**: 256–272.

Miller, R. S. 1967. Pattern and process in competition. *Adv. Ecol. Res.,* 4: 1–74.

Park, T. 1962. Beetles, competition, and populations. *Science* 138: 1369–1375.

Pianka, E. R. 1966. Convexity, desert lizards, and spatial heterogeneity. *Ecology* 47: 1055–1059.

Pianka, E. R. 1973. The structure of lizard communities. *Ann. Rev. Ecol. Syst.* 4: 53–74.

Pimentel, D. 1961. Animal population regulation by the genetic feed-back mechanism. *Amer. Naturalist* 95: 65–79.

Pulliam, H. R., and F. Enders, 1971. The feeding ecology of five sympatric finch species. *Ecology* 52: 557–566.

Rand, A. S. 1964. Ecological distribution in anoline lizards of Puerto Rico. *Ecology* 45: 745–752.

Ricklefs, R. E. 1966. The temporal component of diversity among species of birds. *Evolution* 20: 235–242.

Rosenzweig, M. L. 1966. Community structure in sympatric carnivora. *J. Mammal.* 47: 602–612.

Rosenzweig, M. L., and J. Winakur. 1969. Population ecology of desert rodent communities: habitats and environmental complexity. *Ecology* 50: 558–572.

Schoener. T. W. 1969. Size patterns in West Indian *Anolis* lizards. I. Size and species diversity. *Syst. Zool.* 18: 386–401.

Schoener, T. W. 1970a. Nonsynchronous spatial overlap of lizards in patchy habitats. *Ecology* 51: 408–418.

Schoener, T. W. 1970b. Size patterns in West Indian *Anolis* lizards. II. Correlations with the sizes of particular sympatric species — displacement and convergence. *Amer. Naturalist* 104: 155–174.

Schoener, T. W. 1974. Resource partitioning in ecological communities. *Science* 185: 27–39.

Schoener, T. W., and G. C. Gorman. 1968. Some niche differences in three Lesser Antillean lizards of the genus *Anolis. Ecology* 49: 819–830.

Skellam, J. G. 1951. Random dispersal in theoretical populations. *Biometrika* 38: 196–218.

Smigel, B. W., and M. L. Rosenzweig. 1974. Seed selection in *Dipodomys merriami* and *Perognathus penicillatus. Ecology* 55: 329–339.

Stallcup, P. L. 1968. Spatio-temporal relationships of nuthatches and woodpeckers in ponderosa pine forests of Colorado. *Ecology* 49: 831–843.

Storer, R. W. 1966. Sexual dimorphism and food habits in three North American accipiters. *Auk* 83: 423–436.

Whittaker, R. H. 1967. Gradient analysis of vegetation. *Biol. Rev.* 42: 207–264.

Whittaker, R. H. 1975. *Communities and Ecosystems.* 2nd ed. Macmillan, New York. 385 pp.

9

Reprinted from Amer. Naturalist, **96**(888), 167–174 (1962)

ON BIRD SPECIES DIVERSITY

II. Prediction of Bird Census from Habitat Measurements

ROBERT H. MacARTHUR, JOHN W. MacARTHUR AND JAMES PREER

Department of Zoology, University of Pennsylvania, Marlboro College
and Swarthmore College

A competent bird watcher can look at a habitat and correctly name the bird species which will breed there in abundance. This tells us that some properties of the general appearance of the habitat are sufficient to determine most of the breeding birds; it doesn't tell us just which measurable properties these are. In an earlier paper (MacArthur and MacArthur, 1961) the authors showed that in deciduous forests the diversity of breeding bird species depends upon foliage profile (foliage density plotted against height) and not upon plant species composition. Also, in this paper, it was suggested that each species requires a "patch" of vegetation with a particular profile for its selected habitat, and that the variety of "patches" of vegetation within a habitat determines the variety of bird species breeding there. In the present paper this hypothesis is tested; foliage profiles acceptable to many bird species are measured and compared with those available as patches in various habitats.

Plotting a profile as a point. Although it is doubtless more informative and more accurate to plot a complete profile as was done in the earlier paper, for our purposes it is easier and more picturesque to plot each profile as a point and then show the variety of profiles (representing the variety of "patches") as a cluster of points. This was done as follows: The vegetation was divided into three layers: that 0–2 ft from the ground, that 2–15 ft above ground, and that over 15 ft. (In the early paper 2–25' was used as the second layer; this was changed to 2–15 to improve the accuracy of the technique in brushy fields where most of the current work is being done). The proportion of the total foliage which is in each layer can be plotted in an equilateral triangle (see figure 1). From each point, imagine the three perpendicular lines dropped to the three sides. The lengths of these three lines are the proportions of foliage in the three layers, and the total length of the three combined is independent of the position of the point. Thus, a single point represents the proportions in the three layers and hence gives a crude picture of the profile. A 3-dimensional graph (in a tetrahedron instead of triangle) would allow us to represent a subdivision into four layers giving greater accuracy; in general it is possible with a point to represent the actual profile with any desired degree of accuracy if we are willing to plot (or imagine a plot!) in a space of enough dimension. The pictorial value of our triangle, which can actually be drawn, compensates for the crudeness of its detail. It would be nice to plot the density as well as the proportions, but this, too, would require an extra dimension.

Figures 1–3 show the variety of profiles found in five different habitats in Pennsylvania and Vermont.   Figure 1 contains (as a bar graph) measurements from a recently abandoned Pennsylvania field and (as dots) from a Pennsylvania "slash" plot which is a recently cut forest which has scattered tall trees surrounded by dense ground cover, berries, and young trees. Each habitat was divided into 100 ft squares and each of these is plotted

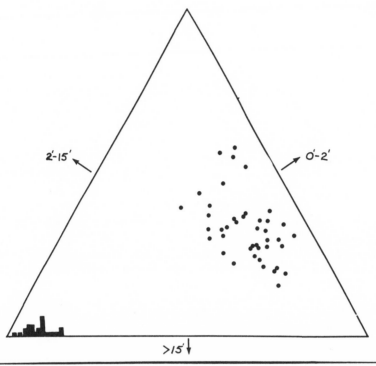

FIGURE 1. Each dot represents the proportions of foliage in each of the three horizontal layers (0–2', 2–15', over 15') in a 100 ft square of Pennsylvania slash plot. Each square (the squares adding into columns of a bar graph) represents the proportions of foliage in the three layers in 100 ft squares of a recently abandoned Pennsylvania field.

as a single point.   Figure 2 shows, as plus signs, the foliage proportions in 100 ft square patches of a dense Pennsylvania second growth forest and, as dots, a brushy field grown up in spots to clumps of young trees up to 40 ft high.   Figure 3 shows, as plus signs, 100 ft square patches of a Vermont second growth forest, and as dots, a Vermont slash plot. Imagine a horizontal line across these figures half way up; above this line the measurements are less accurate than below it.   For, the techniques of measurement (measuring the horizontal distance at which leaves obscure just 1/2 of a board from view — see earlier paper) are most accurate in practice where one can stand on the ground. Hence, measurements in which a substantial part of the vegetation is above 15 ft may be somewhat dis-

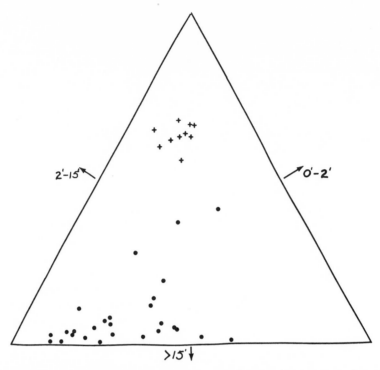

FIGURE 2. The proportions of foliage in three horizontal layers from 100 ft squares of two Pennsylvania habitats. The plus signs are patches of a dense second growth forest, and the dots are a brushy field.

placed in the figures from what they should be, although the cluster size is probably about right.

Clearly there is variability within habitats in the foliage profiles of different 100 ft squares. That is, there are patches in the habitats. And the slash plots and brushy field show much greater variation from patch to patch than do the field or second growth forest. This means that a wider variety of profiles (and hence, perhaps, a wider variety of bird species) are found in the slash and brushy field plots. As a general rule, plots located near the vertices of the triangular graph have little variability; those near the center have great variability. This is no accident; in large part it is due to the greater uncertainty of three approximately equi-probable foliage layers as compared with three layers of which one is overwhelmingly dominant. This is discussed in MacArthur and MacArthur (1961).

Notice that no plots were measured which lay along the middle of the left side of the triangle. That is, no plots had very little foliage in the 2–15 ft layer and yet much in both the 0–2 ft and > 15 ft layers.

BIRD HABITAT SELECTION

During the spring of 1961, breeding bird censuses of a large variety of habitats were made by plotting the territories of singing male birds. Within

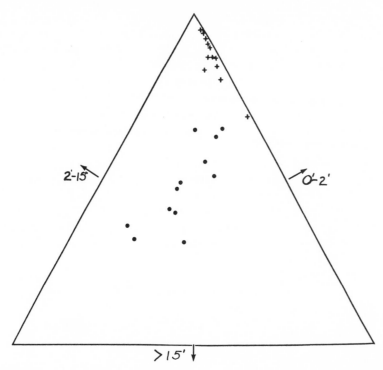

FIGURE 3. The proportions of foliage in three horizontal layers from 100 ft squares of two Vermont habitats. The plus signs are a very dense second growth forest and the dots are a "slash" plot.

each territory, the 100 ft square most used by the bird was selected, measured for foliage profile, and plotted as a point for the bird species which occupied that territory. Figure 4 shows the results of these measurements. (None of the points came from a habitat of which we wished to predict the census.) Notice particularly that the cluster of points for each species is fairly tight; that is, the graph *does* seem to reflect a large part of what the species choose in their habitat selection and each species has a characteristic range of acceptable profiles. (To show that the graph reflects all the bird requires, we must show that every 100 ft square patch whose profile lies in the birds cluster actually has a pair of that species. This is discussed later.) Some locally common species — notably towhee and indigo bunting — are not included. Such species, setting up their territories on an "edge," seem to require two kinds of patch. Pairs of points would be required for each such territory. No such measurements were made. Many species are not included simply because few or none of their territories were measured.

PREDICTION OF BIRD CENSUS

Armed with knowledge of birds' habitat choices and the variety of habitats present in a given area, it is now simple to predict the census of breeding birds, at least roughly. In fact, all we have to do is superimpose the graph of the birds' habitat requirements on the graph of the patch variability of the habitat whose census is to be predicted. And if the habitat contains many 100 ft square patches which are suitable for a bird species (as indicated by figure 4), then we predict that species should be common, and so on. If this prediction fails — if a bird is absent whose preferred vegetation profile is present — then we can conclude that there is some other requirement (other than the vegetation profile) which must be described before we can predict that species accurately. If, on the other hand, the prediction is fairly accurate — if a bird is present whenever patches of vegetation with the appropriate profile are present — then we can conclude that a proper profile alone is sufficient to make that patch occupied by that species.

It is actually possible to make a numerical prediction by counting the number of vegetation patches which lie in the species' zone, but this is unrealistic because it assumes that all species have the same territory size and will settle one pair to a 100 ft square patch, throughout the habitat. For this reason we will be content, at this stage, with a qualitative prediction and will only guess which species should be common, which uncommon and which absent.

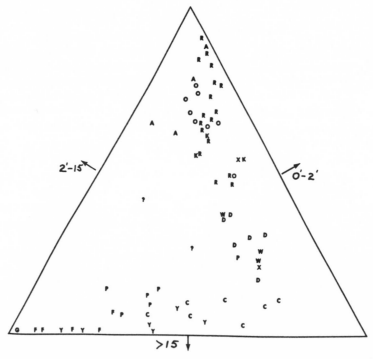

FIGURE 4. Each letter indicates the foliage proportions for a 100 ft square in the territory of a given pair of birds. A = Acadian flycatcher, C = yellow breasted chat, D = catbird, F = field sparrow, G = grasshopper sparrow, K = Kentucky warbler, O = ovenbird, P = prairie warbler, R = red-eyed vireo, W = white-eyed vireo, X = cardinal, Y = Maryland yellowthroat.

Let us now turn to the actual habitats to be predicted, remembering that
the bird habitat preferences were measured on other plots. Accurate censuses of the recently abandoned Pennsylvania field (bar graph in figure 1)
and the Pennsylvania brushy field (dots in figure 2) were made to compare
with the predictions. The recently abandoned field has vegetation patches
lying in the preference zones of grasshopper sparrow, field sparrow and
yellow throat. And, these are the three species which bred in the field —
field sparrow commonly, one pair each of the others. Meadowlark may possibly have bred too, but in any case no data on its habitat profile was
obtained.

Superimposing the bird species preferred profile zones of figure 4 onto
figure 2 which shows the brushy field, we predict a more diverse census:
Field sparrow, yellowthroat, prairie warbler and yellow breasted chat should
all be common, and perhaps a pair of catbird, cardinal or white eyed vireo
should be present. The actual census was: three prairie warbler, three
yellow throat, two yellow breasted chat, two field sparrow, one cardinal,
one catbird, one towhee, one blue-winged warbler, one robin, one yellow-
billed cuckoo. Bearing in mind that the last four species were not plotted
in figure 4 and so could not be predicted, this is remarkably close to the
predicted census.

Finally, this work should not be interpreted as suggesting that all species
use only the foliage profile in choosing their habitat, but merely that there is
a large collection of species whose presence can be predicted from foliage
profile measurements. A few species (for example, acorn woodpecker,
crossbills) may require specific tree species in their habitats; certainly
water birds must use very different criteria. And there is no proof that the
birds themselves use the profile in their choice; all we can say is that the
profile is closely associated with what they use for their choice. P. Klopfer is investigating this last point in his aviaries.

DISCUSSION AND RELATION TO TROPICAL DIVERSITY

As mentioned earlier, perhaps some species in choosing their habitat require profiles of two types. For instance, perhaps a cardinal requires both
woods and open areas, rather than some mean profile. For such species, to
be strictly accurate, we would have to plot not just a cluster of points, but
a cluster of joined pairs of points, the two joined points being the two types
of profile which must be present before the territory is acceptable. For predicting these species the habitats would also have to be plotted as clusters
of joined pairs of points. This would be nice but clearly requires a more
detailed set of measurements than we possess. There is little doubt, however, that this situation is important enough to be a real source of error in
the crude predictions of this paper.

Another source of error lies in the way the clusters of points for habitats
were constructed. Here the habitat was laid out in 100 ft squares with no
reference to what might be "natural patches." But the bird territories were
plotted as they were; although a 100 ft square was chosen within each one,

this square had no reference to an arbitrary preassigned grid. This is altogether proper but causes a few species to be missed in the predictions, since by hunting for an appropriate patch a species may find one which lies across the grid so that it is not reflected in the cluster of points.

In the previous paper we showed that bird species diversity could be predicted from the *mean* foliage profile of the habitat; in fact it was proportional to foliage height diversity. In this paper we are predicting not only the bird species diversity, but also the census, and this is done in terms of the variety of patches within the habitat. In this light, why was the bird species diversity proportional to the foliage height diversity? The bird species diversity actually was best estimated by .46 + 2.01 (foliage height diversity). Now bird species diversity and foliage height diversity were logarithmic measures. For our present purposes we can restate these conclusions in terms of the *number* of equally common species which would give the observed diversity and the number of equally dense layers of foliage which would give the observed diversity. In these terms, the result becomes (using 2 instead of 2.01):

"Number of equally common bird species is proportional to the square of the number of equally dense layers." And in the former paper we guessed that it increased with the square of the number of layers because the number of combinations of layers which a bird could require increases about with the square of the number of layers available (for 1, 2, 3 layers, the number of combinations possible is 1, 3, 7 respectively). How does this result accord with the present paper? It means that the number of bird species whose habitat requirements (for example, figure 4) include the habitat under consideration — that number of bird species increases about with the square of the number of equally dense layers of vegetation. And, since the number of bird species which will breed in a given habitat whose patches of vegetation are plotted as in the clusters of figures 1, 2, 3 is the number of bird species whose habitat preferences (as in figure 4) overlap the habitat cluster, we can conclude that the number of bird species should increase as the area of the habitat cluster increases. More accurately the number of bird species might be proportional to the area of the cluster plus a margin around it of width 1/2 the diameter of the bird species habitat clusters. These areas, imagined roughly from figures 1, 2, 3, do correspond roughly to the number of bird species. Thus a large part of the cause of bird species diversity is the amount of patchiness within the habitat. Another part is caused by many-layered forests being able to support ground species (for example, ovenbird), shrub species (for example, Kentucky warbler) and canopy species (for example, scarlet tanager), but this obvious cause seems much less important than the within-plot variability — the patchiness.

Let us turn, now, to the factors involved in tropical diversity. Three factors could be associated with the increased bird species diversity in the tropics: (1) Tropical habitats could have more internal variability (that is, larger clusters of points). (2) Tropical birds could have more refined habitat selection (that is, smaller clusters of points on that graph). (3) More species could share the same profile (that is, more of the species clusters should overlap). Some evidence for (2) or (3) is gathered in Klopfer and MacArthur (1960, 1961), but this type of graphical analysis should permit better disentangling of the factors associated with tropical diversity. The central question of tropical diversity (given enough time, will the temperate regions support as diverse a fauna as the tropics now have, or has speciation reached a limit?) remains to be answered. The approach of Southwood bears more directly on this problem.

CONCLUSION

(1) A fairly accurate census of breeding birds can be predicted from measurements of the amounts of foliage in three horizontal layers. The abundance of each species is roughly determined by the number of patches of vegetation whose foliage profile is acceptable to that species. This suggests that many species are rare only because their chosen foliage profile is rare.

(2) The main reason one habitat supports more bird species than another is that the first has a greater internal variation in vegetation profile (that is, a greater variety of different kinds of patches). A second reason is of course that a forest with vegetation at many heights above the ground will simultaneously support ground dwellers, shrub dwellers and canopy dwellers. With a few exceptions, the variety of plant species has no direct effect on the diversity of bird species.

(3) Comparable plotting of tropical bird requirements should disentangle three of the possible factors associated with the tropical increase in diversity.

ACKNOWLEDGMENTS

This work was supported by grant G-11575 of the National Science Foundation. Drs. Peter Klopfer, Monte Lloyd and Ernst Mayr have made very useful criticisms.

LITERATURE CITED

Klopfer, P., and R. MacArthur, 1960, Niche size and faunal diversity. Am. Naturalist 94: 293-300.
       1961, On the causes of tropical species diversity. Am. Naturalist 95: 223-226.
MacArthur, R., and J. MacArthur, 1961, On bird species diversity. Ecology 594-598.
Southwood, T. R. E., 1961, The number of species of insect associated with various trees. J. Animal Ecol. 30: 1-8.

10

Reprinted from *Ecol. Monographs,* **37,** 317–319, 331–349 (Fall 1967) with
permission of the publisher, Duke University Press, Durham, N. C.

THE NICHE EXPLOITATION PATTERN OF THE BLUE-GRAY GNATCATCHER[1]

RICHARD B. ROOT[2]

Museum of Vertebrate Zoology, University of California, Berkeley

TABLE OF CONTENTS

INTRODUCTION

The niche concept remains one of the most confusing, and yet important, topics in ecology. Traditionally the concept refers to the functional role, particularly in trophic interactions, of a species within a community. There has been little agreement, however, on what factors adequately define this functional role; in describing niches, various authors have chosen to stress different features of a species' diet, natural enemies, microhabitat, and periods of seasonal or diurnal activity. Much of the misunderstanding between ecologists concerning the competitive exclusion principle and the existence of "vacant" niches can be traced to this ambiguity (see reviews by Udvardy, 1957; Hardin, 1960; and DeBach, 1966).

In 1957, Hutchinson and Macfadyen, writing independently, defined the niche in a new and different way. Both authors cast the niche in terms of the range and combination of environmental conditions that permit a species to exist indefinitely. In other words, the niche is seen as an abstract "space" in the environment which some species must be able to exploit successfully for an extended period. While the Hutchinson-Macfadyen concept is no more helpful than the "role" concept in providing an operational definition for the niche, it serves to direct our attention to new types of investigation. For instance, the same species may occur in several different habitats

or cope with a changing set of conditions within a single habitat. Through a comparative investigation, one may hope to discover features common to these several environments. By this process, one may peel away all but the most critical features, leaving a core of limiting factors that defines the fundamental niche of a particular species. This approach has been attempted in the present study.

The niche may be thought of as composed of several dimensions (Hutchinson, 1957), each corresponding to some requisite for a species. Organisms are usually adapted to exploit only a portion of the requisites that are available in any environment. When the characteristics of these requisites are plotted on a continuous scale (e.g., prey size, position of the habitat in a vegetation continuum), the species exhibits a characteristic "exploitation curve" (Fig. 1). The exploitation curves for all requisites combine to form the species' "exploitation pattern". The shape of the exploitation curve will be determined by the interplay of several selective forces. The population will often respond to interspecific competition by becoming more efficient, through the evolution of specializations, in exploiting a more restricted range of requisites. Intraspecific competition will oppose this tendency toward greater specialization by causing the population to exploit the environment in a more generalized manner, thereby capturing a larger niche space (Svärdson, 1949). Finally, the relative stability of the environment will influence the exploitation pattern (Klopfer, 1962). Species which occupy habitats that fluctuate widely in their suitability for existence must

[1] Manuscript first received January 12, 1967. Accepted for publication June 29, 1967.

[2] Present address: Department of Entomology and Limnology, New York State College of Agriculture; and Section on Ecology and Systematics, Division of Biological Sciences, Cornell University, Ithaca, N. Y.

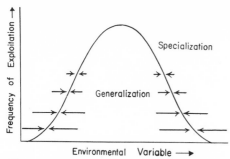

FIG. 1. A hypothetical exploitation curve of a population with respect to one dimension of the niche. The arrows show how selective forces which promote greater specialization or generalization influence the shape of the curve.

either remain highly generalized or possess adaptations, such as the ability to migrate or to become dormant, that permit the population to survive when its specialized requirements fail.

The central theme of this paper is the manner in which the exploitative behavior of the Blue-gray Gnatcatcher (*Polioptila caerulea*) is organized to achieve optimal adaptation in a changing environment.

The gnatcatcher* is a small bird, weighing about 5.8 gm, which feeds exclusively upon arthropods. During the breeding season this species is sexually dimorphic, the male having black feathers in the loral region and a more brightly colored dorsum. The genus *Polioptila* has been divided into nine New World species (Paynter, 1964) which breed allopatrically or in different habitats. Situations where members of the genus come into marginal contact with one another have been described by Brodkorb (1944), Paynter (1955), and Miller and Stebbins (1964). Mayr and Amadon (1951) have placed the genus in the Sylviinae, a subfamily which has undergone its most extensive adaptive radiation in the Old World. The breeding range of *P. caerulea* occupies a major portion of the continental United States and extends southward into Guatemala. The fragmentary literature on the gnatcatcher's breeding behavior is summarized in Bent (1949), Nice (1932), and Root (*In press*). During the winter, the northern populations migrate to the southern United States and Mexico. The food supply and habitat requirements of the species vary on a geographical, seasonal, and yearly basis. By describing the gnatcatcher's behavioral response to these environmental variations and to the increased energy demands associated with raising young, I have attempted to discover how the interplay between specialization and generalization influences the exploitation pattern of a single species.

*Throughout this paper, "gnatcatcher" is used only to designate the species *Polioptila caerulea*.

STUDY AREAS

The principal study area was located on the Hastings Natural History Reservation, situated at the northern end of the Santa Lucia Mountains, in Monterey County, California. The broad-sclerophyll vegetation and physical conditions of the surrounding region have been described by Cooper (1922), Shreve (1927a and 1927b), and Linsdale (1943). Intensive observations were made on a plot which was 56.1 acres (22.7 hectares) in extent and varying from 1700 to 2150 feet in elevation. The study plot was so oriented as to include several plant associations which in this region form a vegetation complex that is expressed variably according to slope exposure (Fig. 6). The vegetation of the generally north-facing slopes consists of a deciduous oak woodland (*Quercus Douglasii* H. & A. with some *Q. lobata* Nee and hybrid oaks). In some areas, these trees develop a spreading life-form and occur in open stands, while more extensive areas are dominated by an oak scrub consisting of small (8 to 24 ft. tall) deciduous oaks (Fig. 2). White (1966) has thoroughly described the deciduous oak stands in this region. The field layer of the oak woodland consists of grasses and forbs characteristic of the California annual type (Heady, 1958). Coast live oaks (*Q. agrifolia* Nee) occur in draws and on the shaded lower slopes. The south-facing slopes are covered with a chaparral which is dominated by chamise (*Adenostoma fasciculatum* H. & A.) on the drier sites (Fig. 3) and buckbrush (*Ceanothus cuneatus* (Hook.) Nutt.) in more mesic situations.

Additional observations were made at the Hastings Reservation in canyons where a mesophytic forest consisting of varying proportions of live oaks, broadleaf maple (*Acer macrophyllum* Pursh.), madrone (*Arbutus Menziesii* pursh.), bay (*Umbellularia californica* (H.A.) Nutt.), and willows (*Salix* spp.) occur.

A wintering concentration of gnatcatchers was studied in a 22 acre woodland of screw-bean (*Prosopis pubescens* Benth.) on the flood-plain of the Colorado River, 2 miles northeast of Yuma, Arizona. A dense layer of naturalized tamarisk (*Tamarix* sp.) and arrowweed (*Pluchea sericea* (Nutt.) Cov.) forms the understory in this association.

The foraging behavior of wintering gnatcatchers was compared with that of their resident congener, the Blacktailed Gnatcatcher (*P. melanura*), at two desert localities near Tucson, Arizona: at about 3000 ft. elevation on the south slope of the Santa Catalina Mountains in Lower Sabino and Lower Bear Canyons, and along the Santa Cruz River on the San Xavier Indian Reservation. An association which is characterized by the presence of saguaro (*Carnegiea gigantea* (Engelm.) Britt. & Rose) and palo verde (*Cercidium microphyllum* (Torr.) Rose & Johnston) covers the rocky slopes at these localities (Fig. 4). In local areas within this vegetation type, bur-sage (*Franseria deltoidea* Torr.), cholla and prickly-pear (*Opuntia* spp.), and ocotillo (*Fouquieria splendens*

Fig. 2. Deciduous oak scrub-woodland at the Hastings Reservation, Monterey County, California.

Fig. 4. Saguaro-palo verde plant association in Lower Sabino Canyon near Tucson, Arizona.

orado and Sonoran deserts of southern California and Arizona.

[*Editors' Note:* Material has been omitted at this point.]

Fig. 3. Chaparral dominated by chamise (*Adenostoma fasciculatum* H. & A.) at the Hastings Reservation.

Engelm.) are common. The saguaro-palo verde vegetation, also called Sonoran desert scrub, and its associated physical conditions in the Tucson region have been described by Shreve (1915) and Whittaker and Niering (1965). A mesquite (*Prosopis juliflora* (Sw.) DC.) woodland, often with an understory of graythorn (*Condalia lycioides* (Gray) Weberb.) and tumbleweeds, occurs on benches at the base of the rocky slopes. The most extensive stands of mesquite woodland were studied on the San Xavier Reservation. Fremont cottonwoods (*Populus Fremontii* Wats.) and willows (*Salix* spp.) form a timbered belt along the stream-course at these localities.

Additional observations of shorter duration were made of *P. melanura* at several localities in the Col-

DIET AND PREY AVAILABILITY

The gnatcatcher's diet was completely restricted to arthropods. (appendix). The three instances when plant fragments were found in the stomachs probably represent cases where plant material was acci-

TABLE 8. The proportion of attack maneuvers directed at prey which was flushed from the foliage.

	Self-Maintenance		Feeding Young	
Sampling period	Total attacks (n)	Attacks at flushed prey (percent)	Total attacks (n)	Attacks at flushed prey (percent)
March 6-May 9	190	9.5		
May 10-June 16	146	0.7	137	5.1
June 26-July 27	122	0.8	215	6.0

TABLE 9. Variation in exploitation patterns of individual Blue-gray Gnatcatchers feeding dependent young in oak woodland. The data are presented as the number of foraging maneuvers per 500 secs. of observation.

Individual*	Secs. of Observation	Foraging Intensity	Tactical Response			Location of Prey			
			glean	hover	hawk	foliage	bark	herbs	air
♀1	736	25.8	10.9	10.2	4.7	15.5	0.7	4.8	4.8
♀4	758	25.6	13.8	5.9	5.9	13.1	3.3	3.3	5.9
♀5	605	28.1	8.3	14.0	5.8	16.5	1.7	4.1	5.8
♀7	729	24.0	13.7	6.2	4.1	17.1	1.4	1.4	4.1
♀12	1096	21.9	5.9	13.7	2.3	13.7	1.4	4.5	2.3
♂5	707	26.1	9.2	9.9	7.0	15.6	0.7	2.8	7.0
♂12	850	20.0	10.0	7.1	2.9	13.5	1.2	2.4	2.9

*Numbers refer to the territory designations in Fig. 11.

TABLE 10. The percent of the stomachs in which major categories of prey occurred.

	Monterey County, California			Yuma, Arizona
	June-August 1959	March-May 1963	June-August 1963	December and January 1963
Homoptera	100.0	56.5	100.0	53.8
Heteroptera	10.0	39.1	26.7	7.7
Coleoptera	80.0	78.2	40.0	92.3
Lepidoptera	25.0	47.8	6.7	15.4
Diptera	50.0	17.4	20.0	23.1
Hymenoptera	80.0	47.8	60.0	15.4
Araneae	50.0	26.1	26.7	53.8
Other	15.0	13.0	13.3	7.7
n	20	23	15	13

dentally ingested with the normal prey. At least 70 different families of arthropods, however, were represented in the diet, reflecting the gnatcatcher's capacity for exploiting a wide variety of situations. Because of the diverse diet, it was often necessary to lump the samples of prey into higher taxonomic units before attempting to draw comparisons. The analysis of diet is based upon the total number of prey individuals, since each item represents a successful attack maneuver and thereby provides a better index to the gnatcatcher's exploitation pattern than such measures as percent volume or weight. The percent of the stomachs in which prey taxa were found is presented (Table 10) as a check upon the importance of each food type to the diet of the entire population: percent occurrence reduces the bias resulting from a few birds feeding upon a concentration of "abnormal" prey.

The behavior of arthropods known to be important prey species of the gnatcatcher was noted throughout the study. Membracid, cicadellid, and mirid bugs, and chrysomelid and mylabrid beetles were usually found on the foliage and small twigs of the trees. Curculionid beetles, ants, and spiders were found on the larger branches as well as in the foliage zone. These insects rarely flushed when approached suddenly. When these winged insects were tapped with a pencil, they either moved to the other side of their perch, or sprang into the air and after falling a few inches, either alit in the foliage below or flew off on a horizontal course. Hemipteran nymphs simply dropped from their perch when struck with a pencil, while ants and spiders fled along the twigs and branches. Depending upon the situation in which such prey was found, gnatcatchers could easily attack these arthropods with gleaning, rushing, or hovering maneuvers.

If the prey was able to escape the initial attack, subsequent tumbling or hawking maneuvers could be employed. Thus the foraging beat and the foraging repertoire of the gnatcatcher are clearly related to the preferred habitats and escape reactions of the major types of prey found in the diet.

The taxonomic composition and the size distribution of the foliage arthropod fauna varied on a seasonal and yearly basis (Figs. 12-14). At the Hastings Reservation, the gnatcatchers foraged at the stations where the foliage arthropod samples were taken during 76.9 to 89.7% of the standard observations (Table

2). The composition of the gnatcatcher's diet also varied in time and space, but these changes in food habits were not related in any simple way to the observed changes in the foliage arthropod fauna. Such differences in the distribution of prey in the diet and in the environment can be a result of either the predator actively selecting prey from the food that is available or of discrepancies between the sampled food resource and that which was available to the predator. Availability is difficult to assess, for it requires more than a measurement of the food within the predator's preferred foraging beat. Because of distasteful char-

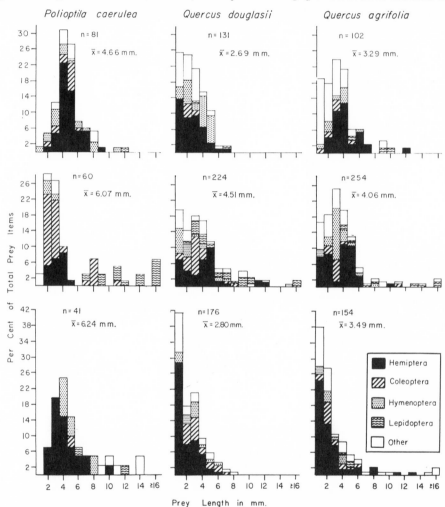

FIG. 12. The size distribution of arthropods in the stomachs of the Blue-gray Gnatcatcher and in the foliage of the deciduous blue oak (*Q. Douglasii*) and the coast live oak (*Q. agrifolia*). The three graphs at the top present the combined samples taken during June-August, 1959; the middle three, March-May, 1963; and the bottom three, June-August, 1963.

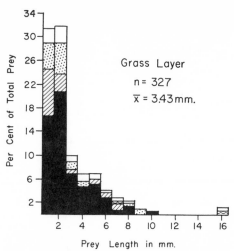

FIG. 13. The size distribution of arthropods in the grass layer of oak woodland. The samples were collected during June-August, 1963. The designation for arthropod taxa are the same as those in Fig. 12.

acteristics, certain types of prey may be avoided completely, e.g., the late instar larvae of the tent caterpillar, *M. constricta* (Root, 1966). For this reason, *M. constricta* larvae were not included in the size distributions of foliage arthropods (Fig. 12). In addition, subtle structural, physiological, and behavioral limitations may prevent the predator from consuming large quantities of certain classes of prey (Tinbergen, 1960). In large part, the avian predator's diet is influenced by a tendency to concentrate on those kinds of prey which the bird can find and ingest most efficiently (see Gibb, 1958 and 1962).

The average length of the prey that was taken by gnatcatchers was always longer than the average length of the arthropods that were present in the preferred foraging stations (Fig. 12-14). The size distribution of prey in the diet at different seasons was relatively stable in comparison with the taxonomic composition of the diet. During the winter, however, when only small arthropods were present in the environment, the mean length of the prey in the diet was also reduced. The mean length of prey in the stomachs of eight dependent young that were shot in June and July, 1963, was 8.86 mm. Since the food of adults shot during this period was somewhat smaller, there is reason to believe that gnatcatchers with young forage selectively for even larger prey.

The major differences between the size distributions of arthropods in the diet and in the environment occurred within the smaller size categories. Gnatcatchers were capable of eating arthropods which were less than 1 mm in length and yet they did so infrequently. This suggests that small prey were "overlooked" because their energy yield was too low to warrant the effort involved in their pursuit. The

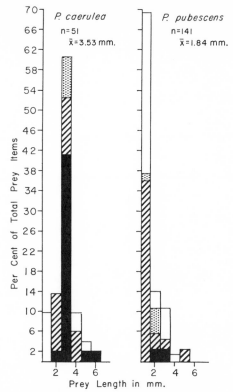

FIG. 14. The size distribution of arthropods in the stomachs of the Blue-gray Gnatcatchers and in the foliage of the screw-bean (*P. pubescens*) near Yuma, Arizona. The graphs present data from combined samples gathered in January and December, 1963. The designations for arthropod taxa are the same as those in Fig. 12.

rarity of items in the diet which exceeded 10 mm in length reflects the scarcity of such prey in the environment. Thus it appears that the distribution of prey lengths in the diet results from a dynamic adjustment between the selection of prey which exceeds a suboptimal size and the degree of scarcity of large prey in the environment.

The upper limit on prey size was probably determined by the bird's inability to dispatch and ingest large insects. Larger prey were dropped frequently and took longer to subdue (above). On two occasions, gnatcatchers were observed to have obvious difficulty in swallowing large adult lepidoptera: the prey was dropped and reoriented in the beak a few times before ingestion was accomplished. On three occasions, similar difficulties were observed in the swallowing of lepidopterous larvae. that exceeded about 18 mm in length (reckoned by comparison with the length of the beak). After swallowing large prey, the birds remained on the perch for about a minute where they

TABLE 11. Changes in relative abundance of certain types of prey in the diet and in the foliage of deciduous oaks and live oaks at the Hastings Reservation. The data are presented as the percent of the total number of individuals (n) in each sample.

	Gnat-catchers	Decid-uous Oaks	Live Oaks
June-August, 1959			
Membracidae	30.7	4.6	5.9
Cicadellidae	13.6	16.0	2.9
Miridae	4.9	4.6	7.7
Lepidopterous larvae	4.9	1.5	0.9
Other	45.9	73.3	82.6
n	81	131	102
March-May, 1963			
Membracidae	3.3	6.7	2.4
Cicadellidae	3.3	5.8	4.7
Miridae	31.7	14.7	24.4
Lepidopterous larvae	18.3	18.3	3.9
Other	43.4	54.5	64.6
n	60	224	254
June-August, 1963			
Membracidae	34.1	1.7	6.4
Cicadellidae	19.5	23.3	21.1
Miridae	4.9	1.1	1.9
Lepidopterous larvae	2.4	4.5	1.9
Other	39.1	69.4	68.7
n	41	176	156

alternated between wiping the base of the beak against the perch and pointing the beak upward while turning the head from side to side. Large prey which nestlings were unable to swallow was eaten by the adults on three occasions.

Closer agreement was seen between the gnatcatcher's foraging response and changes in the abundance of certain types of prey (Table 11). All of the prey individuals included in Table 11 exceeded 2.5 mm in length and possessed characteristics which made them vulnerable to attack by gnatcatchers (above). Mirids and lepidopterous larvae formed a large part of the diet during the spring when these groups were relatively abundant in the foliage. During the summer, membracids and cicadellids became more important in the diet. The higher degree of predation on cicadellids was associated with an increase in the relative abundance of these insects in the environment. The increased proportion of membracids in the summer diet could not be explained on this basis, however. Perhaps gnatcatchers were feeding selectively on the membracids, particularly *Platycotis* spp. and *Stictocephala* sp. which are stout-bodied and exceed 5 mm in length, in response to a seasonal decline in the abundance of large lepidopterous larvae.

THE FOLIAGE-GLEANING GUILD

GUILD CONCEPT

Data were gathered on the exploitation patterns of birds that regularly fed on foliage arthropods in the oak woodlands at the Hastings Reservation. These species are local members of the "foliage-gleaning guild." A *guild* is defined as a group of species that exploit the same class of environmental resources in a similar way. This term groups together species, without regard to taxonomic position, that overlap significantly in their niche requirements. The guild has a position comparable in the classification of exploitation patterns to the genus in phylogenetic schemes.

As with the genus in taxonomy, the limits that circumscribe the membership of any guild must be somewhat arbitrary. To be considered a member of the foliage-gleaning guild in the oak woodland, the major portion of a bird species' diet had to consist of arthropods obtained from the foliage zone of oaks. As a result, birds that occasionally use the foliage zone were excluded even though they may have exerted some influence on the guild's food supply. A species can be a member of more than one guild. For instance, the Plain Titmouse, *Parus inornatus*, while belonging to the foliage-gleaning guild with respect to its foraging habits, is also a member of the hole-nesting guild by virtue of its nest-site requirements.

According to the competitive exclusion principle (Hardin, 1960), there must be some minimal difference in the exploitation patterns of sympatric species. Even if this principle was incorrect, it is obvious that interspecific competition has deleterious effects and that natural selection would tend to favor divergence which reduces the intensity of competition. Most studies of interspecific competition have considered only sympatric members of the same genus on the assumption that species with close taxonomic relationship tend to be the strongest competitors. The work of Brian (1952, 1955) on ants, Hartley (1948) and Maitland (1965) on fish, and Pitelka (1951) on birds, to mention but a few, has demonstrated, however, that intergeneric competition can also be an important factor. One advantage of the guild concept is that it focuses attention on all sympatric species involved in a competitive interaction, regardless of their taxonomic relationship.

The guild concept has additional use. As mentioned earlier, the term niche is used in reference to two quite distinct concepts: the functional role or "occupation" of a species in a community and the set of conditions that permit a species to exist in a particular biotope. This combination of concepts under a single term has led to controversies over the interpretation of data presented to refute the generality of the competitive exclusion principle (cf., Ross 1957, 1958; Savage, 1958). If we restrict our definition of niche to the latter (Hutchinson-Macfadyen) concept, such arguments could be resolved by recognizing that groups of species having very similar ecological roles within a community are members of the same guild, not occupants of the same niche.

Finally, the guild concept may be useful in the comparative study of communities. It is usually impractical to consider at once all species living in a biotope, so ecologists must restrict their analyses to particular taxonomic groups. This procedure frequently obscures the functional relationships within the communities, because species performing several different unrelated roles are considered together. The same guild may be represented in several different communities and can thus serve as the basis for comparing species diversity, degree of character difference, biomass, etc. in different biotopes. In using the guild in these comparisons, we are dealing with a functional or ecological category—one that has been molded by adaptation to the same class of resources and by competition between its local members. Functional classifications based upon "feeding groups" have been used by Salt (1953, 1957) and Turpaeva (1957) in their comparative analyses of natural communities.

The term, guild, has appeared once before in the ecological literature as a translation of "Genossenschaften" in the English edition (1903) of Schimper's "Pflanzengeographie auf physiologischer Grundlage." It was used there to refer to four groups of plants that are dependent upon other plants for their existence—lianes, epiphytes, saprophytes, and parasites. This usage of guild has not become established and can be considered obsolete or "archaic." I am using the word guild in a new sense because it seems to be the most evocative and succinct term for groups of species having similar exploitation patterns.

MEMBERSHIP AT THE HASTINGS RESERVATION

During the breeding season the foliage-gleaning guild in the oak woodlands at the Hastings Reservation includes the Warbling Vireo (*Vireo gilvus*), Hutton's Vireo (*Vireo huttoni*), Orange-crowned Warbler (*Vermivora celata*), Plain Titmouse (*Parus inornatus*), and the gnatcatcher. The titmouse just barely meets the criteria used to delimit the guild, because a large portion of its diet consists of plant material and arthropods living on bark. Occasionally other foliage-gleaning species such as the Common Bushtit (*Psaltriparus minimus*), Black-throated Gray Warbler (*Dendroica nigrescens*), and Yellow Warbler (*Dendroica petechia*) were encountered in the oak woodland adjacent to large stands of chaparral or riparian forest. The House Wren (*Troglodytes aedon*), Bewick's Wren (*Thryomanes bewickii*), Scrub Jay (*Aphelocoma coerulescens*), Oregon Junco (*Junco oreganus*), and Chipping Sparrow (*Spizella passerina*) occasionally foraged for arthropods in the oak foliage but obtained most of their food from the subcanopy or ground.

The Western Flycatcher (*Empidonax difficilis*) is an important avian insectivore in the oak woodland, but its foraging behavior differs greatly from that of the foliage-gleaners. Like other flycatchers, *E. difficilis* hunts by searching a large area from a "sentinel" position on an exposed perch. Most prey is taken in long sweeping aerial attacks on flying insects or arthropods that alight momentarily on the foliage. This concentration on active insects is reflected in the high

proportion of Hymenoptera and Diptera found in the diet (Beal, 1910).

The Warbling Vireos, Orange-crowned Warblers, and gnatcatchers leave the Hastings Reservation in the winter. The resident members of the guild are joined by the Ruby-crowned Kinglet (*Regulus calendula*) and Audubon's Warbler (*Dendroica auduboni*) at this season. During the winter the foliage-gleaning birds range over a wider variety of vegetation types and are less restricted to the foliage zone of trees than during the summer. The resident species begin nesting somewhat earlier than the migrants. A pair of titmice began a nest on 21 February 1963 and I found a Hutton's Vireo nest on 20 March 1963. To the north, in Sonoma County, the last species has been reported to begin nesting in early March (Bent, 1950).

The species diversity of gleaning birds in the oak woodlands reaches a peak in late March and throughout April when transient species and winter visitants overlap in their presence at the Reservation with the breeding members of the guild.

The following discussion considers the niche relations of guild members during the breeding season. Only quantitative data on individuals engaged in self-maintenance foraging during March-August are reported. Qualitative statements about these species are based upon several additional observations that were made at Las Trampas Canyon, Contra Costa County (Root, 1964a) and other localities in central California.

FORAGING BEAT

There were 4 pairs of Warbling Vireos, 2 pairs of Hutton's Vireos, 4 pairs of Orange-crowned Warblers, and 7 pairs of Plain Titmice during the 1963 breeding season on the portion of the study area shown in Fig. 6. The territories of the two vireo species were centered in clumps of evergreen oaks along the draws, while some of the gnatcatcher and titmouse territories extended out on to the ridges and contained almost pure stands of deciduous oak woodland. The Orange-crowned Warblers were found on slopes where there was a well developed shrub layer beneath the oak canopy. Nevertheless, all of the species regularly utilized both deciduous and live oaks.

All of the guild members forage most often in the foliage zone of oaks, but differ in their restriction to this station (Table 12). Hutton's Vireo frequently perches in the subcanopy, but directs most of its attacks "outward" at arthropods in the foliage zone. The titmouse is able to exploit a greater variety of stations within the oak woodland than the other species. Often it is found foraging on the large limbs and trunks in the subcanopy or on open ground (Root, 1964a). During the winter, when the abundance of arthropods drops sharply in the foliage zone of deciduous oaks (Fig. 19), this species is the only guild member to forage regularly in stands of barren trees. My data on the Orange-crowned War-

Table 12. The foraging beat of members of the oak foliage-gleaning guild at the Hastings Reservation. The data are expressed as the percent of standard observations.

		Tree zones		
	n	foliage	sub-canopy	Herb layer
Polioptila caerulea	309	90.0	8.6	1.4
Vireo gilvus	80	90.0	10.0	0.0
Vireo huttoni	60	65.0	33.3	1.7
Vermivora celata	113	84.9	9.7	5.4
Parus inornatus	140	48.6	41.4	10.0

maneuver	position of prey	position of bird
glean	resting on substrate	standing on perch
hover	resting on substrate	in the air
hawk	in the air	in the air

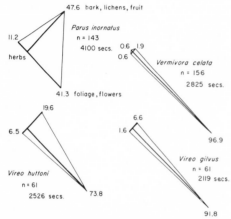

Fig. 15. The substrates where members of the foliage-gleaning guild obtained food in oak woodland. The coordinates are explained in Fig. 10. The length of the timed observations used in computing foraging tactics is given below the total number of maneuvers (n).

bler may be somewhat biased in favor of tree foraging. Occasionally, these birds were observed using twig perches at the base of shrubs and small oaks while foraging on insects in the herb layer. Because of the dense shrub canopy on parts of warbler territories and the relative silence of birds foraging at these lower levels, I may have missed some of their activities. By following the individuals for periods of 30 min or longer, I found that this error is probably minor.

Foraging Maneuvers

With the exception of the Plain Titmouse, the guild members all capture most of their prey on oak foliage (Fig. 15). Both Hutton's Vireo and the Plain Titmouse frequently obtain food from bark surfaces, but during the breeding season most of these attacks are directed at twigs in the foliage zone. The frequencies of basic foraging maneuvers employed by guild members are presented in Fig. 16. These basic maneuvers are defined as follows:

The two vireo species have a similar foraging "style." They remain on a perch for a longer period than the other species before moving rapidly to another vantage point a few feet away. They also seem to search for prey at a greater distance from the perch, often directing attacks at insects 3 ft or more away. Their frequent use of hovering attacks may be related to the wide search radius, because it is probably difficult to alight near prey that is seen at a distance. These hovers differ somewhat from those of the gnatcatcher. Frequently gnatcatchers remain stationary in midair (in hummingbird fashion) while grasping the prey and then fall back to a perch. Vireos do this on occasion, but usually strike the prey while passing by in rapid flight. When extracting prey from dense terminal sprigs of foliage, vireos often grasp the leaves in their feet and hang upside down for a moment. Gnatcatchers normally employ hovering maneuvers to attack prey in this situation. Large prey are mandibulated and battered against the perch by both vireo species. In addition, they occasionally hold large insects against the perch with their foot while tearing them into smaller portions with the beak.

Foraging Orange-crowned Warblers move rapidly from perch to perch as they probe into clusters of leaves with their beaks. Under the best observation conditions it could be seen that many of these probes did not involve the capture of prey: the beak is thrust methodically into places where prey is likely to be found as part of the normal searching routine. Since it was often impossible to distinguish such searching probes from those used in capturing prey, my estimate of foraging intensity for the warblers (Fig. 16) is undoubtedly too high. In contrast with the vireos and gnatcatcher, these warblers employ aerial attacks infrequently. On the other hand, they characteristically stretch, lean, or even hang momentarily from the perch to peck at nearby foliage. Hovering and hawking maneuvers are executed in a manner similar to those of the gnatcatcher. Large insects are mandibulated or battered against the perch.

Plain Titmice are capable of extracting concealed prey from a variety of situations that cannot be exploited by other species in the guild. The titmice often pull apart leaf galls, flowers, curled dead leaves, and lichens with their beaks to expose arthropods. They also hammer apart acorns and other fruits, and pry bark from branches to obtain food. Frequently objects are held down with one foot while the beak is used to tear them apart. In order to compare the foraging tactics of titmice with the other species (Fig. 16), it is necessary to distinguish pecking used to extract prey from that employed in actual capture. As a result, I have reported bouts of pecks at the same object as one gleaning maneuver. The infre-

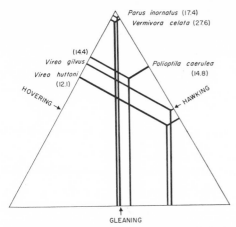

FIG. 16. The foraging tactics of members of the foliage-gleaning guild in oak woodland. The coordinates are explained in Fig. 11. The numbers following each species express its foraging intensity as the number of attack maneuvers/500 sec of observation.

quent hovering and hawking attacks performed by titmice consist of a "clumsy" lunge at nearby prey. Insects at the bottom of terminal foliage clumps or on the underside of twigs are usually captured in hovering maneuvers by the vireos and gnatcatcher. Titmice attack prey in these situations while hanging upside down from a perch. During the 4100 sec of standard observation for this species, the birds hung from perches 30 times (for an estimated total of 120 sec) and remained in an upside position for as long as 15 sec.

DIET

Data on the arthropods in the diets of guild members are summarized in Fig. 17 and Table 13. Only stomachs of birds collected in oak woodland near the Hastings Reservation during the breeding season were examined. Both intact prey and arthropod fragments are included in Table 13. Counting of fragments tends to overestimate taxa with thick exoskeletons, such as the Coleoptera (see Table 1), and therefore fragments could not be used above to compare the frequency of taxa in the gnatcatcher's diet with the abundance of prey on the foliage. By including the fragments in Table 13, however, a more complete impression of relative differences in the diets of guild members can be gained than would be possible if I had counted only the few intact prey that are available.

All of the guild members captured prey within a similar range of sizes, but the two vireo species and the Orange-crowned Warblers tended to capture large caterpillars more frequently than the other species. Hemiptera were a significant item in all diets; the same common species of cicadellids and membracids were eaten to some extent by all of the guild members.

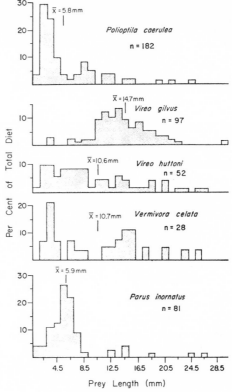

FIG. 17. Size distribution of intact prey in the stomachs of foliage-gleaning birds. Based on birds collected in oak woodland near the Hastings Reservation.

Only the titmice regularly fed upon plant material: seeds and plant fragments were found in 13 of the 16 stomachs examined.

MORPHOLOGICAL ADAPTATIONS

Studies on the functional anatomy of birds have shown that the structure of the beak is closely related to a species' foraging behavior (Engels, 1940; Bowman, 1961; and others). The guild members display a spectrum of bill types (Fig. 18) which, in turn, correspond to gross differences in their foraging tactics and diet. Species that are primarily gleaners (e.g., Orange-crowned Warbler) possess narrow thin-tipped beaks, while aerial foragers, such as the Western Flycatcher (included with guild members in Fig. 18 for comparison), have the base of the beak dorsoventrally compressed. The relatively massive beak of the Plain Titmouse is probably linked to its extensive use of the bill in hammering apart hard objects. By comparison, gnatcatchers seem to exploit two different foraging modes: gleaning arthropods from the foliage is expressed in the sharp tip of the

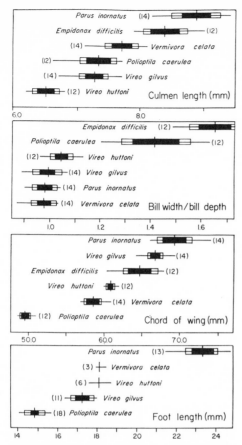

FIG. 18. Morphological variation in the birds of the foliage-gleaning guild. The horizontal lines represent the observed ranges; the rectangles mark one standard deviation on either side of the mean, with the solid portion indicating the 95% confidence limit for the mean. Only male specimens collected during the breeding season, in the Central Coast Range of California were measured. The bill measurements were made with the anterior margin of the nostril as a reference point. Foot length was measured on specially prepared specimens from the end of the hallux to the tip of the middle toe (excluding the claws).

beak, while the compressed base of the bill represents an adaptation for aerial foraging.

These comparisons serve to emphasize that the shape of the beak derives from a mosaic of adaptations suited to a particular set of foraging maneuvers. For this reason the foraging tactics must be considered when selecting dimensions for use in comparing morphological differences between potential competitors. The use of differences only in culmen length to estimate degree of niche overlap in birds (Hutchin-

son, 1959, Klopfer and MacArthur, 1961, and Schoener, 1965) may be inappropriate because the more "active" dimension segregating species would probably be beak width among aerial foragers and beak depth in fruit or seed eaters. In the foliage-gleaning guild, those species that overlap significantly in culmen length have the base of the beak compressed to different degrees (Fig. 18).

Schoener (1965) has shown that there is less difference in beak length between small sympatric birds, particularly in insectivorous families, than between large species. He suggests that this is because the large food morsels captured by large birds are less abundant than small prey. In addition, the larger birds probably have greater energy requirements. As a result, large species can tolerate less overlap than smaller birds that are feeding upon the more abundant small prey. The data on the foliage-gleaning guild supports Schoener's hypothesis in part.

The size distribution of arthropods in the tree foliage (Fig. 12) shows that large insects are relatively less abundant. The weights of guild members (based on breeding males taken near the Reservation) were as follows:

	n	mean wt. in gm
Polioptila caerulea	14	5.7
Vermivora celata	6	9.3
Vireo huttoni	9	11.2
Vireo gilvus	10	11.3
Parus inornatus	13	17.8

These size differences are also reflected in the measurements for the chord of the wing (Fig. 18). The character differences in the series of mean bill lengths (Fig. 18) are as follows:

Species	ϕ
Vireo huttoni	
	1.12
Vireo gilvus	
	1.01
Polioptila caerulea	
	1.05
Vermivora celata	
	1.15
Parus inornatus	

where ϕ is the ratio of the larger to the smaller measurement (Hutchinson and MacArthur, 1959; Hutchinson, 1959). Thus we see that a character difference greater than 1.14 (used by Schoener to identify the "transition zone") separates only the largest species, the titmouse, from the other guild members.

An implicit assumption in Schoener's hypothesis is that beak size (within the same avian family) is directly related to the size of the preferred prey. The two species of *Vireo* appear to fit this assumption (cf., Figs. 17 and 18). For the guild as a whole, however, there is no correlation between bill length and the mean size of prey. Furthermore, while there seem to be specific differences in preferred food size, there is almost complete overlap in the size range of prey

TABLE 13. Arthropods in the diets of foliage-gleaning birds. The data are expressed as the percent of total prey individuals (n) that could be identified.

	Polioptila caerulea	*Vireo gilvus*	*Vireo huttoni*	*Vermivora celata*	*Parus inornatus*
Hemiptera	36.0	10.3	11.9	47.8	13.2
Coleoptera	32.3	15.0	29.8	6.5	55.3
Lepidoptera	7.1	62.0	24.6	37.0	6.5
Hymenoptera	13.8	6.6	22.4	4.3	10.2
Other	10.8	6.1	11.2	4.3	14.8
n	287	213	134	46	81

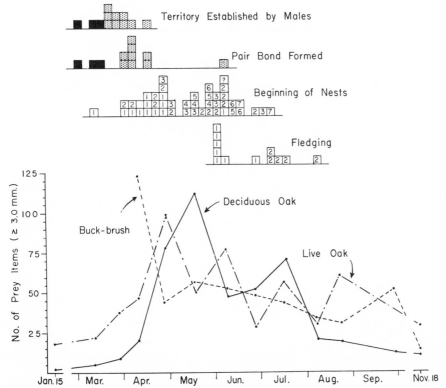

FIG. 19. Seasonal changes in the breeding activities of the Blue-gray Gnatcatcher and the abundance of foliage arthropods at the Hastings Reservation during 1963. In the two histograms at the top, the black squares refer to territories which were centered in chaparral and the stippled squares refer to territories in oak woodland. The numbers in the histogram labeled ''beginning of nests'' refer to the position that each nest occupied in the sequence of nests built by each pair. In the bottom histogram, the numbers designate whether the fledglings were from the first or second successful brood of the season. The points on the graph of prey abundance present the number of arthropods collected in 40 sweeps with a beating net.

captured by the guild members (Fig. 17). Thus it would appear that the shape of the food abundance curve (Fig. 12) restricts selective foraging for prey of different sizes, even in the titmouse, with its relatively large character differences. Partitioning of the guild's food supply is not based upon a single factor, such as prey size, but instead, involves the simultaneous action of differences in "preferred" foraging beat, foraging maneuvers, and diet.

Those guild members that spent most of the time in the foliage zone (Table 12) where small diameter perches abound, have smaller feet than the titmouse

(Fig. 18) which frequently forages from limbs and boles.

ADAPTIVE RESPONSES TO CHANGES IN THE FOOD SUPPLY

The importance of the food supply as a limiting factor of avian populations is well established (Lack, 1954 and 1966, presents extensive reviews). In the temperate and arctic latitudes, and perhaps in the tropics as well, the composition and quantity of this critical resource vary on a seasonal, yearly, and geographical basis. Birds inhabiting such regions must possess adaptations for exploiting the variable food supply. Changes in the abundance of foliage arthropods and the timing of important events in the breeding cycle of the gnatcatchers are summarized in Fig. 19. Since the purpose of Fig. 19 is to compare the foraging conditions at different seasons and in different habitats, only arthropods within the optimal range of prey size for the gnatcatcher (3.0 mm in length and longer) are considered. The following is an account of the gnatcatcher's adaptive responses to these changes in the food supply.

HABITAT SELECTION

The importance of the structural configuration of the vegetation and the topography of the region as determinants of habitat occupancy has been well studied (Lack, 1933, 1949; Svärdson, 1949; MacArthur, MacArthur, and Preer, 1962; Klopfer, 1963; and Wecker, 1963). Indeed, the theories of Lack and Svärdson attempt to explain the habitat selection process almost completely upon the bird's recognition of such structural features of the environment. Both authors point out, however, that structural cues are proximate factors (i.e., "releasers" of the behavioral mechanism) and that ultimate factors, those concerned with survival value, must act through natural selection to delimit the spectrum of cues used by a species in habitat selection. The gnatcatcher's pattern of territory establishment provides an interesting contrast to previous work, in that ultimate factors seem to be directly involved in the habitat selection process.

In other species of migratory birds the "optimal" habitats are selected out first by the early arrivals (Svärdson, 1949). Male gnatcatchers, following their arrival in March and early April, wander over a large area which permits them to assess the favorability of different habitats. Not all habitats within the area are visited. At the Hastings Reservation, gnatcatchers were rarely observed in the broadleaved canyon forest, even though this habitat appeared to offer ideal foraging conditions for this species. The first territories which are defended encompass large areas of chaparral or live oak woodland (Fig. 6). The height and spacing of the dominant plant species in these two vegetation types are quite different: there is a greater similarity in outward structural configuration between the live oak woodland and the deciduous oak woodland, the latter not being attractive to gnatcatchers at this season. Apparently something in addition to these structural cues is involved in the habitat selection process.

The deciduous oak woodlands did not become attractive to gnatcatchers until after the chaparral and live oak habitats were saturated by the early arrivals. Since the deciduous oaks were almost devoid of leaves during this period, the possibility exists that foliage density provides cues eliciting habitat settlement. This possibility could not be evaluated thoroughly in this study because foliage density is clearly related to other important factors, such as the abundance of foliage arthropods, in most field situations. An experimental procedure, such as a comparison of territory establishment between plots where the prey populations had been reduced with insecticides and untreated plots, would be necessary to separate foliage density from other cues that might be involved in habitat selection. Nevertheless, indirect evidence leads me to believe that foliage density is relatively unimportant as a proximate cue for recognition of optimal habitat, at least with the gnatcatcher. In August, when the foliage characteristics of the oak woodlands and chaparral are little changed from their condition in May and June, the gnatcatchers leave the deciduous oak habitats and return to the chaparral and adjacent live oaks. In addition, the western subspecies of the gnatcatcher leaves the evergreen vegetation of California during the winter and migrates to desert habitats where the foliage is relatively sparse.

Nest site characteristics might also provide the gnatcatcher with important cues for the selection of breeding habitats. The inspection of potential nest sites is a dominant activity of both sexes during the period of territory establishment and pair formation (Root, in press). While inspecting nest sites, the gnatcatchers examine the immediate surroundings intently, suggesting that a critical assessment of the habitat is taking place at this time. Klopfer (1962) and Wecker (1963) have suggested that early experiences with the environment may influence an individual's subsequent selection of habitats. Since the surroundings of the nest are the first features of the breeding habitat that each nestling perceives, it would not be surprising if these characteristics were important in habitat recognition. Sargent (1965), experimenting with caged Zebra Finches, has shown that rearing and previous nesting experiences have some effect upon the birds' choice of nesting materials and nest "habitats." Individual gnatcatchers build consecutive nests in very different situations, however. A banded female (from Pair 1) built nests in deciduous oaks, live oaks, and a buck brush shrub; in shaded and unshaded locations; and on vertical and horizontal nest forks. I have thus been unable to discover a feature of the nest fork or its surroundings which appears to limit the individual's choice of habitats.

Differences in the food supply of the habitats in which gnatcatchers occur during the period of territory establishment appears to be the principle variable that can account for the differences in the attractiveness of the habitats. The first territories which are defended include large areas of chaparral and live oak woodland, habitats containing the greatest abundance of foliage arthropods at this season (Fig. 6 and 19). Formation of the pair bond is also related to differences in the food supply. The initial assessment of the habitats by the males is continued by the females during the period when they are consorting with different unmated males (Root, *in press*). The pair bond does not appear to become established until after the abundance of arthropods in the dominant trees and shrubs on the territory exceeds some minimum amount (about 15 large arthropods in 40 sweeps of the foliage).

Svärdson (1949) has proposed that habitat selection by birds is not "released" by any single factor in the environment, but instead, that several environmental stimuli, which can be combined in various ways, reach a "sum" that is sufficient to elicit territory establishment. The habitat selection process of the gnatcatcher seems to be organized in a similar manner. I would suggest that the process takes place in two stages. First, responses which are innate or learned early in life, delimit the general region and the spectrum of habitats within a given region to which the gnatcatchers return in the spring. After their arrival, the second stage begins. While wandering over a series of acceptable habitats, the birds are able to assess the food supply (and perhaps the availability of nest sites) in different areas. Finally, territories are established in those areas which provide the most suitable combination of requisites for survival of both the adults and the young.

Thus the same characteristics of the environment serve as both proximate and ultimate factors in the final stages of the gnatcatcher's selection of breeding habitats. The continued assessment of the environmental requisites during the period prior to the beginning of the first nests assures that the population's initial breeding effort will be concentrated in the most favorable habitats. In years when there is a late spring or some regional catastrophe reduces the favorability of a habitat, the gnatcatchers are able to change their normal pattern of habitat occupancy in accordance with these changes. The gnatcatcher's pattern of territory establishment seems to support Klopfer's expectation (1963) that the habitat selection process of Temperate Zone birds will be so organized as to permit the species to respond opportunistically to such changes in its environment.

Changes in Territory Configuration

Adult gnatcatchers defend a territory which contains the necessary requisites (e.g., food, shelter, and nest sites) for their own survival and rearing young. Territoriality of this sort promotes a "contest" type of intraspecific competition (Nicholson, 1957). The initial contest for territories appears to have a direct influence on the size of the breeding populations. Of the 12 original males that set up territories on the

study area in March and April, 1963, two subsequently moved from the area. These vacancies were filled by two pairs from other areas. The breeding habitats appear to be nearly "saturated" following the period of territory establishment, with each pair claiming an area which contains adequate supplies of the requisites required for successful breeding.

In the weeks that follow the initial settlement of the breeding area, however, there is a reversal in the abundance of food between the habitats encompassed within the original territories. The configuration of the territory boundaries shifts in accordance with these seasonal changes in the dispersion of optimal food resources (Fig. 6). Between May and July, when the abundance of foliage arthropods in the deciduous oaks is high (Fig. 19), nesting activity is focused in the deciduous oak woodlands. As the food supply of the deciduous oaks declines in August, the gnatcatchers move back into the chaparral and live oaks.

Throughout the breeding season, the territorial boundaries are extended to include the dense clumps of evergreen vegetation to which fledglings are led by the adults. Such areas may provide the vulnerable fledglings with a refuge from predators.

An interrelated series of behavioral adaptations facilitates the shifting of territorial boundaries in response to the gnatcatcher's changing demands on its environment, and to the seasonal changes in conditions prevailing in different habitats. During periods when the adults are making frequent trips to the nest with building materials and food for the nestlings, the males can patrol the perimeter of the territories only infrequently. As a result, the pair's most active defense of the territory becomes compressed into a relatively small area around the nest. Adjacent pairs which have lost their attachment to a nest, either through the failure of a nest or the fledging of a brood, are able to move into these vacant areas while the original holder is occupied elsewhere with nesting activities.

The fluid territorial organization of the gnatcatcher is promoted by a lack of synchrony in the nesting cycle of adjacent pairs. The prolonged period over which the birds arrive on the breeding area and begin nesting (Fig. 19) produces asynchronous cycles early in the season. The timing of territory establishment is itself related to changing conditions in the various habitats. In addition, nest predators, through their spotty destruction of nests, interrupt the breeding cycle at different times (Root, *in press*). Since the ability to change the configuration of territories offers obvious advantages for the gnatcatcher, the lack of synchrony has beneficial effects. In this regard, the gnatcatcher differs from those species (cited in Wynne-Edwards, 1962) whose breeding cycles are closely synchronized, apparently in response to seasonal peaks in their food supply. The gnatcatcher's diverse diet, its ability to successively exploit different habitats, and its prolonged breeding season, reduce the necessity for highly synchronous breeding.

CHANGES IN FORAGING BEHAVIOR

Both the abundance and the species composition of the foliage arthropods vary in time and space (Figs. 12-14, Table 11). To exploit such a variable food supply, the gnatcatchers must possess a versatile foraging behavior that can respond quickly to sudden changes in prey availability. Accordingly, the gnatcatcher's capacity to alter its foraging beat, foraging maneuvers, and diet at different seasons has been demonstrated (Figs. 8, 10-12, and 14).

The high species diversity of the arthropod fauna, coupled with the gnatcatcher's selective predation on it, makes it difficult to establish an exact correspondence between changes in the availability of different kinds of prey and changes in the foraging behavior (see section on diet for further discussion of this point). One clear relationship emerges from these data, however. In all of the situations studied, gnatcatchers concentrated their foraging efforts on those arthropods which, by virtue of their large size, high abundance, or sedentary behavior, could be exploited most efficiently. Such opportunistic feeding behavior seems to be characteristic of most true predators: similar trends in prey selection have been observed in gastropods (Paine, 1963), robberflies (Powell and Stage, 1962), fish (Nilsson, 1955; Brooks and Dodson, 1965), and other species of birds (Gibb and Betts, 1963).

ADAPTIVE RESPONSES TO THE BROOD'S FOOD REQUIREMENTS

The energy requirement of the dependent young exerts heavy demands upon the foraging abilities of the adults. Broods of advanced nestlings and young fledglings were fed by the most attentive parent about once every 2-4 min (for further details on the rates at which broods were fed and seasonal changes in the parental roles of both sexes, see Root, *in press*). In addition, the normal self-maintenance energy requirements of the adults must be elevated somewhat by their frantic activity in feeding the young. Thus the interval of about 22 days, when the brood is completely dependent upon the adults for food, obviously constitutes a critical time in the gnatcatcher's utilization of its food supply.

In addition to their adaptive responses to seasonal and spatial changes in the food supply (discussed above), adult gnatcatchers also possess several behavioral adaptations that are primarily related to the increased energy requirements of the brood. While the time and energy budgets (Orians, 1964) were not studied directly, it is obvious that the adult's release from incubation and brooding permits the pair-unit to devote progressively greater proportions of time to foraging as the food requirements of the developing brood increase.

The young gnatcatchers are fed larger arthropods, on the average, than are consumed by the adults. Since large preys are less abundant, their capture requires that pairs with young must search a greater volume of the habitat. This is accomplished by an ob-vious increase in the searching velocity, the number of hunting perches assumed per unit time, by adults with young.

The foraging intensity of the adults increases significantly when they are feeding young (Table 7). The rate at which prey is captured in an interval of time spent foraging is thereby increased. This intensification of the foraging activity is accompanied by shifts in the bird's tactical response. The adults with dependent young capture a greater proportion of their prey in aerial attacks (Fig. 11) and in the pursuit of prey that has been flushed from the foliage (Table 8). Such changes in the foraging tactics reflect a tendency for adults to engage in more strenuous maneuvers in order to capture highly active prey. In comparison, adults which are engaged in self-maintenance activities appear to conserve energy by pursuing mainly those arthropods that can be captured with less effort.

Adults with dependent young also increase their exploitation of arthropods in the herb layer. There is a suggestion that by shifting to the herb layer, these adults are able to capture larger prey, e.g., grasshoppers, than would be possible within the foliage zone. Since the availability of suitable hunting perches is low in the herb layer, it is often necessary for the gnatcatchers to engage in long hovering flights or to flutter constantly while searching from a pliant or upright perch while foraging at this level. Thus it would appear that exploitation of the herb layer requires a greater expenditure of energy than does foraging in the foliage zone of trees.

These tactical responses of adults with young are expressed by exploiting the more peripheral portions of the foraging niche. The magnitude of this niche broadening can be estimated by a comparison of diversity indices for various components of the exploitation pattern. The index, $H = -\Sigma p_i \log_e p_i$, where p_i is the proportion of all behaviors represented by behavior i, was calculated for this purpose (Table 14). The properties of this widely used index have been discussed by MacArthur and MacArthur (1961), MacArthur (1964), Whittaker (1965), and Pielou (1966). Both R. T. Paine (*in litt.*) and L. C. Cole (pers. comm.) have pointed out that to compare H values validly, one must assume that the situation is thoroughly sampled and that further effort will not change the p_is. I found that trial calculations for 50 consecutive maneuvers give fairly consistent results within each sampling category (the difference between extreme H values from 5 such samples was 0.152). This implies that any bias due to sample size is negligible in Table 14 where the H values are calculated for even longer runs of 125 consecutive maneuvers.

Within each sampling period, the adults with young exhibited a greater foraging diversity, both with respect to the substrates where prey was obtained and in the maneuvers employed in capturing prey, than those engaged in self-maintenance activities (Table 14). In each case, the increased diversity was accomplished by foraging outside of the gnatcatcher "adaptive mode": only by performing maneuvers, made strenuous by the lack of primary adaptations, were the adults able to exploit a wider spectrum of resources. Gnatcatchers engaged in self-maintenance foraging employed the same repertoire as those with young, but performed the strenuous maneuvers infrequently. Thus the broadening of the niche during the period of the brood's dependency was produced by a flattening of the exploitation curves (Fig. 1) and not by any important expansion in the limits of the niche.

This response of increasing the foraging diversity to meet the increased energy requirements of the brood may be a common adaptation among birds. Evans (1964) found that nestlings of three sparrow species were fed a more diverse array of prey life-forms from a wider variety of microhabitats than were eaten by the adults. Similarly Kuroda (1963) demonstrated that the foraging behavior of adult Grey Starlings (*Sturnus cineraceus*) became less selective when the normal brood size was experimentally increased.

Orians (1961), Verbeek (1964), and Verner (1965) have shown how the amount of time budgeted by a species to various activities is related to variations in the energy requirements. The changes in the foraging behavior of gnatcatchers with dependent young, however, demonstrate that several additional factors are involved in the response to increasing energy requirements. As a result of the increase in foraging intensity, the selection of larger prey, and the more frequent performance of strenuous maneuvers, the expenditure and capture of energy are very different for a unit of foraging time spent by a bird which is feeding young and one which is engaged solely in self-maintenance.

ORGANIZATION OF THE EXPLOITATION PATTERN

The food supply of the gnatcatcher has been shown to vary widely in time and space. To exploit this resource efficiently, the gnatcatcher's foraging behavior must be plastic enough to respond to variations in the food supply while remaining specialized enough to avoid intense interspecific competition. Put another way, the species must be sufficiently specialized as to compensate for what has been termed the "jack of all trades" principle (MacArthur and MacArthur, 1961) and yet not so specialized that efficiency and survival are reduced in seeking out highly restricted conditions in a diverse and changing environment. The following discussion considers the manner in which the gnatcatcher's exploitation pattern is adapted to meet these conflicting requirements. The terminology and conceptual scheme employed in this discussion have already been considered in the introduction to this paper.

Each dimension of the gnatcatcher's niche is bounded by indistinct limits which are apparently imposed by the birds' inability to exploit the peripheral situations efficiently. Morphological adaptations, considered here in their broadest sense as including modifications of the external features, physiological abilities, and configurations of "neural pathways," are the factors that most likely set these boundaries. Within each set of limits the gnatcatchers are able to alter their behavior in response to changes in the environment. This is accomplished by varying the frequency with which different "adaptive mode" activities are performed. Thus the gnatcatchers changed from chaparral to oak woodland habitats during the late spring (Fig. 6), from twig to foliage insects following bud-burst (Fig. 10), from small arthropods in the winter to larger prey in the summer (Figs. 12 and 14), and from mirids and lepidopterous larvae in March-May to membracids and cicadellids during June-August (Table 11).

The niche dimensions of the birds studied by Crowell (1962) seem to be limited in the same manner as those of the gnatcatcher. Crowell contrasted the foraging behavior of populations living on the mainland with populations of the same species living under conditions of reduced interspecific competition on Bermuda. Morphological divergence between the respective mainland and Bermudian populations was only slight. The exploitation curves of the birds on Bermuda had different modes, but these always fell within the range of activities observed on the mainland. The results suggest that the niches of these species are circumscribed by limits that are conservative and morphologically determined.

The exploitation pattern of a population can be formed by the interplay of two adaptive strategies: either the population can be morphologically variable, so that each individual is restricted to a portion of the total niche, or all of the individuals may possess a similar range of capabilities (cf. Klopfer, 1962; Levins, 1963; and Van Valen, 1965). In most cases, exploitation patterns are probably based upon a mixture of these two extremes. The difference in the foraging behavior of sexually dimorphic birds reviewed by Selander (1966) are representative of a strategy stressing morphological variation. The gnatcatcher seems to emphasize the alternative strategy. Different individuals performed the same maneuvers and obtained food from the same substrates with similar frequencies (Table 9). The observed differences probably reflect the small sample size and the opportunistic response of individuals to the local conditions on their territories. Furthermore, several individuals (not cited in Table 9) from both the winter and breeding grounds were observed to perform the entire foraging repertoire. The limits of the exploitation pattern therefore appear to be similar for all members of a gnatcatcher population.

Gnatcatchers are capable of performing all of the basic foraging maneuvers employed by other members of the foliage-gleaning guild except using the foot to hold prey. Furthermore, they capture prey from all of the major arthropod taxa, and over a similar size range to that exploited by the other insectivorous birds. The gnatcatchers, however, engage in some of these activities (e.g., climbing trunks,

Table 14. Indices of foraging diversity (H) of the Blue-gray Gnatcatcher. H is Shannon's measure for diversity (see the text for a description of its calculation and properties). The values for H are based on samples of 125 consecutive capture maneuvers in each category. The substrates where prey was obtained are classified as tree foliage, bark, herbs, and air; the capture maneuvers as gleans, hovers, hawks, and tumbles.

	Diversity of Substrates		Diversity of Maneuvers	
	Self-maintenance	Feeding Young	Self-maintenance	Feeding Young
May 10-June 16.............................	0.542	1.006	0.804	1.133
June 24-July 27.............................	0.729	1.142	0.923	1.214

hanging beneath perches, or consuming large caterpillars) only on rare occasions, and then with obvious difficulty. The exploitation patterns of other foliage-gleaning birds are expressed in a similar fashion. Thus the niche limits of the guild members appear to overlap broadly, with the major separation between species being achieved through differences in their efficiency in performing the basic maneuvers common to all.

Hinde (1958) has proposed that birds learn to exploit the prey which they can obtain most efficiently. This conclusion has been substantiated in laboratory preference experiments on finches (Kear, 1962). Further support for this hypothesis is found in Hess's (1964) demonstration that the normal pecking preference of newly-hatched chicks can be altered by providing food rewards for different stimulus objects. These results suggest one means by which species can avoid intense interspecific competition while exploiting niches which overlap. When a particular type of prey becomes superabundant, several species that are not highly adapted for its capture can exploit this resource with relative ease. As the food supplies dwindle, the exploitation patterns can contract, with each species concentrating on those situations for which they are best adapted. Thus the exploitation curves of guild members would be expected to alternate between overlap and discreteness in response to seasonal changes in prey availability. Such pulsations in accordance with changes in the food supply have been observed in the feeding behavior of birds (Lack, 1946; Gibb, 1954) and fish (Nilsson, 1960; Lindstrom and Nilsson, 1962) which have similar exploitation patterns.

The members of the foliage-gleaning guild each exhibited a characteristic mixture of preferences in foraging beat, feeding maneuvers, and diet that may be shifted in response to changes in food availability. In this manner, the guild members are apparently able to exploit niches sufficiently distinct to permit co-existence, and yet broad enough to meet the demands of a changing environment. Thus the niche relations in the foliage-gleaning guild fit well with the predictions of other authors; the niche limits are broad, and therefore overlapping, because the environment is highly variable (Klopfer, 1962); and the species are separated by "fine-grained" differences because their prey is small relative to the size of the consumer (MacArthur and Levins, 1964).

Since structural modifications are correlated with these differences in foraging preference, it would appear that a distinctive morphology is an important prerequisite for guild membership. The prevalence of character displacement in closely related sympatric species (Brown and Wilson, 1956; Schoener, 1965) supports this conclusion. The relationship between character displacement and the degree of niche overlap has been considered by Hutchinson (1959), Klopfer and MacArthur (1961) and Klopfer (1962). Their conclusions are based entirely upon differences in one dimension of the trophic apparatus. The results of this investigation, however, demonstrate that consideration of only one morphological feature can yield ambiguous results; two guild members may compensate for a low character difference in one feature by divergence in another (Fig. 18).

The various specializations that limit the gnatcatcher's exploitation pattern are integrated in such a manner that the bird will encounter a large number of foraging situations as it moves through the habitat. Such an arrangement results in what has been termed a "convex" exploitation pattern (MacArthur and MacArthur, 1961). The gnatcatcher's foraging beat is restricted by its inability to use perches that are either pliant or of large diameter. As a result, the birds usually forage in the foliage zone of trees and shrubs where suitable perches abound. A variety of plant species within a given habitat provides perches of suitable size and density for foraging gnatcatchers. The birds are thereby able to enter the foliage zone of most of the plants they encounter while traveling at any level above the herb stratum. Gnatcatchers tend to linger in certain species of trees at different seasons, and in sunlit banks of foliage when the sun is low on the horizon. It seems likely that this results from the bird's "velocity" being reduced by favorable foraging conditions.

The foraging tactics seem to be closely related to the gnatcatcher's relative restriction to the foliage zone. The repertoire is rich in aerial and rushing maneuvers that allow them to capture the small, active insects common amidst the foliage.

The gnatcatcher's diet is limited mainly by the sizes of prey which can be captured efficiently: the energy return from extremely small prey is too low to warrant the effort, and large prey are either too difficult to capture or cannot be swallowed. Prey with effective concealment and escape reactions, and those which are distasteful, are not found in the diet. This still leaves a wide variety of arthropod species whose characteristics are acceptable to gnatcatchers. By having the feeding specializations related to the size and behavorial attributes of the prey, rather than its taxonomic relationships, the gnatcatcher is probably more likely to encounter suitable food in different habitats and at different seasons.

The exploitation patterns of other species of the foliage-gleaning guild are limited in other ways. The Plain Titmouse, for instance, is better able to use a much greater variety of hunting perches. This permits the tits to respond to changes in the dispersion of their food supply by shifting between the foliage zone, the subcanopy, and the ground. On the other hand, the titmice appear to be more highly restricted to the oak woodland habitats at the Hastings Reservation. Similarly, the foraging maneuvers that can be performed easily are less diverse, but the tits are able to exploit plant foods as well as arthropods. Restriction in one dimension of the niche seems to be compensated by increased breadth in another. As a result of such adjustments, it is impossible to judge whether one species is more broadly adapted than another without considering the complete exploitation pattern.

The general organization of the exploitation patterns among the birds of the foliage-gleaning guild is basically similar, although the specific features of the pattern are integrated in different ways. All are able to respond opportunistically to changes in the environment within broad limits, apparently determined by morphological adaptations. Furthermore, the limiting specializations are primarily related to structural features of the habitat and to behavioral and size characteristics of the prey. In this way the birds are less influenced by the spatial and temporal changes that have been observed in the availability of particular prey taxa or the abundance of arthropods associated with particular plant species. Thus the exploitation patterns of foliage-gleaning birds are so organized that each species can remain optimally adapted to the greatest possible variety of opportunities.

SUMMARY

The food supply and habitat of the Blue-gray Gnatcatcher, a small insectivorous bird, vary widely in space and time. This paper considers the manner in which the gnatcatcher's niche exploitation pattern is organized to achieve optimal adaptation to the conflicting demands of a changing environment.

The behavior of the gnatcatcher was studied at the Hastings Reservation in the Central Coast Ranges of California during the breeding season, and at three localities in the deserts of Arizona during the winter. More than 1200 hours were spent in actual observation during three different field seasons. Field observation techniques were developed to quantify changes in the gnatcatcher's foraging beat, feeding tactics, food supply, and diet. Similar data were gathered on other species of sympatric insectivorous birds. In addition, the pattern of habitat occupancy, nest site requirements, and general natural history of the gnatcatcher were studied.

In the early spring gnatcatcher territories are established in chaparral and live oak woodland, where foliage arthropods are abundant. Deciduous oak woodland is not settled until later when the insects living there have begun to increase. Habitat selection by the gnatcatcher can be only partially "released" by structural characteristics of the vegetation: habitats that differ greatly in physiognomy are occupied while some vegetation-types having a similar structure are left vacant. Thus it appears that the abundance of such critical requisites as food and nest sites is directly involved in the habitat selection process.

Gnatcatchers are able to shift their territory boundaries in accordance with seasonal changes in the dispersion of optimal food resources among the various subhabitats. The constriction of the territory during the nest construction and nestling stages of the reproductive cycle leaves vacant areas that may then be colonized by other pairs. The fluid organization of territories is promoted by a lack of synchrony in the nesting cycle of adjacent pairs. As a result, different broods are able to share certain areas well suited for young fledglings by using them at different times.

At their preferred foraging stations, gnatcatchers select the larger and more vulnerable prey. The diet and foraging beat shift in accordance with seasonal changes in the availability of prey.

The energy requirements of the dependent young exert heavy demands upon the foraging abilities of the adults. Pairs fed broods of older nestlings and young fledglings as often as 43 times per hour. Compared with birds engaged in self-maintenance feeding, the food gathered by adults with young was increased by their: (i) devoting a greater proportion of time to foraging, (ii) increasing the foraging intensity, i.e., the frequency of attack maneuvers per unit of foraging time, (iii) increasing the diversity of the foraging tactics and foraging beat, and (iv) capturing larger prey. In meeting the demands of the brood, the adults foraged outside their normal "adaptive mode," a response that probably requires a large expenditure of maintenance energy. The bearing of this result upon the interpretation of "time and energy budgets" is discussed. It appears that the abundance of food is an especially critical requisite during the brief period when the brood is completely dependent upon the adults for sustenance.

A new unit, the ecological "guild," is proposed for groups of species that exploit the same class of environmental resources in a similar way. The utility of the guild concept is discussed. Members of the foliage-gleaning guild, to which the gnatcatcher belongs, overlap in their foraging repertoires, foraging beats, and diets. It appears that niche segregation within the guild is maintained by differences in the efficiency with which portions of a common range of situations can be exploited. These different efficiencies are reinforced by morphological differences.

The exploitation pattern of the gnatcatcher is expressed opportunistically within limits that are largely set by morphology. The limiting specializations are related to structural features of the habitat and to behavioral and size characteristics of the prey; these are factors that tend to change less in space and time than do other features of the environment.

ACKNOWLEDGMENTS

I have profited greatly from the guidance and encouragement of Frank A. Pitelka throughout this investigation. A National Science Foundation graduate fellowship and a grant from Francis S. Hastings to the Museum of Vertebrate Zoology provided the necessary financial support for the research. I am particularly indebted to John Davis and the personnel of the Hastings Reservation for many courtesies extended to me. Joe T. Marshall, Jr., and his students at the University of Arizona greatly facilitated my research in the Tucson area.

Larry L. Wolf lent valuable assistance throughout the investigation. In addition, George E. Chaniot, John Davis, Arnthor Gardarsson, John Tramantano, Laidlaw O. Williams, and Edwin O. Willis assisted me during certain phases of the field work. James Bell gave me permission to collect specimens on his ranch. Paul D. Hurd made the determinations of the stomach contents. Keith L. Dixon has generously made some of his unpublished data available for inclusion in this paper.

In addition to the people already mentioned, I have profited from discussion about the research with Alden H. Miller, LaMont C. Cole, Keith L. White, Robert T. Paine, Richard T. Holmes, my wife Elizabeth, and many others. Frank A. Pitelka, Peter R. Marler, Herbert G. Baker, and William L. Brown, Jr., have made critical comments on the manuscript.

LITERATURE CITED

Andrew, R. J. 1956. Intention movements of flight in certain passerines, and their use in systematics. Behaviour 10: 179-204.

Beal, F. E. L. 1907. Birds of California in relation to the fruit industry. U.S. Dept. Agric. Biol. Survey Bull. 30: 1-100.

———. 1910. Birds of California in relation to the fruit industry. Part II. U.S. Dept. Agric. Biol. Survey Bull. 34: 1-96.

Bent, A. C. 1949. Life histories of North American thrushes, kinglets, and their allies. U.S. Natl. Mus. Bull. 196.

———. 1950. Life histories of North American wagtails, shrikes, vireos, and their allies. U.S. Natl. Mus. Bull. 197.

Borror, D. J. & D. M. DeLong. 1960. An introduction to the study of insects. Holt, Rinehart, and Winston: New York. 1030 pp.

Bowman, R. I. 1961. Morphological differentiation and adaptation in the Galapagos finches. Univ. Calif. Publ. Zool. 58: 1-326.

Brian, M. V. 1952. The structure of a dense natural ant population. J. Anim. Ecol. 21: 12-24.

———. 1955. Food collection by a Scottish ant community. J. Anim. Ecol. 24: 336-351.

Brodkorb, P. 1944. The subspecies of the gnatcatcher, Polioptila albiloris. Jour. Wash. Acad. Sci. 34: 311-316.

Brooks, J. L. and S. I. Dobson. 1965. Predation, body size, and composition of plankton. Science 150: 28-35.

Brown, W. L. & E. O. Wilson. 1956. Character displacement. Systematic Zool. 5: 49-64.

Brues, C. T., A. L. Melander & F. M. Carpenter. 1954. Classification of insects. Bull. Mus. Comp. Zool. 108.

Burleigh, T. D. 1958. Georgia birds. Univ. Okla. Press: Norman. 747 pp.

Chamberlin, C. 1901. Some architectural traits of the Western Gnatcatcher. Condor 3: 33-36.

Chemsak, J. A. 1957. Use of polyethylene bags for inactivating sweep samples. J. Econ. Ent. 50: 523.

Cooper, W. S. 1922. The broad-sclerophyll vegetation of California. Carnegie Inst. of Washington Publ. 319: 1-124.

Crowell, K. L. 1962. Reduced interspecific competition among the birds of Bermuda. Ecology 43: 75-88.

Dansereau, P. 1958. A universal system for recording vegetation. Contrib. Inst. Bot. Univ. de Montreal 72: 1-58.

DeBach, P. 1966. The competitive displacement and coexistence principles. Ann. Rev. Entom. 11: 183-212.

Dennis, C. J. 1964. Observations on treehopper behavior (Homoptera, Membracidae). Amer. Midl. Natur. 71: 452-459.

Dixon, K. L. 1962. Notes on the molt schedule of the Plain Titmouse. Condor 64: 134-139.

Eaton, E. H. 1914. Birds of New York. New York State Mus. Mem. 12.

Engels, W. L. 1940. Structural adaptations in thrashers (Mimidae: genus Toxostoma) with comments on interspecific relationships. Univ. Calif. Publ. Zool. 42: 341-400.

Evans, F. C. 1964. The food of Vesper, Field and Chipping Sparrows nesting in an abandoned field in Southeastern Michigan. Amer. Midl. Natur. 72: 57-75.

Ficken, M. S. 1962. Maintenance activities of the American Redstart. Wilson Bull. 74: 153-165.

———, & R. W. Ficken. 1962. The comparative ethology of the wood warblers: A review. The Living Bird (1962): 103-121.

Forbush, E. H. 1929. Birds of Massachusetts and other New England states. Vol. 3. Norwood Press: Mass. 466 pp.

Gibb, J. 1954. Feeding ecology of tits, with notes on treecreeper and goldcrest. Ibis 96: 513-543.

———. 1958. Predation by tits and squirrels on the eucosmid Ernarmonia conicolana (Heyl.). J. Anim. Ecol. 27: 375-396.

———. 1962. L. Tinbergen's hypothesis on the role of specific search images. Ibis 104: 106-111.

———, & M. M. Betts. 1963. Food and food supply of nestling tits (Paridae) in Breckland pine. J. Anim. Ecol. 32: 489-533.

Goss, N. S. 1891. History of the birds of Kansas. Crane & Co.: Topeka. 692 pp.

Grinnell, J. 1921. The principle of rapid peering in birds. In Joseph Grinnell's philosophy of nature. (1943). Berkeley: Univ. Calif. Press. 237 pp.

———, & A. H. Miller. 1944. The distribution of the birds of California. Pac. Coast Avifauna 27: 1-608.

Hardin, G. 1960. The competitive exclusion principle. Science 131: 1292-1298.

Hartley, P. H. T. 1948. Food and feeding relationships

in a community of fresh-water fishes. J. Anim. Ecol. 17: 1-14.

Heady, H. F. 1958. Vegetational changes in the California annual type. Ecology 39: 402-416.

Hess, E. H. 1964. Imprinting in birds. Science 146: 1128-1139.

Hinde, R. A. 1959. Food and habitat selection in birds and lower vertebrates. Proc. 14th Intern. Congr. Zool.: 808-810.

Horvath, O. 1964. Seasonal differences in Rufous Hummingbird nest height and their relation to nest climate. Ecology 45: 235-241.

Hutchinson, G. E. 1957. Concluding remarks. Cold Spring Harbor Symposia on Quant. Biol. 22: 415-427.

————. 1959. Homage to Santa Rosalia or why are there so many kinds of animals? Amer. Natur. 93: 145-159.

————, & R. H. MacArthur. 1959. A theoretical ecological model of size distributions among species of animals. Amer. Natur. 93: 117-125.

Kear, J. 1962. Food selection in finches with special reference to interspecific differences. Proc. Zool. Soc. Lond. 138: 163-204.

Kilham, L. 1965. Differences in feeding behavior of male and female Hairy Woodpeckers. Wilson Bull. 77: 134-145.

Klopfer, P. H. 1962. Behavioral aspects of ecology. Prentice-Hall: New Jersey. 172 pp.

————. 1963. Behavioral aspects of habitat selection: The role of early experience. Wilson Bull. 75: 15-22.

————, & R. H. MacArthur. 1961. On the causes of tropical species diversity: Niche overlap. Amer. Natur. 95: 223-226.

Kuroda, N. 1963. Adaptive parental feeding as a factor influencing the reproductive rate in the Grey Starling. Res. Population Ecol. 5: 1-10.

Lack, D. 1933. Habitat selection in birds. J. Anim. Ecol. 2: 239-262.

————. 1946. Competition for food by birds of prey. J. Anim. Ecol. 15: 123-129

————. 1949. The significance of ecological isolation, pp. 299-308. In G. L. Jepsen, G. Simpson, and E. Mayr (ed.), Genetics, paleontology, and evolution. Princeton Univ. Press: Princeton.

————. 1954. The natural regulation of animal numbers. Oxford Univ. Press: Oxford. 343 pp.

————. 1966. Population studies of birds. Clarendon Press: Oxford. 341 pp.

Levins, R. 1963. Theory of fitness in a heterogeneous environment II. Developmental flexibility and niche selection. Amer. Natur. 97: 75-90.

Lindstrom, T. & Nils-Arvid Nilsson. 1962. On the competition between whitefish species. In E. D. Le Cren and M. W. Holdgate (ed.), The exploitation of natural animal populations. Symposia Brit. Ecol. Soc. 2: 326-340.

Linsdale, J. M. 1943. Work with vertebrate animals on the Hastings Natural History Reservation. Amer. Midl. Natur. 30: 254-267.

MacArthur, R. H. 1958. Population ecology of some warblers of the northeastern coniferous forest. Ecology 39: 599-619.

————. 1964. Environmental factors affecting bird species diversity. Amer. Natur. 98: 387-397.

————, & R. Levins. 1964. Competition, habitat selection, and character displacement in a patchy environment. Proc. Nat. Acad. Sci. 51: 1207-1210.

————, & J. W. MacArthur. 1961. On bird species diversity. Ecology 42: 594-598.

————, ————, & J. Preer. 1962. On bird species diversity II. Prediction of bird census from habitat measurements. Amer. Natur. 96: 167-174.

Macfadyen, A. 1957. Animal ecology: Aims and methods. Pitman & Sons: London. 264 + xx pp.

Maitland, P. S. 1965. The feeding relationship of salmon, trout, minnows, stone loach and three-spined sticklebacks in the River Endrick, Scotland. J. Anim. Ecol. 34: 109-133.

Marshall, J. T., Jr. 1957. Birds of pine-oak woodland in southern Arizona and adjacent Mexico. Pac. Coast Avifauna 32: 1-125.

Mayr, E. & D. Amadon. 1951. A classification of Recent Birds. American Mus. Novitates No. 1360.

Miller, A. H. 1951. An analysis of the distribution of the birds of California. Univ. Calif. Publ. Zool. 50: 531-644.

———— & R. C. Stebbins. 1964. The lives of desert animals in Joshua Tree National Monument. Univ. of Calif. Press: Berkeley. 452 + vi pp.

Nice, M. M. 1932. Observations on the nesting of the Blue-gray Gnatcatcher. Condor 34: 18-22.

Nicholson, A. J. 1957. The self-adjustment of populations to change. Cold Spring Harbor Symposia Quant. Biol. 22: 153-173.

Nickell, W. P. 1956. Vertical nest placement in the Blue-gray Gnatcatcher. Wilson Bull. 68: 159-160.

Nilsson, Nils-Arvid. 1955. Studies on the feeding habits of trout and char in North Swedish lakes. Rept. Inst. Freshwater Res. Drottningholm 36: 163-225.

————. 1960. Seasonal fluctuations in the food segregation of trout, char, and whitefish in 14 North-Swedish lakes. Rept. Inst. Freshwater Res. Drottningholm 41: 185-205.

Nolan, V., Jr. 1963. Reproductive success of birds in a deciduous scrub habitat. Ecology 44: 305-313.

Orians, G. H. 1961. The ecology of blackbird (Agelaius) social systems. Ecol. Monogr. 31: 285-312.

Paine, R. T. 1963. Trophic relationships of 8 sympatric predatory gastropods. Ecology 44: 63-73.

Paynter, R. A., Jr. 1955. The ornithogeography of the Yucatan Peninsula. Peabody Mus. Nat. Hist., Yale Univ. Bull. 9.

————. 1964. Subfamily Polioptilinae. In J. E. Peters, Checklist of birds of the world. Vol. 10: 443-455.

Pielou, E. C. 1966. Shannon's formula as a measure of specific diversity: its use and disuse. Amer. Natur. 100: 463-465.

Pitelka, F. A. 1951. Ecologic overlap in hummingbirds. Ecology 32: 641-661.

Powell, J. A. & G. I. Stage. 1962. Prey selection by robberflies of the genus Stenopogon, with particular observations of S. engelhardti Bromley (Diptera: Asilidae). Wasmann J. Biol. 20: 139-157.

Root, R. B. 1964a. Ecological interactions of the Chestnut-backed Chickadee following a range extension. Condor 66: 229-238.

————. 1964b. Niche organization in the Blue-gray Gnatcatcher (Polioptila caerulea). Ph.D. Thesis. Univ. of Calif. Berk. (No. 65-3074) 146 p. Univ. Microfilms. Ann Arbor, Mich.

————. 1966. Avian response to a population outbreak of the tent caterpillar, Malacosoma constrictum. Pan-Pacific Entomol. 42: 48-53.

————. 1968. The behavior and reproductive success of the Blue-gray Gnatcatcher. Condor, in press.

———— and R. M. Yarrow. A predator-decoy method for capturing insectivorous birds. Auk 84: 423-424.

Ross, H. H. 1957. Principles of natural coexistence indicated by leafhopper populations. Evolution 11: 113-129.

——. 1958. Further comments on niches and natural coexistence. Evolution 12: 112-113.

Salt, G. W. 1953. An ecologic analysis of three California avifaunas. Condor 55: 258-273.

——. 1957. An analysis of avifaunas in the Teton Mountains and Jackson Hole, Wyoming. Condor 59: 373-393.

Sargent, T. D. 1965. The role of experience in the nest building of the Zebra Finch. Auk 82: 48-61.

Savage, J. M. 1958. The concept of ecologic niche, with reference to the theory of natural coexistence. Evolution 12: 111-112.

Schaldach, W. J., Jr. 1963. The avifauna of Colima and adjacent Jalisco, Mexico. Proc. Western Found. Vert. Zool. 1: 1-100.

Schimper, A. F. W. 1903. Plant-geography upon a physiological basis. English Ed. Clarendon Press: Oxford. 839 + xxx pp.

Schoener, T. W. 1965. The evolution of bill size differences among sympatric congeneric species of birds. Evolution 19: 189-213.

Selander, R. K. 1966. Sexual dimorphism and differential niche utilization in birds. Condor 68: 113-151.

Shreve, F. 1915. The vegetation of a desert mountain range as conditioned by climatic factors. Carnegie Inst. Wash. Publ. 217.

——. 1927a. The vegetation of a coastal mountain range. Ecology 8: 27-44.

——. 1927b. The physical conditions of a coastal mountain range. Ecology 8: 398-414.

Smith, W. P. 1954. Breeding bird census. —Upland mixed forest. Audubon Field Notes 8: 367.

Sprunt, A., Jr. 1954. Florida bird life. Coward-McCann: New York. 527 pp.

Svärdson, G. 1949. Competition and habitat selection in birds. Oikos 1: 157-174.

Tinbergen, L. 1960. The natural control of insects in pinewoods I. Factors influencing the intensity of predation by songbirds. Arch. Neerl. Zool. 13: 265-343.

Turpaeva, E. P. 1957. [Food interrelationships of dominant species in marine benthic biocoenoses.] Akad. nauk SSSR. Trans. Inst. Okeanologii v.20. f.171. (Amer. Inst Biol. Sci.: Washington, D. C. Transl. 1959. pp. 137-148

Udvardy, M. D. F. 1957. An evaluation of quantitative studies in birds. Cold Spring Harbor Symposia Quant. Biol. 22: 301-311.

Van Valen, L. 1965. Morphological variation and width of ecological niche. Amer. Natur. 99: 377-390.

Verbeek, N. A. M. 1964. A time and energy budget study of the Brewer Blackbird. Condor 66: 70-74.

Verner, J. 1965. Time budget of the male Long-billed Marsh Wren during the breeding season. Condor 67: 125-139.

Webster, J. D. 1959. Breeding bird census—Beech-maple forest. Audubon Field Notes 13: 462.

Wecker, S. C. 1963. The role of early experience in habitat selection by the prairie deer mouse, *Peromyscus maniculatus bairdi*. Ecol. Monogr. 33: 307-325.

White, K. L. 1966. Structure and composition of foothill woodland in central coastal California. Ecology 47: 229-237.

Whittaker, R. H. 1965. Dominance and diversity in land plant communities. Science 147: 250-260.

—— & W. A. Niering. 1965. Vegetation of the Santa Catalina Mountains, Arizona: A gradient analysis of the south slope. Ecology 46: 429-452.

Wynne-Edwards, V. C. 1962. Animal dispersion in relation to social behavior. Oliver and Boyd: Edinburgh. 653 pp.

[*Editors' Note:* Material has been omitted at this point.]

11

Reprinted from *Science*, **179**, 767 (Feb. 23, 1973)

DISTRIBUTIONAL ECOLOGY OF NEW GUINEA BIRDS

Jared M. Diamond

[*Editors' Note:* In the original, material precedes this excerpt.]

Nonspatial Segregating Mechanisms

Spatial overlap of closely related species is possible if they separate on the basis of time, diet, or foraging techniques.

Infrequently, closely related bird species segregate by occupying the same space at different times of day or of the year. The kingfisher *Melidora macrorhina* is nocturnal, whereas other kingfishers are diurnal. The south New Guinea savanna near Merauke is alternately occupied by two marsh hawks, *Circus approximans* in the dry season and *C. spilonotus* in the wet season.

Differences in body size provide the commonest means by which closely related species can take the same type of food in the same space at the same time (*15*). Larger birds can take larger food items than can smaller birds, but smaller birds can perch on more slender branches than can larger birds. One can frequently see a bird foraging out along a branch up to the point where the branch begins to bend under its weight. In a tree occupied by birds of many species the larger birds are often concentrated toward the main branches, the smaller birds toward the periphery (Fig. 9). Among congeners sorting by size in New Guinea, the ratio between the weights of the larger bird and the smaller bird is on the average 1.90; it is never less than 1.33 and never more than 2.73. Species with similar habits and with a weight ratio less than 1.33 are too similar to coexist locally (that is, to share territories) and must segregate spatially. For instance, the cuckoo-shrikes *Coracina tenuirostris* and *C. papuensis* segregate by habitat on New Guinea, where their average weights are 73 grams and 74 grams, respectively, but they often occur together in the same tree on New Britain, where their respective weights are 61 g and 101 g. New Guinea has no locally coexisting pairs of species with similar habits and with a weight ratio exceeding 2.73, presumably because a medium-sized bird of relative weight $\sqrt{2.73} = 1.65$ can coexist successfully with both the large species and with the small species. Thus one finds a sequence of three or more species rather than just two species of such different sizes. For example, the eight

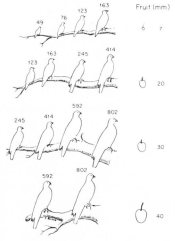

Fig. 9. Schematic representation of niche relations among the eight species of *Ptilinopus* and *Ducula* fruit pigeons in New Guinea lowland rain forest. On the right is a fruit of a certain diameter (in millimeters), and on the left are pigeons of different weights (in grams) arranged along a branch. Each pigeon weighs approximately 1.5 times the next pigeon. Each fruit tree attracts up to four consecutive members of this size sequence. Trees with increasingly large fruits attract increasingly large pigeons. In a given tree the smaller pigeons are preferentially distributed on the smaller, more peripheral branches. The pigeons having the weights indicated are: 49 g, *Ptilinopus nanus*; 76 g, *P. pulchellus*; 123 g, *P. superbus*; 163 g, *P. ornatus*; 245 g, *P. perlatus*; 414 g, *Ducula rufigaster*; 592 g, *D. zoeae*; 802 g, *D. pinon*.

fruit pigeons of the genera *Ptilinopus* and *Ducula* coexisting in the lowland forests of New Guinea form a graded size sequence over a 16-fold range in weight (Fig. 9).

May and MacArthur (*6*) predicted on theoretical grounds that species segregating along a single niche dimension in a fluctuating environment must maintain a certain minimum niche difference. This minimum spacing seems in fact to have been reached in nature by those bird species that segregate according to size. Thus, on Pacific islands with 30 to 50 bird species, the ratio between the weights of pairs of birds segregating by size is approximately 4, but this ratio has already been compressed to 2 on islands with

100 species. On New Guinea (513 species) the average value of this ratio is compressed no further, and the extra species are accommodated by expanding the size sequences to smaller or larger birds or else by finer subdivision of space or foraging techniques.

Similarly sized species that take the same food may overlap spatially if they harvest the food in different ways. For instance, small insectivorous birds differ in tactics according to whether insects are caught in midair by sallying, are pounced on and plucked off surfaces, are gleaned off surfaces, pried out of bark, taken from flowers, or are extracted from epiphytes and accumulations of dead leaves. Species with a given type of strategy further differ in the ratio between traveling time and stationary time in the frequency of movements, and in their average rate of travel (*23*). Thus, the montane flycatchers *Pachycephala modesta* and *Poecilodryas albonotata* differ in that the former remains perched for an average of 2 seconds, the latter for an average of 30 seconds between moves; and in that the former travels 1 m, and the latter 12 m, per move. *Pachycephala modesta* could be described as a quick and cursory searcher, *Poecilodryas albonotata* as a slow and selective searcher.

Finally, related species may segregate by different diets. For example, the whistler *Pachycephala leucostigma* eats mainly fruit while other whistlers eat mainly insects.

[*Editors' Note:* Material has been omitted at this point.]

REFERENCES

6. R. M. May and R. H. MacArthur, *Proc. Nat. Acad. Sci. U.S.A.* **69**, 1109 (1972).
15. J. M. Diamond, *Avifauna of the Eastern Highlands of New Guinea* (Nuttall Ornithological Club, Cambridge, Mass., 1972).
23. M. L. Cody, *Amer. Natur.* **102**, 107 (1968).

12

Reprinted from *Amer. Naturalist,* **100**(910), 65-75 (1966)

FOOD WEB COMPLEXITY AND SPECIES DIVERSITY

ROBERT T. PAINE

Department of Zoology, University of Washington, Seattle, Washington

Though longitudinal or latitudinal gradients in species diversity tend to be well described in a zoogeographic sense, they also are poorly understood phenomena of major ecological interest. Their importance lies in the derived implication that biological processes may be fundamentally different in the tropics, typically the pinnacle of most gradients, than in temperate or arctic regions. The various hypotheses attempting to explain gradients have recently been reviewed by Fischer (1960), Simpson (1964), and Connell and Orias (1964), the latter authors additionally proposing a model which can account for the production and regulation of diversity in ecological systems. Understanding of the phenomenon suffers from both a specific lack of synecological data applied to particular, local situations and from the difficulty of inferring the underlying mechanism(s) solely from descriptions and comparisons of faunas on a zoogeographic scale. The positions taken in this paper are that an ultimate understanding of the underlying causal processes can only be arrived at by study of local situations, for instance the promising approach of MacArthur and MacArthur (1961), and that biological interactions such as those suggested by Hutchinson (1959) appear to constitute the most logical possibilities.

The hypothesis offered herein applies to local diversity patterns of rocky intertidal marine organisms, though it conceivably has wider applications. It may be stated as: "Local species diversity is directly related to the efficiency with which predators prevent the monopolization of the major environmental requisites by one species." The potential impact of this process is firmly based in ecological theory and practice. Gause (1934), Lack (1949), and Slobodkin (1961) among others have postulated that predation (or parasitism) is capable of preventing extinctions in competitive situations, and Slobodkin (1964) has demonstrated this experimentally. In the field, predation is known to ameliorate the intensity of competition for space by barnacles (Connell, 1961b), and, in the present study, predator removal has led to local extinctions of certain benthic invertebrates and algae. In addition, as a predictable extension of the hypothesis, the proportion of predatory species is known to be relatively greater in certain diverse situations. This is true for tropical vs. temperate fish faunas (Hiatt and Strasburg, 1960; Bakus, 1964), and is seen especially clearly in the comparison of shelf water zooplankton populations (81 species, 16% of which are carnivores) with those of the presumably less productive though more stable Sargasso Sea (268 species, 39% carnivores) (Grice and Hart, 1962).

In the discussion that follows no quantitative measures of local diversity are given, though they may be approximated by the number of species represented in Figs. 1 to 3. No distinctions have been drawn between species within certain food categories. Thus I have assumed that the probability of, say, a bivalve being eaten is proportional to its abundance, and that predators exercise no preference in their choice of any "bivalve" prey. This procedure simplifies the data presentation though it dodges the problem of taxonomic complexity. Wherever possible the data are presented as both number observed being eaten and their caloric equivalent. The latter is based on prey size recorded in the field and was converted by determining the caloric content of Mukkaw Bay material of the same or equivalent species. These caloric data will be given in greater detail elsewhere. The numbers in the food webs, unfortunately, cannot be related to rates of energy flow, although when viewed as calories they undoubtedly accurately suggest which pathways are emphasized.

Dr. Rudolf Stohler kindly identified the gastropod species. A. J. Kohn, J. H. Connell, C. E. King, and E. R. Pianka have provided invaluable criticism. The University of Washington, through the offices of the Organization for Tropical Studies, financed the trip to Costa Rica. The field work in Baja California, Mexico, and at Mukkaw Bay was supported by the National Science Foundation (GB-341).

THE STRUCTURE OF SELECTED FOOD WEBS

I have claimed that one of the more recognizable and workable units within the community nexus are subwebs, groups of organisms capped by a terminal carnivore and trophically interrelated in such a way that at higher levels there is little transfer of energy to co-occurring subwebs (Paine, 1963). In the marine rocky intertidal zone both the subwebs and their top carnivores appear to be particularly distinct, at least where macroscopic species are involved; and observations in the natural setting can be made on the quantity and composition of the component species' diets. Furthermore, the rocky intertidal zone is perhaps unique in that the major limiting factor of the majority of its primary consumers is living space, which can be directly observed, as the elegant studies on interspecific competition of Connell (1961a,b) have shown. The data given below were obtained by examining individual carnivores exposed by low tide, and recording prey, predator, their respective lengths, and any other relevant properties of the interaction.

A north temperate subweb

On rocky shores of the Pacific Coast of North America the community is dominated by a remarkably constant association of mussels, barnacles, and one starfish. Fig. 1 indicates the trophic relationships of this portion of the community as observed at Mukkaw Bay, near Neah Bay, Washington (ca. 49° N latitude). The data, presented as both numbers and total calories consumed by the two carnivorous species in the subweb, *Pisaster ochraceus*,

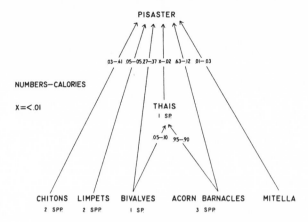

FIG. 1. The feeding relationships by numbers and calories of the *Pisaster* dominated subweb at Mukkaw Bay. *Pisaster*, N = 1049; *Thais*, N = 287. N is the number of food items observed eaten by the predators. The specific composition of each predator's diet is given as a pair of fractions; numbers on the left, calories on the right.

a starfish, and *Thais emarginata*, a small muricid gastropod, include the observational period November, 1963, to November, 1964. The composition of this subweb is limited to organisms which are normally intertidal in distribution and confined to a hard rock substrate. The diet of *Pisaster* is restricted in the sense that not all available local food types are eaten, although of six local starfish it is the most catholic in its tastes. Numerically its diet varies little from that reported by Feder (1959) for *Pisaster* observed along the central California coastline, especially since the gastropod *Tegula*, living on a softer bottom unsuitable to barnacles, has been omitted. *Thais* feeds primarily on the barnacle *Balanus glandula*, as also noted by Connell (1961b).

This food web (Fig. 1) appears to revolve on a barnacle economy with both major predators consuming them in quantity. However, note that on a nutritional (calorie) basis, barnacles are only about one-third as important to *Pisaster* as either *Mytilus californianus*, a bivalve, or the browsing chiton *Katherina tunicata*. Both these prey species dominate their respective food categories. The ratio of carnivore species to total species is 0.18. If *Tegula* and an additional bivalve are included on the basis that they are the most important sources of nourishment in adjacent areas, the ratio becomes 0.15. This number agrees closely with a ratio of 0.14 based on *Pisaster*, plus all prey species eaten more than once, in Feder's (1959) general compilation.

A subtropical subweb

In the Northern Gulf of California (ca. 31° N.) a subweb analogous to the one just described exists. Its top carnivore is a starfish (*Heliaster kubiniji*), the next two trophic levels are dominated by carnivorous gastropods, and the main prey are herbivorous gastropods, bivalves, and barnacles. I have

collected there only in March or April of 1962-1964, but on both sides of
the Gulf at San Felipe, Puertecitos, and Puerta Penasco.   The resultant
trophic arrangements (Fig. 2), though representative of springtime conditions and indicative of a much more stratified and complex community, are
basically similar to those at Mukkaw Bay.   Numerically the major food item
in the diets of *Heliaster* and *Muricanthus nigritus* (a muricid gastropod),
the two top-ranking carnivores, is barnacles; the major portion of these
predators' nutrition is derived from other members of the community, primarily herbivorous mollusks.   The increased trophic complexity presents
certain graphical problems.   If increased trophic height is indicated by a
decreasing percentage of primary consumers in a species diet, *Acanthina
tuberculata* is the highest carnivore due to its specialization on *A. angelica*,
although it in turn is consumed by two other species.   Because of this, and
ignoring the percentages, both *Heliaster* and *Muricanthus* have been placed
above *A. tuberculata*.   Two species, *Hexaplex* and *Muricanthus* eventually
become too large to be eaten by *Heliaster*, and thus through growth join it
as top predators in the system.   The taxonomically-difficult gastropod
family Columbellidae, including both herbivorous and carnivorous species
(Marcus and Marcus, 1962) have been placed in an intermediate position.

The Gulf of California situation is interesting on a number of counts.   A
new trophic level which has no counterpart at Mukkaw Bay is apparent, in-
terposed between the top carnivore and the primary carnivore level.   If
higher level predation contributes materially to the maintenance of diversity, these species will have an effect on the community composition
out of proportion to their abundance.   In one of these species, *Muricanthus*,

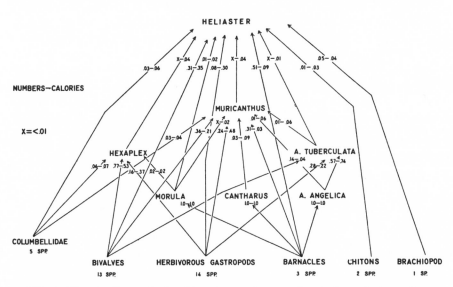

FIG. 2. The feeding relationships by numbers and calories of the *Heliaster*
dominated subweb in the northern Gulf of California.   *Heliaster*, N = 2245; *Muricanthus*, N = 113; *Hexaplex*, N = 62; *A. tuberculata*, N = 14; *A. angelica*, N = 432;
Morula, N = 39; *Cantharus*, N = 8.

the larger members belong to a higher level than immature specimens (Paine, unpublished), a process tending to blur the food web but also potentially increasing diversity (Hutchinson, 1959). Finally, if predation operates to reduce competitive stresses, evidence for this reduction can be drawn by comparing the extent of niche diversification as a function of trophic level in a typical Eltonian pyramid. *Heliaster* consumes all other members of this subweb, and as such appears to have no major competitors of comparable status. The three large gastropods forming the subterminal level all may be distinguished by their major sources of nutrition: *Hexaplex*—bivalves (53%), *Muricanthus*—herbivorous gastropods (48%), and *A. tuberculata*—carnivorous gastropods (74%). No such obvious distinction characterizes the next level composed of three barnacle-feeding specialists which additionally share their resource with *Muricanthus* and *Heliaster*. Whether these species are more specialized (Klopfer and Mac-Arthur, 1960) or whether they tolerate greater niche overlap (Klopfer and MacArthur, 1961) cannot be stated. The extent of niche diversification is subtle and trophic overlap is extensive.

The ratio of carnivore species to total species in Fig. 2 is 0.24 when the category Columbellidae is considered to be principally composed of one herbivorous (*Columbella*) and four carnivorous (*Pyrene, Anachis, Mitella*) species, based on the work of Marcus and Marcus (1962).

A tropical subweb

Results of five days of observation near Mate de Limon in the Golfo de Nocoya on the Pacific shore of Costa Rica (approx. 10° N.) are presented in Fig. 3. No secondary carnivore was present; rather the environmental resources were shared by two small muricid gastropods, *Acanthina brevidentata* and *Thais biserialis*. The fauna of this local area was relatively simple and completely dominated by a small mytilid and barnacles. The co-occupiers of the top level show relatively little trophic overlap despite the broad nutritional base of *Thais* which includes carrion and cannibalism. The relatively low number of feeding observations (187) precludes an accurate appraisal of the carnivore species to total web membership ratio.

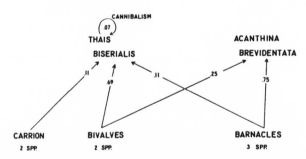

FIG. 3. The feeding relationship by numbers of a comparable food web in Costa Rica. *Thais*, N = 99; *Acanthina*, N = 80.

CHANGES RESULTING FROM THE REMOVAL OF THE TOP CARNIVORE

Since June, 1963, a "typical" piece of shoreline at Mukkaw Bay about eight meters long and two meters in vertical extent has been kept free of *Pisaster*. An adjacent control area has been allowed to pursue its natural course of events. Line transects across both areas have been taken irregularly and the number and density of resident macroinvertebrate and benthic algal species measured. The appearance of the control area has not altered. Adult *Mytilus californianus*, *Balanus cariosus*, and *Mitella polymerus* (a goose-necked barnacle) form a conspicuous band in the middle intertidal. The relatively stable position of the band is maintained by *Pisaster* predation (Paris, 1960; Paine, unpublished). At lower tidal levels the diversity increases abruptly and the macrofauna includes immature individuals of the above, *B. glandula* as scattered clumps, a few anemones of one species, two chiton species (browsers), two abundant limpets (browsers), four macroscopic benthic algae (*Porphyra*-an epiphyte, *Endocladia*, *Rhodomela*, and *Corallina*), and the sponge *Haliclona*, often browsed upon by *Anisodoris*, a nudibranch.

Following the removal of *Pisaster*, *B. glandula* set successfully throughout much of the area and by September had occupied from 60 to 80% of the available space. By the following June the *Balanus* themselves were being crowded out by small, rapidly growing *Mytilus* and *Mitella*. This process of successive replacement by more efficient occupiers of space is continuing, and eventually the experimental area will be dominated by *Mytilus*, its epifauna, and scattered clumps of adult *Mitella*. The benthic algae either have or are in the process of disappearing with the exception of the epiphyte, due to lack of appropriate space; the chitons and larger limpets have also emigrated, due to the absence of space and lack of appropriate food.

Despite the likelihood that many of these organisms are extremely long-lived and that these events have not reached an equilibrium, certain statements can be made. The removal of *Pisaster* has resulted in a pronounced *decrease* in diversity, as measured simply by counting species inhabiting this area, whether consumed by *Pisaster* or not, from a 15 to an eight-species system. The standing crop has been increased by this removal, and should continue to increase until the *Mytilus* achieve their maximum size. In general the area has become trophically simpler. With *Pisaster* artificially removed, the sponge-nudibranch food chain has been displaced, and the anemone population reduced in density. Neither of these carnivores nor the sponge is eaten by *Pisaster*, indicating that the number of food chains initiated on this limited space is strongly influenced by *Pisaster*, but by an indirect process. In contrast to Margalef's (1958) generalization about the tendency, with higher successional status towards "an ecosystem of more complex structure," these removal experiments demonstrate the opposite trend: in the absence of a complicating factor (predation), there is a "winner" in the competition for space, and the local system tends toward simplicity. Predation by this interpretation interrupts the successional process and, on a local basis, tends to increase local diversity.

No data are available on the microfaunal changes accompanying the gradual alteration of the substrate from a patchy algal mat to one comprised of the byssal threads of *Mytilus*.

INTERPRETATION

The differences in relative diversity of the subwebs diagrammed in Figs. 1-3 may be represented as Baja California (45 spp.) >> Mukkaw Bay (11 spp.) > Costa Rica (8 sp.), the number indicating the actual membership of the subwebs and not the number of local species. All three areas are characterized by systems in which one or two species are capable of monopolizing much of the space, a circumstance realized in nature only in Costa Rica. In the other two areas a top predator that derives its nourishment from other sources feeds in such a fashion that no space-consuming monopolies are formed. *Pisaster* and *Heliaster* eat masses of barnacles, and in so doing enhance the ability of other species to inhabit the area by keeping space open. When the top predator is artificially removed or naturally absent (i.e., predator removal area and Costa Rica, respectively), the systems converge toward simplicity. When space is available, other organisms settle or move in, and these, for instance chitons at Mukkaw Bay and herbivorous gastropods and pelecypods in Baja California, form the major portions of the predator's nutrition. Furthermore, *in situ* primary production is enhanced by the provision of space. This event makes the grazing moiety less dependent on the vagaries of phytoplankton production or distribution and lends stability to the association.

At the local level it appears that carnivorous gastropods which can penetrate only one barnacle at a time, although they might consume a few more per tidal interval, do not have the same effect as a starfish removing 20 to 60 barnacles simultaneously. Little compensation seems to be gained from snail density increases because snails do not clear large patches of space, and because the "husks" of barnacles remain after the animal portion has been consumed. In the predator removal area at Mukkaw Bay, the density of *Thais* increased 10- to 20-fold, with no apparent effect on diversity although the rate of *Mytilus* domination of the area was undoubtedly slowed. Clusters (density of 75–125/m²) of *Thais* and *Acanthina* characterize certain rocks in Costa Rica, and diversity is still low. And, as a generality, wherever acorn barnacles or other space-utilizing forms potentially dominate the shore, diversity is reduced unless some predator can prevent the space monopoly. This occurs in Washington State where the shoreline, in the absence of *Pisaster*, is dominated by barnacles, a few mussels, and often two species of *Thais*. The same monopolistic tendencies characterize Connell's (1961a,b) study area in Scotland, the rocky intertidal of northern Japan (Hoshiai, 1960, 1961), and shell bags suitable for sponge settlement in North Carolina (Wells, Wells, and Gray, 1964).

Local diversity on intertidal rocky bottoms, then, appears directly related to predation intensity, though other potential factors are mentioned below. If one accepts the generalizations of Hedgpeth (1957) and Hall

(1964) that ambient temperature is the single most important factor influencing distribution or reproduction of marine invertebrates, then the potential role of climatic stability as measured by seasonal variations in water temperature can be examined. At Neah Bay the maximum range of annual values are 5.9 to 13.3 C (Rigg and Miller, 1949); in the northern Gulf of California, Roden and Groves (1959) recorded an annual range of 14.9 to 31.2 C; and in Costa Rica the maximum annual range is 26.1 to 31.7 C (Anon., 1952). Clearly the greatest benthic diversity, and one claimed by Parker (1963) on a regional basis to be among the most diverse known, is associated with the most variable (least stable) temperature regimen. Another influence on diversity could be exercised by environmental heterogeneity (Hutchinson, 1959). Subjectively, it appeared that both the Mukkaw Bay and Costa Rica stations were topographically more distorted than the northern Gulf localities. In any event, no topographic features were evident that could correlate with the pronounced differences in faunal diversity. Finally, Connell and Orias (1964) have developed a model for the organic enrichment of regions that depends to a great extent on the absolute amount of primary production and/or nutrient import, and hence energy flowing through the community web. Unfortunately, no productivity data are available for the two southern communities, and comparisons cannot yet be made.

PREDATION AND DIVERSITY GRADIENTS

To examine predation as a diversity-causing mechanism correlated with latitude, we must know why one environment contains higher order carnivores and why these are absent from others. These negative situations can be laid to three possibilities: (1) that through historical accident no higher carnivores have evolved in the region; (2) that the sample area cannot be occupied due to a particular combination of *local* hostile physiological effects; (3) that the system cannot support carnivores because the rate of energy transfer to a higher level is insufficient to sustain that higher level. The first possibility is unapproachable, the second will not apply on a geographic scale, and thus only the last would seem to have reality. Connell and Orias (1964) have based their hypothesis of the establishment and maintenance of diversity on varying rates of energy transfer, which are determined by various limiting factors and environmental stability. Without disagreeing with their model, two aspects of primary production deserve further consideration. The animal diversity of a given system will probably be higher if the production is apportioned more uniformly throughout the year rather than occurring as a single major bloom, because tendencies towards competitive displacement can be ameliorated by specialization on varying proportions of the resources (MacArthur and Levins, 1964). Both the predictability of production on a sustained annual basis and the causation of resource heterogeneity by predation will facilitate this mechanism. Thus, per production unit, greater stability of production should be correlated with greater diversity, other things being equal.

The realization of this potential, however, depends on more than simply the annual stability of carbon fixation. Rate of production and subsequent transfer to higher levels must also be important. Thus trophic structure of a community depends in part on the physical extent of the area (Darlington, 1957), or, in computer simulation models, on the amount of protoplasm in the system (Garfinkel and Sack, 1964). On the other hand, enriched aquatic environments often are characterized by decreased diversity. Williams (1964) has found that regions of high productivity are dominated by few diatom species. Less productive areas tended to have more species of equivalent rank, and hence a greater diversity. Obviously, the gross amount of energy fixed by itself is incapable of explaining diversity; and extrinsic factors probably are involved.

Given sufficient evolutionary time for increases in faunal complexity to occur, two independent mechanisms should work in a complementary fashion. When predation is capable of preventing resource monopolies, diversity should increase by positive feedback processes until some limit is reached. The argument of Fryer (1965) that predation facilitates speciation is germane here. The upper limit to local diversity, or, in the present context, the maximum number of species in a given subweb, is probably set by the combined stability and rate of primary production, which thus influences the number and variety of non-primary consumers in the subweb. Two aspects of predation must be evaluated before a generalized hypothesis based on predation effects can contribute to an understanding of differences in diversity between *any* comparable regions or faunistic groups. We must know if resource monopolies are actually less frequent in the diverse area than in comparable systems elsewhere, and, if so, why this is so. And we must learn something about the multiplicity of energy pathways in diverse systems, since predation-induced diversity could arise either from the presence of a variety of subwebs of equivalent rank, or from domination by one major one. The predation hypothesis readily predicts the apparent absence of monopolies in tropical (diverse) areas, a situation classically represented as "many species of reduced individual abundance." It also is in accord with the disproportionate increase in the number of carnivorous species that seems to accompany regional increases in animal diversity. In the present case in the two adequately sampled, structurally analalgous, subwebs, general membership increases from 13 at Mukkaw Bay to 45 in the Gulf of California, a factor of 3.5, whereas the carnivore species increased from 2 to 11, a factor of 5.5.

SUMMARY

It is suggested that local animal species diversity is related to the number of predators in the system and their efficiency in preventing single species from monopolizing some important, limiting, requisite. In the marine rocky intertidal this requisite usually is space. Where predators capable of preventing monopolies are missing, or are experimentally removed, the systems become less diverse. On a local scale, no relationship between latitude (10° to 49° N.) and diversity was found. On a geographic scale, an increased stability of annual production may lead to an increased capacity for systems to support higher-level carnivores. Hence tropical, or other, ecosystems are more diverse, and are characterized by disproportionately more carnivores.

LITERATURE CITED

Anon. 1952. Surface water temperatures at tide stations. Pacific coast North and South America. Spec. Pub. No. 280: p. 1-59. U. S. Coast and Geodetic Survey.

Bakus, G. J. 1964. The effects of fish-grazing on invertebrate evolution in shallow tropical waters. Allan Hancock Found. Pub. 27: 1-29.

Connell, J. H. 1961a. Effect of competition, predation by *Thais lapillus*, and other factors on natural populations of the barnacle *Balanus balanoides*. Ecol. Monogr. 31: 61-104.

————. 1961b. The influence of interspecific competition and other factors on the distribution of the barnacle *Chthamalus stellatus*. Ecology 42: 710-723.

Connell, J. H., and E. Orias. 1964. The ecological regulation of species diversity. Amer. Natur. 98: 399-414.

Darlington, P. J. 1957. Zoogeography. Wiley, New York.

Feder, H. M. 1959. The food of the starfish, *Pisaster ochraceus*, along the California coast. Ecology 40: 721-724.

Fischer, A. G. 1960. Latitudinal variations in organic diversity. Evolution 14: 64-81.

Fryer, G. 1965. Predation and its effects on migration and speciation in African fishes: a comment. Proc. Zool. Soc. London 144: 301-310.

Garfinkel, D., and R. Sack. 1964. Digital computer simulation of an ecological system, based on a modified mass action law. Ecology 45: 502-507.

Gause, G. F. 1934. The struggle for existence. Williams and Wilkins Co., Baltimore.

Grice, G. D., and A. D. Hart. 1962. The abundance, seasonal occurrence, and distribution of the epizooplankton between New York and Bermuda. Ecol. Monogr. 32: 287-309.

Hall, C. A., Jr. 1964. Shallow-water marine climates and molluscan provinces. Ecology 45: 226-234.

Hedgpeth, J. W. 1957. Marine biogeography. Geol. Soc. Amer. Mem. 67, 1: 359-382.

Hiatt, R. W., and D. W. Strasburg. 1960. Ecological relationships of the fish fauna on coral reefs of the Marshall Islands. Ecol. Monogr. 30: 65-127.

Hoshiai, T. 1960. Synecological study on intertidal communities III. An analysis of interrelation among sedentary organisms on the artificially denuded rock surface. Bull. Marine Biol. Sta. Asamushi. 10: 49-56.

————. 1961. Synecological study on intertidal communities. IV. An ecological investigation on the zonation in Matsushima Bay concerning the so-called covering phenomenon. Bull. Marine Biol. Sta. Asamushi. 10: 203-211.

Hutchinson, G. E. 1959. Homage to Santa Rosalia or why are there so many kinds of animals? Amer. Natur. 93: 145–159.

Klopfer, P. H., and R. H. MacArthur. 1960. Niche size and faunal diversity. Amer. Natur. 94: 293–300.

———. 1961. On the causes of tropical species diversity: niche overlap. Amer. Natur. 95: 223–226.

Lack, D. 1949. The significance of ecological isolation, p. 299–308. *In* G. L. Jepsen, G. G. Simpson, and E. Mayr [eds.], Genetics, paleontology and evolution. Princeton Univ. Press, Princeton.

MacArthur, R., and R. Levins. 1964. Competition, habitat selection, and character displacement in a patchy environment. Proc. Nat. Acad. Sci. 51: 1207–1210.

MacArthur, R. H., and J. W. MacArthur. 1961. On bird species diversity. Ecology 42: 594–598.

Marcus, E., and E. Marcus. 1962. Studies on Columbellidae. Bol. Fac. Cienc. Letr. Univ. Sao Paulo 261: 335–402.

Margalef, R. 1958. Mode of evolution of species in relation to their place in ecological succession. XVth Int. Congr. Zool. Sect. 10, paper 17.

Paine, R. T. 1963. Trophic relationships of 8 sympatric predatory gastropods. Ecology 44: 63–73.

Paris, O. H. 1960. Some quantitative aspects of predation by muricid snails on mussels in Washington Sound. Veliger 2: 41–47.

Parker, R. H. 1963. Zoogeography and ecology of some macro-invertebrates, particularly mollusca in the Gulf of California and the continental slope off Mexico. Vidensk. Medd. Dansk. Natur. Foren., Copenh. 126: 1–178.

Rigg, G. B., and R. C. Miller. 1949. Intertidal plant and animal zonation in the vicinity of Neah Bay, Washington. Proc. Calif. Acad. Sci. 26: 323–351.

Roden, G. I., and G. W. Groves. 1959. Recent oceanographic investigations in the Gulf of California. J. Marine Res. 18: 10–35.

Simpson, G. G. 1964. Species density of North American recent mammals. Syst. Zool. 13: 57–73.

Slobodkin, L. B. 1961. Growth and regulation of Animal Populations. Holt, Rinehart, and Winston, New York.

———. 1964. Ecological populations of Hydrida. J. Anim. Ecol. 33 (Suppl.): 131–148.

Wells, H. W., M. J. Wells, and I. E. Gray. 1964. Ecology of sponges in Hatteras Harbor, North Carolina. Ecology 45: 752–767.

Williams, L. G. 1964. Possible relationships between plankton-diatom species numbers and water-quality estimates. Ecology 45: 809–823.

13

Reprinted from *Brit. Ecol. Soc. Jubilee Symp.*, A. Macfayden and P. J. Newbould, eds., Suppl. *J. Ecol.*, **52**, *J. Animal Ecol.*, **33**, pp. 131–140, 147–148 (1964)

EXPERIMENTAL POPULATIONS OF HYDRIDA

By L. B. SLOBODKIN

Department of Zoology, Ann Arbor, Michigan

INTRODUCTION

Experimental population studies are usually conducted by placing a small number of animals in some suitable medium in a closed container. These animals and their descendants constitute the population. The population is free to adjust its numbers, age distribution and occasionally its species composition. The containers are examined periodically, while the medium is renewed and the animals counted or subjected to some other treatment.

This type of experiment has repeatedly demonstrated that the reproductive and survival characteristics of single individuals are more or less radically dependent on the kind and number of animals in the experimental containers. While the precise mechanisms by which these changes are brought about are of intrinsic interest their occurrence is no longer of great theoretical significance.

It has been possible, however, to use population experiments to generate tentative hypotheses with a claim to phylogenetic or physiological generality. The experiments with hydrids were designed to test the validity of two such hypotheses that have grown out of work in my laboratory.

As discussed in detail elsewhere, experimental studies of *Daphnia* populations have demonstrated a damped oscillatory approach to population equilibrium, a linear relation between terminal population size and food supply and a very simple relation between population size and a function of predation (*cf*. Slobodkin 1961). In addition, if ecological efficiency is defined as the caloric value of the animals removed on a sustained yield basis from a population divided by the calories of food consumed by that population, it is found that the maximal experimentally achieved values in the *Daphnia* experiments are in fair agreement with existing field estimates (*cf*. Slobodkin 1962).

There is no obvious connection between the taxonomic and ecological position of *Daphnia* and the conclusions derived from *Daphnia* population experiments. It therefore seemed of great interest to do subsequent population experiments on animals as different as possible from *Daphnia*.

Technical considerations impose severe restrictions on the choice of experimental animals for population studies (Slobodkin 1961). Within these restrictions hydrids seemed of great interest since they are taxonomically and ecologically very distinct from *Daphnia* but can be readily cultured, are free of genetic complexity and have a relatively short generation time. Preliminary experiments were conducted using *Hydra oligactis* Pallas 1776 from the Huron River, Ann Arbor, Michigan. There was a high variance in the results but as far as they went they tended to confirm both the simple relation between predation and population size and the order of magnitude of maximal ecological efficiency as found in *Daphnia* (Slobodkin 1962).

The present paper is concerned with a more ambitious and extensive repetition of these experiments. Two species of hydrids were used to test the applicability of the above

119

hypotheses to a multi-species situation. Among the ancillary questions touched on are the competitive interaction between brown and green hydrids, the role of light in the nutrition of green hydrids and the relation between the sizes of individuals, predators and prey.

The most noticeable ecological differences between hydrids and *Daphnia* are: (1) Hydrids are generally attached to a substrate, while *Daphnia* move freely in three dimensions. (2) Hydrids can change size and shape readily, while *Daphnia* can only increase their size with time and have a relatively rigid exoskeleton.

In addition, green hydrids have presumably symbiotic algae in their endoderm.

Three experiments were conducted, which will be discussed in detail below. Orientation might be easier if they are briefly outlined here.

The first experiment was based on the hypothesis that the likelihood of joint survival of two ecologically different species in the same restricted space would be enhanced if the environment were such as to emphasize these differences. Since the primary ecological distinction between brown and green hydrids is in the possibility of algal photosynthesis, competing populations of brown and green hydra were set up in the light at various levels of 'predatory intensity' (defined below). A control population of the two species was also established in the dark. As it developed, in the presence of light, green hydrids eliminated brown hydrids, unless the predatory intensity was very high.

The second experiment was in part a control for the competition in light experiments and consisted of maintaining brown *Hydra* in the light at two food levels. At each food level the populations were started with either one, five or 200 animals per 30 ml container. The different inoculations were used as a consequence of the research of Armstrong (1960). Using populations of *Dugesia tigrina* (Girard), Armstrong found that a population started with a very heavy initial flatworm inoculation, stabilized at a higher numerical level and smaller size of individuals than a flatworm population started with a lower inoculum. Since the size of hydra is as flexible as that of flatworms it seemed possible that the hydra populations would also show a permanent effect of the size of inoculum. In addition, the damped oscillatory approach to population equilibrium found in *Daphnia* population experiments has different patterns depending on the size and age structure of the inoculum (*cf.* Slobodkin 1954; Frank 1957).

The third experiment was designed to determine the efficiency of two species systems. Darwin conjectured that a mixed species system should have a higher yield than a one species system, although this had primary reference to a land crop system rather than to a self-regulating population. The experimental and field efficiency estimates, on the other hand, lead to the tentative conclusion that a multi-species system would have the same ecological efficiency as a single species system. This experiment involved twenty-four populations, including two clones of *Hydra littoralis* Hyman 1931, one clone of *Chlorohydra viridissima* (Pallas) 1766, three food levels and four predatory levels.

METHODS

The populations were maintained in Petri dishes containing 30 ml of synthetic pond water. For counting, populations were placed over millimetre ruled paper with white lines on a black background, under a movable binocular microscope. All of the Petri dish was scanned by the experimenter ticking off the number of each kind of animal on a rack of hand counters. The species and number of buds per animal were recorded. The animals were not removed from the Petri dishes during the counts. There was some movement of hydra during the counts but no pile-up occurred at any point in the Petri dishes during the counts. The data used in all analyses are means over extended numbers of censuses, more than twenty-five in every case, so that small sampling errors that may exist in individual counts are probably not significant in any of the conclusions to be drawn.

The census data will be discussed in terms of total number of mouths present. That is, an animal consisting of one central stalk and three feeding buds is counted as four mouths. The rationale for this procedure is that buds may stay attached to their parent for some time after they are in all other respects independent organisms. Once a bud begins to feed it enters into the general competition for food.

The hydrids were fed brine shrimp (*Artemia* sp.) hatched from eggs vacuum packed by the San Francisco Aquarium Association. The hatching and collecting procedure was copied from that of W. F. Loomis (personal communication). Usually 5 ml of saturated table salt solution was added to 500 ml of tap water in a shallow baking dish. Approximately a quarter teaspoonful of *Artemia* eggs was scattered on the water surface. At *c.* 20° C the eggs hatched usually in 24 hours. Occasionally different batches of eggs had slightly different hatching requirements. As much as 10–15 ml of saturated salt solution was necessary for some of them. When a directional light was used, many of the nauplii would collect in the region nearest the light, from which they could be removed, free of eggshells and unhatched eggs by use of a plastic gravy baster. They were then washed through a silk plankton net to remove salt and were re-suspended in a graduated centrifuge tube in artificial pond water. When this tube was placed in an ice water bath the nauplii settled to the bottom. A live settled volume of 0·5 ml of nauplii was then re-suspended in 10 ml of artificial pond water. One ml of this suspension constituted one food ration. All of the hydrid populations were fed and washed daily but counted only every 4 days.

The feeding procedure consisted of adding the appropriate number of food rations to the Petri dish, swirling gently to achieve dispersion and leaving the dishes under a towel, to prevent aggregation of the *Artemia*. After half an hour the culture medium was poured off into a finger bowl. The Petri dishes were immediately flooded with new medium to prevent drying of the adhering hydrids. *Hydra* that had been poured off with the old medium were retrieved from the uneaten *Artemia* and casts, using a medicine dropper. This process was then repeated until no *Artemia* remained. Every fourth day all the hydrids were gently pushed from the glass with a rubber 'policeman' (spatula) and poured into a clean Petri dish. Towards the end of the experiment sterile disposable Petri dishes made of wettable plastic were substituted for glass ones, at a saving in time, money and an improvement in apparent consistency of results.

My laboratory was moved during these experiments. The new and more spacious laboratory apparently had something in its plumbing (possibly mercury). In any case many of the populations either died completely or precipitously declined immediately after the move. For these populations data collected after the laboratory move have not been included in the analyses.

The amount of food provided was estimated by direct counts of the number of *Artemia* in a food ration. An uninhabited Petri dish would be 'fed' a food ration in 30 ml of medium. The *Artemia* were then counted and removed one at a time. This was done only as time permitted. One hundred and eleven such counts were made between August 1959 and July 1960 by four different technicians. The mean count was 612·0 with 95% confidence intervals of $\pm 86\cdot6$ per food ration (see Fig. 1). Populations were maintained in various experiments at $\frac{1}{2}$, 1, 2 and 4 times this ration per day.

The artificial pond water for the *Hydra* was also taken from W. F. Loomis. Ten milli-litres of each of two stock solutions were added to a gallon jug (4 l. approx.), filled to its neck with demineralized water. One stock solution contained 3·8 g NaHCO₃/l. The other contained 3 g CaCl₂/l.

Counts were also made of the number of *Artemia* in the used medium that was poured off. The variance of these counts is very high, varying with the size of each population. At times as much as 98% of the food was eaten but more usually around 90% would be a better estimate using the empty dish count as an estimate of the food particles provided.

When I refer to predation I mean removal of animals by a technician. Animals re-moved typically had no buds. The pattern of predation was fixed for each population as part of the experimental design and always consisted of a number of animals proportional to the increase in the population since the last census. For example, when I refer to a predation rate of 90%, I mean that the number of mouths in the previous census has been subtracted from the number of mouths in the current census and a number of budless

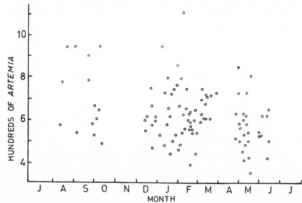

FIG. 1. The distribution of the counts of *Artemia* in one food ration over a period of 1 year. There is a high variance but no trend. The estimate of energy income is derived from the mean of these counts.

animals equal to nine-tenths of this difference is to be removed. All fractional animals were saved and removed at subsequent censuses. That is, if 12½ animals were to be re-moved at one census and 8½ at the subsequent census, 12 would be taken at the first census and 9 at the second. No animals were removed from declining populations. In the *Daphnia* studies (see Slobodkin 1959) an estimate of mortality between censuses was made from the corpses found in the culture vessels. Hydrids disintegrate almost im-mediately so that this procedure could not be attempted.

In two species situations the animals are removed in proportion to the relative numbers of the two species present at the time of the census, regardless of which species was responsible for the population increase. That is, if one increased, members of both species would be removed, even though the other species had not increased or had even declined.

For the efficiency studies it was necessary to estimate the caloric equivalent of both the hydrids and their food. At various times animals removed by predation and whole pop-ulations at the termination of experiments had been weighed dry in a tared beaker using a Mettler semimicro one pan automatic balance. Forty-one weighings were made of brown *Hydra* from competing populations in the dark giving a mean weight per animal of $2\cdot116\times10^{-5}$ g with 95% confidence intervals of $\pm0\cdot341\times10^{-5}$ g. Ninety-nine weighings of green hydrids from these populations gave a mean weight per animal of $7\cdot19\times10^{-6}$ g with 95% confidence intervals $\pm0\cdot653\times10^{-6}$ g. The weight of a single *Artemia* nauplius was taken as 2×10^{-6} g, from the data of Armstrong (1960). The calorific value of *Artemia* nauplii were taken as 6400 cal/g. This is derived from the value of Slobodkin & Richman (1961) for calories per ash-free gramme and an assumed ash content of 5%. All subsequent discussion will assume a value of $1\cdot28\times10^{-2}$ calories per *Artemia*. Calorific determinations of *Hydra oligactis* and *Chlorohydra viridissima* were not significantly different from each other. The mean value for these two species is 5607 ($S_{\bar{x}} = 89\cdot7$) cal/g, which is assumed valid for all hydrids. Combining the weight and caloric data provides an estimate of $0\cdot119$ cal/*Hydra littoralis* and $0\cdot0403$ cal/*Chlorohydra viridissima*. Combining the weights, caloric estimates and *Artemia* counts provides an estimate of $31\cdot32$ calories of food per food ration per 4 days.

Other assumptions, computational details and variations of procedure will be discussed in connection with particular experiments.

It will be noted that several of the above assumptions have either large or unknown errors associated with them. These are the best we can do at the moment. The consistency of most of the data to be discussed, both within these experiments and in comparison with other, quite independent, studies made by myself and others lends confidence to the initial assumptions. There is the possibility that bias of one sort or another determined by my intellectual, and therefore emotional, commitment to certain apparent ecological conclusions has in some way altered the analyses. With this in mind I have made it a point to use technical assistants for most of the population experiments, choosing careful, intelligent, people without formal training in zoology. The data could not have been gathered with my predispositions in mind since in general the technicians were unaware of the conclusions or theoretical considerations involved in this or in previous population experiments.

In addition I can only offer the invitation to ecologists in general to test experimentally the conclusions I will draw. I personally do not intend to do any more experimental efficiency studies using self-regulating populations for fear that in the absence of external verification I will fall into a solipsist trap.

It must be emphasized that these are self-regulating populations, not cohorts or individual animals selected by the investigator.

RESULTS

(1) *The relation between size of inoculum, the number of food particles provided per day and population growth*

To determine the effect of initial population size on growth pattern and terminal population size six *Hydra littoralis* populations were started, three at the base food level and three at half the base food level. At each food level one population was started with one animal, one with five animals and one with 200 animals. The temperature varied from 19 to 20° C.

From Figs. 2(a) and 2(b) it can be seen that regardless of the size of initial inoculum the populations tend towards the same final size. From Figs. 3(a), 3(b) and 3(c) it can be seen that the growth patterns at each inoculation level is linearly related to food supply, at least over the range $\frac{1}{2}$ to 1 food level.

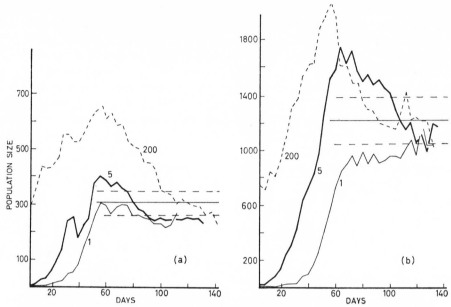

FIG. 2. The growth curves of three *Hydra littoralis* populations all at the same food level but started with an initial inoculum of one, five or 200 animals. The horizontal line indicates the mean number of *Artemia* per day provided. The dashed parallel lines indicate the 95% confidence intervals for the mean number of *Artemia* provided. (a) Populations with half food ration. (b) Populations with full food ration.

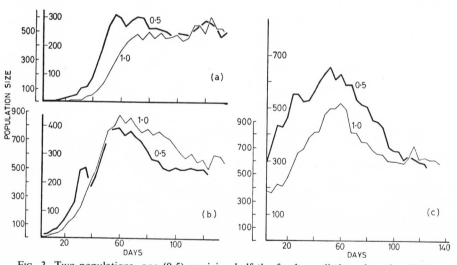

FIG. 3. Two populations, one (0·5) receiving half the food supplied to the other (1·0). The ordinate for the high food population is doubled. (a) One animal was used as the inoculum in both these populations. (b) Five animals used in inoculum. (c) 200 animals used in inoculum.

In Fig. 2 the horizontal line indicates the mean number of *Artemia* provided, the dashed lines representing the 95% confidence limits. The terminal population sizes seem to be approaching a situation in which each *Hydra* receives on the average one *Artemia*, at least, per day.

The pattern of population growth varies with the size of initial inoculum. The reproductive rate per individual is lower in high inoculum populations than in low inoculum populations at the same food level (Fig. 4). The percentage of the total mouths that were associated with buds, as either the main mouth or the bud's mouth, was taken as an indication of reproductive rate. We can tentatively conclude that the relation between increased peak size and inoculum was the result of a greater number of reproducing animals, despite the fact that in the high inoculum populations the reproductive rate per animal was lower. The reproductive rate at an inoculum of 200 was lower than that at

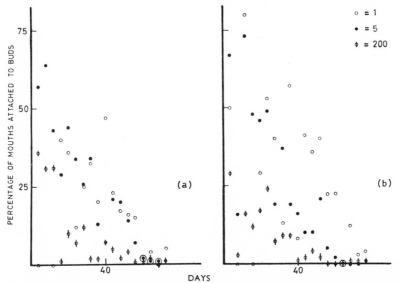

Fig. 4. The percentage of all mouths attached to budding individuals or to buds at two food levels (a at 0·5 level, b at 1·0 level) and three inoculum levels as a function of time since the initiation of the experiment. There is significantly less budding when the inoculum is 200 animals.

inoculum of one or of five but not 200 or even 40 times lower, as it would have to be if peak sizes were to stay the same. Comparing the populations with corresponding inoculum and different food levels, no significant difference in reproductive rate (as measured above) is found. Since the low food level populations can be thought of as inoculated twice as heavily with respect to their food supply as the high food level populations we can conclude that a twofold difference in inoculation rate makes no significant difference in reproductive rate while a fivefold or greater difference in inoculation rate does make a significant difference.

(2) *The effect of predation on* Hydra littoralis *in the light*

Five *H. littoralis* populations were set up with an initial inoculum of seventy-five animals and maintained at approximately 20° C under fluorescent lamps in a controlled temperature room. These populations were designed to act as a control to competition experiments (to be discussed below), and to test the applicability of a simple equation relating population size to predatory rate which had been successful in analysing the relation between *Daphnia* populations and predation (Slobodkin 1959). They were all maintained at a feeding rate of one food ration per day and were counted every 4 days. At the time of counting four of the populations had animals removed at the rates of $F = 25\%$, 50%, 75% or 90% as described in the 'Methods' section (see above). All five populations were maintained for approximately 300 days. The population growth curves are shown in Fig. 5. The cumulative mean population size as of the seventy-seventh census will be taken as the best estimate of the mean value, since after the seventy-seventh census the laboratory was moved.

D. *pulex* populations were found to conform reasonably well to the equation:

$$P_f = P_0\left(1 - \frac{f}{2-f}\right) \qquad (1)$$

in which P_f is the number of animals in a population subjected to predation at rate f, and P_0 is the control population size. f is the fraction of the newborn removed as yield. It differs from F as used in the non-*Daphnia* experiments since F is the fraction of the

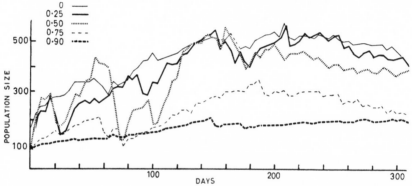

FIG. 5. The population growth curves of five *Hydra littoralis* populations subjected to five levels of predation.

population increase removed. The rapid disintegration of dead hydrids makes a direct estimate of *newborn* as such impossible. Nevertheless, the mean value of P_F for an experiment involving sixteen populations of *Hydra oligactis* was satisfactorily predicted by the equation:

$$P_F = P_0\left(1 - \frac{F}{2-F}\right) \qquad (2)$$

The *H. oligactis* data are presented graphically in Slobodkin (1962). The corresponding attempt to fit the present data to equation 2 is presented in Fig. 6 and Table 1.

The calculations in this table were used to compute from each F and the observed P_F that value of P_0 which would provide complete agreement with equation 2. The mean of these P_0 estimates was then fitted into equation 2 and used in combination with the particular values of F to compute the expected values of P_F. In general, the deviation between expectation and observation is sufficiently great not to lend support to the applicability of equation 2. The difference in definition between the predation systems of equations 1 and 2 leaves some residual question as to the applicability of equation 1. The horizontal line in Fig. 6 again indicates the estimated mean number of *Artemia* provided per day.

(3) *Competition between* Hydra littoralis *and* Chlorohydra viridissima

Four Petri dishes were inoculated with equal numbers of *Hydra littoralis* and *Chlorohydra viridissima* and maintained under the same conditions as the *Hydra littoralis* populations discussed above. Predation levels of 0, 25, 50 and 90% were applied to the mixed populations. In all dishes except that at 90% predation the brown *Hydra* became extinct. The time required for extinction increased with predation rate. The population growth patterns are presented in Fig. 7.

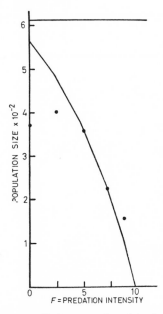

Fig. 6. An unsuccessful attempt to fit the mean size of the five populations shown in Fig. 3(b) to an elementary equation describing the interaction between population size and predation intensity. Horizontal line represents estimated number of *Artemia* supplied per day.

Table 1. *Effect of different predation rates on* Hydra oligactis

F is the predation rate. The observed P_F is the mean of the first seventy-seven censuses for each of five *H. littoralis* populations. From F and equation 2 an expected P_0 was calculated from each observed P_F. From the mean expected P_0 (564·8) the expected P_F was calculated from equation 2.

F	Observed P_F	Expected P_0	Expected P_F
0	371·3	371·3	564·8
0·25	403·3	470·5	484·1
0·50	359·8	539·7	376·5
0·75	229·6	574·0	225·9
0·90	157·9	868·5	102·7

Mean 564·8

When a population was kept in darkness, by being wrapped in black cloth, both species persisted (Fig. 8).

This data is of interest in two respects. First, it demonstrates the significance of light for interspecific competition. Second, it is the first experimental demonstration of a theoretical situation postulated by Gause (1934) in which predation stabilizes the interaction between two competing species.

Note that predation lowers the total number of hydrids so that at $F = 90\%$ the total number of hydrids does not exceed the estimated number of *Artemia* per day provided for any period longer than 40 days. Mr Thomas Griffing (personal communication) has found that isolated *Hydra littoralis* can live 40 days or longer without any food at all. Another set of competing populations maintained in darkness will be discussed below.

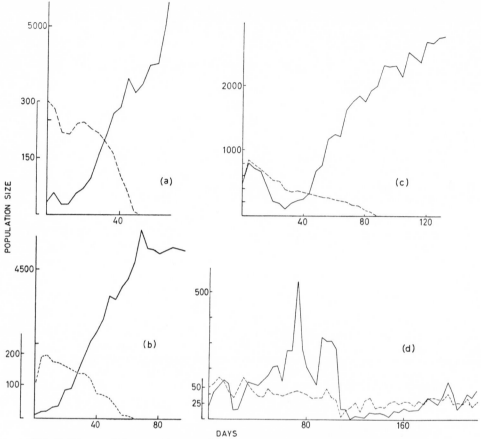

FIG. 7. The population growth curve of a two species population of *Hydra littoralis* (- - -), and *Chlorohydra viridissima* (———) kept in the light. (a) Without predation, inoculum of 300 individuals of each species (the ordinate is enlarged for *Hydra littoralis*). (b) Predation at $F = 25\%$, inoculum of 100 individuals of each species. (c) Predation at $F = 50\%$, inoculum of 500 individuals of each species. (d) Predation at $F = 90\%$, inoculum of 50 individuals of each species.

[*Editors' Note:* Material has been omitted at this point.]

REFERENCES

Anderson, R. L. & Bancroft, T. A. (1952). *Statistical Theory in Research.* New York.

Armstrong, J. (1960). *The dynamics of* Dugesia tigrina *populations and of* Daphnia pulex *populations as modified by immigration.* Ph.D. dissertation, Department of Zoology, University of Michigan, Ann Arbor.

Engelmann, M. D. (1961). The role of soil arthropods in the energetics of an old field community. *Ecol. Monogr.* **31**, 221–38.

Fogg, G. E. (1963). Remarks in *Marine Biology. I, First International Symposium of Marine Biology* (Ed. by G. Riley). Washington.

Frank, P. W. (1957). Coactions in laboratory populations of two species of *Daphnia. Ecology,* **38,** 510–19.

Gause, G. F. (1934). *The Struggle for Existence.* Baltimore.

Gohar, H. A. F. (1940). *Publ. Mar. biol. Sta. Ghardaqa (Red Sea),* No. 2, pp. 25–120.

Goreau, T. F. & Goreau, N. I. (1960). Distribution of labelled carbon in reef-building corals with and without zooxanthellae. *Science,* **131,** 668.

Loomis, W. F. (1954). Experimental factors controlling growth in *Hydra. J. Exp. Zool.* **126,** 223–34.

Muscatine, L. & Hand, C. (1958). Direct evidence for the transfer of materials from symbiotic algae to the tissues of a Coelenterate. *Proc. Nat. Acad. Sci., Wash.* **44,** 1259–63.

Slobodkin, L. B. (1954). Population dynamics in *Daphnia obtusa* Kurz. *Ecol. Monogr.* **24,** 69–88.

Slobodkin, L. B. (1959). Energetics in *Daphnia pulex* populations. *Ecology,* **40,** 232–43.

Slobodkin, L. B. (1960). Ecological energy relationships at the population level. *Amer. Nat.* **94,** 213–36.

Slobodkin, L. B. (1961). *Growth and Regulation of Animal Populations.* New York.

Slobodkin, L. B. (1962). Energy in animal ecology. *Advanc. Ecol. Res.* **1,** 69–101.

Slobodkin, L. B. & Richman, S. (1961). Calories/gm in species of animals. *Nature, Lond.* **191,** 299.

Yonge, C. M. (1944). Experimental analysis of the association between invertebrates and unicellular algae. *Biol. Rev.* **19,** 68–80.

14

Reprinted from *Ecology*, **34**(2), 301–307 (1953)

INTERSPECIFIC COMPETITION BETWEEN TWO SPECIES OF BEAN WEEVIL [1]

SYUNRO UTIDA

Entomological Laboratory, College of Agriculture, Kyoto University, Kyoto, Japan

INTRODUCTION

Lotka (1934) studied mathematically the problem of the competition between two species belonging to the same ecological niche, and reached the following conclusions. When two species having the same needs or habits compete with each other in a given environment, four possible conditions of population equilibrium are deduced mathematically:

(1) Each species inhibits its own potential increase more than that of the other and both continue to exist together; (2) the first species inhibits the potential increase of the second species and drives out the second species from the given space; (3) the second species drives out the first; (4) the inhibition of each species by the other is greater than its own inhibition, and it is the initial densities that determine chiefly which species could survive.

Experimental verification of the existence of these four cases was attempted by Gause (1934) in protozoan populations and by Crombie (1945, 1946) and Park (1950) in populations of several graminivorous insects. Recently, Park (1950) investigated the competition between the flour beetles, *Tribolium castaneum* and *Tribolium confusum,* and obtained an interesting result; in one case *T. castaneum* drives out *T. confusum* and in another case the result is vice versa. He believed that this incompatibility is caused by the disturbance resulting from the infection with the protozoan parasite, *Adelina tribolii.*

The present investigation is planned with a similar purpose: to verify the interspecific competition, using two species of bean weevil, the azuki bean weevil *Callosobruchus chinensis* and the southern cowpea weevil *C. quadrimaculatus.* In addition, an examination is made of the influence of parasitization upon the outcome of competition between species belonging to the same ecological niche. The parasite used is the pteromalid wasp *Neocatolaccus mamezophagus,* which prefers both of these bean weevils as hosts without distinguishing between them.

Here, the writer wishes to express his hearty thanks to Dr. C. Harukawa for the invaluable advice given him. His thanks are also due to Miss H. Kakemi for her patient and faithful work in making the monotonous counts.

MATERIALS AND METHODS

Two species of the bean weevil, the azuki bean weevil *Callosobruchus chinensis* and the southern cowpea weevil *Callosobruchus quadrimaculatus,* were used for the experiments. Both species had been reared continuously under constant conditions in the laboratory for several years.

The experiments were carried out under constant environmental conditions. In a dark incubator electrically regulated at a constant temperature of about 30° C, petri dishes about 8.5 cm in diameter and 1.8 cm in height were set. Atmospheric moisture in the dish was regulated at a constant relative humidity of about 75 per cent by means of a supersaturated solution of sodium chloride. The variety *Dainagon* of the azuki bean was used for the food of weevils. The water content of the bean was almost 15.0 per cent. The technique used was almost iden-

[1] Contributions from the Entomological Laboratory, Kyoto University, No. 212.

tical with that described in a previous work on experimental populations of the azuki bean weevil (Utida 1941). Here, only its outline will be described. Four pairs of newly emerged weevils of both species were put in the dish prepared. Counts of the populations of both species took place at one week intervals, dead and living individuals, both female and male, being identified separately. Living individuals were lightly narcotised with ether, counted and put back in the dish. Ten-gram lots of beans were supplied as food, and were renewed every 3 weeks, a period which corresponds closely to the average time required by the weevils for completion of a generation. At each renewal of the food the remnants of food were taken away from the dish.

Method of calculating density

The number of individuals emerging between two successive counts can be obtained by a simple calculation using the number of living individuals and that of dead individuals in the two consecutive counts. The total number of adult weevils emerging in each generation can be obtained by summing up the number of emerging weevils in 3 consecutive counts, for both species of weevil require about 3 weeks for the completion of a generation. In this way, the change of population density which is caused by emergence and death of adult weevils

belonging to a single generation can be smoothed out, and the resulting figures give the changes of density from generation to generation in the populations of both species.

Single-Species Equilibria

To understand the behavior of the mixed and competing populations, we must observe in advance the trend of population growth and the level of the equilibrium state in each single-species population. Experiments yielding this information for the azuki bean weevil have been conducted previously under conditions similar to those of the present experiment and have been discussed in detail (Utida 1941). The steady density of the equilibrium state in the population of the azuki bean weevil was about 400; in one population it was 390.5 and in another it was 418.9.

The trend of growth and the steady state of the equilibrium in the population of *C. quadrimaculatus* are given in Table I and represented graphically in Figure 1. Two identical populations start with 8 pairs of weevils. In the second generation the population density increases to about 150. In the third generation it reaches about 300. Thereafter, the density of population fluctuates around the mean density of the equilibrium state,

TABLE I. *Trend of population of* Callosobruchus quadrimaculatus, *as represented by the number of emerged individuals in each generation*

Number of generation	Population A		Population B	
	♂	♀	♂	♀
1	8	8	8	8
2	104	63	63	87
3	166	148	245	133
4	143	102	135	94
5	171	125	155	95
6	63	92	103	91
7	166	122	160	132
8	168	128	134	142
9	154	140	148	111

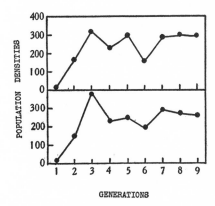

FIG. 1. Trend of population of *C. quadrimaculatus,* as represented by the population densities in each successive generation.

TABLE II.   *Trend of the population when* Callosobruchus chinensis *and* C. quadrimaculatus
compete with each other

It is represented by the number of emerged individuals in each generation.

Number of generation	Population C				Population D			
	C. chinensis		*C. quadrimaculatus*		*C. chinensis*		*C. quadrimaculatus*	
	♂	♀	♂	♀	♂	♀	♂	♀
1	4	4	4	4	4	4	4	4
2	49	56	21	36	62	66	38	30
3	35	41	142	98	69	76	138	101
4	2	7	51	53	7	3	44	52
5	0	0	212	112	0	1	193	135
6	0	0	162	103	0	0	127	160
7	0	0	172	134	0	0	133	111
8	0	0	122	128	0	0	130	104
9	0	0	104	93	0	0	110	114

which is about 250 in both populations.

Comparing the level of the equilibrium state and the trend of population growth in the two species of weevil, the following points become clear. The trends of growth are similar in *Callosobruchus chinensis* and *C. quadrimaculatus,* but the *C. chinensis* population grows slightly more rapidly than does that of *C. quadrimaculatus,* since the second-generation densities are 258 and 287 in the former population and 170 and 150 in the latter population. The equilibrium density is higher for *C. chinensis* than for *C. quadrimaculatus.* The fluctuation in the equilibrium state is almost identical in its mode in both populations.

COMPETITION WITHOUT PARASITISM

The initial populations consisted of 4 pairs of *C. chinensis* mixed with the same number of pairs of *C. quadrimaculatus.* Two such mixed populations were run in parallel to obtain the changes of the population density when both species compete with each other under the given environmental conditions. The results are given in Table II and represented in Figure 2.

In the second generation the density of *C. chinensis* became higher than that of *C. quadrimaculatus,* but in the third generation the relative proportions were

reversed. After the third generation the density of *C. chinensis* decreased sharply and in the fourth generation it fell to zero and the population of *C. chinensis* was completely extinct.

On the other hand, the population of *C. quadrimaculatus* grew rapidly and attained the equilibrium state in the third generation. *C. quadrimaculatus* drives out *C. chinensis* completely in the fourth generation, and consequently the equilibrium state composed of both species of weevils can not be established. The level of the steady state in the population of *C. quadrimaculatus* was about 250 in both of the two parallel populations. It was almost equal to the value obtained

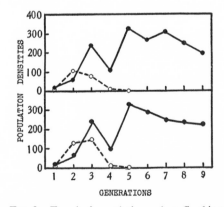

FIG. 2.   Trend of population, when *C. chinensis* and *C. quadrimaculatus* compete.   --○--: *C. chinensis,* —●—: *C. quadrimaculatus.*

in the homogeneous (control) population of *C. quadrimaculatus.*

COMPETITION WITH PARASITISM

A pteromalid wasp, *Neocatolaccus mamezophagus,* can parasitize the full-grown larva of *C. quadrimaculatus* as well as that of *C. chinensis.* When *Neocatolaccus* attacks the interacting population of two species of weevil, the relationships between the three species are very complicated.

Two populations were set up, consisting of different combinations of numbers of the host species. One population started under the dominance of *C. quadrimaculatus* and thereafter a certain definite

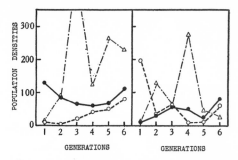

FIG. 3. Competition between *C. chinensis* and *C. quadrimaculatus,* when both species are parasitized by *Neocatolaccus mamezophagus.* Left figure: *C. quadrimaculatus* initially dominant. Right figure: *C. chinensis* initially dominant. --○--: *C. chinensis,* —●—: *C. quadrimaculatus,* —·—△—·—: *N. mamezophagus.*

number of the parasitic wasp was liberated in this population. Another population started under the dominance of *C. chinensis,* and thereafter a certain definite number of *Neocatolaccus* was liberated.

In both cases, the results are almost similar as represented in Table III and Figure 3. The parasite population was established and maintained relatively high density. The populations of both host species existed together and held rela-

TABLE III. *Competition between* C. chinensis *and* C. quadrimaculatus, *when both species are parasitized by a common parasite,* Neocatolaccus mamezophagus.

The population density of both host species is represented by the number of emerged individuals in each generation and that of the parasite is represented by the number of individuals emerged during a host generation.

| Number of generation | Population E | | | | | |
| | *C. chinensis* | | *C. quadri-maculatus* | | *Neocatolaccus mamezophagus* | |
	♂	♀	♂	♀	♂	♀
1	4	4	64	64	4	4
2	1	3	42	43	34	46
3	12	10	34	29	297	253
4	20	20	32	26	72	52
5	23	28	46	30	136	129
6	42	37	51	61	138	92
	Population F					
1	96	96	4	4	4	4
2	16	18	12	19	81	47
3	38	23	31	30	21	37
4	3	3	30	21	143	134
5	3	8	9	13	19	28
6	36	25	41	37	6	20
	Population G					
1	—	—	86	86	4	4
2	—	—	60	81	55	144
3	4	4	33	50	178	92
4	6	12	52	39	147	44
5	36	21	39	29	122	69
6	42	37	24	20	122	105

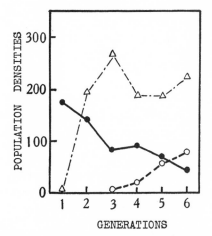

FIG. 4. Competition between two species of host and their common parasite, when *C. chinensis* is introduced after equilibrium is established between *C. quadrimaculatus* and *N. mamezophagus.* --○--:*C. chinensis,* —●—: *C. quadrimaculatus,* —·—△—·—: *N. Mamezophagus.*

tively low levels of density. Both species of weevil competed with each other, but existed together in the same universe without becoming extinct.

To confirm the present result, a further experiment was made. After the balance was established between the populations of *C. quadrimaculatus* and *N. mamezophagus,* 8 pairs of *C. chinensis* were introduced into the population. The result is given in Table III and represented graphically in Figure 4. The population of *C. chinensis* became established and existed together with the population of *C. quadrimaculatus*. The result is the same as in the former experiment, when both species of weevil were introduced at the same time with the common parasite, *Neocatolaccus*.

DISCUSSION

The present problem has been fully investigated by Park, Crombie and others, using granary beetles such as *Tribolium, Calandra,* and *Rhizopertha,* and some mathematical postulates have been justified and important findings have been made. Although the present investigation is a preliminary one and suffers from an insufficient number of replicates compared to Park's extensive work on *Tribolium* (Park 1950), it is clear-cut in its result.

It is demonstrated that the competition between *C. chinensis* and *C. quadrimaculatus* corresponds to the second or third case of Lotka's system of competition mentioned above. It has already been shown experimentally in protozoan populations by Gause and in granary insects by Crombie that when two species with an identical niche compete with each other in a limited environment, one will drive out the other. Experimental animals used by Crombie were *Rhizopertha, Sitotroga, Tribolium,* and *Oryzaephilus,* which belong to different genera, families and even orders, while the weevils used in the present work belong to the same genus *Callosobruchus*. Competition between *Rhizopertha* and *Sitotroga* is limited in its operation to the larval stages and the mode of life of the adults of these insects is quite different, while in the present case both species of weevil compete with each other continuously all through the life cycle from the time of egg-laying to the emergence of the adult weevil. Aspects of interference include competition for oviposition site, mutual interruption of rhythmic behavior patterns by males and females, prevention of completion of copulation by individuals of the same or the other species, egg mortality caused by mechanical interference of adult individuals, and competition for food materials in the larval stage. Consequently, it is to be supposed that the process of extinction is very rapid and the end result is clear-cut, when such incompatibility obtains between species belonging to the same genus and having the same needs and habits. This conclusion seems not to hold good in the results obtained by Park on interaction between *Tribolium confusum* and *T. castaneum,* which also belong to the same genus and have entirely similar modes of life. This disagreement may be due to the fact that in the *Tribolium* populations parasitization by microorganisms, which is frequently observed, markedly influences the competition.

On the basis of this principle of interspecific competition, a theoretical interpretation was given to Motomura's formula of geometric progression in an animal community (Utida 1943). This formula, which was empirically obtained by Motomura, has proved to be of wide application. Williams' formula describing the relation between the number of species and the number of individuals in an animal population can be derived mathematically from Motomura's formula (Utida, unpublished).

As stated by Lack (1949), sympatric and allopatric speciation may be explainable by this principle of interspecific competition. In North America, the geographical distribution of the two weevils used in the present experiments is separated allopatrically into northern and southern districts.

Hutchinson (1948) pointed out two possible exceptions to this principle, (1) where external factors may act to rarefy the mixed population, so that the environmental possibilities are not completely exploited, and (2) where continual chance oscillations of the environmental variables may continually reverse the direction of competition, so that no equilibrium can be established (Allee *et al.* 1949).

Lack (1949) also claimed that two species might be able to live in the same habitat feeding on the same food without effective competition, provided that their numbers were controlled by parasites or predators. Against this is the theory of Nicholson (1933) that two species cannot persist together when both are controlled by the same predator or parasite. The claim of Lack may conform to the first term of Hutchinson's proposition, the action of parasites or predators being one of the external factors controlling the mixed population. But when mixed populations are controlled by the action of predator or parasite as an external factor, the mixed populations persist together even if the environmental capacity is almost completely utilized by the competing species. This depends upon the assumption that the parasite searches at random and parasitizes both host species without discriminating between them.

Lack's claim has not been experimentally proved as yet, except for the present experiment. When two species of bean weevil are attacked by the parasite *Neocatolaccus*, the equilibrium densities of both bean weevil populations decrease but neither species becomes completely extinct; both exist continuously, all through the experiment. This situation seems frequent and important in the host-parasite and prey-predator systems. It is likely that the mixed population of *Tribolium confusum* and *T. castaneum*

parasitized by the protozoan *Adelina* provides another example. The theory of Nicholson is probably correct only when two host species compete with each other severely for the determination of their adult densities after both are attacked by a common parasite. In this case, the random searching by the parasite is not operative in regulating the equilibrium state, but Gause's general principle of competition for food between two host species is operative.

It can be concluded from the result of the present experiment that predation or parasitism plays a more important role in regulating the number of host or prey individuals than does competition between two or more species of host or prey belonging to the same ecological niche. Perhaps this conclusion may also apply when extended to the comparison between predation and intra-specific competition.

In a natural community species interact with each other as host and parasite or as predator and prey, as well as by way of interspecific competition for the same niche requirements. Neither type of interaction can be ignored, and both kinds together provide the warp and the weft of community organization. The regulatory action of inter- or intraspecific competition can be transformed or outweighed by such higher-order interspecific interactions as predation and parasitism. Accordingly it is no mistake to conclude that the abundance of a given species in nature is mainly controlled by its predators or parasites and that its competitors of the same or related species are less important than has been assumed.

Pursuing this line of reasoning, we reach the theoretical conclusion that the distinction between allopatric and sympatric species does not exist in the lower strata of community organization, but comes in only at the higher levels of carnivory. In other words, the greatest number of species (as well as individuals) should be found at the lowest ecological rank in the food-chain hierarchy.

SUMMARY

In this paper, interspecific competition was studied experimentally between species belonging to the same ecological niche. Two species of bean weevil, *Callosobruchus chinensis* and *C. quadrimaculatus,* were reared under constant laboratory conditions and counts of weevils were made every generation at regular intervals. Unispecific control populations of *C. chinensis* and *C. quadrimaculatus* grow and attain the steady state of population after two or three generations under the experimental conditions. The steady population density of *C. quadrimaculatus* maintains a higher level than that of *C. chinensis*. When both species compete within the same microcosm, the population of *C. chinensis* is depressed and falls to zero after several generations. When a paratitic wasp, *Neocatolaccus mamezophagus,* is introduced into the mixed population of competing species, the population of the wasp is established and the populations of the two weevils are reduced markedly below their level of steady state, but they exist together in the same microcosm. The same result occurs regardless of the population densities of the host species relative to each other. These results are discussed on the general background of population theory.

REFERENCES

Allee, W. C., A. E. Emerson, O. Park, T. Park, and K. D. Schmidt. 1949. Principles of animal ecology. Philadelphia: Saunders.

Crombie, A. C. 1945. On competition between different species of graminivorous insects. Proc. Roy. Soc. London, **132B**: 362–395.

——. 1946. Further experiments on insect competition. Proc. Roy. Soc. London, **133B**: 76–109.

Gause, G. F. 1934. The struggle for existence. Baltimore: Williams & Wilkins.

Hutchinson, G. 1948. Teleological mechanisms: Circular causal systems in ecology. Ann. New York Acad. Sci., 50: 221–246.

Lack, David. 1949. The significance of ecological isolation. *In* "Genetics, paleontology and evolution," pp. 299–308. Princeton: Princeton Univ. Press.

Lotka, A. J. 1934. Theorie analytique des associations biologiques. Actualités Scientifiques et Industrielles, **187**: 1–45.

Nicholson, A. J. 1933. The balance of animal populations. J. Anim. Ecol., 2: 132–178.

Park, Thomas. 1948. Experimental studies of interspecies competition. 1. Competition between populations of the flour beetles, *Tribolium confusam* Duval and *Tribolium castaneum* Herbst. Ecol. Monogr., 18: 265–308.

Utida, Syunro. 1941. Studies on experimental population of the azuki bean weevil *Callosobruchus chinensis*. I. The effect of population density on the progeny populations. Mem. Col. Agr. Kyoto Imp. Univ., **48**: 1–30.

——. 1941. Ibid. V. Trend of population density at the equilibrium position. Mem. Coll. Agr. Kyoto Imp. Univ., 51: 27–34.

——. 1943. Some considerations on "the law of geometrical progression in animal community" proposed by Dr. Motomura. Seitaigaku-Kenkyu (Ecol. Rev.), 9: 173–178 (in Japanese).

15

COMMUNITIES AND ECOSYSTEMS

R. H. Whittaker

[*Editors' Note:* In the original, material precedes this excerpt.]

Niche Space and Species Importance

Some niche characteristics, such as prey size, can be quantified. We
may consider growth-forms in the Sonoran desert of the lower mountain slopes in southeastern Arizona, a spectacular high desert of giant
cactus, ocotillo, palo verde, and a great variety of plants (Figure 2·6).
Occupation of niche space in this community involves first the relation
of species to the vertical height axis. The heights at which plant species bear the greater part of their foliage range from near-zero in herbs
with stems and leaves on the ground surface, to a few centimeters in
other herbs, to a few decimeters in most of the semishrubs, to 0.5 to
2.0 meters in true shrubs of different species, to 2 to 5 meters in ar-
borescent shrubs (ocotillo, palo verde, mesquite), whereas the giant
cactus or saguaro has a photosynthetic surface to the top of its stem,
up to 6 to 9 meters. Average position of the buds or tissues that survive
unfavorable seasons and from which the foliage develops, from the
ground surface up (or in some herbs below the ground surface), provides a convenient expression of plant height.

A second set of niche relationships involve seasonal time. There are
two rainy seasons in this desert, one in winter and one in late summer,
separated by dry seasons, and two waves of plant growth correspond

Figure 2·6. A Sonoran desert community of highly mixed growth-form composition. Taken near Tucson, Arizona, with the Santa Catalina Mountains in the background. The giant cacti are the saguaro, *Carnegiea gigantea*. [Courtesy of and copyright by W. A. Niering.]

to the rainy seasons. Among the perennial plants most utilize the moisture of both rainy seasons, but in different patterns of leaf and stem photosynthesis, patterns expressed in growth-forms and the seasonal behavior of foliage. The plants may again be arranged along a gradient, from those with persistent, evergreen leaves, through semi-deciduous species with leaves (or the leaf-bearing twigs of the semishrubs) persistent through less severe dry seasons but not more severe ones, to the deciduous mesquite, palo verde, and semishrubs. These in turn grade into forms like ocotillo, with short-lived leaves quickly produced and soon lost after rains, to the cacti, which lack leaves. In plants of the latter part of this sequence the stems and branches are green and photosynthesize to supplement the photosynthesis by the leaves; in the cacti photosynthesis is wholly by stem and branches. The plants form, then, a gradient of decreasing leaf persistence and increasing stem and branch photosynthesis, and we shall assume that differences along the gradient are significant in relation to plant competition. The gradient is also one of adaptive patterns in relation to water shortage—different solutions of the plant's problem of how to photosynthesize enough while also conserving water sufficiently to stay alive in a desert.

The two gradients of height and foliage persistence may now become the axes of a chart (Figure 2·7). They define a two-dimensional niche space or surface in which plant species have located themselves in evolution. The plant species are scattered in this space as the principle of Gause would lead us to expect: each species has its own distinctive niche area. Even though niche areas overlap, the centers of these areas are dispersed in the niche space. We have chosen only two niche gradients for this illustration, but there are many other niche characteristics that can be treated as gradients (along with some that are difficult to treat thus). For these many niche characteristics we may generalize our two-dimensional niche surface into an *n*-dimensional niche hyperspace. Each species occupies some part, or hypervolume, of the niche hyperspace, and the centers of these species hypervolumes are dispersed in the niche hyperspace.

The niche hyperspace concept, developed by G. E. Hutchinson, is most useful for interpreting the way the principle of Gause works out in practice among the species of a natural community. In particular it provides us with an approach to quantitative relations between the species we are interested in.

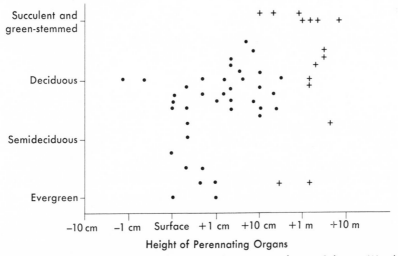

Figure 2·7. Growth-form differentiation in the Sonoran desert, Arizona. Woody plant species of the desert are plotted by the seasonal relations of their foliage, on the vertical axis, and plant height in relation to ground surface (on a logarithmic scale) on the horizontal axis. The larger woody plants of the community are armed with spines, as indicated by the crossed points. [Whittaker and Niering, 1965.]

[*Editors' Note:* Material has been omitted at this point.]

LITERATURE CITED

Whittaker, R. H., and W. A. Niering. 1965. Vegetation of the Santa Catalina Mountains, Arizona. (II) A gradient analysis of the south slope. Ecology **46**: 429–452.

16

Reprinted from *Science,* **159,** 1432–1437 (Mar. 29, 1968)

Population Regulation and Genetic Feedback

Evolution provides foundation for control of herbivore, parasite, and predator numbers in nature.

David Pimentel

Population Characteristics

Although within a relatively short period man has learned how to put himself into space, he still is not certain how the numbers of a single plant or animal population are naturally controlled. Aspects of this problem have been investigated since Aristotle's time, they were given important consideration in Darwin's *Origin of Species,* and yet the unknowns far outweigh the discoveries. If we knew more about natural regulation of population, we would be in a better position to devise more effective and safer means of control for important populations of plant and animal pests. We might also be better able to limit the growth of human populations, although that problem is exceedingly complex because of the social activities and nature of man.

Before considering how populations in nature are regulated, we should review various characteristics of animals and plants—as individuals and as populations. Do populations of animals in nature fluctuate severely or are they relatively constant? Stability and constancy have been proposed as characteristics of natural populations. Speaking about birds, Lack (*1*) says, "of the species which are familiar to us in England today, most were familiar to our Victorian great-grandparents and many to our medieval ancestors; and the known changes in numbers are largely attributable to man." He continues, "All the available censuses confirm the view that, where conditions are not disturbed, birds fluctuate in numbers between very restricted limits. Thus, among the populations considered above, the highest total recorded was usually between two and six times, rarely as much as ten times, the lowest. This is a negligible range compared with what a geometric rate of increase would allow." Discussing the stability in animal populations in general, MacFadyen (*2*) writes: "it is generally agreed that the same species are usually found in the same habitats at the same seasons for many years in succession, and that they occur in numbers which are of the same order of magnitude."

Further evidence for the thesis that species populations are relatively constant is found in a study of the changes in the fauna of Ontario, Canada (*3*). When Snyder (*4*) evaluated the bird fauna, he found that, over a period of about 70 years, two species became extinct, 23 species increased in number, and six species decreased in number. This represents a total change of only 9 percent of 351 bird species found in Ontario (*5*) and agrees favorably with an 11-percent change (*6*) for 149 species of birds over a 50-year period in Finland. These data suggest that there is relative constancy in the abundance of species populations. The word "relative" must be emphasized because changes in numbers must be related to a species' real potential for fluctuations; to para-

The author is chairman of the department of entomology and limnology, Cornell University, Ithaca, New York.

phrase Lack (*1*), the changes observed are mere "ripples" compared to the possible "waves." Although in geological time 99 percent of all species have become extinct, during periods of 100 years or more constancy is the rule.

There are exceptions to this rule of constancy. What are the population characteristics of plants and animals newly introduced on islands and continents? Typically when a species population enters a new biotic community in which no ecological barrier exists, outbreaks occur in these populations. The following examples of introductions into the United States illustrate this point: Japanese beetle *(Popilla japonica);* European gypsy moth *(Porthetria dispar);* South American fire ant *(Salenopsis saevissima);* Asiatic chestnut blight *(Endothia parasitica)* (fungus); European starling *(Sturnus vulgaris);* and the English sparrow *(Passer domesticus).* Outbreak of chestnut blight was so severe that for all practical purposes it destroyed its host, the American chestnut tree *(Castanea dentata).* After increases in the number of Japanese beetles, a bacterial pathogen epidemic spread through the population and is now effectively controlling the numbers of the beetle.

In nature the numbers of many herbivore, parasitic, and predaceous species are limited by resistant factors inherent in the host. Are resistant factors which limit or prevent pest attack commonly found in plants and animals? Various kinds of resistant factors exist in plants and animals in nature and appear to be quite prevalent. The spines occurring in many kinds of plants, such as cacti, gorse, and hawthorn, prevent feeding by browsing animals. Toxins or growth inhibitors which occur in many kinds of plants limit animal feeding, for example, tannins in oak leaves (*7*), cyanide in bird's-foot trefoil *(Lotus corniculatus)* (*8*), and nepetalactone in catnip (*9*). To prevent predator attack (*10*), poisonous sprays are ejected from many insects and other arthropods, such as acetic acid by whip scorpion (Pedipalpidas), formic acid by ants (Formicidae), and *p*-benzoquinones by flour beetles (Tenebrionidae). Repellent sprays from glands in some vertebrates, such as the skunk *(Mephitis* spp.), the Indian mongoose *(Herpestes auropunctatus),* and the toad *(Bufo marinus),* ward off attacking enemies. Nutritional changes in certain plants prevent the multiplication of attacking insects, for example, aphids on corn (carotene) (*11*) and leafhoppers on beets (linoleic acid) (*12*). A kind of

armor plating protects various animals (armadillos, turtles, and certain beetles) from their attackers. Such physiological defense mechanisms as specific antibodies and phagocytosis are present in many kinds of animals (man and other vertebrates) and effectively control pathogen and parasite infections.

When these natural resistant factors in plants and animals are successful, they prevent the uncontrolled increase of the feeding species. Are animal numbers abundant or rare? Rarity, like constancy, is relative. Numbers of a given species can be related to the numbers of another species, to the unit area occupied, or to the food resources of the species. Andrewartha and Birch (*13*) noted that "the truth is that the vast majority of species are rare, by whatever criterion they are judged." In the *Origin of Species,* Darwin wrote, "rarity is the attribute of a vast number of species of all classes, in all countries." In enumerating the number of insects abundant enough to be considered pests, Smith (*14*) warned "that such species form only an insignificant fraction of the total number of phytophagous insects." Of the 240 species of nocturnal Lepidoptera collected by Williams (*15*), 35 species were represented by a single individual each; 85 (including the 35 above) were represented by five or fewer individuals; 115 by ten or fewer; and 205 by 100 or fewer individuals; therefore, there were only 35 species with over 100 individuals. Further data in support of rarity is found in Dunn's (*16*) work with Panamanian snakes; he reports that "about 1/10 of the species make up 1/2 of the individuals in the snake populations."

Many of the abundant species would not be classed as abundant if they were compared to their food source. For example, many species of insects that are easily captured in the field are rare if they are sought on their host plant or if their biomass is compared with the biomass of the plant or animal upon which they feed.

One of the dynamic relationships in the community and ecosystem is the food chain, because animals must seek food to live. Elton (*17*) stated that the "whole structure and activities of the community are dependent upon questions of food-supply." What proportion of animals feed on living, as opposed to nonliving, matter? In nature, the majority of all animals may be classified as herbivore (grazer on living plants), parasite, or predator; few species are truly saprophytic. Though many animals are

associated with dead plant matter, these animals are not saprophytes but are herbivores feeding on bacteria, fungi, and other minute organisms in the decaying matter. Jacot (*18*) stated, "I am quite certain that perhaps as much as one half of the Oribatoidea are not saprophytic. Their function is feeding on fungi." Overgaard (*19*) reported that the evidence suggested that nematodes feed not upon humus but upon plant roots, fungi, bacteria, and other animals. Speaking similarly about saprophagous insects, Chapman (*20*) noted that, "These are usually designated as those feeding upon decaying and fermenting matter. It is evident at the start that these insects live in media which may be teeming with microorganisms, and that the decaying material is the medium upon which the microorganisms live." *Drosophila* depend upon yeasts and other microorganisms present in decaying fruit (*21*).

Genetic Feedback

Stability and constancy are characteristics of natural populations; in many hosts there are resistant factors that limit any severe attack of feeding species, and most animals feed on living matter. These seemingly diverse factors are related and are the foundation of the mechanism for population regulation which I termed "genetic feedback" (*22*). Population numbers (herbivore, parasite, or predator) are regulated in this way: high herbivore densities create strong selective pressures on their host-plant populations; selection alters the genetic makeup of the host population to make the host more resistant to attack; this in turn feeds back negatively to limit the feeding pressure of the herbivore. After many such cycles, the numbers of the herbivore populations are ultimately limited, and stability results.

Through the functioning of the genetic feedback mechanism, resistant factors in a given plant can be used to control a parasite which feeds on it. For example, on a susceptible plant genotype the animal population feeding heavily may be reproducing at a rate of two offspring per individual in the population. Under these conditions, the animal population would increase rapidly and would soon cause severe damage to the plant population by overfeeding. On the other hand, if resistant genes were concentrated in the plant population so that only resistant genotypes dominated, animal reproduction might be at a rate of one-half offspring per individual in the population. Then the animal population would decrease, and the damage to the plant population would be kept to a minimum.

An example of this type of change in a plant population took place in the Kansas wheat crop which was susceptible to the Hessian fly *(Phytophaga destructor)*. As a result of the low resistance of the wheat to attack, Hessian fly populations increased, and the wheat crop suffered damage. With R. H. Painter's *(23)* development of a resistant variety of wheat, reproduction on the resistant wheat dropped to less than one per individual, and soon the fly population declined to a low level. Thus, by manipulation of the genotypes found in the wheat and not of the quantity of wheat, the fly population was controlled.

Although the interactions of wheat and Hessian flies can be considered to be man-made, we find evidence that under natural conditions biotic communities develop their own controls. In fact, populations in nature are usually regulated by several mechanisms that operate interdependently. These include not only genetic feedback but competition, parasitism, predation, and environmental heterogeneity.

This can be illustrated by a study of what might happen when a new animal species is introduced into a biotic community and becomes established on a plant. At first, the animal increases rapidly on its new plant host and reaches outbreak level. Under these conditions, competition for food among the animals is intense. In addition, the severe feeding pressure tends to eliminate many of the plants; this results in an altered distribution of the plants. With the plant hosts more sparsely distributed, the animal has increased difficulty in locating hosts, and some hosts have time to grow, reproduce, and maintain themselves.

Thus, at the early stages of interaction between animal and plant, competition and environmental heterogeneity along with the pressure from parasites and predators frequently limits the numbers of the animal and prevents the complete destruction of the host. If these factors are successful, then slower-acting genetic change and evolution can take place.

Genetic change in the plant takes several generations because plant response to selective pressure exerted by the animal is slow. When a large animal popu-

lation exerts severe feeding pressure on the plant population, large numbers of plants are destroyed. The first plants destroyed are primarily those most susceptible to the feeding pressure of the animal; the surviving plants generally carry one or more resistant genes. Under natural conditions, the evolution that occurs in the plant would rarely be caused by mutation but would be due to a recombining and concentrating of the genes already existing at low frequencies in the plant population. Resistance in the host is generally polygenic, and evolution proceeds slowly as genes are recombined in individuals and concentrated in the population. For example, there might be 20 loci (two genes per locus) in the host plants for some resistant character such as hardness. As susceptible genes at each locus are slowly replaced with resistant genes, the amount of resistance gradually increases.

When we look at the problem from a different angle, we find that the change and resistance in the plant can be measured as a response in the survival of the animal; that is, the number of eggs produced, the rate of development and growth, mortality, and longevity of the animal might all be influenced by increasing the concentration of resistant genes in the plant host. At a critical level of resistance in the host, the low birth rate and high death rate in the animal population would result in a significant decrease in numbers, and eventually the population would be sparse. Then with animal numbers rare in relation to those of the plant host, the animal population would only be removing "interest" (excess individuals or energy, or both) from the plant population, and relative equilibrium would exist between plant and animal. The animal would no longer be removing "capital" (those individuals or that energy, or both, needed for maintenance of the plant population). Evolution of this kind with a balance between supply and demand is possible with the genetic feedback mechanism.

Feeding pressure of herbivores, parasites, and predators on their plant or animal host may be limited by various protective mechanisms in their host, but there are examples of subtle genetic changes that significantly affect the survival of the animal that uses the host plant or host animal for food. For instance, when young pea aphids *(Acyrthosiphum pisum)* were placed on a common crop variety of alfalfa *(Medicago sativa)*, they produced a

mean of 290 offspring in 10 days, whereas the same number of aphids for a similar period on a resistant alfalfa variety produced a mean of only two offspring *(24)*. In another example, the mean rate of oviposition (eggs per generation) of the chinch bug *(Blissus leucopterus)* on a susceptible strain of sorghum *(Sorghum vulgare)* was about 100, whereas on a resistant strain the mean oviposition was less than one *(25)*. In both, reproduction in the animals feeding on the resistant plant hosts decreased more than 99 percent. This reduced reproduction obviously would have dramatic effects on the population dynamics of the feeding animal populations.

Resistance is effective in limiting animal numbers, and evidence suggests that it plays a dominant role in controlling populations in nature. If so, this would explain why population outbreaks occur frequently in newly introduced species. With little or no resistance, the new species increases rapidly on its susceptible food hosts. Until resistance in the plant host gradually increases, both outbreaks and intense fluctuations will occur. When relative stability is eventually reached and resistance is fully effective, animal numbers will be low. This is one reason why most animals are rare and especially rare relative to their food resource.

In addition, the relative stability and responsiveness of living systems are believed to account in part for the fact that most animals feed on living matter. The interaction between eating and eaten species and genetic feedback within the community form a complex but fully responsive system. Living systems, of course, respond to change and can evolve to provide a functional system whereby careful control of supply and demand can be achieved within the community as a whole.

The adaptation of supply by the plant and demand by the animal evolves and in time attains a state of relative balance. The plant host responds and evolves to its attacking animal only if the numbers of the animal are sufficient to exert some selective pressure on the host. This means that the trophic interactions between herbivore and plant, parasite and host, and predator and prey are important in determining the structure of the community. Based on this knowledge, Elton's statement that "the whole structure and activities of the community are dependent upon questions of food supply" takes on great significance in population control.

Parasite-Host Systems

The validity of genetic feedback functioning as a regulatory mechanism in populations was investigated under controlled laboratory conditions (*26*). The premise of the first experiment was that the numbers of the feeding species would be controlled as genetic resistance evolved in the host population. The housefly (*Musca domestica*) was the host species, and a wasp (*Nasonia vitripennis*) was the parasite or feeding species (Fig. 1). These two species were allowed to interact in the experimental unit for 1004 days while host numbers were kept constant and parasite numbers were allowed to vary. The control unit was similar in design, except that hosts for the parasite population came from a population of houseflies that had not been exposed to the parasite. Hosts that survived exposure to the control parasites were destroyed to prevent the control host population from evolving. In both the control and experimental units, all parasites that emerged from their host types were saved and were returned to their respective population cages.

During the period of study, measurable evolution took place in both the host and parasite populations in the experimental unit. The experimental host population became more resistant to the parasite, as evidenced by a drop in the average reproduction of 135 to 39 progeny per experimental female parasite and a decrease in longevity from about 7 to 4 days. Concurrently, the parasite population evolved some avirulence toward its host. As the experiments progressed, selective pressure on the experimental host population declined, and density of the parasite population declined to about one-half that of the control (about 3700 for the control and 1900 for the experimental). The amplitude of the fluctuations of experimental population (Fig. 2) was significantly less than those experienced by the control.

The ecology and evolution of this same parasite and host were investigated in another experiment during which both parasite and host density were allowed to vary. A specially designed cage, consisting of 30 plastic cells joined together to make a multicelled structure, provided space-time structure for normal parasite-host interactions. With this cage, the population characteristics exhibited by the control or newly associated parasite-host system were compared with those of the first

Fig. 1. Wasp parasite and housefly host pupa.

experimental system in which some ecological balance had evolved.

During the 581-day period for the control system and 322-day period for the experimental system, parasite numbers averaged 118 per cell in the control system and only 32 in the experimental system. Host numbers averaged 172 per cell in the control and 462 in the experimental (Fig. 3). Population fluctuations in the control system were severe, whereas in the experimental system they were dampened. The greater stability which the experimental system had already attained enabled it to make efficient use of its environmental resources and in this way increase its chances for survival.

One of the outstanding examples of the genetic feedback functioning in a natural population is the relationship of myxomatosis virus and European rabbits in Australia. After its introduction there in 1859, the European rabbit (*Oryctolagus cuniculus*) population increased to outbreak levels within the following 20 years (*27*). To reduce the density of the rabbit to a harmless level, the myxomatosis virus obtained from South American rabbits was introduced into the rabbit population. In essence, this action was analogous to introducing a new virus species into another community, for the myxomatosis virus and European rabbit had never been associated before. The virus spread rapidly in the rabbit population and immediately reached outbreak levels. During the

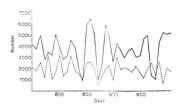

Fig. 2. Population trends of parasite populations for the last 254 days of the 1004-day period of two laboratory parasite-host systems. Solid line, control parasite; dashed line, experimental parasite.

first epidemic, myxomatosis was fatal to about 98 percent of the rabbits; the second epidemic resulted in about 85 percent mortality; and by the sixth epidemic, mortality was about 25 percent (*28*). Today the virus is less effective than it had been but is still taking its toll of rabbits. Fenner summarized the situation by stating, "We could then envisage a climax association in which myxomatosis still caused moderately severe disease with an appreciable mortality, much as smallpox does in human communities. The reproductive capacity of the rabbit is such that this sort of disease need not seriously interfere with its population size."

In this adjustment between virus and rabbit, attenuated genetic strains of virus evolved by mutation and tended to replace the virulent strains (*29*). In addition, passive immunity to myxomatosis is conferred to kittens born of immune does (*30*). Finally, a genetic change has occurred in the rabbit population, and this has provided intrinsic resistance to the myxomatosis virus (*31*). This clearly illustrates the alternate functioning of the feedback of density, selection, and genetic change which has in turn altered the density of both populations. There was some similarity between the virus-rabbit relationship, the laboratory wasp-fly relationship, and the type of evolution which took place. In the virus-rabbit association, most of the evolution occurred in the parasite, whereas in the wasp-fly association most of the evolution took place in the host.

Transmission of the myxomatosis virus depends upon mosquitoes (*Aedes* and *Anopheles*) that feed only on living animals (*32*). Rabbits infected with the virulent strain of virus live for a shorter period of time than those infected with the less virulent strain. Because rabbits infected with the less virulent strain live for longer periods of time, mosquitoes have access to that virus for longer periods of time. This gives the avirulent strain a competitive advantage over the virulent strain. In addition, in regions where the avirulent strain is located, rabbits are more abundant, and this allows more total virus to be present than in a comparable region infected with the virulent. Thus, the virus with the greatest rate of increase and density within the rabbit is not the virus selected for, but the virus with demands balanced against supply has survival value in the ecosystem.

Another example of how population regulation evolves from one dominant mechanism to another can be found in

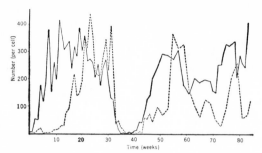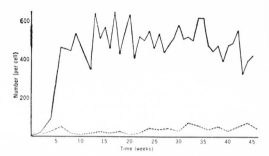

Fig. 3. Population trends of control parasite-host system (left) and experimental system (right) that has evolved some stability or balance between the interacting species. Solid line, mean number of hosts; dashed line, mean number of parasites per cell in 30-cell population cages.

a comparison of the results of the initial interaction of the parasite-host study with the results of these populations after they had interacted for 1004 days (26). Initially, parasite density in the experimental system averaged about 3700. Although the density of the parasite population fluctuated, the mean reproduction of a parasite pair at the carrying capacity of the environment would have to be two or a pair with births equaling deaths. Because the experimental parasites produced about 135 progeny per female, 133 of these would have to die each generation to leave a single parasite pair surviving to replace the parent pair. Early in the experimental system, competition was primarily responsible for limiting parasite numbers and causing the death of 133 of the 135 offspring produced per female. The decline from 135 to 39 progeny per female of the experimental parasite meant that the loss of 96 progeny was due to changes brought about by genetic feedback. To maintain the population at this lowered reproductive rate, only 37 of the progeny could be lost to competition. Thus, competition in the beginning was the dominant control mechanism operating in the experimental system, but genetic feedback became dominant with time and through evolution.

Competition and Coexistence

When we consider how the genetic feedback mechanism functions, it seems logical to apply it to situations in which competing species might evolve to occupy the same niche. Competition here refers to species at the same time and place which share the same essential resource in short supply (2). Niche is defined as an animal's "place in the biotic environment, its relationship to

food and enemies" in the community (17).

Competing species seeking the same plant, prey, or host can coexist if their numbers are controlled by genetic feedback. For example, let us assume that two aphid populations feed on sap from the same plant species. The two aphid species can coexist because the more abundant aphid species will eventually be controlled through the processes of genetic feedback. The amount of change that occurs in the characteristics of the plant for protection against the feeding pressure of the animal is dependent on density. Because more plants are selectively destroyed by the abundant aphid, the resistant polygenic factors effective against the abundant aphid would increase in the plant population. This means that the abundant aphid ultimately will be more limited by changes in the plant than the sparse aphid will. Thus, the numbers of both competing aphid populations are controlled by differential evolution of the plant relative to each population. Results of field studies with two aphid species that attack alfalfa (33) suggest that two competitive animal species seeking the same food host can be differentially influenced by evolution in the plant.

Genetic feedback may also operate in yet another way to enable two species to coexist and utilize the same thing (food, space, and so on) in the ecosystem. In this case, let us assume that both species are fairly evenly balanced in their competitive ability and that species A is only slightly superior to species B. As the numbers of A are increasing, the numbers of B will be declining and becoming sparse. The abundant individuals of species A must contend principally with intraspecific competitive selection because there is a greater

chance for individuals of this species to interact with their own kind. Haldane (34) pointed out that intraspecific competitive selection is frequently biologically disadvantageous for the species. At the same time, individuals of species B are contending primarily with interspecific competitive selection. Thus, under this selection species B would evolve and improve its ability to compete with its more abundant cohort species A. As species B improves as a competitor, its numbers increase, and finally B becomes the more abundant species. Then the dominant kind of competition (interspecific or intraspecific) affecting each species is reversed. After many such oscillations and with each oscillation decreasing in intensity, a state of relative stability should result.

This idea—that intraspecific selection on the dominant species and interspecific selection on the sparse species favors the sparse species—was tested successfully with the housefly and blowfly (*Phaenicia sericata*) in a multicelled cage (35). In another population system (surviving for 160 weeks or 80 fly generations), there was a persistent alternation of dominance of first the blowfly and then the housefly. A genetic check on the fly populations showed that the currently dominant species remained genetically static, while the sparse species or "underdog" evolved to become the better competitor and dominant species. Although there has been an oscillation in dominance, no damping of the fluctuation has been noted to date.

Conclusion

The importance of the genetic feedback mechanism as a regulatory system in communities is substantiated by its wide application to such diverse interacting population systems as herbivore and plant, parasite and host, predator and prey, and interspecific competitor systems. The real significance of this mechanism for population regulation lies in the fact that it has its foundation in evolution. Population regulation by genetic feedback supports Emerson's (*36*) view that evolution in natural populations is toward homeostasis (balance) within populations, communities, and ecosystems.

Students of population ecology and especially of parasitology and epidemiology generally accept the fact that evolutionary trends in relationships of parasite and host are toward balance. The deductive basis for this generalization rests on the ecological principle that disharmony results in serious losses to both parasite and host. Large numbers of fatal infections in the host population eventually lead to host extinction which in turn brings about the extinction of the parasite. The success of any living population is measured by its relative abundance and distribution as well as its ability to survive in time.

Homeostasis, in herbivore-plant, parasite-host, and predator-prey species and among other community members in general, results in improved survival of the community system. The evolved balance in supply and demand achieved by the feeding species and its host establishes a sound economy for the community. This, of course, enables the community to make effective use of the resources available to it.

Increased species diversity in a community is due in part to community homeostasis. The genetic integration of interspecific competitors which makes possible the use of the same resource by competing species and enables them to occupy the same niche contributes to greater species diversity. The increased network of interactions within the community, resulting from a greater number of species present, further contributes to community homeostasis.

With more knowledge concerning the regulation of natural populations, man will be in a better position to control the pests on his food crops and the parasitic diseases of mankind. This will also help conserve the millions of living species which are vital for the functioning of the vast living system of which he is a part.

References and Notes

1. D. Lack, *The Natural Regulation of Animal Numbers* (Clarendon Press, Oxford, 1954).
2. A. MacFadyen, *Animal Ecology* (Pitman, London. 1957).
3. F. A. Urquhart, *Changes in the Fauna of Ontario* (Univ. of Toronto Press, Toronto, 1957).
4. L. L. Snyder, in *Changes in the Fauna of Ontario*, F. A. Urquhart, Ed. (Univ. of Toronto Press, Toronto 1957), pp. 26–42.
5. ———, *Ontario Birds* (Clarke, Erwin, Toronto, 1951).
6. O. Kalela, *Bird-Banding* **20**, 77 (1949).
7. P. P. Feeny, thesis, Oxford University (1966).
8. J. M. Kingsbury, *Poisonous Plants of United States and Canada* (Prentice-Hall, Englewood Cliffs, N.J., 1964).
9. T. Eisner, *Science* **146**, 1318 (1964).
10. ——— and J. Meinwald, *ibid.* **153**, 1341 (1966).
11. B. F. Coon, R. C. Miller, L. W. Aurant, *Pennsylvania Agricultural Experiment Station Report* (1948).
12. J. H. Pepper and E. Hastings, *Montana Agricultural Experiment Station Technical Bulletin 413* (1943).
13. H. G. Andrewartha and L. C. Birch, *The Distribution and Abundance of Animals* (Univ. of Chicago Press, Chicago, 1954).
14. H. S. Smith, *Econ. Entomol.* **28**, 873 (1935).
15. R. A. Fisher, A. S. Corbet, C. B. Williams, *J. Anim. Ecol.* **12**, 42 (1943).
16. E. R. Dunn, *Ecology* **30**, 39 (1949).
17. C. Elton, *Animal Ecology* (Sigwick and Jackson, London, 1927).
18. A. P. Jacot, *Ecology* **17**, 359 (1936).
19. C. Overgaard, *Natura Jutlandica* **2**, 1 (1949).
20. R. N. Chapman, *Animal Ecology* (McGraw-Hill, New York, 1931).
21. M. Demerec, *Biology of Drosophila* (Wiley, New York, 1950).
22. D. Pimentel, *Amer. Natur.* **95**, 65 (1961).
23. R. H. Painter, *Insect Resistance in Crop Plants* (Macmillan, New York, 1951).
24. R. G. Dahms and R. H. Painter, *J. Econ. Entomol.* **33**, 482 (1940).
25. R. G. Dahms, *J. Agr. Res.* **76**, 271 (1948).
26. D. Pimentel and R. Al-Hafidh, *Ann. Entomol. Soc. Amer.* **56**, 676 (1963).
27. D. G. Stead, *The Rabbit of Australia* (Winn, Sydney, Australia, 1935).
28. F. Fenner, in *The Genetics of Colonizing Species*, H. G. Baker and G. L. Stebbins, Eds. (Academic Press, New York, 1965), pp. 485–499.
29. H. V. Thompson, *Ann. Appl. Biol.* **41**, 358 (1954).
30. F. Fenner, *Cold Spring Harbor Symp. Quant. Biol.* **18**, 291 (1953).
31. I. D. Marshall, *J. Hyg.* **56**, 288 (1958).
32. M. F. Day, *J. Australian Inst. Agr. Sci.* **21**, 145 (1955).
33. R. H. Painter, *Proc. Int. Congr. Entomol. 12th* **1964**, 531 (1964).
34. J. B. S. Haldane, *The Causes of Evolution* (Longmans, Green, New York, 1932).
35. D. Pimentel, E. H. Feinberg, P. W. Wood, J. T. Hayes, *Amer. Natur.* **99**, 97 (1965).
36. A. E. Emerson, in *Principles of Animal Ecology*, W. C. Allee, A. E. Emerson, O. Park, T. Park, and K. P. Schmidt, Eds. (Saunders, Philadelphia, 1949), pp. 640–695.
37. Supported in part by environmental biology grant GB-4567 from NSF.

Reprinted from *Science*, **162**, 1453–1459 (Dec. 27, 1968)

Genotype, Environment, and Population Numbers

Animal numbers are regulated by the genetic composition
of the population and by environmental factors.

Francisco J. Ayala

In his work *On the Origin of Species,* Darwin wrote that "the causes which check the natural tendency of each species to increase in numbers are most obscure. . . . We know not exactly what the checks are even in a single instance." The regulation of population numbers is of major importance for the understanding of natural selection and biological evolution. It has implications of economic interest, particularly for the control of animal pests. Finally, it is a major problem for modern man who has become aware that the quality of human life is seriously threatened by the so-called "population explosion."

Population biology is concerned with the distribution and abundance of organisms. The factors considered when studying a particular population are the relationship of the animals to their food, to the places where they live, to the weather, and to other animals that share the same food or place to live, that prey on them, or that are related to them in any way. Unfortunately the genetic constitution of the population is usually not given sufficient attention. Populations of a species are treated as if they were genetically homogeneous in space and in time. Yet, to understand the causes which regulate animal numbers, both genetic and environmental factors must be considered.

Students of natural populations of animals encounter many difficulties, particularly in the estimation of adult numbers and the causes of mortality (*1*). Some problems can more easily be approached in laboratory studies, which permit control of the more important factors, while one or a few variables are manipulated at a time. Models can be produced; the validity of which must, of course, be ultimately tested in the field.

Drosophila flies are particularly favorable organisms for laboratory studies of some population problems. They multiply rapidly in cultures which are easy to maintain at moderate expense. Moreover, much is known about their biology, since they have been intensively studied for the last 60 years. I now describe some experimental approaches using drosophila that have provided information on the factors which regulate population numbers.

Innate Capacity for Increase

All components of the life cycle of drosophila are influenced by the genetic constitution of the flies. Genetic variation has been found to affect fertility of females and hatchability of eggs (*2*), fertility and mating activity of males (*3*), rate of development (*4, 5*), longevity (*6*), and others. The ability of a population to increase in numbers or to maintain a certain size is related to these properties of the flies. However, it is not clear how they interact with each other to determine reproductive capacity. A statistic variously named the Malthusian parameter, intrinsic rate of natural increase, or innate capacity for increase has been proposed which

The author is assistant professor at Rockefeller University, New York, New York.

incorporates the significant components of the life cycle into a single value (7–9).

The innate capacity for increase in numbers, r_m, may be defined as the maximum rate of increase attained by a population at any particular combination of quality of food, temperature, humidity, and so forth, when the quantity of food, space, and other animals of the same species are kept at an optimum, and other organisms of other species are excluded. The population is assumed to have a constant age schedule of births and deaths. Essentially, the ability of a population to increase in numbers depends upon the birth rate and the survival rate of the animals. The number of animals will increase, remain constant, or decrease depending on whether the birth rate is greater than, equal to, or smaller than the death rate. The rate of change depends on the magnitude of the difference between birth rate and death rate. To estimate the innate capacity for increase of a population the distribution of ages must be considered since the expectation of births and the probability of death vary with age. For a population with a stable age distribution, that is, with a constant age schedule of births and deaths, the innate capacity for increase is connected with the schedules of fecundity and mortality by the expression:

$$\int_0^{\infty} e^{-r_m x}\, l_x m_x \delta x = 1$$

where e is the base of natural logarithms, l_x the probability at birth of being alive at age x, m_x the number of female offspring produced per unit time by a female aged x, and 0 to ∞ the life-span. To estimate r_m the integration is replaced with a summation over discrete time intervals:

$$\sum_0^t e^{-r_m x}\, l_x m_x = 1$$

The detailed procedure for estimating r_m from this equation is given by Andrewartha and Birch (8). The accurate solution of the equation requires data not easily obtained, as well as some laborious calculations, although various approximations which simplify the operations are usually acceptable. Sometimes it is preferable to use a related statistic, the finite rate of increase ($\lambda = \text{antilog}_e\ r_m$). While r_m measures the infinitesimal rate of increase, λ can express the rate of increase per day, per week, or for any time interval. Genetically different populations of

Table 1. Mean productivity and population size after equilibrium of various geographic strains of two related drosophila species. Measurements are 44 for productivity and 17 for population size.

Population	Temperature (°C)	Productivity (No./food unit)	Population size
Drosophila serrata			
Sydney	25	550 ± 17	1782 ± 76
Cooktown	25	568 ± 20	2221 ± 80
Popondetta	25	477 ± 13	1828 ± 90
Sydney	19	483 ± 13	1803 ± 87
Cooktown	19	486 ± 12	2017 ± 84
Popondetta	19	357 ± 8	1580 ± 52
Drosophila birchii			
Cairns	25	351 ± 16	1262 ± 83
Popondetta	25	152 ± 9	469 ± 49
Cairns	19	324 ± 11	1091 ± 66
Popondetta	19	121 ± 5	428 ± 33

D. pseudoobscura have different innate capacities for increase in numbers (10). Populations polymorphic for certain chromosomal arrangements, and therefore carrying more genetic variability, had greater capacity for increase than monomorphic populations. *Drosophila pseudoobscura* flies with irradiated genetic material have a lower capacity for increase than nonirradiated controls (11). Geographic races of the same species also have different innate capacities for increase (12). Given any genetic constitution, the statistic r_m is very sensitive to differences in temperature (10–12) and in quality of food (13).

The statistic r_m has been criticized as being an abstraction far removed from nature. Indeed, while it provides some information about certain physiological characteristics of a population, its relevance to natural conditions is limited. Natural populations do not have stable age-distributions. Estimates of r_m obtained for any particular population are applicable only for the conditions of temperature, humidity, quality of food, and so forth, under which the experiments are carried out. In Tantawy's experiments (11) the population with the highest capacity for increase at 15°C had the lowest at 25°C. Finally, r_m measures the capacity for increase when the quantity of food and space are kept at an optimum and there are no competitors. In nature, food and a place to live may be limited or inaccessible; the chances of an animal to survive and multiply depend on its ability to compete with other animals.

Role of the Genotype

The regulation of population numbers may be studied in the laboratory under conditions more nearly approximating those existing in the field. One approach consists of introducing a genetically defined population into a restricted environment with a limited supply of food provided at regular intervals. Environmental factors like moisture and temperature are also controlled. After a few generations population numbers usually reach an equilibrium and thereafter oscillate about a mean equilibrium level. At equilibrium the rate of births equals the rate of deaths. The effect of genetic constitution on population numbers may be studied by comparison of the performance of genetically different populations which are treated identically. Conversely, treatment of genetically identical populations may be modified to ascertain the effect of selected environmental components on population size.

A variant of this method is the "serial transfer" technique, which has proven to be useful in the experimental study of population regulation in drosophila. Adult flies are introduced into an experimental "cage," usually a glass jar, with a measured amount of food and allowed to lay eggs for a specified period of time, usually 2 or 3 days. At regular intervals the flies are transferred to new cages with fresh food. When adult flies begin to emerge in the cages where the eggs were laid, they are collected, counted and weighed under anesthesia, and then added to the cage containing the adult population. The ovipositing adult flies are thus always in a single cage with fresh food, while a number of cages contain eggs, larvae, pupae, and newly emerged adults. The adult population is anesthetized and censused at regular intervals. The technique also provides information about the number of births, that is the number of flies emerging, per unit food or per unit time. From the rate of birth and the population size, estimates may be obtained of the average longevity of adult flies.

The serial transfer technique was employed to study the performance of several geographic strains of *D. serrata* and *D. birchii*, two sibling species common in eastern Australia and New Guinea (14). Strains of *D. serrata*, collected at Popondetta, New Guinea; Cooktown, Queensland; and about 200 kilometers north of Sydney, New South Wales, were used, while *D. birchii* was collected at Cairns, Queensland, and at Popondetta, New Guinea. These strains represent populations adapted to very different climates, and may therefore be rather different genetically. The Popondetta and Sydney strains were collected at about 9° and 33° south latitude, respectively, separated by 2700 kilometers.

Experimental populations (300 flies each) were established in the laboratory with descendants of the flies collected in the field. Two populations were started with each strain, one kept at 25°C and the other at 19°C. They were observed for a year, which corresponds to about 17 and 10 generations at 25°C and 19°C, respectively. The populations increased rapidly, reaching by the third generation an equilibrium size around which they oscillated thereafter (15). The mean number of flies produced per food unit and the mean population size from the fourth generation on are given in Table 1. Standard erorrs for the means indicate the amplitude of the oscillations around the levels of equilibrium.

There are striking differences among the genetically different strains in their ability to exploit the experimental environment. The Cooktown strain of *D. serrata* has the largest population size at either temperature. At 25°C, the Popondetta and Sydney strains have approximately equal size, but the Sydney population is larger at 19°C, which suggests that Sydney flies may be adapted to live in a climate considerably colder than that of Popondetta.

There is no strict correspondence between rate of birth and average population numbers. The number of flies emerging per unit time is approximately the same in the Sydney and Cooktown populations at either temperature, but the Cooktown population is larger in average size. This means that under the experimental conditions the average longevity of the Sydney flies is smaller. In *D. birchii*, the Cairns populations have greater productivity and size than those from Popondetta. The considerable difference in adaptation to the experimental environment of these two strains reflects the large genetic differences existing between them (14).

Genetic Variability

Carson (16) detected differences in the productivity and size of experimental populations of *D. robusta* derived from different geographic localities.

Table 2. Mean productivity and population size after equilibrium of various hybrid populations of drosophila. Measurements are 41 for productivity and 15 for population size.

Population	Temperature (°C)	Productivity (No./food unit)	Population size
Drosophila serrata			
Sydney × Cooktown	25	593 ± 16	2360 ± 74
Sydney × Popondetta	25	622 ± 18	2541 ± 117
Cooktown × Popondetta	25	540 ± 18	2419 ± 76
Sydney × Cooktown	19	554 ± 19	2418 ± 171
Sydney × Popondetta	19	572 ± 14	2448 ± 86
Cooktown × Popondetta	19	479 ± 12	2227 ± 172
Drosophila birchii			
Cairns × Popondetta	25	342 ± 26	1331 ± 123
Cairns × Popondetta	19	303 ± 10	1203 ± 55

The performance of a strain collected at the center of the geographic distribution of the species (central population) was superior to that of another strain collected at the margin of its distribution. This seems to be the case also for *D. serrata*. Central populations of drosophila have been observed to possess greater genetic variability than marginal populations. The superior performance of central populations in the laboratory has been explained by arguing that populations with greater genetic variability are more efficient in adapting to a new environment.

The role of genetic variability in the adaptation of a population to a new environment can be directly approached in the laboratory. "Hybrid" populations can be produced by mass-crossing two strains. Females of, say, strain A are mated with males of strain B, and females of strain B with males of strain A. If a large number of parents are used the progenies of the two reciprocal crosses will contain most of the genetic variability present in both parental strains. Mass-crosses between each two strains were made among the three strains of *D. serrata* and the two of *D. birchii*. Experimental populations established with progenies of the mass-crosses were studied for about a year at 25°C and 19°C. Table 2 shows the mean productivity and size of these hybrid populations. Hybrid populations have larger size, and, generally, also greater productivity, than the corresponding parental populations (Tables 1 and 2). Comparison of the mean of the hybrid populations with the means of the two parental populations with the Student's *t*-test showed that the productivity and size of the hybrid populations are always significantly greater (15).

In serial transfer experiments with *D. pseudoobscura*, the average size and average productivity of a population polymorphic for two chromosomal arrangements were greater than those of either one of two monomorphic populations (17). The fitness of certain chromosomal arrangements of *D. pseudoobscura* was positively correlated with the initial amount of genetic variability (18). In experiments where *D. pseudoobscura* and *D. serrata* were competing for food and living space, the average numbers of *D. pseudoobscura* were greater in the populations with more genetic variability. In *D. melanogaster* the average population size of a strain carrying several mutant genes was about one-third the size of a wild-type population. A hybrid population, having both wild-type and the mutant genes, was superior to both parental populations (19).

Evolutionary changes occur in the adaptation of a population to an experimental environment by natural selection. The observed superior performance of populations with greater initial genetic variability is likely to result from natural selection being more efficient in those populations where more genotypes are available for selection. In the experiments, selection is intensive since food and space are quite restricted. The amount of food available is sufficient for the development of probably less than 1 percent of the eggs laid. The adult flies also compete intensively for food and space in the extremely crowded cages. The average longevity of *D. serrata* in the experiments is about 9 and 13 days at 25°C and 19°C, respectively (15). Under optimum conditions their average longevity is about 25 and 45 days at 25°C and 20°C, respectively (12).

If the populations are adapting gradually to the experimental environment as a result of evolutionary changes, it might be possible to observe this process and measure it in some way. The adaptation of the populations to the environment may be measured by their numbers. Larger population size implies greater efficiency in transforming the limited resources of food and space into biomass, or living matter. The progressive adaptation of a population to an experimental environment could be measured by a gradual increase in population numbers while the environmental conditions remain constant.

The performance of two populations, one established with the Popondetta strain and the other with the progenies of Sydney × Popondetta mass-crosses, was studied for 18 months at 25°C and at 19°C (*20*) (Fig. 1). A statistic, the coefficient of regression of population numbers on time, was used to evaluate the apparent increase in population size. The coefficients of regression are positive, and statistically different from zero, in the four populations (Fig. 1). The populations have gradually shown greater adaptation to the experimental environment.

The Popondetta populations increased at an average rate of 10.5 flies per week at 25°C and 8.4 flies per week at 19°C. The corresponding rates of increase per week are 19.5 and 20.4

for the hybrid populations. The rate of increase in the hybrid populations is approximately double that in the Popondetta populations. The experiment shows measurable evolution over a short time span in *D. serrata*. Moreover, it provides one of the few available biological illustrations of Fisher's (*9*) fundamental theorem of natural selection: "The rate of increase in fitness of any organism at any time is equal to its genetic variance in fitness at that time." The rate of evolution is considerably larger in the hybrid populations, which have greater genetic variance, than in the Popondetta populations.

The gradual increase in population size may conceivably be due to environmental, rather than to genetic, changes. Although temperature, amount and quality of food, space, and all other components of the physical environment were kept constant within measurable limits, it is possible that undetected modifications of the environment may have occurred. This possibility was tested by comparison of the performances of control populations directly established from the laboratory stocks with populations derived

from the experimental populations described above (*21*). The experimental populations had greater productivity and larger average size than the controls demonstrating their superior genetic adaptation to the experimental environment.

Radiation and Selection

The process of mutation ultimately furnishes the materials for adaptation to changing environments. Genetic variations which increase the reproductive fitness of a population to its environment are preserved and multiplied by natural selection. Deleterious mutations are eliminated more or less rapidly depending on the magnitude of their harmful effects. High-energy radiations, such as x-rays, increase the rate of mutation (*22*). Mutations induced by radiation are random in the sense that they arise independently of their effects on the fitness of the individuals which carry them. Randomly induced mutations are usually deleterious. In a precisely organized and complex system like the genome of an organism, a random change will most frequently decrease, rather than increase, the orderliness or useful information of the system. A potentially beneficial mutation induced by radiation is likely to have occurred in the past history of a population. If the population has lived for a long time in the same environment mutations beneficial in that environment may already be incorporated in the gene pool of the population. Mutational changes are more likely to be beneficial to the population when the environment changes.

Experiments can be made to test whether an increase in genetic variance induced by x-radiation might result in an increase in the reproductive fitness of the population. Certain conditions must be fulfilled for such experiments to succeed. Large numbers of individuals must be irradiated to make more probable the induction of some favorable mutations. The dosage should not be too large lest the potential selection of favorable induced mutations be more than counteracted by deleterious mutations. Finally, the population should be exposed to a new environment, as different as possible from that to which they are adapted, to increase the chance of a measurable rate of evolutionary change during the experimental period.

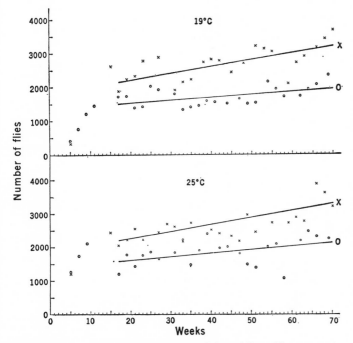

Fig. 1. Population size of four experimental populations of *Drosophila serrata* at two temperatures. *X*, Hybrid population; *O*, Popondetta population.

Experiments designed to meet these requirements were carried out with *D. serrata* flies at 25°C and at 19°C (*23*). At each temperature two populations were irradiated and a third one was the control. The males of the experimental populations were given 1000 roentgen of x-rays in each of three consecutive generations and mated to the nonirradiated females. The populations were maintained by the serial transfer technique.

Figure 2 shows the changes in population numbers at 25°C. The two irradiated populations decreased in size during the first 6 weeks, owing presumably to the elimination of carriers of deleterious mutations. Thereafter there was a rapid increase in the size of the two experimental populations which became considerably larger than the control from week 10 until the experiment was terminated. Comparable results were observed at 19°C. Table 3 contains the mean productivity and size of the six populations from weeks 16 to 41.

Natural selection resulted in better adapted genotypes in the irradiated than in the control populations, as measured by either their size or their productivity. The increase in the rate of evolution of the irradiated populations can be measured as follows. The differences in size between each irradiated population and the control are obtained for each measurement throughout the experimental period and the coefficient of regression of the differences on time is calculated. The two irradiated populations at 25°C increased in size at a rate of 26 and 41 flies per week faster than the controls. At 19°C the increase in the rate of evolution of the irradiated populations over the control was 36 and 46 flies per week.

Comparable results were obtained with *D. birchii* (*23*), and with two irradiated populations of *D. serrata* studied for nearly 3 years (*24*). Carson did not observe any sustained increase in the size of irradiated populations of *D. melanogaster* (*25*). His experiments, however, were conducted under conditions different in several important respects (*23*). Radiation-induced mutations in *D. pseudoobscura* increased the fitness of certain chromosomes when natural selection was operative, but not otherwise (*26*). The conclusion is that the genetic variability induced by high frequency radiation may result, after several generations of strong natural selection, in more efficient adapta-

Table 3. Mean productivity and population size of four irradiated populations of *Drosophila serrata* and their controls.

Population	Temperature (°C)	Productivity (No./food unit)	Population size
Control	25	198 ± 7	1294 ± 50
Exp. 1	25	317 ± 10	1955 ± 65
Exp. 2	25	378 ± 9	2558 ± 98
Control	19	118 ± 14	498 ± 110
Exp. 1	19	250 ± 10	1358 ± 102
Exp. 2	19	259 ± 9	1515 ± 75

tion of a population to a new environment. Needless to say, this conclusion has no application to man with his long generation time and limited reproductive capacity. Moreover, human values would hardly allow for the enormous price in lives and physical misery that the species would have to pay for such hypothetical improvement of its adaptation to the environment.

Temperature, Food, Space

The size of a population living in a certain environment depends upon its genetic constitution. I have expounded at some length the genetic aspect of the regulation of population numbers because in discussions of this topic genetic considerations are often ignored altogether. Animal populations, however, do not live in a vacuum. The effect of the genotype depends on the environment of the organism. I shall now consider several experiments with drosophila flies which indicate the ef-

fect of various components of the environment on the size of animal populations.

Animals can survive and multiply only within certain temperature ranges. Within the survival range, temperature affects various properties of the organisms. In drosophila temperature influences fertility (*2, 27*), speed of development and longevity (*10, 12*). Humidity also affects reproductive efficiency (*28*). The innate capacity for increase in numbers of various geographic strains of *D. serrata* and *D. birchii* is nearly double at 25°C that at 20°C (*12*). *Drosophila pseudoobscura* of various genetic compositions had an average capacity for increase more than double at 25°C that at 16°C (*10*).

"The amount of food for each species of course gives the extreme limit to which it can increase." There is little argument about this statement from Darwin's *On the Origin of Species*. There is, however, considerable debate about whether food is a major check of animal numbers. According to Lack (*1*) the numbers of many species of birds, and also of many other animals, are limited by food. Andrewartha and Birch (*8*) think instead that shortage of material resources, such as food, is probably the least important among the possible ways in which the numbers of a population may be limited. Crowding and food limitation, however, affect longevity, as well as fecundity and speed of development of drosophila (*29*). Quality as well as quantity of food is important (*13, 30*).

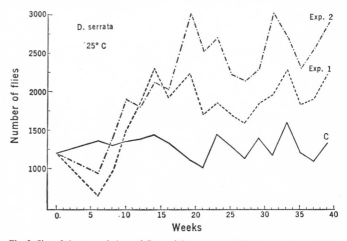

Fig. 2. Size of three populations of *Drosophila serrata* at 25°C; *C*, control population; *Exp-1* and *Exp-2*, irradiated populations.

Table 4. Effects of temperature (25°C versus 10°C), transfers per week (2 versus 3), and genetic composition (hybrid versus Popondetta strain) in eight populations of *Drosophila serrata*. The standard errors for the mean and the treatments are the same within each column. The effects are indicated as average deviations from the mean.

	Productivity per food unit			Population size	
	No.	Biomass (mg)	Indiv. weight (mg)	No.	Indiv. weight (mg)
Mean	535.3	296.0	0.560	2670	0.627
Temp.	+ 24.8*	+ 2.4	− 0.016*	− 370*	− 0.025*
Transfers per week	+ 20.9*	+ 14.8*	− 0.005	− 208*	+ 0.009*
Genetic composition	+ 66.8*	+ 39.6*	− 0.004	+ 394*	− 0.021*
Standard error	5.8	5.7	0.003	45	0.003

* The effect is statistically significant, $P < .05$.

When a population is growing with a limited amount of food and space, the number of animals per unit of food and space rises. Eventually, an equilibrium must be reached when the number of births per unit time equals the number of deaths. The average size of a population, however, need not be proportional to the rate of births. Of two populations with the same number of births per unit time, one will be twice as large as the other if the average longevity of the animals in the first population is also double. The large increase in human numbers during recent times has been due more to increase in average longevity than to increase in rate of births.

An experiment with *D. serrata* illustrates this point. Temperature and amount of food were each studied at two levels. The temperature was either 25°C or 19°C, the amount of food was either two or three food units per week. Genetic composition was also a variable, the populations being either the Popondetta strain or the hybrid, Sydney × Popondetta. All possible combinations among the variables were made (factorial design). Thus, there was a Popondetta population at 25°C with three food units per week; a second at 25°C with two food units; a third at 19°C with three food units; and a fourth at 19°C with two food units. Four parallel Sydney × Popondetta populations were also established. The populations were maintained by the serial transfer technique, and therefore the living space for the adult flies was equal in the eight populations (*31*).

The hybrid populations have considerably greater productivity and size than the Popondetta populations (Table 4). This confirms again that a greater initial genetic variance results, with natural selection, in better adapted genotypes.

The effects of temperature are clear. The number of flies produced is slightly larger at 25°C than at 19°C. The flies developed at the higher temperature are, however, smaller; in fact, the biomass produced per food unit is the same at either temperature. The average population number at 19°C is about 30 percent larger than at 25°C. Since somewhat fewer flies are added per unit time to the populations at 19°C, the average longevity of these flies must be considerably greater, about 40 percent, than in the 25°C populations (*31*) (Table 4).

The effects of food amount (Table 4) are most interesting. Populations receiving three food units per week are only about 17 percent larger than populations transferred twice a week. The number of flies born per food unit is nearly equal for both populations. Nearly 50 percent more flies are added per week to the populations with three food units, but their average numbers are only 17 percent greater. The increase in amount of food, and therefore in numbers of births, results in a decrease in average longevity with but a small increase in population size. The

Table 5. Performance of three genetically different populations of *Drosophila pseudoobscura* in competition with *Drosophila serrata*. The numbers are means of six replicates calculated from weeks 6 to 47. CH and AR are two chromosomal arrangements.

	Flies emerging (No./per week)	Old flies (No.)
Population 1		
D. pseudoobscura (CH)	60.4	64.5
D. serrata	264.1	299.6
Population 2		
D. pseudoobscura (AR)	107.0	144.6
D. serrata	138.8	150.7
Population 3		
D. pseudoobscura (CH and AR)	123.8	152.6
D. serrata	144.4	151.1

limiting factor is presumably living space. The available space in the cages is about 400 cubic centimeters, a very restricted space indeed for the average 2670 flies living in it.

It is unlikely that living space is ever so limited for natural populations of drosophila. The lesson is that rate of births may not be strictly correlated with average population size; and that factors other than food may exercise the primary control of population numbers even when the number of individuals developing to maturity depends on the amount of food. The numbers in the experimental populations are simultaneously limited by both food and space. Food is a limiting factor, for an increase in amount of food produces an increase in number. Since the increase in number is not proportional to the increase in food, space is postulated as the additional limiting factor. In nature, predators or other components of the environment may take the limiting role played by living space in the experiments.

Population density may be simultaneously limited by more than one factor. This question was further pursued in an experiment with *D. serrata* in which the amount of space as well as the amount of food was varied (*32*). A factorial design was used with a total of 18 populations. The mean number of flies for all populations was 1145. An increase or decrease in the amount of food of 33 percent produced an increase or decrease of 26 percent in population size, respectively. A change of 33 percent in the amount of space resulted in a 10 percent change in population size. Again both food and space are limiting factors of population numbers, although within the range studied food plays a greater role than space. Comparable results have been obtained with several geographic strains of *D. serrata*, and with *D. birchii* (*33*), *D. melanogaster* and *D. pseudoobscura* (*34*).

Cooperation and Competition

Other animals of the same species are not always competitors for the available resources of food and space. Laboratory studies have shown that there exist various forms of cooperation among individuals of the same species. The most obvious form of cooperation is, of course, the need for sexually reproducing organisms of finding a mate. But undercrowding as well as overcrowding is frequently harmful. In drosophila, the viability of the larvae (4), longevity, and other properties of the flies (29) generally have an optimum at intermediate densities. The presence of larvae, or their metabolic products, of different genetic constitution enhances the larval viability of some strains but not of others (4, 35, 36). Particularly interesting is the discovery that female flies prefer to overposit near the places where eggs have already been laid by other females (37). Del Solar has demonstrated that this "gregarious" tendency is genetically controlled.

Interactions among animals of related species influence their numbers. Larval viability of drosophila can be enhanced or handicapped by the presence of larvae of other drosophila species in the same cultures (36, 38). In laboratory populations, if two species share limited resources of food and space, one species will generally eliminate the other within a few generations (39). However, I have demonstrated that with the serial transfer technique two species of drosophila can coexist if the environment is properly adjusted (40). For instance, at 25°C D. nebulosa eliminated D. serrata in four or five generations; at 19°C both species coexisted until the experiment was terminated after some 60 generations.

Coexistence of two species in a relatively uniform and constant environment seems to contradict the so-called "competitive exclusion principle." If two related species share at least one essential resource available in limited quantity, one species will be at advantage in the exploitation of the limited resource. According to the competitive exclusion principle, the relative advantage of one species will accumulate over time with the eventual elimination of the other species. The serial transfer technique, which permits the study of competition independently among the adult and among the larvae, makes possible an explanation for the coexistence of two species. It seems that one species is at an advantage in the larval stage while the other species is at advantage in the adult stage. Or, looking at it differently, one species is at an advantage in the exploitation of one limited resource—food, and the other species is at an advantage in the exploitation of a different limited resource—living space for the adults. The selection pressure is different among the larvae than among the adults. The relative advantages can-

cel each other at certain relative frequencies of the two species. The equilibrium frequencies are observed to be stable. It seems that an increase in the frequency of one species results in a net disadvantage for that species until the equilibrium is restored. Thus, the phenomenon of frequency-dependent selection previously observed at the intraspecific level (41) apparently occurs also at the interspecific level.

When two or more species share the same environmental resources, the numbers of each species depend on its genetic constitution as well as on the genetic constitution of the competing species. This can be illustrated by an experiment with D. serrata and D. pseudoobscura (42). There were three types of populations depending on the genotypes of D. pseudoobscura. The genetic constitution of D. serrata was the same in all populations; D. pseudoobscura was either polymorphic for two chromosomal arrangements (CH and AR), or monomorphic for CH or AR. There are six replicates for each type, with a total of 18 populations. The populations were started with 300 flies of each species and kept at 23.5°C for 47 weeks. D. pseudoobscura flies decreased in frequency from their original 50 percent during the first few weeks. By the third generation an equilibrium was reached and the relative frequency of each species remained approximately constant thereafter until the experiment was terminated (Table 5).

The numbers of D. pseudoobscura are greater in the monomorphic AR than in the monomorphic CH population, and larger in the polymorphic than in either monomorphic population. Although the genetic composition of D. serrata was the same in all populations, the average numbers of this species depend on the genetic composition of D. pseudoobscura.

Summary

The abundance of animals is regulated by factors both internal and external to the animals. Of major importance is the genetic constitution of the population, which has too often been ignored in ecological studies. Laboratory experiments with drosophila flies show that populations with greater genetic variability have larger population sizes. The rate of evolution of a population becoming adapted to a new environment is positively correlated

with the initial amount of genetic variability in the population.

Temperature, humidity, and other climatic factors affect population numbers. Food and a place to live may jointly limit the maximum size that a population can reach. Finally, the biotic components of the environment influence the size of animal populations.

References and Notes

1. D. Lack, *Population Studies of Birds* (Clarendon Press, Oxford, 1966).
2. W. W. Alpatov, *J. Exp. Zool.* **63**, 85 (1932); W. S. Stone, F. D. Wilson, V. L. Gerstenberg, *Genetics* **48**, 1089 (1963); D. Marinkovic, *ibid.* **57**, 701 (1967).
3. E. B. Spiess, B. Langer, L. D. Spiess, *Genetics* **54**, 1139 (1966).
4. R. C. Lewontin, *Evolution* **9**, 27 (1955).
5. E. B. Spiess, *ibid.* **12**, 234 (1958); P. Bentvelzen, *Genetica* **34**, 229 (1963).
6. B. Wallace, *Evolution* **6**, 333 (1952); M. Vetukhiv, *ibid.* **11**, 348 (1957).
7. A. J. Lotka, *Elements of Physical Biology* (Williams and Wilkins, Baltimore, 1925); P. H. Leslie and R. M. Ranson, *J. Anim. Ecol.* **9**, 27 (1940).
8. H. G. Andrewartha and L. C. Birch, *The Distribution and Abundance of Animals* (Univ. of Chicago Press, Chicago, 1954).
9. R. A. Fisher, *The Genetical Theory of Natural Selection* (Clarendon Press, Oxford, 1930).
10. Th. Dobzhansky, R. C. Lewontin, O. Pavlovsky, *Heredity* **19**, 597 (1963).
11. A. O. Tantawy, *Genetica* **34**, 34 (1963).
12. L. C. Birch, Th. Dobzhansky, P. O. Elliott, R. C. Lewontin, *Evolution* **17**, 72 (1963).
13. S. Ohba, *Heredity* **22**, 169 (1967).
14. F. J. Ayala, *Evolution* **19**, 538 (1965).
15. ——, *Genetics* **51**, 527 (1965).
16. H. L. Carson, *ibid.* **46**, 553 (1961).
17. Th. Dobzhansky and O. Pavlovsky, *Heredity* **16**, 169 (1961).
18. M. W. Strickberger, *Genetics* **51**, 795 (1965).
19. H. L. Carson, *Proc. Nat. Acad. Sci. U.S.* **44**, 1136 (1958); *Evolution* **15**, 496 (1961).
20. F. J. Ayala, *Science* **150**, 903 (1965).
21. ——, *Evolution* **22**, 55 (1968).
22. H. J. Muller, *Science* **66**, 84 (1927).
23. F. J. Ayala, *Genetics* **53**, 883 (1966).
24. ——, *Proc. Nat. Acad. Sci. U.S.* **58**, 1919 (1967).
25. H. L. Carson, *Genetics* **49**, 521 (1964).
26. Th. Dobzhansky and B. Spassky, *Evolution* **1**, 191 (1947).
27. E. C. Hammond, *Quart. Rev. Biol.* **14**, 35 (1939); A. O. Tantawy and M. Vetukhiv, *Amer. Natur.* **94**, 395 (1960).
28. M. J. Heuts, *Proc. Nat. Acad. Sci. U.S.* **33**, 210 (1947); H. Kalmus, *J. Genet.* **47**, 58 (1945); D. D. Sameoto and R. S. Miller, *Ecology* **49**, 177 (1968).
29. R. Pearl, *The Biology of Population Growth* (Knopf, New York, 1926); ——, J. R. Miner, S. L. Parker, *Amer Natur.* **61**, 289 (1927); L. C. Birch, *Evolution* **9**, 389 (1955).
30. A. B. DaCunha, *ibid.* **5**, 395 (1951); F. W. Robertson, *Proc. Nutr. Soc.* **21**, 169 (1962).
31. F. J. Ayala, *Amer. Natur.* **100**, 333 (1966).
32. ——, *Bull. Ecol. Soc. Amer.* **49**, 76 (1968).
33. ——, *Ecology* **49**, 562 (1968).
34. ——, *ibid.* **48**, 67 (1967).
35. M. Dawood and M. Strickberger, *Genetics* **50**, 999 (1964).
36. D. R. Weisbrot, *ibid.* **53**, 427 (1966).
37. E. Del Solar, *ibid.* **58**, 275 (1968); —— and H. Palomino, *Amer. Natur.* **100**, 127 (1966).
38. A. Sokoloff, *Ecol. Monogr.* **25**, 387 (1955); R. C. Lewontin and Y. Matsuo, *Proc. Nat. Acad. Sci. U.S.* **49**, 270 (1963); R. S. Miller, *Amer. Natur.* **98**, 221 (1964).
39. J. A. Moore, *Evolution* **6**, 407 (1952); J. S. F. Barker, *Genetics* **51**, 747 (1965); *Evolution* **21**, 299 (1967).
40. F. J. Ayala, *Amer. Natur.* **100**, 81 (1966); *Genetics* **56**, 542 (1967).
41. C. Petit, *Bull. Biol. Fr. Belg.* **92**, 248 (1958); L. Ehrman, B. Spassky, O. Pavlovsky, Th. Dobzhansky, *Evolution* **19**, 337 (1965); K. Kojima and K. M. Yarbrough, *Proc. Nat. Acad. Sci. U.S.* **57**, 645 (1967).
42. F. J. Ayala, *Genetics*, in press.
43. Supported by PHS career development award 1K3 GM37265-01 from the National Institute of General Medical Sciences.

18

Reprinted from *Amer. Naturalist*, 95(882), 137–145 (1961)

THE PARADOX OF THE PLANKTON*

G. E. HUTCHINSON

Osborn Zoological Laboratory, New Haven, Connecticut

The problem that I wish to discuss in the present contribution is raised by the very paradoxical situation of the plankton, particularly the phytoplankton, of relatively large bodies of water.

We know from laboratory experiments conducted by many workers over a long period of time (summary in Provasoli and Pintner, 1960) that most members of the phytoplankton are phototrophs, able to reproduce and build up populations in inorganic media containing a source of CO_2, inorganic nitrogen, sulphur, and phosphorus compounds and a considerable number of other elements (Na, K, Mg, Ca, Si, Fe, Mn, B, Cl, Cu, Zn, Mo, Co and V) most of which are required in small concentrations and not all of which are known to be required by all groups. In addition, a number of species are known which require one or more vitamins, namely thiamin, the cobalamines (B_{12} or related compounds), or biotin.

The problem that is presented by the phytoplankton is essentially how it is possible for a number of species to coexist in a relatively isotropic or unstructured environment all competing for the same sorts of materials. The problem is particularly acute because there is adequate evidence from enrichment experiments that natural waters, at least in the summer, present an environment of striking nutrient deficiency, so that competition is likely to be extremely severe.

According to the principle of *competitive exclusion* (Hardin, 1960) known by many names and developed over a long period of time by many investigators (see Rand, 1952; Udvardy, 1959; and Hardin, 1960, for historic reviews), we should expect that one species alone would outcompete all the others so that in a final equilibrium situation the assemblage would reduce to a population of a single species.

The principle of competitive exclusion has recently been under attack from a number of quarters. Since the principle can be deduced mathematically from a relatively simple series of postulates, which with the ordinary postulates of mathematics can be regarded as forming an axiom system, it follows that if the objections to the principle in any cases are valid, some or all the biological axioms introduced are in these cases incorrect. Most objections to the principle appear to imply the belief that equilibrium under a given set of environmental conditions is never in practice obtained. Since the deduction of the principle implies an equilibrium system, if such sys-

*Contribution to a symposium on Modern Aspects of Population Biology. Presented at the meeting of the American Society of Naturalists, cosponsored by the American Society of Zoologists, Ecological Society of America and the Society for the Study of Evolution. American Association for the Advancement of Science, New York, N. Y., December 27, 1960.

153

tems are rarely if ever approached, the principle though analytically true, is at first sight of little empirical interest.

The mathematical procedure for demonstrating the truth of the principle involves, in the elementary theory, abstraction from time. It does, however, provide in any given case a series of possible integral paths that the populations can follow, one relative to the other, and also paths that they cannot follow under a defined set of conditions. If the conditions change the integral paths change. Mere failure to obtain equilibrium owing to external variation in the environment does not mean that the kinds of competition described mathematically in the theory of competitive exclusion are not occuring continuously in nature.

Twenty years ago in a Naturalists' Symposium, I put (Hutchinson, 1941) forward the idea that the diversity of the phytoplankton was explicable primarily by a permanent failure to achieve equilibrium as the relevant external factors changed. I later pointed out that equilibrium would never be expected in nature whenever organisms had reproductive rates of such a kind that under constant conditions virtually complete competitive replacement of one species by another occurred in a time (t_c), of the same order, as the time (t_e) taken for a significant seasonal change in the environment. Note that in any theory involving continuity, the changes are asymptotic to complete replacement. Thus ideally we may have three classes of cases:

1. $t_c \ll t_e$, competitive exclusion at equilibrium complete before the environment changes significantly.
2. $t_c \simeq t_e$, no equilibrium achieved.
3. $t_c \gg t_e$, competitive exclusion occurring in a changing environment to the full range of which individual competitors would have to be adapted to live alone.

The first case applies to laboratory animals in controlled conditions, and conceivably to fast breeding bacteria under fairly constant conditions in nature. The second case applies to most organisms with a generation time approximately measured in days or weeks, and so may be expected to occur in the plankton and in the case of populations of multivoltine insects. The third case applies to animals with a life span of several years, such as birds and mammals.

Very slow and very fast breeders thus are likely to compete under conditions in which an approach to equilibrium is possible; organisms of intermediate rates of reproduction may not do so. This point of view was made clear in an earlier paper (Hutchinson, 1953), but the distribution of that paper was somewhat limited and it seems desirable to emphasize the matter again briefly.

It is probably no accident that the great proponents of the type of theory involved in competitive exclusion have been laboratory workers on the one hand (for example, Gause, 1934, 1935; Crombie, 1947; and by implication Nicholson, 1933, 1957) and vertebrate field zoologists (for example, Grinnell, 1904; Lack, 1954) on the other. The major critics of this type of approach, notably Andrewartha and Birch (1954), have largely worked with insects in the field, often under conditions considerably disturbed by human activity.

DISTRIBUTION OF SPECIES AND INDIVIDUALS

MacArthur (1957, 1960) has shown that by making certain reasonable assumptions as to the nature of niche diversification in homogeneously diversified[1] biotopes of large extent, the distribution of species at equilibrium follows a law such that the rth rarest species in a population of S_s species and N_s individuals may be expected to be

$$\frac{N_s}{S_s} \sum_{i=1}^{r} \frac{1}{S_s - i + 1} .$$

This distribution, which is conveniently designated as type I, holds remarkably well for birds in homogeneously diverse biotopes (MacArthur, 1957, 1960), for molluscs of the genus Conus (Kohn, 1959, 1960) and for at least one mammal population (J. Armstrong, personal communication). It does not hold for bird faunas in heterogeneously diverse biotopes, nor for diatoms settling on slides (Patrick in MacArthur, 1960) nor for the arthropods of soil (Hairston, 1959). Using Foged's (1954) data for the occurrence of planktonic diatoms in Braendegård Sø on the Danish island of Funen, it is also apparent (figure 1) that the type I distribution does not hold for such assemblages of diatom populations under quite natural conditions either.

MacArthur (1957, 1960) has deduced two other types of distribution (type II and type III) corresponding to different kinds of biological hypotheses. These distributions, unlike type I, do not imply competitive exclusion. So far in nature only type I distributions and a kind of empirical distribution which I shall designate type IV are known. The type IV distribution given by diatoms on slides, in the plankton and in the littoral of Braendegård Sø, as well as by soil arthropods, differs from the type I in having its commonest species commoner and all other species rarer. It could be explained as due to heterogeneous diversity, for if the biotope consisted of patches in each one of which the ratio of species to individuals differed, then the sum of the assemblages gives such a curve. This is essentially the same as Hairston's (1959) idea of a more structured community in the case of soil arthropods than in that of birds. It could probably arise if the environment changed in favoring temporarily a particular species at the expense of other species before equilibrium is achieved. This is, in fact, a sort of temporal analogue to

[1]A biotope is said to be *homogeneously diverse* relative to a group of organisms if the elements of the environmental mosaic relevant to the organism are small compared to the mean range of the organisms. A *heterogeneously diverse biotope* is divided into elements at least some of which are large compared to the ranges of the organisms. An area of woodland is homogeneously diverse relative to most birds, a large tract of stands of woodland in open country is heterogeneously diverse (Hutchinson, 1957, 1959).

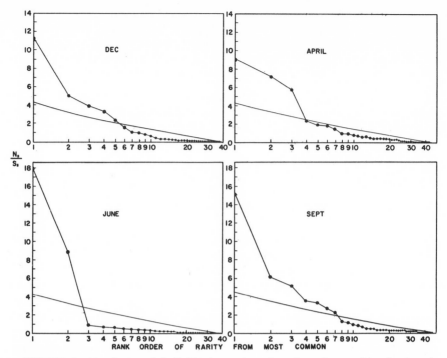

FIGURE 1. Abundance of individual species plotted against rank order for the planktonic diatoms of Braendegård Sø, for the four seasons, from Foged's data, showing type IV distributions. The unmarked line gives the type I distribution for a like number of species and individuals. The unit of population for each species is the ratio of total number of individuals (N_s) to total number of species (S_s).

heterogeneous diversity. Existence of the type IV distribution does not necessarily imply non-equilibrium, but if we assume niches are separated out of the niche-hyperspace with any boundary as probable as any other, we may conclude that either non-equilibrium in time or unexpected diversity in space are likely to underlie this type of distribution.

APPLICATION TO THE PLANKTON

Before proceeding to inquire how far plankton associations are either never in equilibrium in time or approach heterogeneous diversity in space in a rather subtle way, it is desirable to inquire how far ordinary homogeneous niche diversification may be involved. The presence of a light gradient in all epigean waters by day does imply a certain diversification, but in the epilimnia of lakes the chances of any organism remaining permanently in a particular narrow range of intensities is small in turbulent water. By day the stability of the epilimnion may well never be zero, but since what has to be explained is the presence of many species of competitors in a small volume of water, the role of small vertical variations is probably insignificant. A few organisms may be favored by peculiar chemical conditions at the surface film, but again this hardly seems an adequate explanation. The Langmuir spirals in the wind drift might also separate motile from non-motile forms or organisms of different densities to some extent but again the effect is likely to be small and transitory. It is hard to believe that in turbulent open water many physical opportunities for niche diversification exist.

SYMBIOSIS AND COMMENSALISM

The mathematical theory of competition permits the treatment of commensal and symbiotic relations by a simple change in sign of one or both of the competition functions. It can be shown (Gause and Witt, 1935) that under some conditions commensal or symbiotic species can occupy the same niche. There is a little evidence that occasionally water in which one alga has been growing may be stimulatory to another species (Lefevre, Jacob and Nisbet, 1952; see also Hartman, 1960) though it is far more likely to be inhibitory. Since some phytoplankters require vitamins and others do not, a more generally efficient species, requiring vitamins produced in excess by an otherwise less efficient species not requiring such compounds, can produce a mixed equilibrium population. It is reasonably certain that this type of situation occurs in the phytoplankton. It is interesting to note that many vitamin-requiring algae are small and that the groups characteristically needing them (Euglenophyta, Crytophyceae, Chrysophyceae, and Dinophyceae) tend to be motile. The motility would give such organisms an advantage in meeting rare nutrient molecules, inorganic or organic. This type of advantage can be obtained by non-motile forms only by sinking in a turbulent medium (Munk and Riley, 1952) which is much more dangerous than even random swimming.

ROLE OF PREDATION

It can be shown theoretically, as Dr. MacArthur and I have developed in conversation, that if one of two competing species is limited by a predator, while the other is either not so limited or is fed on by a different predator, co-existence of the two prey species may in some cases be possible. This should permit some diversification of both prey and predator in a homogeneous habit.

RESULTS OF NON-EQUILIBRIUM CONDITIONS

The possibility of synergistic phenomena on the one hand and of specific predation on the other would probably permit the development of a somewhat diversified equilibrium plankton even in an environment that was essentially boundaryless and isotropic. It may, however, be doubted that such phenomena would ever permit assemblages of the order of magnitude of tens of species to co-occur. At least in homogeneous water in the open ocean there would seem to be no other alternative to a non-equilibrium, or as MacArthur (1960) would term it, an opportunistic community.

The great difficulty inherent in the opportunistic hypothesis is that since, if many species are present in a really variable environment which is controlling their competition, chance extinction is likely to be an important aspect of the process.[2] That this is not an important aspect of the problem, at least in some cases, is shown by the continual presence of certain dominant species of planktonic diatoms as microfossils in sediments laid down under fairly uniform conditions over periods of centuries or millenia. This is, for instance, clear from Patrick's (1943) study of the diatoms of Linsley Pond, in which locality *Stephanodiscus astrea*, *Melosira ambigua* and certain species of Cyclotella must have co-occurred commonly for long periods of time. It is always possible to suppose that the persistent species were continually reintroduced from outside whenever they became extinct locally, but this does not seem a reasonable explanation of the observed regularity.

IS THE PHYTOPLANKTON A VALID CONCEPT?

In view of the paradoxical nature of the phytoplankton, perhaps it is justifiable to inquire to what extent the concept itself has validity. In the ocean it is reasonably certain that the community is a self-perpetuating one, but in lakes it has long been regarded as largely an evolutionary derivative of the littoral benthos (for example, Wesenberg-Lund, 1908, pp. 323-325) and in recent years much evidence has accumulated to suggest that the derivation in some cases is not an evolutionary process in the ordinary sense of the word, but a process occurring annually, some individuals of a benthic flora moving at times into plankton. The remarkable work of Lund (1954, 1955) on Melosira indicates that the planktonic species of this genus become benthic, though probably in a non-reproductive condition, when turbulence is inadequate to keep them afloat. Brook (1959) believes that some of the supposed planktonic varieties of littoral-benthic desmids are non-genetic modifications exhibited by populations annually derived from the littoral. If most of the phytoplankton consisted of species with well-defined, if somewhat restricted, benthic littoral niches, from which at times large cultures in the open water were developed but perhaps left no descendants, much of our paradox would disappear. In the sea we should still apparently have to rely on synergism, predation and opportunism or failure to achieve equilibrium, but in fresh waters we might get still more diversity from transitory invasions of species which in the benthos probably occupy a heterogeneously diverse biotope like the soil fauna studied by Hairston (1959).

---

[2]The chance of extinction is always finite even in the absence of competition, but for the kind of population under consideration the arguments adduced, for instance, by Cole (1960) appear to the writer to be unrealistic. In a lake of area 1 km^2 or 10^6 m^2, in a layer of water only one meter deep, any organism present at a concentration of one individual per litre, which would be almost undetectibly rare to the planktologist using ordinary methods, would have a population N_0 of 10^9 individuals. If the individuals divided and the two fission products had equal chances of death or reproduction, so that in the expected case the population remained stable, the probability of random extinction (Skellam, 1955) is given by $p_e = [t/(1 + t)]^{N_0}$ where t is measured in generations. For large values of N_0 and t we may approximate by $t = -N_0/\ln p_e$. In the lake in question p_e would reach a value of 0.01 in 2.2×10^8 generations which for most phytoplankters would be a period of over a million years. Less than half a dozen lakes are as old as this, and all these are vastly larger than the hypothetical lake of area 1 km^2.

The available data appear to indicate that in a given lake district there is no correlation between the area of a lake and the number of species comprising its phytoplankton. This is apparent from Järnefelt's (1956) monumental study of the lakes of Finland, and also from Ruttner's (1952) fifteen Indonesian lakes. In the latter case, the correlation coefficient of the logarithm of the numbers of phytoplankton species on the logarithm of the area (the appropriate quantities to use in such a case), is -0.019, obviously not significantly different from zero.

It is obvious that something is happening in such cases that is quite different from the phenomena of species distribution of terrestrial animals on small islands, so illuminatingly discussed by Dr. E. O. Wilson in another contribution to this symposium. At first sight the apparent independence indicated in the limnological data also may appear not to be in accord with the position taken in the present contribution. If, however, we may suppose that the influence of the littoral on the species composition decreases as the area of the lake increases, while the diversity of the littoral flora that might appear in the plankton increases as the length of the littoral, and so its chances of diversification, increases, then we might expect much less effect of area than would initially appear reasonable. The lack of an observed relationship is, therefore, not at all inconsistent with the point of view here developed.

CONCLUSION

Apart from providing a few thoughts on what is to me a fascinating, if somewhat specialized subject, my main purpose has been to show how a certain theory, namely, that of competitive exclusion, can be used to examine a situation where its main conclusions seem to be empirically false. Just because the theory is analytically true and in a certain sense tautological, we can trust it in the work of trying to find out what has happened to cause its empirical falsification. It is, of course, possible that some people with greater insight might have seen further into the problem of the plankton without the theory that I have with it, but for the moment I am content that its use has demonstrated possible ways of looking at the problem and, I hope, of presenting that problem to you.

LITERATURE CITED

Andrewartha, H. G., and L. C. Birch, 1954, The distribution and abundance of animals. XV. 782 pp. Univ. of Chicago Press, Chicago, Ill.

Brook, A. H., 1959, The status of desmids in the plankton and the determination of phytoplankton quotients. J. Ecol. 47: 429–445.

Cole, L. C., 1960, Competitive exclusion. Science 132: 348.

Crombie, A. C., 1947, Interspecific competition. J. Animal Ecol. 16: 44–73.

Feller, W., 1939, Die Grundlagen der Volterraschen Theorie des Kampfes von Dasein in Wahrscheirlichkeitstheoretischer Behandlung. Acta Biotheoret. 5: 11–40.

Foged, N., 1954, On the diatom flora of some Funen Lakes. Folia Limnol. Scand. 5: 1–75.

Gause, G. F., 1934, The struggle for existence. IX. 163 pp. Williams and Wilkins, Baltimore, Md.

——— 1935, Vérifications experimentales de la théorie mathematique de la lutte pour la vie. Actual. scient. indust. 277: 1–62.

Gause, G. F., and A. A. Witt, 1935, Behavior of mixed populations and the problem of natural selections. Amer. Nat. 69: 596–609.

Grinnell, J., 1904, The origin and distribution of the chestnut-backed Chickadee. Auk 21: 364–382.

Hairston, N. G., 1959, Species abundance and community organization. Ecology 40: 404–416.

Hardin, G., 1960, The competitive exclusion principle. Science 131: 1292–1298.

Hartman, R. T., 1960, Algae and metabolites of natural waters. In The Pymatuning symposia in ecology: the ecology of algae. Pymatuning Laboratory of Field Biology, Univ. Pittsburgh Special Publ. No. 2: 38–55.

Hutchinson, G. E., 1941, Ecological aspects of succession in natural populations. Amer. Nat. 75: 406–418.

——— 1953, The concept of pattern in ecology. Proc. Acad. Nat. Sci. Phila. 105: 1–12.

——— 1957, Concluding remarks. Cold Spring Harbor Symp. Quant. Biol. 22: 415–427.

——— 1959, Il concetto moderno di nicchia ecologica. Mem. Ist. Ital. Idrobiol. 11: 9–22.

Järnefelt, H., 1956, Zur Limnologie einiger Gewässer Finnlands. XVI. Mit besonderer Berücksichtigung des Planktons. Ann. Zool. Soc. "Vancimo" 17(1): 1–201.

Kohn, A. J., 1959, The ecology of Conus in Hawaii. Ecol. Monogr. 29: 47–90.

——— 1960, Ecological notes on Conus (Mollusca: Gastropoda) in the Trincomalee region of Ceylon. Ann. Mag. Nat. Hist. 13(2): 309–320.

Lack, D., 1954, The natural regulation of animal numbers. VIII. 343 pp. Clarendon Press, Oxford, England.

Lefèvre, M., H. Jakob and M. Nisbet, 1952, Auto- et heteroantagonisme chez les algues d'eau dounce. Ann. Stat. Centr. Hydrobiol. Appl. 4: 5–197.

Lund, J. W. G., 1954, The seasonal cycle of the plankton diatom, Melosira italica (Ehr.) Kütz. subsp. subarctica O. Müll. J. Ecol. 42: 151–179.

——— 1955, Further observations on the seasonal cycle of Melosira italica (Ehr.) Kütz. subsp. subarctica O. Müll. J. Ecol. 43: 90–102.

MacArthur, R. H., 1957, On the relative abundance of bird species. Proc. Natl. Acad. Sci. 45: 293–295.

——— 1960, On the relative abundance of species. Amer. Nat. 94: 25–36.

Munk, W. H., and G. A. Riley, 1952, Absorption of nutrients by aquatic plants. J. Mar. Research 11: 215–240.

Nicholson, A. J., 1933, The balance of animal populations. J. Animal Ecol. 2(suppl.): 132–178.

——— 1957, The self-adjustment of populations to change. Cold Spring Harbor Symp. Quant. Biol. 22: 153–173.

Patrick, R., 1943, The diatoms of Linsley Pond, Connecticut. Proc. Acad. Nat. Sci. Phila. 95: 53–110.

Provasoli, L., and I. J. Pintner, 1960, Artificial media for fresh-water algae: problems and suggestions. *In* The Pymatuning symposia in ecology: the ecology of algae. Pymatuning Laboratory of Field Biology, Univ. Pittsburgh Special Publ. No. 2: 84–96.

Rand, A. L., 1952, Secondary sexual characters and ecological competition. Fieldiana Zool. 34: 65–70.

Ruttner, F., 1952, Plankton studien der Deutschen Limnologischen Sunda-Expedition. Arch. Hydrobiol. Suppl. 21: 1–274.

Skellam, J. G., 1955, The mathematical approach to population dynamics. *In* The numbers of man and animals, ed. by J. G. Cragg and N. W. Pirie. Pp. 31–46. Oliver and Boyd (for the Institute of Biology), Edinburgh, Scotland.

Udvardy, M. F. D., 1959, Notes on the ecological concepts of habitat, biotope and niche. Ecology 40: 725–728.

Wesenberg-Lund, C., 1908, Plankton investigations of the Danish lakes. General part: the Baltic freshwater plankton, its origin and variation. 389 pp. Gyldendalske Boghandel, Copenhagen, Denmark.

Part IV

NICHE AND HABITAT DIMENSIONS

Editors' Comments
on Papers 19 Through 30

NICHE VERSUS HABITAT

The preceding section developed the concept of niche axes as directions of differentiation among the species of a community. We concern ourselves now with axes we regard as the most critical ones for the relationships among species. Measurements of species' responses to these axes describe species' *niche dimensions;* and similar measurements of species' responses to critical habitat axes describe species' habitat dimensions.

What are the relevant, quantitative questions to ask about niches? To begin, some measure of the population's response to environment is necessary. Different measures serve different purposes. The most familiar population measure for relating species to habitat axes is density (number of individuals per unit area); but this is a static measure, giving only part of the information. Full information on the species response to environment includes its behavior and population dynamics, and we may use measures such as fertility, growth rate, and frequency of different behaviors, among others. For example, we can relate birds to the vertical height and food-size axes in a forest by the relative frequency of prey captured at different heights, or the relative frequency of capture of prey of different sizes.

Having made a choice of population measure, the following questions suggest themselves:

1. What is the form of the population response?
2. What are the relative widths along the axis for different species?
3. How much do the species overlap?
4. What are the relative positions of the species along the axis?

We need to pursue our questions further in two directions. Although it is true that some gradients cannot be neatly classified as uniquely intercommunity (extensive) or intracommunity (intensive), many can. The above four questions can be asked about species' responses to both kinds of factors, although the choice of response measure will likely differ. Study of species responses to extensive (habitat) factors developed first historically, and we consequently treat this problem first; but parallel questions apply to niche analysis.

It will be apparent that species are relating to many different axes at once, some habitat, some niche, some a little of each. We shall arbitrarily include the latter axes among the niche axes. Then the n niche axes by which species in a given community are related define an n-dimensional abstract space that we term the "niche hyperspace," in which each species has its position. The m habitat axes relating species to extensive environmental factors in an area form an m-dimensional abstract space, the habitat hyperspace, in which each species has its place. Our four questions can thus be asked both about one-dimensional relations of species along a single axis, and multidimensional relations in a "space" of many axes. The multidimensional concept of habitat relations developed first; it was formulated by Ramensky (1924, 1930) and became the basis of the extensive development of gradient analysis in plant ecology (Whittaker, 1967, see Paper 21, 1973; McIntosh, 1967). The multidimensional concept of niche was one of the great contributions of G. E. Hutchinson and was formulated in a major article given as one of the last three in this volume. That article has often been misinterpreted on the assumption that the environmental factors Hutchinson was describing as niche axes were habitat factors. As the introduction to the study of niche and habitat dimensions we give a shorter passage in which Hutchinson develops the multidimensional concept of niche, and with this distinguishes between intensive factors (those relating species within a given biotope or community, and hence characterizing niches) and extensive factors (those relating species along gradients connecting different biotopes and communities, hence characterizing habitats).

REFERENCES

McIntosh, R. P. 1967. An index of diversity and the relations of certain concepts to diversity. *Ecology* **48**: 392–404.

Ramensky, L. G. 1924. Die Grundgesetsmässigkeiten im Aufbau der Vegetationsdecke (in Russian). *Vêstnik Opŷtnogo Dêla, Voronezh,* pp. 37–73. [Abstract in *Bot. Centralblatt* (n.f.) 7: 453–455, 1926.]

Ramensky, L. G. 1930. Zur Methodik der vergleichenden Bearbeitung und Ordunng von Pflanzenlisten und anderen Objekten, die durch mehrere, verschiedenartig wirkende Faktoren bestimmt werden. *Beitr. Biol. Pflanz.* **18**: 269–304.

Whittaker, R. H. 1967. Gradient analysis of vegetation. *Biol. Rev.* **42**: 207–264.
Whittaker, R. H. (ed.). 1973. Ordination and Classification of Communities, Vol. 5
of *Handbook of Vegetation Science.* Junk, The Hague.

HABITAT DIMENSIONS

For one of the significant early analyses of habitat we can return to
Joseph Grinnell. In "Animal Life in the Yosemite," by Grinnell and
Storer, the distributions of some 270 species of birds, mammals, rep-
tiles, and amphibians were diagrammed in relation to elevation in the
Sierra Nevada, California. Grinnell and Storer were interested in the relations of species to elevation belts as life zones in the sense of Merriam
(1898, 1899). They were, however, too good as scientists to make their
data fit the life zones. Study of their diagram is revealing for (1) the
manner in which they used tapered figures, suggesting that a species
population has a center along a habitat gradient, and on each side of
this center decreases to limits that are tapering, not abrupt, and (2) the
fact that the centers and limits of species appear to be located at random
along the elevation gradient. The latter relates to the "principle of
species individuality" stated by Ramensky (1924) and Gleason (1926):
Each species is distributed according to its own genetic, physiological,
and population characteristics and its own way of relating to environmental factors (including other species); hence no two species are distributed alike.

The forms of species distributions along habitat gradients, suggesting
Gaussian curves, and the individuality of species are illustrated in a study
by Pennak of copepods in relation to tide levels on sandy beaches (Paper
20). Similar results were obtained for insect species in relation to elevation and moisture gradients by Whittaker (1952), and in a number of
studies of plant populations along environmental gradients (Whittaker,
1951, 1956, 1960; Curtis and McIntosh, 1951; Brown and Curtis, 1952;
Curtis, 1959). From work with plant populations, Fig. 8 of Paper 21
illustrates the bell-shaped, apparently Gaussian curves, and the scattering or "individuality" of the species distributions, along the topographic
moisture gradient in two mountain ranges.

Some species have broader distributions than others in Fig. 8, and
some are bimodal (having two population peaks along the gradient; see
curve *C*). The forms of the curves suggest that relative habitat width of
different species should be measured as dispersions—standard deviations,
say, expressing the relative spread of the population on each side of its

mode or center. McNaughton and Wolf (1970) used standard deviations in this way, as does Fig. 1 in Paper 41. Whittaker (1956, 1960) observed the correlation of wide habitat ranges and bimodal distributions with morphological differentiation in plant populations; it is natural to assume that genetic differentiation within a species contributes to determining its habitat width.

Measures expressing species overlap have been used much more widely. There is an extensive literature on association and correlation of species in sets of samples (Goodall, 1952, 1973; Whittaker, 1967; Dagnelie, 1960). "Association" can be used for measures of how similar two species are in their occurrence in a set of samples, considering their presence and absence only; percentage coocurrence, in the selection that follows, is one among many such measures. "Correlation" of species refers here to the degree to which they are alike in quantitative representation in a set of samples; percentage similarity of distribution, in the following selection, is perhaps the simplest expression of this kind. With such measures we are not asking the probability that two species are alike in distribution. Given the principle of species individuality, we are interested only in *degrees* of overlap and relative similarity, versus dissimilarity of distribution. Papers 22 and 23 both proceed from discussion of distributional measurement (question 3) to use of such measurement for the arrangement of species in relation to one another (question 4).

With these selections we have already moved from single axes to arrangements of species in relation to multiple axes. The arrangement of species, or samples, or other ecological entities in relation to one or more environmental gradients or axes is termed *ordination* (Ramensky, 1930; Goodall, 1954; Whittaker, 1967). From results of gradient analysis using ordinations on two or more axes, we can approach our four questions in a multidimensional context:

1. The form of population responses: In two dimensions a bell-shaped curve becomes a bell-shaped surface, or a population hill or ridge. A population response surface in relation to two habitat gradients is illustrated in Fig. 2, Paper 41, the final article in this volume. The response is more difficult to visualize in three or more dimensions. It can be conceived, however, as a population cloud with a center of maximum density in the multidimensional space, and with density tapering (according to Gaussian curves) in all directions away from that center. Such can be termed an "atmospheric" distribution (Bray and Curtis, 1957).

2. Relative breadth of species distributions: The standard deviation is the natural expression of habitat width along one axis, but it is difficult to apply to several axes when the relations of these axes to one

another may be quite unclear. A change of approach is possible, given a set of samples representing communities from a range of different environments affected by more than one habitat gradient. The percentages of samples in which two species occur express their relative widths of occurrence—but only over the particular range of environments included in those samples. Different species can thus be compared within the same sample set; but they cannot (without special precautions on sample choice) be compared between different sample sets. One may also wish to consider relative evenness of occurrence in the various samples as an aspect of habitat breadth. A species that occurred in all 100 samples with the same number of individuals would seem to have a "broader" habitat than one that had most of its individuals concentrated in a single sample, and one or a few individuals in each of the other samples. We can use the Simpson (1949) index, $C = \Sigma p^2$ as a simple expression of the relative concentration of a population in one or a few samples of the set (p is a fraction, the number of individuals of a species in a given sample divided by the total for that species in the sample set). The reciprocal of the Simpson index, $1/C$, is then an expression of breadth of occurrence in the sample set. It can be interpreted as expressing an equivalent number of equally common occurrences— that is, samples with the same number of individuals of the species.

If we want one measurement to express both the number or proportion of samples a species occurs in, and its relative evenness in these, then there is no simple answer on how these questions should be weighted when they are combined in one measure. Both the reciprocal of the Simpson index and the Shannon-Wiener index, $H' = -\Sigma p \log p$, are possibilities. Both these are measures used in the study of species diversity; in that use p is the fraction of the individuals in a given sample that belong to a given species. For use in characterizing habitat breadth we have turned these measures around; instead of computing p for the different species in a single sample (for a diversity measure), we compute p for a single species in different samples. The logarithmic basis of the Shannon-Wiener index damps its expression of increased range of samples; if the number of samples in which a species occurs is doubled, the H' value will not double but will increase by a much smaller fraction. The antilog of H' can also be used, however, as an equivalent number of equally common occurrences. Both $1/C$ and the antilog of H' can be divided by the number of samples in the set to give measures of relative breadth that have 1.0 as their upper limit. It should be clear that either of these measures will be strongly affected by the manner in which the set of samples has been chosen, including the size of the set and the range of communities included and the possible bias in representation of particular communities. Problems of breadth measurement as applied to niches are discussed further in selections by Levins (Paper 28) and Colwell and Futuyma (Paper 29) in the second half of this section.

Measures of habitat breadth have given rather few research rewards so far. A principal observation is that species differ widely in habitat breadth; but this is little more than a kind of corollary of the principle of species individuality. There has been extensive study of genetic characteristics of species, including reproductive systems and ecotypic differentiation, that relate to habitat breadth. It is not clear that there is any real correlation of habitat width or breadth with relative importance, or dominance, of species in communities (Gauch and Whittaker, 1972). There may be some correlations of habitat breadth with environment. Islands tend to have fewer species, and those species broader habitats, than equivalent areas on the continent. Some very distinctive environments (such as salt lakes and serpentine soils) have many species narrowly restricted to such habitats. There is some evidence, however, that along climatic gradients from the lowland tropics to high latitudes and altitudes, average habitat breadths of species may increase. Paper 24 by Terborgh considers the question.

3. Species overlap: Cases are observed (Beauchamp and Ullyott, 1932; Hairston, 1951; Connell, 1961; Jaeger, 1974) in which animals that are close competitors, with closely similar niches, exclude one another at boundaries where they meet. Their populations virtually non-overlap. Such cases seem uncommon among animals and rare among plants, for which Fig. 8 is typical (see, however, Barber, 1965). We should conceive the habitat hyperspace as being not compartmented among species excluding one another at boundaries, but as containing population clouds of different breadths and degrees of overlap with one another, The measures of overlap already discussed are appropriate for more than one axis, and will be discussed further below.

4. Relative positions of species: De Vries' "constellation" (Paper 23) represents one of the significant findings of gradient analysis: species populations are dispersed in habitat hyperspace (Whittaker, 1956. 1967). Although they may be adapted to one another when they occur together, species do not evolve to form distinct groups with closely similar distributions, groups that would appear as clusters in habitat hyperspace. As indicated in the paragraph from Whittaker (Paper 21) already given, evolution toward difference in habitat, by which competition is reduced, is an extension of the principle of species individuality and relates this to the principle of Gause. Paper 25 by James gives effective illustrations of the dispersion of species positions in habitat hyperspace.

We conclude our discussion of habitats with two recent articles on birds, one concerned with their relations to altitude as a single major habitat axis, and the other with multidimensional treatment. Terborgh (Paper 24) sought, with the results of gradient analysis of plants in mind, the extent to which bird populations fit the same patterns. Much of the study concerns "ecological amplitudes," as he terms habitat widths along the elevation gradient. A result of much interest is the observation that birds of higher elevations, where the number of species is smaller, have wider habitat ranges (see Fig. 9 and pp. 34ff.). Of interest also are the comments on habitat overlap. Ecotones (boundaries between different kinds of communities) account for less than 20 percent of distributional limits; approximately one-third of all limits are ascribed to competitive exclusion (p. 36). Terborgh considers the latter probably an underestimate; overestimate would also be possible, for observations of distributions along gradients give no clear indication of the extent of competition effects on these distributions. Paper 26 by May and MacArthur later in this section bears on these interpretations, for it suggests that competing species may limit one another with overlapping bell-shaped distributions, without forming any recognizable boundary between their populations. Terborgh's observations on the controls of distributions are thus not conclusive, although the questions are important ones. His conclusion (p. 32) that, with a single exception, faunal overturn is continuous with elevation is consistent with a second principle stated by Ramensky (1924) and with other research in gradient analysis: Along continuous habitat gradients undisturbed natural communities mostly intergrade continuously (Ramensky, 1924; Gleason, 1926; Whittaker, 1967; McIntosh, 1967; Beals, 1969). A species then occurs over a range of habitats, and the ecotope becomes the correct level from which to view even localized exclusions. This problem is dealt with in Levin (1974). Terborgh also discusses the manner in which sample similarity (or "faunal congruity" as he terms it) changes along the elevation gradient; this question has recently been analyzed by Gauch (1973).

The study of bird habitats by Frances James (Paper 25) is of special interest for the manner in which it links together habitat ordination and implications of vegetation structure for bird niches (as developed in the second half of this section). James characterizes the habitat requirements of birds as structural "niche gestalts" (pp. 217–221), an idea that seems a direct descendant of Grinnell's original niche concept. Ten vegetational variables are subjected to a principal-components analysis to derive abstract axes defining a habitat hyperspace (pp. 221–225). Bird species can then be located, or ordinated, in this habitat hyperspace (Fig. 7, p. 226). A second ordination is based on discriminant functions (omitted in our reproduction of the article). Principal components analysis is no longer considered the best technique of ordination for ecological purposes, as suggested on page 234. Given curvilinear and nonmonotonic relations among species, such as illustrated in Paper 21, Fig. 8, principal-components analysis based on species representation in samples gives distorted ordination of samples (Austin and Noy-Meir, 1972; Gauch and Whittaker, 1972; Beals, 1973). The wider the range of habitats represented, the greater the distortion and the less interpretable the axes may be (Jeglum et al., 1971; Gauch and Whittaker, 1972). James' application, however, is an appropriate one, for she is dealing not with species representation but with vegetational characteristics as monotonic variables. The result (Fig. 7) shows how bird species, with their different "niche-gestalts," are arranged in a habitat hyperspace.

REFERENCES

Austin, M. P., and I. Noy-Meir. 1972. The problem of non-linearity in ordination: experiments with two-gradient models. *J. Ecol.* **59**: 763–774.

Barber, H. N. 1965. Selection in natural populations. *Heredity* **20**: 551–572.

Beals, E. W. 1969. Vegetational change along altitudinal gradients. *Science* **165**: 981–985.

Beals, E. W. 1973. Ordination: mathematical elegance and ecological naiveté. *J. Ecol.* **61**: 23–35.

Beauchamp, R. S. A., and P. Ullyott. 1932. Competitive relationships between certain species of fresh-water triclads. *J. Ecol.* **20**: 200–208.

Bray, J. R., and J. T. Curtis. 1957. An ordination of the upland forest communities of southern Wisconsin. *Ecol. Monogr.* **27**: 325–349.

Brown, R. T., and J. T. Curtis. 1952. The upland conifer-hardwood forests of northern Wisconsin. *Ecol. Monogr.* **22**: 217–234.

Connell, J. H. 1961. The influence of interspecific competition and other factors on the distribution of the barnacle *Chthalamus stellatus. Ecology* **42**: 710–723.

Curtis, J. T. 1959. *The Vegetation of Wisconsin: An Ordination of Plant Communities.* University of Wisconsin, Madison, Wis. 657 pp.

Curtis, J. T., and R. P. McIntosh. 1951. An upland forest continuum in the prairie-forest border region of Wisconsin. *Ecology* **32**: 476–496.

Dagnelie, P. 1960. Contribution à l'étude des communautés vegetales par l'analyse factorielle (English summary). *Bull. Serv. Carte Phytogéogr.* **B5**: 7–71, 93–115.

Gauch, H. G., Jr. 1973. The relationship between sample similarity and ecological distance. *Ecology* **54**: 618–622.

Gauch, H. G., Jr., and R. H. Whittaker. 1972. Comparison of ordination techniques. *Ecology* **53**: 868–875.

Gleason, H. A. 1926. The individualistic concept of the plant association. *Bull. Torrey Bot. Club* **53**: 7–26.

Goodall, D. W. 1952. Quantitative aspects of plant distribution. *Cambridge Phil. Soc., Biol. Rev.* **27**: 194–245.

Goodall, D. W. 1954. Objective methods for the classification of vegetation. III. An essay in the use of factor analysis. *Austral. J. Bot.* **2**: 304–324.

Goodall, D. W. 1973. Sample similarity and species correlation, pp. 105–156. In R. H. Whittaker (ed.), Ordination and Classification of Communities, Vol. 5 of *Handbook of Vegetation Science.* Junk, The Hague.

Grinnell, J., and T. I. Storer. 1924. *Animal life in the Yosemite*. University of California Press, Berkeley, Calif. 752 pp.

Hairston, N. G. 1951. Interspecies competition and its probable influence upon the vertical distribution of Appalachian salamanders of the genus *Plethodon*. *Ecology* **32**: 266–274.

Jaeger, R. G. 1974. Competitive exclusion: comments on survival and extinction of species. *BioScience* **24**: 33–39.

Jeglum, J. K., C. F. Wehrhahn, and J. M. A. Swan. 1971. Comparisons of environmental ordinations with principal component vegetational ordinations for sets of data having different degrees of complexity (French summary). *Can. J. For. Res.* **1**: 99–112.

Levin, S. A. 1974. Dispersion and population interactions. *Amer. Naturalist* **108**: 207–228.

McNaughton, S. J., and L. L. Wolf, 1970. Dominance and the niche in ecological systems. *Science* **167**: 131–139.

McIntosh, R. P. 1967. An index of diversity and the relations of certain concepts to diversity. *Ecology* **48**: 392–404.

Merriam, C. H. 1898. Life zones and crop zones of the United States. *Bull. U.S. Biol. Surv.* **10**: 1–79.

Merriam, C. H. 1899. Results of a biological survey of Mount Shasta, California. *North Amer. Fauna* **16**: 1–179.

Ramensky, L. G. 1924. Die Grundgesetsmässigkeiten im Aufbau der Vegetationsdecke (in Russian). *Vêstnik Opŷtnogo Dêla, Voronezh*, pp. 37–73. [Abstract in *Bot. Centralblatt* (n.f.) **7**: 453–455, 1926.]

Ramensky, L. G. 1930. Zur Methodik der vergleichenden Bearbeitung und Ordnung von Pflanzenlisten und anderen Objekten, die durch mehrere, verschiedenartig wirkende Faktoren bestimmt werden. *Beitr. Biol. Pflanz.* **18**: 269–304.

Whittaker, R. H. 1951. A criticism of the plant association and climatic climax concepts. *Northwest Sci.* **25**: 17–31.

Whittaker, R. H. 1952. A study of summer foliage insect communities in the Great Smoky Mountains. *Ecol. Monogr.* **22**: 1–44.

Whittaker, R. H. 1956. Vegetation of the Great Smoky Mountains. *Ecol. Monogr.* **26**: 1–80.

Whittaker, R. H. 1960. Vegetation of the Siskiyou Mountains, Oregon and California. *Ecol. Monogr.* **30**: 279–338.

Whittaker, R. H. 1967. Gradient analysis of vegetation. *Biol. Rev.* **42**: 207–264.

NICHE DIMENSIONS

Habitat axes, as environmental variables with spatial, intercommunity extension, seem to us relatively clear as concepts, and they are limited in number. Although the axes and the habitat space they form are abstract concepts, our questions of habitat dimensions have fairly definite geometric meaning. Niche axes, in contrast, are less clear as such and are essentially innumerable. There is no great difficulty in regarding vertical height above the ground and depth in the soil as axes, or in treating diurnal and seasonal time as axes in relation to which species behave differently. If a community has a microtopography of small rises and depressions that repeat themselves every 2 or 3 meters, this microtopographic gradient can be treated as an axis of intracommunity pattern. The more we go beyond these, however, to more purely biological niche relations, the more difficult axis definition becomes. Biologically defined axes predominate, for niche evolution is toward refinement in biological relationships.

Prey sizes form an easily conceived niche axis. But what about the subtleties of differing prey and predator behavior that become directions of differentiation among these? What about plant and prey chemistry? We believe that most or all plants, and many animals, are at least partially protected against consumption by chemicals that make them relatively unpalatable and in some cases poisonous. The range of kinds of chemicals is large, and different consumers are adapted to tolerate (or inactivate) different protective chemicals and feed on different plants or animal prey. There are thus many directions of differentiation in chemical protection among plants and prey animals that become directions of niche differentiation for the animals consuming them. We can determine something about the occurrence of protective chemicals, and who eats whom, but we cannot really make a set of niche axes out of what we know.

Also how are we to treat the many limiting factors that affect species populations? We consider limiting factors to be the most essential ways niches differ (see Paper 8 by Levin); but these factors are diverse, interrelated in various ways as they affect particular species, and difficult to recognize in field research. What, in fact, is a factor and how finely can it be divided? Need a single prey species or a single resource necessarily constitute a single factor, or can these be effectively subdivided in a stable community without violating the principle of competitive exclusion? How different need factors be to be different and independent? It seems clear that a single prey species need not correspond to a single limiting factor for predators. If eleven different prey species are being controlled by seven different predators (plus interactive effects of food limitation and palatability, availability of shelter, and behavioral difference), how are we to make one or more niche axes out of these population controls? At a more sophisticated level, species will differ in responses to several different kinds of interactive gradients with a number of kinds of other species, often including both competitors and predators as well as prey or resources. This is multidimensional interaction, indeed; but it is not easy to make a coordinate system, with a definite number of axes, in which species responses can be measured and arranged as in our discussion of habitat dimensions.

These comments do not diminish the significance of Hutchinson's

(Papers 19 and 39) multidimensional concept of the niche. The conceptual and theoretical value of his insight remains, and it makes possible effective research in favorable cases. We can measure niche dimensions along some of those axes that are most distinct. It is well to recognize, though, how small a part of the wealth of niche relationships in communities has become accessible to such research. The limitations on what we can do may suggest also a particular effort toward maximum clarity in our questions about niches, and caution on the kinds of measures we apply and the interpretations we derive.

The selections that follow bear on the four questions about dimensions that we have already used to discuss habitats:

1. Form of population distributions: A bell-shaped or Gaussian curve of population response to a niche axis is generally assumed, but need not always apply. As we have observed, for niche characterization we often apply this form to frequency of life-history events or different kinds of behavior rather than to densities of individuals.

2. Niche width and breadth: Gaussian curves imply the standard deviation as an appropriate measure of niche width along a given axis, and this is applied in Papers 10 and 26 in this volume by Root and by May and MacArthur. Treatment of niche breadths in a multidimensional context is difficult, but Levins (Paper 28) and Colwell and Futuyma (Paper 29) discuss the problem in this section.

3. Niche overlap: These selections also discuss niche overlap, using in part measures already discussed for habitat. Many authors (see, for example, Papers 26 and 28 by May and MacArthur and by Levins) use overlap measures and competition coefficients interchangeably; but this may not be justified.

4. Niche arrangement: In principle the concept of ordination—the arrangement of entities in relation to one or more gradients or axes—can apply as well to niche as to habitat axes. We do, in fact, arrange species along particular niche axes—see Root (Paper 10) and Price (1972) as well as May and MacArthur (Paper 26). Given the difficulties of defining niche space, however, multidimensional niche ordinations are considerably more primitive than habitat ordinations. The article by MacArthur et al. (Paper 9) in Part III is an approach to niche ordination, as is the arrangement of desert plants by Whittaker (Paper 15).

5. Niche shape: To these four questions we add a fifth raised by Maguire (1973). Of two hypothetical birds, one feeds on a limited prey size, 5–10 mm in length, at all vertical heights, 1–30 m, in a forest, whereas the other feeds on a wide range of prey, 1–30 mm, but only in the low-tree stratum between 5 and 10 m height. Their niches differ in "shape" or "direction". (We avoid saying whether they are "the same" in breadth.) Both differ from a third hypothetical species with a

"squarer" niche feeding on prey 10–20 mm in length 10–20 m above the ground. Such questions of shape are possible, but we know of no research effectively asking them.

Our first selection on niche dimensions is a recent and significant theoretical study by May and MacArthur (Paper 26). These authors interrelate our first four questions by (1) assuming Gaussian form, for which (2) niche widths measured as standard deviations are (3) related to niche overlap by the ratio of the deviation to distance apart along the axis, with the result that (4) species are found to be arranged in sequence with a spacing approaching deviation/distant = 1. Some of their examples are species along habitat axes, suggesting that the same principle of spacing for species with closely similar requirements applies to both niche and habitat axes. Their study, with its inherent mathematical assumptions, provides only the beginnings of a full theory. But it is a substantial beginning, one that seems to fit well the gross features of observed distributions, and one that is to some extent testable. It remains to be seen how such studies can be extended to multidimensional situations, such as Harrison's (1962) classification of rain-forest birds according to food type and height in canopy; but the qualitative answers should be the same. Through evolutionary time additional species can be fitted into the sequence along a niche axis, or into positions in the niche space, but as they do so the niches of competing species will be narrowed. Finally, as species migration patterns evolve, the proper evolutionary context is not intracommunity, and these ideas may bear extension to multidimensional situations combining niche and habitat axes, hence to species' ecotopes (Paper 41).

There *are* problems with May and MacArthur's approach in Paper 26, and we would be remiss in not pointing them out. In the sense that the work of May and MacArthur is an updating of the earlier work of MacArthur and Levins (1967) and Levins (Paper 28), it is useful to examine the paper with regard to the critical restrictions summarized by Kohn (1971) in his poignant critique of the earlier theory. The restrictions as identified by Kohn (1971) are:

1. "The niche dimensions considered are those that serve to separate species." When separation is less than predicted, the species may be assumed to be sorting out along several dimensions. When separation is greater than predicted, the environment is unsaturated. Thus the theory is to this extent untestable, in the same sense as the competitive exclusion principle. If correct, however, it could provide a tool by which to organize research.

2. "Overlap in resource utilization is equivalent to competition." This is a very serious restriction, unlikely to be justified, as pointed out by Dayton (1973). Anemones and starfish, for example, have very high dietary overlap, but are far from being competitors. Anemones survive largely on the scraps from the starfish's table (Dayton 1973). More-

over even in an equilibrium community, the per-capita competition co-efficients a_{ij} of the Lotka–Volterra theory are unlikely to be symmetric. In many systems the opposite extreme—preemptive competition — is the case.

3. "Usable resources are equally available and rapidly renewed." The implication that displacement along a resource spectrum is without cost cannot be justified.

4. "Species abundances are identical or differences are unimportant." This detail cannot really be removed by a renormalization of units to equate all species levels, because it is the competitive effect that is at issue.

5. "Additional species considered for entrance into the community matrix have the same mean values of a and the same covariance of a_{ij} and a_{ji} as those already present." Since the May and MacArthur work replaces the question of invasibility by one of coexistence, this restriction becomes essentially the one raised earlier of symmetric competition.

To this list, we add a sixth difficulty:

6. Parameters derived locally may not apply to a large collection of local environments. Indeed, classical theory asks only, "who eventually wins?", giving short shrift to the colonist which makes its way from patch to patch. Barnacles in the intertidal lose out in competition to mussels but exist by their ability to recolonize temporarily open habitats. In such cases, species may be packing along a successional gradient; but overlap is not a measure of competition. Indeed, one species may make the environment more desirable for its successors.

There is, moreover, no reason to assume that species distributions are centered at their physiological optima. When excluded from desirable habitats, species may displace their distributions, for example by shifting the dispersal distribution of larvae; but this will be at some cost. Even with evolutionary adaptation, the species is unlikely to recover all that it loses.

How far will a species displace its distribution when faced with competition? Our view is that in the preponderance of cases the eventual displacement within a particular habitat represents essentially a balance between intra- and interspecific competition. Environmental unpredictability sets a limit to tolerable competition, but this upper limit may never be approached in many environments. Another view of niche overlap theory arises from an article by Pianka (Paper 27).

The ideas of May and MacArthur owe much to Richard Levins, who discusses related ideas in the three excerpts from his book, given as Paper 28, as well as to earlier work of Hutchinson [1957 (Paper 39), 1959], MacArthur (1970), and MacArthur and Levins (1967). Especially relevant here are Levins' suggestions on the measurement of niche breadth (pp. 41–45) and overlap (pp. 50–55). In the selection on niche breadth Levins uses for measurement the two indices we have already discussed for habitat breadth—breadth, *B,* is either the reciprocal of the Simpson index (formula 3.2) or the antilog of the Shannon–Wiener index (formula 3.1). It should be noted that Levins is using these for both niche and habitat axes. Seasonal niche occurrence (Table 3.3) and microhabitat and food-choice range in a given community (Table 3.2) represent niche breadth. Distributions of species over tide levels, or altitudes (p. 42), are of course habitat width rather than niche breadth as we define it. The statement (p.44) that the abundant species are usually the ones with broad niches should be regarded with caution. There are two questions here: (1) Are abundant species abundant because they have broad niches, using a broad range of resources, within a given community? (2) Is there a correlation between abundance in a given community, and width of habitat range? We believe that (2) is likely to be much less important than (1), and that the very different question (1) is not yet adequately answered.

Levins continues (pp. 50–55) with a discussion of niche overlap. For this he uses not the simple expressions such as discussed under habitat overlap (see also Paper 29), but the "alpha" competition coefficients as approximated by $a_{ij} = \sum_h p_{ih} p_{jh}/B_j$, in which B_j is niche breadth for species *i.* As May and MacArthur observe, the system stability for a community and hence permissible niche overlap hinge on the matrix of such coefficients. A competition coefficient, a_{ij}, it will be recalled, expresses the effect of species *j* on the population growth of species *i*; the larger the coefficient, the more intense the competition and, presumably, the greater the niche overlap. In principle, if competition coefficients could be determined for a set of interacting species in a community, inferences could be drawn on the niche overlaps of these species and on the organization of the species into an interacting system.

A note of caution on such use is in order. To have real meaning, alpha values must be determined by experiments testing species in competition with one another; calculation of these values from field data is quite another matter (Kohn, 1971; Dayton, 1973). Suppose that population densities are measured in different microhabitats, or in different seasons, in a given community (Paper 28, Table 3.7), The differences in density are sometimes used to calculate alpha values. There is, however, no way of knowing the relative contributions of competition versus environmental fluctuation, changing resource limits, and predation in determing these densities. It is consequently uncertain whether, or to what degree, the alphas as calculated will express competition. Even more perilous would be a calculation of alphas based on species densities in a number of different communities, with these densities affected by population responses such as illustrated in Fig. 8, Paper 21. With these cautions, Levins' treatment (pp. 50–55) is offered nevertheless as an original approach whose influence is clear in many works, including the preceding selection by May and MacArthur (Paper 26). In general, however, the validity of any such approach is tied to the validity of an underlying set of equations, and this caveat must always be borne in mind.

Before the selection, we offer some paragraphs of mathematical comment, leading up to the key mathematical statement (p. 55, 1.1–2): "In order for a community to be stable the determinant of its matrix must be positive." This derives from the theorem, one of the most familiar of mathematical stability theory, that for the system of equations (where the F_i are "nice" functions)

$$\frac{dx_i}{dt} = F_i(x_1, ..., x_n), \quad i = 1, ..., n,$$

an equilibrium $x_1^0, ..., x_n^0$ [defined by $F(x_1^0, ..., x_n^0) = 0$] is stable provided the *eigenvalues* of a certain matrix (the *Jacobian* matrix) all have negative real parts; and is unstable if any has positive real part. What does all this mean, and how does it apply to the present situation?

First, what is this Jacobian matrix? Very simply, the Jacobian matrix for the functions $F_1(x_1, ..., x_n), ..., F_n(x_1, ..., x_n)$ is the $n \times n$ matrix of partial derivatives

$$\begin{pmatrix} \dfrac{\partial F_1}{\partial x_1} & \dfrac{\partial F_1}{\partial x_2}, ..., & \dfrac{\partial F_1}{\partial x_n} \\ \vdots & \vdots & \vdots \\ \dfrac{\partial F_n}{\partial x_1} & \dfrac{\partial F_n}{\partial x_2}, ..., & \dfrac{\partial F_n}{\partial x_n} \end{pmatrix}$$

evaluated at the equilibrium. This matrix is important because it gives the instantaneous rate of change in every relevant function—that is, it contains information on the dependence of each growth rate on each of the variables. These derivatives contain the most important information about a function *near* the equilibrium.

The *eigenvectors* of a matrix **M** are nontrivial vectors which are changed only in magnitude when multiplied by the matrix; i.e., an eigenvector V is such that

$$MV = \lambda V,$$

where λ is a scalar. λ is known as an *eigenvalue*. One of the most familiar eigenvectors in ecological theory is the stable age distribution.

To compute the eigenvalues, one usually rewrites the above equation as

$$(M - \lambda I)\, V = 0,$$

where I is the identity matrix. Such a system can have a nontrivial solution if and only if the determinant,

$$\det(M - \lambda I) = 0;$$

and this determines a polynomial equation for λ.

Now, assume that the equations

$$\frac{dx_i}{dt} = \frac{r_i x_i (K_i - x_i - \Sigma\, a_{ij} x_j)}{K_i}$$

are applicable to the community. Here

$$F_i(x_i, ..., x_n) = \frac{r_i x_i (K_i - x_i - \Sigma\, a_{ij} x_j)}{K_i};$$

and the equilibrium $x_1^0, ..., x_n^0$ is determined as in Levins, by setting $F_i(x_1^0, ..., x_n^0) = 0$. If the coefficients a_{ij} and the parameters r_i and K_i are constant, the quoted theorem is valid. If the parameters vary [as in Paper 26, Paper 28, and Vandermeer (1970, 1972)], according to some probability distribution, a slightly modified approach is required. The criterion for stability is no longer that the eigenvalues simply have negative real parts—rather, they must be bounded away from zero by an amount that depends upon the variance in the environmental parameters (see Paper 26).

Assume now that all parameters are fixed. Applying the above technique, one finds that the Jacobian matrix may be written compactly as

$$
J = -
\underbrace{
\begin{pmatrix}
\dfrac{r_1 x_1}{K_1} & 0 & . & . & . & 0 \\[2ex]
0 & \dfrac{r_2 x_2}{K_2} & & & & . \\[2ex]
. & & & & & \\
. & & & & & \\
. & & & \dfrac{r_n x_n}{K_n} & & \\[2ex]
0 & . & . & . & &
\end{pmatrix}
}_{\Lambda}
\underbrace{
\begin{pmatrix}
1 & a_{12} & a_{13} & . & . & . & a_{1n} \\[1ex]
a_{21} & 1 & a_{23} & . & . & . & a_{2n} \\[1ex]
. & & & & & & \\
. & & & & & & \\
a_{n1} & a_{n2} & a_{n3} & . & . & . & 1
\end{pmatrix}
}_{A}
$$

or, even more simply,

$$J = -\Lambda \cdot A,$$

where Λ is the diagonal matrix above and A is the *community matrix* (Paper 28). It is well known that if an $n \times n$ matrix has all eigenvalues with negative real parts, its determinant has sign equal to $(-1)^n$. But $(-1)^n$ is also the sign of the determinant of $-\Lambda$; and since $\det J = \det(-\Lambda) \cdot \det A$, stability would require that $\det A > 0$, as claimed by Levins. To use this as a necessary and sufficient condition for stability is completely inappropriate, however, and this is an extremely important point which has often been missed or passed over in the literature.

Indeed, as Strobeck (1973) points out, in general the community matrix alone does not contain enough information to determine necessary and sufficient conditions for stability; one must look to the other parameters as well.

Before leaving Levins' work, we give his development of another idea that relates to niche breadth, the concept of *fitness set*, which expresses vectorially the range of phenotypes available in a mixture to two environments in which these phenotypes have different responses and optima (pp. 14–20). Critical to the theory, as developed by Levins, is the convexity or concavity of one aspect of the boundary of the set; and the form of the "adaptive function," which gives the overall fitness in the mixed environment in terms of the fitnesses in the individual environments. The "environments" of the fitness set might refer either to different biotopes and communities, or to different microenvironments within a given community, such as the different components of community grain as referred to on page 19. In the latter case, which appears to be Levins' intent, the fitness set expresses the phenotypes tolerance of difference in intracommunity environment, or niche. The fitness set should thus express something about niche breadth; and Levins' use of it (p. 19) is to consider evolutionary strategy, expecially whether the niche breadth is based on a single phenotype or more than one phenotype.

A good, short discussion of the fitness set concept is given by Wilson and Bossert (1971). We introduce again some points of caution. Fitness (relative survival of different genotypes) is difficult if not impossible to measure for the range of phenotypes encountered in field conditions; and fitness values for field populations are at best imperfect. The concavity versus convexity of fitness sets is vulnerable to details of curvature of the responses curves in Fig. 2.lb and sample error, not simply determined by relative separation of the curves. It seems likely that judgments of concavity will often be unreliable and that Fig. 2.2 will be a less effective way of showing the relationships than

Fig. 2.Ib. A more fundamental question is the usefulness of concavity, if it can be determined, as evidence. A population comprises a range of genotypes and phenotypes. If change in environment is continuous, a fitness set for a particular population sample may not adequately predict the range of phenotypes occupying the full range of environments and whether that range is occupied by continuous or discontinuous population variation. For macroenvironmental variation (habitat and climatic gradients) fitness sets cannot reliably detect whether species adaptation is continuous (forming a cline) or relatively discontinuous (forming ecotypes or subspecies). For discontinuous microenvironmental or niche variables, such as two different food plants in the same community, the appropriate questions involve movements and genetic exchange within the population as well as phenotypic specialization. Perhaps most crucial, the fitness-set approach does not lend itself to consideration of the coevolution of species and environment. These limitations much restrict the usefulness of the fitness-set concept for evolutionary inference. Having identified the drawbacks of the theory, we now present Levins' original statement of it (on pages 14–20 of Paper 28).

The last sentence of this selection refers us back to our concerns with more direct approaches to niche breadth, considered in Paper 29 by Colwell and Futuyma. The first part of the article reviews Levins' use of the Simpson and Shannon–Wiener indices for niche breadth. Equation 3 for niche overlap is the same as that given for habitat overlap by Whittaker and Fairbanks (Paper 22) and that termed "proportional similarity" in Price's article at the end of this chapter,

$$C_{ih} = 1 - \tfrac{1}{2}\Sigma \, |p_{ij} - p_{hj}| = \Sigma \, \min \, (p_{ij} \text{ or } p_{hj}).$$

Schoener (1970) and Huey et al. (1974) also have used this measure for niche overlap. Difficulties for these simple measures resulting from properties of sample sets are discussed by Colwell and Futuyma, and more exacting measures are developed for niche breadth (equation 21) and niche overlap (equation 24). We are not necessarily advocating the use of these measures; the quality of information available to ecologists may often not justify their use. They suffer also from the limitations of single measures applied to multidimensional relationships. There is sometimes advantage in use of the simplest and clearest measures available with constant consciousness of their limitations; more complex measures can give a sense of confidence that data from communities do not justify. The paper is, however, an effective discussion of the problem and its relation to the niche concept (p. 575). The second and third paragraphs from the end are relevant to our discussion of inferences from competition coefficients calculated for field populations.

Other discussions and approaches to the measurement of niches are Horn (1966), MacArthur and Levins (1967), Maguire (1967), Orians and Horn (1969), Pianka (1969, 1973), Terborgh and Diamond (1970), Kohn (1971), Power (1971), Gallopin (1972), Baker and Baker (1973), Pielou (1972), and Roughgarden (1972). In a recent article, Roughgarden (1974) distinguishes the use of such measures as applied to resource utilization and as applied to morphological traits. Roughgarden thus introduces a somewhat different approach to niche measurement.

It is appropriate to close this part, heavy with theory and the limitations of measurements, with a successful application. Peter Price's (Paper 30) examination of niche width and overlap for a guild of six parasitic wasps serves admirably. The reader may well be interested in related papers by Price (1970, 1972); in the latter, as in Heatwole and Davis (1965), length of ovipositor is studied as an axis of niche difference among parasitic wasps.

REFERENCES

Baker, M. C., and A. E. M. Baker, 1973. Niche relationships among six species of shorebirds on their wintering and breeding ranges. *Ecol. Monogr.* **43**: 193–212.

Dayton, P. K. 1973. Two cases of resource partitioning in an intertidal community: making the right prediction for the wrong reason. *Amer. Naturalist* **107**: 662-670.

Gallopin, G. C. 1972. Trophic similarity between species in a food web. *Amer. Midland Naturalist* **87**: 336–345.

Harrison, J. L. 1962. Distribution of feeding habits among animals in a tropical forest. *J. Anim. Ecol.* **31**: 53–63.

Heatwole, H., and D. M. Davis, 1965. Ecology of three sympatric species of parasitic insects of the genus *Megarhyssa* (Hymenoptera, Ichneumonidae). *Ecology* **46**: 140–150.

Horn, H. S. 1966. Measurement of "overlap" in comparative ecological studies. *Amer. Naturalist* **100**: 419–424.

Huey, R. B., E. R. Pianka, M. E. Egan, and L. W. Coons. 1974. Ecological shifts in sympatry: Kalahari fossorial lizards *(Typhlosaurus). Ecology* **55**: 304–316.

Hutchinson, G. E. 1959. Homage to Santa Rosalia, or why are there so many kinds of animals? *Amer. Naturalist* **93**: 145–159.

Kohn, A. J. 1971. Diversity, utilization of resources, and adaptive radiation in shallow-water marine invertebrates of tropical oceanic islands. *Limnol. Oceanogr.* **16**: 332–348.

MacArthur, R. H. 1970. Species packing and competitive equilibrium for many species. *Theoret. Pop. Biol.* **1**: 1–11.

MacArthur, R. H., and R. Levins. 1967. The limiting similarity, convergence, and divergence of coexisting species. *Amer. Naturalist* **101**: 377–385.

Maguire, B., Jr. 1967. A partial analysis of the niche. *Amer. Naturalist* **101**: 515–523.

Maguire, B., Jr. 1973. Niche response structure and the analytical potentials of its relationship to the habitat. *Amer. Naturalist* **107**: 213–246.

Orians, G. H., and H. S. Horn. 1969. Overlap in foods and foraging of four species of blackbirds in the Potholes of central Washington. *Ecology* **50**: 930–938.

Pianka, E. R. 1969. Sympatry of desert lizards *(Ctenotus)* in western Australia. *Ecology* **50**: 1012–1030.

Pianka, E. R. 1973. The structure of lizard communities. *Ann. Rev. Ecol. Syst.* **4**: 53–74.

Pielou, E. C. 1972. Niche width and niche overlap: a method for measuring them. *Ecology* **53**: 687–692.

Power, D. M. 1971. Warbler ecology: diversity, similarity, and seasonal differences in habitat segregation. *Ecology* **52**: 434–443.

Price, P. W. 1970. Characteristics permitting coexistence among parasitoids of a sawfly in Quebec. *Ecology* **51**: 445–454.

Price, P. W. 1972. Parasitoids utilizing the same host: adaptive nature of differences in size and form. *Ecology* **53**: 190-195.

Roughgarden, J. 1972. Evolution of niche width. *Amer. Naturalist* **106**: 683–718.

Roughgarden, J. 1974. Niche width: biogeographic patterns among *Anolis* lizard populations. *Amer. Naturalist* **108**: 429–442.

Schoener, T. W. 1970. Nonsynchronous spatial overlap of lizards in patchy habitats. *Ecology* **51**: 408–418.

Strobeck, C. 1973. *N* species competition. *Ecology* **54**: 650-654.

Terborgh, J., and J. M. Diamond. 1970. Niche overlap in feeding assemblages of New Guinea birds. *Wilson Bull.* **81**: 29–52.

Vandermeer, J. H. 1970. The community matrix and the number of species in a community. *Amer. Naturalist* **104**: 73–83.

Vandermeer, J. H. 1972. Niche theory. *Ann. Rev. Ecol. Syst.* **3**: 107–132.

Wilson, E. O., and W. H. Bossert. 1971. *A Primer of Population Biology.* Sinauer, Stamford, Conn. 192 pp.

19

Reprinted from G. E. Hutchinson, *A Treatise on Limnology: Vol. II.
Introduction to Lake Biology and the Limnoplankton,*
Wiley, New York, 1967, pp. 232–233

A TREATISE ON LIMNOLOGY: VOL. II.
INTRODUCTION TO LAKE BIOLOGY AND THE
LIMNOPLANKTON

G. E. Hutchinson

[*Editors' Note:* In the original, material precedes this excerpt.]

Within any biotope we may also recognize a series of habitats which may be characterized in terms of the various species present. The *habitat* of a species, within the geographical range, may be regarded as operationally defined by specifying those parts of the ecosystem that must be present in a biotope in order for the species to occur. The habitat is regarded as having spatial extension.

The *niche* of a species is defined purely intensively. It is assumed (Hutchinson 1957, 1959b) that all the variation of the factors required to define a habitat can be ordered linearly on the axes of an *n*-dimensional coordinate system. If the species S_1 requires that the variable X' have values between X_1' and X_2' . . . we can define a hyperspace N_1; any point within N_1 corresponds to values of the variables X', X'' · · · which permit the species to occur. This hyperspace is called the *fundamental niche* of the species. The space of which the niche is a part is called the niche space or, symbolically, the **N**-space.

The habitats of two species, being two volumes of the physical space of the biotope (**B**-space) defined by the presence within them of certain parts of the ecosystem, can overlap or be coextensive. It is considered by many ecologists, including myself (see pages 356–357), that under ordinary circumstances at equilibrium two species cannot co-occur in the same niche (*principle of competitive exclusion*). For example, if we consider two species (S_1, S_2) of planktonic copepod or rotifer, one feeding on large and one on small algal cells, and both requiring an identical range of temperature and chemical composition, the habitat of one species would be a volume of water in ordinary **B**-space of the required physicochemical properties, with small algae suspended in it; the habitat of the other could be the same water with larger algae. Any volume of water might contain both algae and so be a habitat for both species of zooplankton. In the ideal niche diagram, however, the hyperspaces N_1 and N_2, defining the fundamental niches of the two species, would be separated by the different values of the food size on the axis along which food size is measured. More elaborate, if still two-dimensional cases, are easily envisioned (Fig. 71).

[*Editors' Note:* Material has been omitted at this point.]

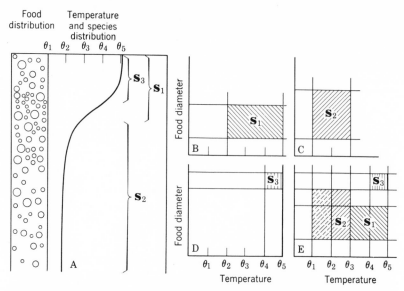

FIGURE 71. Analysis of niche space occupied by three hypothetical species differing in their temperature tolerances and the size of food taken. *A.* The vertical distribution of temperature, food organisms, and the three species S_1, S_2, and S_3. The tolerance of S_1 by itself, between θ_2 and $\theta_5°$, eating small food. *C.* The same for the cold stenotherm but more efficient S_2 between θ_1 and $\theta_3°C$., eating small and intermediate food. *D.* The same for the warm stenotherm S_3 between θ_4 and $\theta_5°C$., eating large food. *E.* This shows only the realized niches; S_2, overlaps S_1 in the niche space S_1, excluding it from the hypolimnion, but is unable to realize its whole potential niche as there is no water cooler than θ_2 in the lake; S_1 and S_3 can coexist as they eat different food.

REFERENCES

Hutchinson, G. E. 1957. See paper 43 this volume.
Hutchinson, G. E. 1959b. Il concetto moderno di nicchia ecologica. Memorie Ist. Ital. Idrobiol. 11: 9–22.

COMPARATIVE ECOLOGY OF THE INTERSTITIAL FAUNA OF FRESH-WATER AND MARINE BEACHES

Robert W. Pennak

[*Editors' Note:* In the original, material precedes this excerpt.]

One of the most interesting ecological phenomena associated with the marine beach copepods is the striking preference of each species for a particular zone of the intertidal region. As shown in Figs. 7 and 8, for example, three species *(Psammotopa vulgaris, Adelopoda ramabula,* and *Arenosetella*

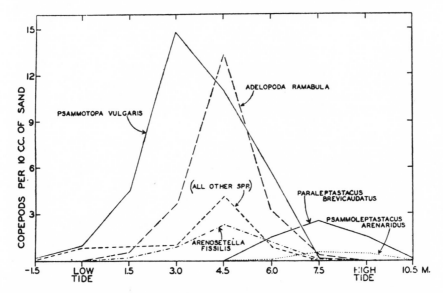

Fig. 7. — Horizontal distribution of interstitial copepods in Nobska Beach, near Woods Hole, Massachusetts, in July, 1939. Data for top 16 cm. of sand (Modified from Pennak, 1942).

[*Editors' Note:* Material has been omitted at this point.]

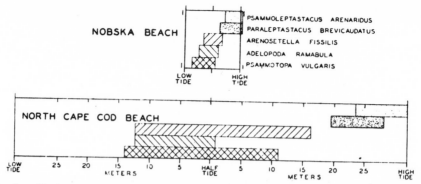

Fig. 8. — Horizontal distribution of five common species of intertidal copepods in two sandy beaches near Woods Hole, Massachusetts. Each bar represents the horizontal extent of the median 80 per cent of the total population. Data for the top 16 cm. of sand (Modified from Pennak, 1942).

fissilis) attain their maximum abundance in the half tide region, but *Paraleptastacus brevicaudatus* and *Psammoleptastacus arenaridus* are restricted to a zone near the high tide mark where the sand is covered with water for only a short time each day.

If it is true that the horizontal distribution of beach copepods is correlated with the amount of water in the sand (or period of submergence), wide beaches should show greater horizontal ranges for a particular species than narrow beaches. This phenomenon is clearly indicated in Fig. 8. Five common species were found in both Nobska and North Cape Cod beaches in Massachusetts, where the tidal amplitudes differed greatly and where the distances between tide marks were about 11 and 65 meters, respectively. Each species had a similar horizontal distribution in the two beaches with reference to the low, mid, and high tide marks, but each species also showed a much wider horizontal distribution in North Cape Cod beach than in the narrow Nobska beach.

[*Editors' Note:* Material has been omitted at this point.]

LITERATURE CITED

Pennak, R. W. 1942. Ecology of some copepods inhabiting intertidal beaches near Woods Hole, Massachusetts. Ecology 23: 446–456.

21

Reprinted from *Biol. Rev.*, **42**, 229 (1967)

GRADIENT ANALYSIS OF VEGETATION

R. H. Whittaker

[*Editors' Note:* In the original, material precedes this excerpt.]

Species can avoid competition also by occupying different habitats; that is, different positions along environmental gradients and in environmental hyperspace. Implications of species interactions for their distributions do not really support the assumption that they should evolve toward natural clusters of species with closely similar distributions (Whittaker, 1962). Instead, an extension of the principle of Gause implies that species should evolve toward dispersion of their distributional centres in environmental hyperspace (Whittaker, 1965). Since the species which occur together are also niche-differentiated, their populations do not form boundaries of mutual exclusion but overlap freely. The structure of compositional gradients—broadly overlapping population distributions mostly of binomial form, with centres scattered along the environmental gradient—is thus a consequence of species evolution toward both niche and habitat diversification.

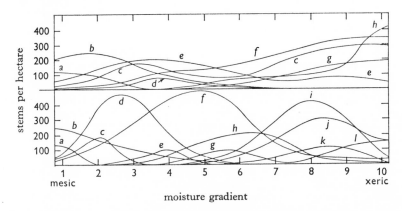

Fig. 8. Contrasts of beta diversities of vegetation along topographic moisture gradients. Above, moderately low beta diversity (less change in composition along the gradient, with widely dispersed population curves) at 460–470 m. elevation, Siskyou Mountains, Oregon (Whittaker, 1960). Below, high beta diversity (narrower population curves, greater change in composition along the gradient) at 1830–2140 m., Santa Catalina Mountains, Arizona (Whittaker & Niering, 1965). Half-change values expressing relative change in composition in the ten-step transects were 1·1 for the tree stratum of the Siskyou transect, 3·4 for that of the Catalina transect. Species, above, *a, Taxus brevifolia*; *b, Chamaecyparis lawsoniana*; *c, Castanopsis chrysophylla*; *d, Abies concolor*; *e, Pseudotsuga menziesii*; *f, Lithocarpus densiflora* (×0·5); *g, Quercus chrysolepis*; *h, Arbutus menziesii*. Species, below, *a, Abies concolor*; *b, Quercus rugosa*; *c, Pseudotsuga menziesii*; *d, Pinus ponderosa*; *e, Arbutus arizonica*; *f, Quercus hypoleucoides* (×0·5); *g, Pinus chihauhuana*; *h, Quercus arizonica*; *i, Arctostaphylos pringlei*; *j, Pinus cembroides*; *k, Garrya wrightii*; *l, Quercus emoryi*.

[*Editors' Note:* Material has been omitted at this point.]

REFERENCES

Whittaker, R. H. (1960). Vegetation of the Siskiyou Mountains, Oregon and California. *Ecol. Monogr.* **30**, 279–338.

Whittaker, R. H. (1962). Classification of natural communities. *Bot. Rev.* **28**, 1–239.

Whittaker, R. H. (1965). Dominance and diversity in land plant communities. *Science* **147**, 250–60.

Whittaker, R. H. & Niering, W. A. (1964). Vegetation of the Santa Catalina Mountains, Arizona. I. Ecological classification and distribution of species. *J. Ariz. Acad. Sci.* **3**, 9–34.

22

Reprinted from *Ecology*, **39**, 57–59 (1958)

A STUDY OF PLANKTON COPEPOD COMMUNITIES IN THE COLUMBIA BASIN, SOUTHEASTERN WASHINGTON

R. H. Whittaker and C. Warren Fairbanks

[*Editors' Note:* In the original, material precedes this excerpt.]

DISTRIBUTIONAL INTERRELATIONS OF SPECIES

Distributional relations of species may also be investigated through the matrix and plexus approach. A larger number of means of measuring the "association" of two species are available to choose from (see also Goodall 1952). Some of these (Forbes 1907, 1925; Shelford 1915; Dice 1945, 1952; Cole 1949, 1957) are based on direct comparison of the number of samples in which two species occur together, in a given sample series, with the number of occurrences together which would be expected by chance. Others employ coefficients of correlation in one form or another (Iljinski & Poselskaja 1929; Stewart & Keller 1936; Tuomikoski 1942, 1948; Cole 1949; Kontkanen 1949; Nash 1950; Goodall 1953, 1954a; de Vries *et al.* 1954; de Vries 1953; Damman & de Vries 1954). Still others are simple expressions of the percentage occurrence together of species in a set of samples (Agrell 1945; Dice 1945, 1952; Whittaker 1952). Distributional associations of plant species have been studied also by Ramensky (1930), Katz (1930), Scheygrond (1932), Iversen (1936, 1954), Matuszkiewicz (1948), Motyka *et al.* (1950), Gardner (1951), Gilbert & Curtis (1953), Hale (1955), Greig-Smith (1952), Bray (1956), and McIntosh (1957), those of animal species by Backlund (1945), Kontkanen (1950a), and Webb (1950).

It is doubtful that correlation in the usual, product-moment, sense is what should be measured in problems like those of the present study (*cf.* Cole 1949, and Bray 1956). It is further doubtful that measurement should be based on comparison of actual co-occurrence with that expected by chance. If, in a given series of lake samples, species *a* occurs in 20 of the 40 and *b* in 30 of the 40, co-occurrence expected by chance is

$$\frac{20}{40} \times \frac{30}{40} = \frac{3}{8},$$ or 15 lakes. Occurrence together

of the two species in 15 lakes is, by this standard, non-association; and occurrence together in 10 is negative association. But occurrence together in 10 or 15 lakes may be a matter not of chance, but of distributional amplitudes in relation to environment; there may be 10 or 15 lakes of the 40 in which environmental conditions permit both species to maintain populations.

It is considered that the relation to be measured is not co-occurrence relative to chance expectation, or product-moment correlation, but *relative distributional overlap*. Figure 4 illustrates two species with distributional curves overlapping along an environmental gradient (*cf.* Fig. 1); community samples are assumed to be evenly spaced along the gradient. Two approaches to measurement are possible: (1) The number of samples in which both occur may be compared with the number in which one or both occur; in effect lengths of the gradient representing overlap and total

amplitudes are being compared—$\dfrac{b-c}{a-d}$. (2) The

cross-hatched overlap area may be compared with one-half the total areas of the two curves (without assuming these to be of binomial form). If the area under a given curve is treated as the sum of the heights of lines for samples such as are marked out between *a* and *b* (*i.e.* the relative numbers of individuals of the species taken in samples along the gradient), and the sum of these heights for a given species is 100%, then the relation of the cross-hatched area to the total areas of the curves becomes a percentage. Measurements

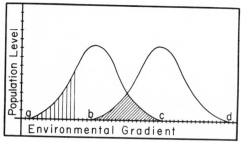

Fɪɢ. 4. Distributional overlap of two species in relation to an environmental gradient.

in terms of means and dispersions are easily conceived, but are thought unsuitable to the kind of material with which ecologists must usually deal.

For computation, the first measurement becomes simply the percentage of samples in which both species occur, among samples in which one or both occur: $PC = \dfrac{c}{a+b-c}$, when *a* is the

number of samples containing the first species, *b* the number containing the second, and *c* the number containing both. The most effective means of computing the second measurement is to total the population of a given species in all samples in the series. Numbers of that species in particular samples are then converted to percentages of that total. When all species have been thus treated, a table of percentage distributions corresponding to Table I, but based on vertical computation of totals and percentages, results. From this table the percentage similarities of species distributions may be simply computed for all pairs of species by the same formula as for percentage similarity of community samples.

Of these two measurements the first, which may be termed *percentage co-occurrence,* is analogous to coefficient of community and the second to percentage similarity of community samples. As is the case with the other measurements, they give somewhat different results, and each has its advantages and limitations under different circumstances. Both are, clearly, very strongly influenced by the distribution of samples relative to the total field of occurrence of the species. Rarely are samples taken along a gradient as in Figure 4; often they may be concentrated in the overlap $b - c$ or in one of the non-overlaps $a - b$ and $c - d$. Species populations occur and overlap in relation to many environmental gradients, not one; and rarely can the whole field of occurrence of both be sampled. The measurements cannot in general express the distributional relation of species in the field, only their distributional relation in a particular set of samples. Granting the conspicuous limitations of such measurements, they may still reveal relations of species in samples which are not apparent from simple inspection of tables. As measurements, they are devoid of statistical sophistication, but free also from assumptions about statistical properties of the material and what is to be measured that are embodied in some more formal procedures. For the present, early stage of development of these techniques, they may be preferred as simple and direct expressions of that which an ecologist regards as appropriate to measurement in the distributional relations of species in his samples.

Percentage co-occurrences and percentage similarities of species distributions are given in Table IV. Copepod species occurring in fewer than three water bodies are excluded; distributions of these are discussed in the following section. It is impossible to manipulate the matrices so that all higher values are concentrated along the diagonal, and multi-directional relations among species are indicated by this fact.

TABLE IV. Matrix for relative distributional overlaps of copepod species. Values above the diagonal are percentage co-occurrences, those below are percentage similarities of species distributions. (x = less than 0.5%)

	Epischura nevadensis	Diaptomus ashlandi	Eucyclops agilis	Cyclops bicuspidatus	Diaptomus sicilis	Cyclops vernalis	Macrocyclops albidus	Cyclops varicans	Diaptomus sanguineus	Diaptomus shoshone	Diaptomus novamexicanus
Epischura nevadensis..		75	14	13	..	..	..	..	..	..	..
Diaptomus ashlandi...	29		13	12	..	..	..	..	..	..	..
Eucyclops agilis......	x	x		40	13	15	13	12	11	..	6
Cyclops bicuspidatus..	4	11	11		29	14	4	17	15	8	17
Diaptomus sicilis.....	..	..	5	6		21	..	12	..	..	13
Cyclops vernalis......	..	..	3	1	9		27	15	8	..	..
Macrocyclops albidus..	..	..	2	1	..	16		29	22	..	..
Cyclops varicans......	..	..	30	10	16	2	22		20	14	..
Diaptomus sanguineus..	..	..	1	1	..	1	18	11		25	..
Diaptomus shoshone...	..	..	..	1	..	..	..	21	19		..
Diaptomus novamexicanus......	..	..	x	58	2	..	..	..	..		

Of the two measurements, percentage co-occurrence is regarded as providing the better results in the present material; these values are diagrammed as a plexus in Figure 5. The relations cannot be represented effectively in two dimensions. The heavy *E. agilis-C. bicuspidatus* line should be regarded as an axis from which three plane surfaces radiate outward—*D. ashlandi* and *E. nevadensis* in one direction, *C. vernalis, D. sicilis,* and *D. novamexicanus* in a second, and *Macrocyclops albidus, C. varicans, D. sanguineus,* and *D. shoshone* in a third. *C. bicuspidatus* and *E. agilis* are the most widely distributed species in non-saline and mildly saline water bodies; they are consequently distributionally related to every other species in the plexus (except *E. agilis* vs. *D. shoshone*). The planes extending outward from the *E. agilis-C. bicuspidatus* axis are formed by species which are characteristic of more distinctive environments. The extreme members of two of the planes, *D. ashlandi* and *E. nevadensis,* and *D. sanguineus* and *D. shoshone,* are almost wholly unrelated to species on the other planes; *C. vernalis* and *D. sicilis* are related, however, to *M. albidus* and *C. varicans,* and these relations are not shown. For the most part, positions toward the center of the plexus are occupied by common and widely distributed or cosmopolitan cyclopids;

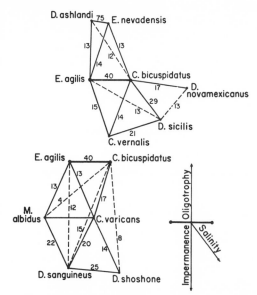

FIG. 5. Plexus of distributional relations of copepod species, based on measurements of percentage co-occurrence (Table IV). Percentage co-occurrence values are indicated by lines connecting species; solid lines are drawn to scale for 100 minus percentage co-occurrence, broken lines are not drawn to scale. The heavy line connecting *E. agilis* and *C. bicuspidatus* is the axis from which radiate three planes for groups of distributionally related species: *D. ashlandi* and *E. nevadensis* in oligotrophic conditions, *C. vernalis* and *D. sicilis* in saline conditions, *M. albidus, C. varicans, D. sanguineus,* and *D. shoshone* in unstable or temporary ponds.

distal positions are occupied by diaptomids (and the temorid *E. nevadensis*) of more restricted American or western distributions (see also Fig. 3).

Interpretation of the plexus in terms of environmental relations is not difficult. *D. ashlandi* and *E. nevadensis* are the species of oligotrophic lakes, *D. novamexicanus* is a species of eutrophic lakes. *C. vernalis* is a species of slightly to moderately saline ponds, *D. sicilis* a species of saline lakes and ponds. *M. albidus* and *C. varicans* are species of semi-permanent ponds; *D. sanguineus* and *D. shoshone* are species of temporary ponds. The plexus thus represents a central complex of widely distributed species most characteristic of "intermediate" conditions—eutrophic, non-saline, relatively stable water bodies—and radiating from this three planes for species of three types of extreme conditions—oligotrophy, salinity, and seasonal instability. The three planes thus correspond to the three axes of the community-type plexus (Fig. 3).

[*Editors' Note:* Material has been omitted at this point.]

REFERENCES

Agrell, I. 1945. The collemboles in nests of warmblooded animals with a method for sociological analysis. K. Fysiogr. Sallsk. i Lund, Handl., N.F., **56**(19) : 1–19.

Backlund, H. O. 1945. Wrack fauna of Sweden and Finland, ecology and chorology. Opuscula Entom., Suppl., **5**: 1–236.

Bray, J. R. 1956. A study of mutual occurrence of plant species. Ecology, **37**: 21–28.

Cole, L. C. 1949. The measurement of interspecific association. Ecology, **30**: 411–424.

———. 1957. The measurement of partial interspecific association. Ecology, **38**: 226–233.

Damman, A. W. H. and D. M. de Vries. 1954. Testing of grassland associations by combinations of species. Biol. Jaarb. [Antwerpen], **21**: 35–46.

Dice, L. R. 1945. Measures of the amount of ecologic association between species. Ecology, **26**: 297–302.

———. 1952. Natural communities. Ann Arbor: Univ. Mich. Press. 547 pp.

Forbes, S. A. 1907. On the local distribution of certain Illinois fishes: an essay in statistical ecology. Ill. State Lab. Nat. Hist. (Nat. Hist. Surv.), Bull. 7: 273–303.

———. 1925. Method of determining and measuring the associative relations of species. Science, N. S., **61**: 524.

Gardner, J. L. 1951. Vegetation of the creosotebush area of the Rio Grande valley in New Mexico. Ecol. Monog., **21**: 379–403.

Gilbert, Margaret L. and J. T. Curtis. 1953. Relation of the understory to the upland forest in the prairie-forest border region of Wisconsin. Wisc. Acad. Sci. Arts & Letters, Trans., **42**: 183–195.

Goodall, D. W. 1952. Quantitative aspects of plant distribution. Cambridge Phil. Soc., Biol. Rev., **27**: 194–245.

———. 1953. Objective methods for the classification of vegetation. I. The use of positive interspecific correlation. Austral. Jour. Bot., **1**: 39–63.

———. 1954a. Objective methods for the classification of vegetation. III. An essay in the use of factor analysis. Austral. Jour. Bot., **2**: 304–324.

Greig-Smith, P. 1952. Ecological observations on degraded and secondary forest in Trinidad, British West Indies. II. Structure of the communities. Jour. Ecol., **40**: 316–330.

Hale, M. E., Jr. 1955. Phytosociology of corticolous cryptogams in the upland forests of southern Wisconsin. Ecology, **36**: 45–63.

Iljinski, A. P. and M. A. Poselskaja. 1929. A contribution to the question of the associability of plants. (Russ. with Engl. summ.) Trudy Prikl. Bot., Genet., i Selek. (Bull. Appl. Bot., Genet., & Plant-Breeding), **20**: 459–474.

Iversen, J. 1936. Biologische Pflanzentypen als Hilfsmittel in der Vegetationsforschung.—Ein Beitrag zur ökologischen Charakterisierung und Anordnung der Pflanzengesellschaften. Skalling-Lab. [København], Meddel., **4**: 1–224.

———. 1954. Über die Korrelation zwischen den Pflanzenarten in einem grönländischen Talgebiet. Vegetatio, **5/6**: 238–246.

Katz, N. J. 1930. Die grundlegenden Gesetzmässigkeiten der Vegetation und der Begriff der Assoziation. Beitr. Biol. der Pflanz, **18**: 305–333.

Kontkanen, P. 1949. On the determination of affinity between different species in synecological analyses. (Finnish summ.) Ann. Entom. Fenn., **14** (Suppl.): 118–125.

———. 1950a. Quantitative and seasonal studies on the leafhopper fauna of the field stratum on open areas in North Karelia. (Finnish summ.) Soc. Zool.-Bot. Fenn. Vanamo, Ann. Zool., **13** (8): 1–91.

McIntosh, R. P. 1957. The York Woods, a case history of forest succession in southern Wisconsin. Ecology, **38**: 29–37.

Matuszkiewicz, W. 1948. Roślinność lasów okolic Lwowa. (Engl. summ.) Univ. Mariae Curie-Skłodowska, Lublin, Ann., Sect. C, **3**: 119–193.

Motyka, J., B. Dobrański, and S. Zawadzki. 1950. Wstępne badania nad łąkami poludniowo-wschodniej Lubelszczyzny. (Russ. & Engl. summs.) Univ. Mariae Curie-Skłodowska, Lublin, Ann., Sect. E, **5**: 367–447.

Nash, C. B. 1950. Associations between fish species in tributaries and shore waters of western Lake Erie. Ecology, **31** : 561–566.

Ramensky, L. G. 1930. Zur Methodik der vergleichenden Bearbeitung und Ordnung von Pflanzenlisten und anderen Objekten, die durch mehrere, verschiedenartig wirkende Faktoren bestimmt werden. Beitr. Biol. der Pflanz., **18**: 269–304.

Scheygrond, A. 1932. Het plantendek van de Krimpenerwaard IV. Sociographie van het hoofd-associatie-complex Arundinetum-Sphagnetum. Nederland. Kruidk. Arch., **1932**: 1–184.

Shelford, V. E. 1915. Principles and problems of ecology as illustrated by animals. Jour. Ecol., **3**: 1–23.

Stewart, G. and W. Keller. 1936. A correlation method for ecology as exemplified by studies of native desert vegetation. Ecology, **17**: 500–514.

Tuomikoski, R. 1942. Untersuchungen über die Vegetation der Bruchmoore in Ostfinnland. I. Zur Methodik der pflanzensoziologischen Systematik. (Finnish summ.) Soc. Zool.-Bot. Fenn. Vanamo, Ann. Bot., **17** (1): 1–203.

———. 1948. Entomologian synekologisista tilastoista ja hyönteisyhteisöjen typologiasta. (Germ. summ.) Ann. Entom. Fenn., **14**: 101–115.

de Vries, D. M. 1953. Objective combinations of species. Acta Bot. Neerland., **1**: 497–499.

de Vries, D. M., J. P. Baretta, and G. Hamming. 1954. Constellation of frequent herbage plants, based on their correlation in occurence. Vegetatio, **5/6**: 105–111.

Webb, W. L. 1950. Biogeographic regions of Texas and Oklahoma. Ecology, **31**: 426–433.

Whittaker, R. H. 1952. A study of summer foliage insect communities in the Great Smoky Mountains. Ecol. Monog., **22**: 1–44.

23

Reprinted from *Vegetatio*, **5-6**, 105-111 (1954)

CONSTELLATION OF FREQUENT HERBAGE PLANTS, BASED ON THEIR CORRELATION IN OCCURRENCE*)

(with 1 table and 2 figures)

by

D. M. DE VRIES

With collaboration of Miss J. P. BARETTA and G. HAMMING.

(Central Institute for Agricultural Research, Wageningen).

INTRODUCTION.

Dominancy and high frequency on one side, as well as combinations of species on the other side, are important from sociological and ecological point of view. Predominancy of a species can be looked upon as an expression of its success in the struggle for life. After all concurrence, habitat and priority of species determine the composition of plant communities. Further more the predominant species often cause bigger or smaller variations in the structure of the plantcover, such as landscape types, zones and mosaics. Moreover, occurring abundantly, species can contribute to a large extent, in changing the original habitat. And last of all the rate of indication of environmental factors by plant species is often very much dependant on the abundance of those species. On the other hand a combination of species will limit the habitat more than species will do this separately. A fixed species for instance will have a smaller pH-amplitude, while others would be more sensitive to humidity, P-status of the soil, grazing, severe frost or shadow.

In view of the preceding, we see both ways of distinction of plant communities, namely those according to the predominancy and those according to the combination of species, have sense and significance. An objection to the use of dominance communities (6) which are distinguished according to the abundance of the leading species, is situated in the possibility, that predominant species will change in different seasons or different years, under the influence of divergent weather conditions. But this objection is not invincible, if one will only count with the season and abnormal weather conditions. In distinguishing combinations of species a big difficulty is situated in the fact, which of the many on the spot occurring species one will choose. Up till now the defining of plant communities, based on the characteristic combination of species, has been more or less subjective. Objectivity in this respect is desirable, though the value of intuition must not be rated too low. The ecology of a combination of species shall only find clearly expression then, when the joining species not only occur, but are frequent too. Most species are *frequency indicators* which means to say, they are species with only a narrow ecological amplitude in the case of a high frequency (specific frequency = frequency of occurrence). Out of our grassland research, it became evident, that *presence indicators* (6), these are species which in small numbers already give a good indication of an environmental factor, are very sparse. If in grassland there are presence-indicators for a complex of environmental factors, like the habitat is, in short to say, whether there really exist exclusive characteristic species (School of BRAUN-BLANQUET), seems to me doubtful.

The conclusion is simple: It is desirable to use both the combination of species and the frequency of occurrence to define plant communities. If the size of the reading is taken small enough, highly frequent species after all can be looked upon as potential

---

*) Received for publication 1.XI.1953.

dominants, therefore we prefer combinations of frequent species which in my opinion form the backbone of the associations. By comparing the presence percentages and the average frequencypercentages of all species which form the composition of suchlike combinations of species, one can decide later, which of these species are more or less faithful to one special combination or a group of combinations.

PROBLEMATIC.

Which of the sometimes many species, which can occur frequently at the same time, come into consideration to be included in the characteristic combination of species? Those that join strongest in nature, showing the biggest coupling. What is the most objective way to find this out? By starting a survey from areas which are not selected for special combinations of species, but are as homogeneous as possible. Further more one calculates in mathematically statistical way correlation coefficients (r-values) of two species, every time. In this way one comes to know, whether these species are found together or are absent together, less often. The extreme r-values are $+1$ and -1. $+1$ means: two species are always found together while -1 means, that one is always present, if the other is absent. Species which show the biggest positive values, may be joined to typical combinations. In view of this, one can still ask a question of general importance. Shall the typical combinations be parted more or less clearly, or shall they be chained? In other words, does one speak of discontinuity or continuity of the chain? In the first case there would indeed be plant communities which can be seen as sharply defined combinations of species.

BASIC MATERIAL.

The compilation refers to 1000 analyses (for some species from 1478) of old permanent grasslands. These permanent grasslands were examined for the greatest part since 1940, they are scattered all over the Netherlands and belong to different types. It was conditioned, that the sampled areas, because of the abundance of the species, should make a homogeneous impression. They were examined for characterizing by the combined 25 cm^2 specific frequency and order method (2, 3). From each grassland were taken about a 100 handfuls of grass (it being tall herbage) or as many borings (it being a lower herbage). Each of the handfuls or borings was cut after a fixed number of steps in front of the toe of the shoes. A 100 handfuls or borings per ha was enough, to get fairly reliable frequency percentages (1), for the smaller areas one could take less. It was not necessary, to follow the way of the ideal, equal distribution of the readings over the field, by means of a diagonal and 4 or 6 lines running parallel to it, it not being the purpose to get an exact picture of a plot as a whole. To make it more simple, sampling was mostly done by means of two diagonals.

METHOD OF COMPILATION.

The 1000 frequencydeterminations of old grasslands which were available at the beginning of the correlation calculation, contained in total 325 plant species. If one should have to determine the correlation of all these species, two by two, it would mean 52650 calculations. As the usual calculation of the product moment correlation coefficient requires a lot of time, it appeared, to be practically impossible to determine the r of all plant species mutual, in this way. Not only we had to limit the amount of species, but also the calculation must be simplified considerably, if possible. Dr Ir G. HAMMING took care of the last. He considered it an error to use the product moment correlation coefficient, because the distribution of the F%'s is often far from normal. He provided us with two formula's which made it possible to calculate the r quickly. They follow below.

Put the case, that the r must be determined of the species X and Y. In the here following diagram (Figure 1) A, B, P, etc. mean:

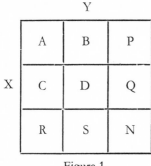

Figure 1.

A equals the number of times X and Y occur together;
P equals the number of times Y occurs in total;
R equals the number of times X occurs in total;
P — A = B equals the number of times Y occurs without X;
R — A = C equals the number of times X occurs without Y;
D equals the number of times X nor Y occur at all;
N equals the total numbers of samples;
S = N — R = B + D;
Q = N — P = C + D.

Consequently one determines A, P, R and N, and by subtraction, one can find the resting figures. The first formula which is being used, and which has been found by Dr Ir G. HAMMING in an empirical way, runs: $r = -0.6 \log \frac{BC}{AD}$. If we find however, that the r is higher than 0.75, this formula is not right and in this case the quantile correlation coefficient of HAMMING must be calculated by the following formula:

$r = \sin (T \times 90°)$, in which $T = \dfrac{AD - BC}{\sqrt{P \times Q \times R \times S}}$ (rank correlation coefficient of

M. G. KENDALL). Considered afterwards, it seems to Dr HAMMING the best to use always the last formula.

In order to get results as exact as possible, one must take account of the centre of distribution of the 25 cm² frequency percentages. In the case of frequently occurring species which are found in the material of our Dutch grasslands the biggest number of times in frequencies of about 25, 50 or 75%, a species is only counted to be present, if it has a frequency higher than the above mentioned limits. For details see p. 4. In all other cases the presence or absence of a species in a sample, is defined according to common parlance. We count an r unreliable, if R or P is lower than 30, which means, one of the mentioned species occurs less than 30 times. Many species are excluded in this way for the correlation calculation, but too many are still left.

All the positive and negative correlations of the 45 most general species, these are the species that were found in more than 200 of the 1000 samples taken of Dutch grasslands, are calculated; these amount in total to 990 r-values. They are the following species, in alphabetic order and with their abbreviations between brackets:

Achillea millefolium L. *(Ach)*, Agropyron repens P.B. *(Agr)*, Agrostis canina L. *(Ac)*, A. stolonifera L. *(As)*, A. tenuis Sibth. *(At)*, Alopecurus geniculatus L. *(Ag)*, A. pratensis L. *(Ap)*, Anthoxanthum odoratum L. *(Ao)*, Bellis perennis L. *(Be)*, Bromus mollis L. *(Bm)*, Cardamine pratensis L. *(Car)*, Carex disticha Huds. *(Cd)*, C. hirta L. *(Ch)*, C. nigra Reichard. *(Cn)*, Cerastium caespitosum Gilib. *(Cer)*, Cirsium arvense Scop. *(Cir)*, Cynosurus cristatus L. *(Cy)*, Dactylis glomerata L. *(D)*, Deschampsia caespitosum (L.) P.B. *(Dc)*, Festuca pratensis Huds. *(Fp)*, F. rubra L. *(Fr)*, Glechoma hederacea L. *(Gle)*,

Glyceria fluitans R.Br. *(Gf)*, Hoicus lanatus L. *(Hl)*, Hordeum secalinum Schreb. *(Hs)*, Leontodon autumnalis L. *(Leo)*, Lolium perenne L. *(Lp)*, Luzula campestris D.C. *(L)*, Lysimachia nummularia L. *(Ln)*, Phleum pratense L. *(Phl)*, Plantago lanceolata L. *(Pl)*, P. major L. *(Pm)*, Poa annua L. *(Pa)*, P. pratensis L. *(Pp)*, P. trivialis L. *(Pt)*, Potentilla anserina L. *(Pot)*, Prunella vulgaris L. *(Pr)*, Ranunculus acer L. *(Ra)*, R. repens L. *(Rr)*, Rumex acetosa L. *(Ru)*, Stellaria graminea L. *(Sg)*, Taraxacum officinale Web. *(Ta)*, Trifolium pratense L. *(Tp)*, T. repens L. *(Tr)* and Trisetum flavescens (L.) P.B. *(Tri)*.

Amongst these species, the following are only counted to be present, if their 25 cm²
F% amounts to more than 25: *Agropyron repens* P.B., *Agrostis tenuis* SIBTH., *Anthoxanthum odoratum* L., *Cardamine pratensis* L., *Cynosurus cristatus* L., *Festuca pratensis* HUDS.,
Holcus lanatus L., *Phleum pratense* L., *Ranunculus acer* L., *A. repens* L., *Rumex acetosa* L.
and *Taraxacum officinale* WEB.

The limit of the following species is situated at 50 F%: *Agrostis stolonifera* L..
Festuca rubra L., *Lolium perenne* L., *Poa pratensis* L., and *Trifolium repens* L., and *Poa trivialis*
L. at 75%.

Up till now the r-values of the following 38 chosen species have been calculated
too, their figures are mentioned between brackets like their abbreviations: Achillea
ptarmica L. *(Ach p)* (14), Arrhenatherum elatius (L.) J. et C. Presl *(Ar)* (55), Avena
pubescens Huds. *(Av)* (23), Calamagrostis canescens (Web.) Roth *(Calam)* (14), Caltha
palustris L. *(Cal)* (53), Carex panicea L. *(Cp)* (59), Carum carvi L. *(Caru)* (17), Cirsium
dissectum (L.) Hill *(Ci)* (47), Comarum palustre L. *(Com)* (14), Crepis biennis L. *(Cre)*
(21), Daucus carota L. *(Dau)* (13), Eriophorum angustifolium Honckeny *(Eri)* (14),
Festuca ovina L. *(Fo)* (54), Filipendula ulmaria (L.) Maxim *(F)* (58), Galium verum
L. *(Gv)* (17), Glyceria maxima (Hartm.) Holmb. *(Gm)* (47), Heracleum sphondylium
L. *(He)* (13), Hieracium pilosella L. *(Hp)* (16), Hydrocotyle vulgaris L. *(Hyd)* (13),
Hypochoeris radicata L. *(Hyp)* (15), Lychnis flos-cuculi L. *(Lfc)* (53), Lysimachia
vulgaris L. *(Lv)* (14), Medicago lupulina L. *(Ml)* (13), Molinia coerulea (L.) Moench
(M) (59), Nardus stricta L. *(N)* (13), Phalaris arundinacea L. *(Pha)* (55), Pimpinella
saxifraga L. *(Pimp)* (7), Plantago media L. *(Pl me)* (4), Potentilla erecta (L.) Räuschel
(Pe) (53), Ranunculus bulbosus L. *(Rb)* (11), Rumex acetosella L. *(Rla)* (12), Sanguisorba
minor Scop. *(San)* (10), Sieglingia decumbens (L.) Bernh. *(S)* (53), Succisa pratensis
Moench *(Suc)* (17), Thalictrum flavum L. *(Thal)* (14), Trifolium fragiferum L. *(Tf)*
(4), Valeriana dioica L. *(Vd)* (13), V. officinalis L. *(Vo)* (14) and Viola palustris L.
(Vio) (13).

RESULTS.

One will find the positive and negative correlations of the 45 most general species
in T a b l e I (look for their abbreviations on p. 107).

The r-values seem to be rather low as a rule which is not surprising, as these are
the most spread species of the Dutch grasslands. If still more attention had been paid
to homogeneity and smaller areas had been chosen, it would not have been impossible,
to get higher correlations. If this method had been followed, there would have been
less chance of also taking some divergent small areas in sampling by the frequency
method over the total area. This involved a gathering of species in a collective sample
which did not really belong together.

The highest postitive r-values are: 0.61 of *Dactylis glomerata* L. and *Trisetum flavescens*
(L.) P.B., 0.53 of *Alopecurus geniculatus* L. and *Glyceria fluitans* R.Br., and 0.52 of *Anthox-
anthum odoratum* L. and *Rumex acetosa* L. The highest negative r-values are: —0.57 of
Agropyron repens P.B. and *Agrostis tenuis* SIBTH., —0.47 from *Agrostis canina* L. and
Lolium perenne L., —0.41 from *Agrostis canina* L. and *Dactylis glomerata* L., —0.40 from
Alopecurus geniculatus L. and *Festuca rubra* L., and from *Anthoxanthum odoratum* L. and
Phleum pratense L., —0.35 from *Festuca rubra* L. and *Lolium perenne* L., and —0.34 from
Festuca rubra L. and *Poa trivialis* L. Negative correlations of species can point to very
different claims on the environment. They can point also to very active competition.
The last for instance is done by means of a growth checking product. It is maintained,
the roots of *Festuca rubra* L. and *Lolium perenne* L. can produce this. Notwithstanding it
all being ordinary herbage, a considerable difference in correlative possibility will attract
attention after looking more closely at the numbers. Species like *Bellis perennis* L.,
Deschampsia caespitosa (L.) P.B. and *Poa pratensis* L. do not care much with which species
they grow together. Their average r values only are: 0.10 (positive average 0.11 and negative

T A B L E   I

Positive and negative (in italics) correlation coefficients (multiplied by 100) between the 45 commonest herbage plants. For the abbreviations see p.107

see p.107

	Ach	Agr	Ac	As	At	Ag	Ap	Ao	Be	Bm	Car	Cd	Ch	Cn	Cer	Clr	Cy	D	Dc	Fp	Fr	Gle	Gf	Hl	Hs	Leo	Lp	L	Ln	Phl	Pl	Pm	Pa	Pp	Pt	Pot	Pr	Ra	Rr	Ru	Sg	Ta	Tp	Tr	Tri
Ach																																													
Agr	9																																												
Ac	24	28																																											
As	8	11	25																																										
At	31	57	1	15																																									
Ag	24	7	1	5	23																																								
Ap	8	18	17	0	17																																								
Ao	8	27	15	29	22	0																																							
Be	5	9	38	22	0	8	8																																						
Bm	11	18	18	5	12	8	9	22																																					
Car	18	5	15	8	21	13	4	13	18																																				
Cd	18	10	23	10	20	7	20	7	33	17																																			
Ch	5	2	6	1	14	1	1	3	29	7	17																																		
Cn	23	6	44	11	1	30	30	11	1	7	3																																		
Cer	17	19	15	8	16	1	7	10	16	28	26	49	26																																
Clr	20	6	12	7	11	5	11	6	3	1	8	1	4	8	1																														
Cy	11	14	17	11	3	8	21	21	10	8	4	30	3	19	8	5																													
D	22	22	12	7	26	11	29	29	33	14	21	13	8	21	8	8	16																												
Dc	7	2	41	9	8	8	33	8	11	11	30	17	3	22	5	6	8	5																											
Fp	0	0	1	20	12	12	8	3	8	10	6	33	27	5	11	9	16	26	9																										
Fr	23	14	2	4	4	18	30	40	5	10	13	22	4	7	5	16	4	15	10	5																									
Gle	12	24	5	17	2	23	53	8	9	7	31	24	8	8	14	11	6	6	8	3	19																								
Gf	29	31	11	5	11	1	1	7	46	9	8	13	22	16	15	24	15	26	13	17	30	24																								
Hl	13	14	36	18	25	10	18	11	5	12	13	6	30	3	1	11	4	15	5	26	4	2	3																							
Hs	17	3	4	1	12	5	16	5	5	25	5	1	22	6	11	24	9	6	4	9	17	10	1	30																						
Leo	22	25	47	32	9	2	14	16	17	3	31	24	3	32	10	15	0	5	16	23	39	20	31	1	25																					
Lp	22	1	8	19	11	33	4	1	3	2	7	3	8	4	5	11	16	5	17	4	20	31	36	7	4	4																			
L	11	3	25	3	32	11	21	4	1	8	34	7	2	8	16	11	14	16	7	23	5	17	5	8	35	13	22																		
Ln	21	16	28	30	11	12	23	15	12	20	11	5	9	7	7	9	8	4	10	19	20	16	14	9	4	15	21	32	5																
Phl	11	8	12	1	32	10	9	5	9	7	5	15	12	6	1	5	8	18	15	5	9	25	17	7	7	1	24	35	19																
Pl	21	16	31	24	23	23	9	24	4	4	12	7	5	5	11	5	8	4	9	1	30	9	16	8	14	11	4	7	18	24	1														
Pm	11	11	27	10	1	10	31	17	23	24	22	16	2	4	20	11	2	18	17	12	12	13	9	7	14	8	21	5	13	4	15	40													
Pa	15	11	31	26	24	23	31	7	9	9	33	10	5	21	24	21	11	11	10	18	25	33	5	5	1	11	13	25	13	3	42	26	8												
Pp	20	15	27	10	1	10	4	31	6	9	1	42	18	9	15	2	20	4	13	11	24	25	6	8	6	17	17	26	3	18	4	6	6	2												
Pt	6	1	21	26	17	12	15	5	7	24	34	12	4	10	17	3	19	16	17	15	28	6	3	5	1	11	13	2	18	4	30	3	25	18	6											
Pot	18	12	26	15	16	15	26	6	7	3	12	18	12	3	29	11	26	4	1	13	22	28	14	20	8	1	21	25	4	30	14	4	10	13	5	31										
Pr	26	3	7	21	26	10	14	3	8	17	33	9	0	20	24	20	17	16	4	18	28	11	19	36	11	8	15	2	27	13	30	13	24	9	10	0	13									
Ra	17	9	13	24	26	7	17	2	5	5	1	4	11	17	2	24	19	4	6	13	13	6	4	1	22	11	17	10	24	4	13	17	24	12	9	28	24	22								
Rr	17	3	18	5	24	13	16	5	35	26	12	0	10	10	23	26	22	16	15	1	33	15	30	7	17	21	20	31	29	30	35	7	6	10	3	4	23	24	15							
Ru	22	18	25	1	1	10	52	4	20	0	33	11	11	17	20	9	26	22	2	24	5	6	10	1	22	22	24	4	31	4	1	13	13	13	23	14	14	0	1	22						
Sg	13	14	33	11	26	26	26	37	19	19	26	11	1	7	24	9	9	14	5	5	28	33	2	5	17	15	27	27	31	35	31	30	29	12	27	31	35	14	5	19	6					
Ta	18	1	1	21	30	4	4	30	20	17	42	9	23	1	23	38	24	14	20	4	13	25	5	2	17	20	4	10	13	1	4	13	22	22	13	13	14	4	2	3	23				
Tp	12	11	4	11	1	19	1	9	19	0	1	11	11	10	9	8	9	14	4	1	11	2	13	13	22	0	27	27	7	7	35	30	13	15	10	4	14	0	19	3	13	19			
Tr	12	7	26	15	9	2	11	19	7	14	9	2	10	1	19	23	24	14	1	4	15	5	20	5	22	7	4	4	13	2	1	22	8	13	22	12	35	2	5	1	3	13	13	16	
Tri	32	2	14	18	9									24				14	61	1	28	24	24	39	20	0	7	4	8	34	31	20	20	22	18	12	39	31	35	7	7	19	43	16	7
	Ach	Agr	Ac	As	At	Ag	Ap	Ao	Be	Bm	Car	Cd	Ch	Cn	Cer	Clr	Cy	D	Dc	Fp	Fr	Gle	Gf	Hl	Hs	Leo	Lp	L	Ln	Phl	Pl	Pm	Pa	Pp	Pt	Pot	Pr	Ra	Rr	Ru	Sg	Ta	Tp	Tr	Tri

average 0.08), 0.08 (+0.09 and −0.06) and 0.10 (+0.09 and −0.11) respectively, while their highest r-values are fairly low too, +0.35 (and −0.22), +0.26 (and −0.16) and −0.33 (+0.29). On the contrary species like *Agrostis canina* L., *Anthoxanthum odoratum* L., and *Trisetum flavescens* (L.) P.B. are connected to more typical combinations. Not only are their average r-values twice as high: 0.29 (+0.18 and −0.20), 0.19 (+0.20 and −0.18) and 0.18 (+0.21 and −0.14) respectively, but their highest r-values begin to look like something too. They are +0.44 (and −0.47), +0.52 (and −0.40) and +0.61 (and −0.41). It can be noted here, the highest r-value, that of *Trisetum* +0.61, would have been still higher, if it had not deserted its companion *Dactylis* in the North provinces of the Netherlands.

It was to be expected, that higher r-values would be found amongst the 38 less common species, than amongst the 45 commonest grassland plants. Amongst these less common species, several are important in sociological respect. Being in want of space, we shall not be able to give them all here, but shall suffice by only giving the values which will amount to at least +0.60. Here they follow in order from high to low, in brackets behind the abbreviations (look for these on p. 107):

M-S (0.91), S-Pe (0.86), M-Fo, M-Ci, S-Ci, S-Fo (0.84), M-Cp (0.83), M-Pe (0.81), S-Cp (0.80), M-Suc (0.79), Cp-Eri (0.77), Pe-Ci (0.76), Cp-Ci (0.75), Ar-D, Com-Eri (0.74), N-Fo (0.73), Cp-Vio, M-Lv (0.72), Cal-Gm, Com-F, Cp-Cn (0.70), Ach p-F, Cp-Lv, Cp-Pe, Ci-Fo, M-Vio (0.69), Ar-Dau, Ar-Tri, Cala-Cal, Hp-L, Hp-Rla, Hyd-Vio (0.68), Ar-Cre, Av-He, Av-Rb, Av-San (0.67), Ach p-Lv, Cp-Suc, Eri-M, Fo-Hyd, F-Vio, Gm-Lfc (0.66), Ac-Cp, Ar-He, Cal-Lfc, Cn-Vd, Cp-Com, Caru-Cre, Caru-Tri, Com-Vio, Eri-Vd, Eri-Vio, Hp-Gv, Lv-Suc, S-Vio (0.65), Ao-Cre, Cn-Lv, Dau-He, Dau-Ml, M-Vd (0.64), Ach p-Cala, Ac-M, Ar-Av, Av-D, Pha-Gm, Pe-N, Rb-Tri (0.63), Cn-Hyd, Cn-Vio, Cp-Fo, Cre-Dau, F-Hyd, F-S, F-Vd (0.62), Cp-F, F-Fo, He-Tri, Lfc-Thal (0.61), Ar-Fr, Av-Gv, Fr-Gv, Lfc-Vo (0.60).

To get a clear picture of the mutual correlation of the most important herbage plants, F i g u r e 2 was drawn. In doing this, use was made of the r-values of coupled species. In establishing the place of the species, the aim was to the put most coupled species nearest together, and the negative correlated ones far apart. It was not possible to get a quite clear picture in this way, as for that the figure should have been at least three dimensional. In this way one sees Agr and Ap upper right as well as below right and At below right as well as upper left, positively correlated with other species. Because not all the distances between the abbreviations could be inversely proportional to the positive correlations of species, they were indicated by lines which clearly show the class of correlation. Moreover the r-values in these lines were placed there, multiplied by a hundred. In the first place, the most ordinary herbage plants (their abbreviations are mentioned on p. 107) except for Ach, Bm, Ch, Cer, Dc, Fp, Gle, Leo and Pp which are only little chained to other species, were inserted in the picture. Also the following 16 less common species which are interesting out of sociological point of view (their abbreviations mentioned on p. 108): Ar, Av, Cal, Cp, Ci, Fo, F, Gv, Gm, Hp, Lfc, M, Pha, Pe, Rla and S. When this figure was drawn, there were not sampled grasslands enough, where the out of sociological point of view important *Nardus stricta* L. occurred. In the mean time this was rectified and the following correlations of N could be calculated: with Ac +37, At +36, Ao +35, Cp +45, Fo +73, Fr +29, Hyp +8, L +59, M +22, Pe +63, Rla +26, S +22 and Sg +43. Out of this follows that N should be placed in the upper left corner of the figure, as was to be expected.

The 52 species are not spread equally over the figure; but now and again one finds a lot of them drawn together. Out of these combinations of frequent species, one can often recognize communities from the School of B R A U N - B L A N Q U E T. Yet one sees practically everywhere connections, though they are not quite so pronounced everywhere. This does not indicate, in my opinion, that plant communities, conceived as combinations of species, are sharply limited in nature.

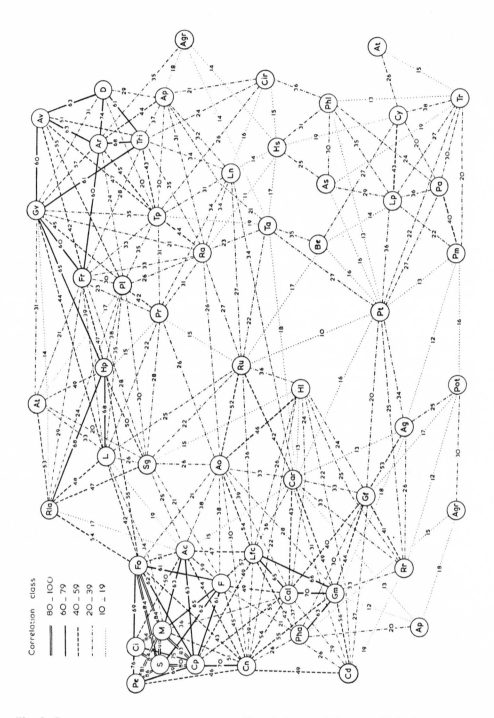

Fig. 2. CONSTELLATION OF HERBAGE PLANTS. (Correlation coefficients multiplied by 100. For the abbreviations see p. 107)

This constellation figure, drawn on a base of mathematically-statistical data, gives a big surprise in showing its ecological background so clearly. A big complex of species, that can endure drought, is to be found in the upper part of the figure. The most right situated group, with the triangle combination Ar-D-Tri as a centre, consists of lime loving species while the group situated left in this complex consists of acid loving species, such as Rla, L, Sg, At and Hp. Further more we know, that the species Fr which acts here as a link between right and left, can be subdivided in an acid loving and a lime loving variety. The combination Fo-N forms, what could be described as a transition to the wet complex left in the figure. The mutual strong chained species M, S, Cp, Ci and Pe (upper left) are most frequent on the poorest and most acid hayfields. The triangle combination Cal-Lfc-Gm (middle left) occurs more on less poor, but still too wet hay-fields, while during the same circumstances, but used for grazing purposes, Ag, Gf and Rr will become frequent (below left). The triangle combination Ao-Ru-Hl in the centre is characteristic for the typical transitional hayfields. Below right in the figure one lastly can fined the not too dry and not too wet grasslands, where the species Pt, Lp, As, Phl, Cy and Tr predominate. The central combination here is Lp-Cy, while the most fertile combinations are Lp-Pt and Lp-Phl, the poorest is Cy-At and the overgrazed those of the plants Pa and Pm, with Lp.

It often occurs one finds in the same grassland two or more combinations of two species together which do not connect in the figure. This can be very useful for the indication of environmental factors. If for instance the combinations Lp-Cy and D-Tri occur together, it does not only indicate one has a moderately fertile pasture, but more-over, that it is fairly dry and not acid.

In choosing the combinations promising to be most successful, one must not consider the strongest coupling only, but also the commonness and the frequency of the participating species. Moreover combinations with a bigger spreading are in preference, for instance those that occur in the mountains abroad. In view of this, the combination D-Tri will suffice better, than D-Ar, notwithstanding the r of D-Tri ($+0.61$) being lower than with Ar ($+0.74$). The first is more common. Not only is Ar more sensitive to severe frost, but it endures grazing badly. The result is, that above a certain height limit, and in more grazed grasslands, the combination D-Ar will not be found. This is the reason why in the Netherlands, with its intensive grasslandculture and where only seldom the real hayfields are found, Ar is less common than Tri. On the other hand the combination D-Tri which is found from the Northsea right up to the high mountains, represents in our opinion the Arrhenatheretum as well as the Trisetetum. In this way the combination M-Cp is prefered to M-S, because Cp is more abundant and Cal-Lfc to Cal-Gm, because already in the mountainous country of the Eifel Gm discontinues to come.

REFERENCES

1. Nielen, G. C. J. F. et J. G. P. Dirven, 1950 — De nauwkeurigheid van de plantensociolo-gische ¼ dm² frequentie-methode. (Summary: The accuracy of the 25 cm² specific frequency method.) *Versl. Landbouwk. Onderz.*, **56**, 13, p. 1—27, 's-Gravenhage.
2. Vries, D. M. de, 1937 — Methods of determining the botanical composition of hayfields and pastures. *Rep. Fourth Intern. Grassl. Congr.*, Great-Britain, p. 474—480, Aberystwyth.
3. —, 1938 — The plant sociological combined specific frequency and order method. *Chron. Bot.*, **4**, 2, p. 115—117, Leiden.
4. —, 1953 — Objective combinations of species. *Acta Bot. Neerl.*, **1**, 4, p. 497—499, Amsterdam.
5. —, et J. P. Baretta, 1952 — De constellatie van graslandplanten, (Summary: The constellation of herbage plants.) *Versl. C.I.L.O.* over 1951, p. 26—28, Wageningen.
6. —, et G. C. Ennik, 1953 — Dominancy and dominance communities. *Acta Bot. Neerl.*, **1**, 4, p. 500—505, Amsterdam.

24

Reprinted from *Ecology,* **52**(1), 26–36 (1971), with permission of the publisher, Duke University Press, Durham, N. C.

Distribution on Environmental Gradients: Theory and a Preliminary Interpretation of Distributional Patterns in the Avifauna of the Cordillera Vilcabamba, Peru

John Terborgh

[*Editors' Note:* In the original, material precedes this excerpt.]

THEORY

Given the setting described above, that of a sharply zoned vegetation superimposed on a continous physical gradient, one may inquire into the various factors that may enter into the determination of animal distributions. In this section we shall examine the implications of three plausible mechanisms, each of which has been evoked by other authors in less explicit form in attempts to explain particular distributional limits. Each mechanism is mutually exclusive of the other two and the three are complementary. While it is possible to think of other models to add to the set, in practice we have found it necessary to limit the number to three and to treat them in such a way as to include all observed distributions within their scope. The manner in which this is accomplished will be explained at an appropriate place later in the paper.

A simple statement can express the premise that underlies each model. The statements carry a number of implications which can be deduced logically or graphically. The set of implications that follows from each statement is unique; that is, the sum of the implications that forms any single model predicts a pattern of distribution that is distinguishable on one or more grounds from the patterns predicted by the other models. An exposition of the models follows seriatim.

Model I. Distributional limits of species on a gradient are determined by factors in the physical or biological environment that vary continuously and in parallel with the gradient.—Implicit in this statement are the absence of exclusion reactions between competing species and effects of discontinuities in the habitat, i.e., ecotones. Each species will show a characteristic optimum position on the gradient where it reaches maximum abundance. On either side of the optimum, abundance will decline more or less rapidly depending on the ecological amplitude of the species. Since optima will fall at random on the gradient, faunal turnover will be highly regular and will be determined only by the mean ecological amplitude of the several species present (Fig. 4a). A derivation of the form of what shall be called the congruity curve is given in the appendix. By eco-

logical amplitude we mean the range of conditions included within the limits of a species' occurrence on an environmental gradient, as measured in units of the gradient variable. For example, when the gradient is one of elevation, a species that is found between 1,500 and 2,500 m has an amplitude equal to 1,000 m.

Included within the scope of this model are all relevant features of the environment that vary in the specified manner with the gradient (elevation). These features may be physical, such as temperature, degree of cloudiness, etc., or biological. Of the many conceivably important biological variables, we can offer as examples three that appear to us to vary in the correct manner with elevation: 1) net annual plant productivity, 2) density of insects and 3) importance of epiphytic plants in the vegetation. The first and second of these decrease and the third increases upwards on the gradient.

Model II. Distributional limits of species are determined by competitive exclusion.—This model is simply a restatement of the Gaussian theorem as applied to spatial distributions. When the ecological requirements of closely related species are sufficiently similar, their coexistence will be unstable and their populations will be forced to occupy mutually exclusive domains (Fig. 4b). The abundances of two such species will fall off sharply in the zone of contact instead of trailing off gradually as is implied by the first model. We shall refer to the resulting truncation of the population density curves as a repulsion interaction. The points on a gradient at which excluding species meet will fall at ecotones only coincidentally.

MODEL I

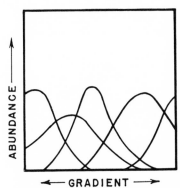 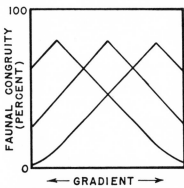

Fig. 4a. Properties of model I. The left-hand drawing illustrates that the abundances of different species and their ecological amplitudes on an environmental gradient are independent of one another. The righthand drawing portrays the expected random or smooth turnover of species along the gradient. Each "congruity curve" represents the degree of faunal similarity found in samples taken at higher and lower values of the gradient in comparison to a central or reference sample (see the Appendix for further explanation).

MODEL II

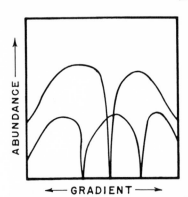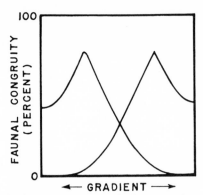

FIG. 4b. Properties of model II. Population density curves are truncated where competitors meet (repulsion interaction). Mean amplitude of the species in replacing series of three is less than that of species in series of two (amplitude compression). Since species are expected to replace each other at random points along the gradient, smooth congruity curves are predicted.

Hence the turnover in a whole fauna will be random and give rise to smooth congruity curves. Closely related species may replace one another sequentially in series of two, three or more. Mutual exclusion on a gradient must eventually lead to crowding in the larger series since all natural gradients encompass a finite range of conditions. Because of this spatial restriction, the amplitudes of species in the larger series should, on the average, be reduced in relation to species that are not repulsed by an adjacent competitor. This effect will be termed amplitude compression.

The reader may have noticed that the construction of this model is based on an extremely narrow concept of competition. Only a single type of interaction is included; that which leads to spatial exclusion of congeneric species. While this is an important and easily recognized form of competition, it is but one of many possible kinds of competitive interactions (in the broadest sense) that could restrict a species' ecological amplitude.

A form of competition which is undoubtedly of importance, especially in diverse tropical faunas, may be termed diffuse competition; that which comes from all related and nonrelated coexisting species that share a common pool of limited resources (Terborgh and Diamond 1970). Further, the more relaxed definitions of biological competition encompass most types of interspecific interactions, including predation and parasitism. Presence of the latter sorts of competitors on an environmental gradient could confine a species to the region of maximum fitness near its environmental optimum. Wherever diffuse competition

or the inroads of predators or parasites is more intense at one end of a species' amplitude than at the other end, it could lead to truncation of the abundance curve towards the former end.

Since, at the empirical level, it is difficult at best to implicate these kinds of interactions as probable causal mechanisms in limiting the distributions of species, we have found it necessary to exclude them from the competition model as presently constituted. Where distributional limits are imposed by the effects of one of these kinds of interaction, the instances will not be discriminated by our criteria and in practice will be relegated to one of the other two models, usually Model I. The problem of surmounting some of these weaknesses of the competition model will be the subject of future papers.

Model III. Distributional limits are determined by habitat discontinuities (ecotones).—If the spread of species populations is blocked by habitat discontinuities there will be massive faunal turnover at ecotones. Were all the species on a gradient so constrained, the faunal congruity curves would be nearly flat except at ecotones where sharp breaks would occur (Fig. 4c). Population density curves would thus tend to assume a square wave shape since available evidence indicates that the total density of individuals is nearly constant across ecotones.

A resumé of the properties and implications of these models is given in Table 1. Species whose distributions are restricted by competitive exclusion are predicted to be replaced abruptly by congeners, and to show repulsion interactions and

MODEL III

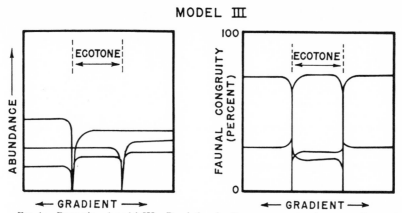

Fig. 4c. Properties of model III. Population density curves are sharply truncated at ecotones, giving rise to square-wave rather than bell-shaped curves. High rates of species turnover at ecotones produce steep, nearly discontinuous congruity curves.

TABLE 1. Predictions of three models of species distribution on environmental gradients. Predictions unique to the given model are in italics

	Model		
Distributional feature	Gradient	Competition	Ecotone
1. Population density curve.........................	± normal	*repulsion interaction*	*truncation*
2. Mutual exclusion.................................	none	*yes*	none
3. Amplitude compression...........................	none	*yes*	none
4. Congruity curves................................	smooth, symmetrical	smooth, symmetrical	*dicontinuous*
5. Amplitude distribution curve......................	± normal	*skewed right*	variable
6. No species near terminus.........................	*reduced*	not reduced	not reduced
7. Species loss near terminus........................	*not reduced*	reduced	reduced
8. Species gain near terminus........................	reduced	reduced	reduced
9. Mean amplitude near terminus.....................	*reduced*	constant	constant

amplitude compression, all features unique to the competition model. Where ecotones impose an insurmountable ecological barrier, the population density curves of the affected species should be truncated at the ecotones, and if appreciable numbers of species are involved the faunal congruity curves will contain conspicuous discontinuities. These predictions establish criteria by which species responding in accordance with two of the models can be recognized. For the time being the remaining model will have to serve as a repository for all data that is not explained by the other two. Thus we recognize that the present scheme yields only a first approximation to a complete solution to the interpretation of distributions.

The ecological amplitudes of the species inhabiting a gradient can often be expressed in terms of some convenient unit e.g., meters of elevation or depth, parts per thousand of salinity, etc. When the distributions of all the species in a group are known, their ecological amplitudes can be plotted as a frequency diagram (cf. Fig. 5). Contempla-

tion of the three models leads to the conclusion that each predicts a different frequency plot. The amplitudes of species that are responding only to the gradual change in conditions along a gradient will cluster approximately normally about the mean value as each species is similarly exposed to the opportunity of expanding into regions of higher and lower intensity of the controlling factor. The extent to which any species may become specialized or generalized with respect to any one gradient, however, is subject to obvious limitations. Excessive specialization that leads to a very narrow amplitude must increasingly expose the population to the risk of extinction. Stochastic fluctuations in population density will impose an ultimate limit on specialization. Empirically we find that this limit falls between 100 and 200 vertical meters for the species-rich Vilcabamba avifauna. At the opposite end of the adaptational spectrum is found the extreme generalist. Provided that successful existence at different points on a gradient requires some degree of divergent adaptation, the maximum

extent to which any interbreeding population may
enlarge its distribution will be regulated by the
rate of gene flow within the population. This, in
turn, is dependent on a number of factors of which
the most important is a species' dispersal tendency.
The amount of adaptation required for an incre-
ment of increase in amplitude may also be expected
to vary. For a few species such as swifts which
freely use the airspace over the entire 3,000-m
Vilcabamba slope, the gradient poses no great
adaptational challenge. For the great majority,
however, it does, as we note in the fact that only
a handful of species occurs over an elevational
differential of more than 2,000 m. Since diametri-
cally opposed evolutionary strategies appear to be
involved in the determination of ecological amplitude in species responding to a gradient, a high
variance in amplitude is to be anticipated.

Within a group of species whose distributional
limits are prescribed by competitive exclusion interactions, the frequency distribution of amplitudes
could take on a variety of forms, but whenever
there are a number of compressed species the
curve will be skewed to the right. There is no
predicting the form of the amplitude distribution
curve of species limited by ecotones since these
boundaries may fall at any point on a gradient.

Although the remaining predictions included in
Table 1 have so far been of little use in practice,
we include them for their intrinsic theoretical interest. Heretofore we have regarded gradients as
continua, and so they may be treated in their middle sections, but we shall now consider the reper-
cussions of their eventual termination. To species
responding to ecotones, a terminus is simply another ecotone and has no special significance.

Series of mutually excluding species, whose
members are compressed within the confines of a
limited gradient, represent another trivial case because the termini effectively act as unbreachable
ecotones. However, some distinctive patterns can
be expected in a set of species whose distributional
limits are determined by the gradient itself, because of the disadvantage of possessing an environmental optimum that lies close to a terminus.
Adaptation to an extreme position on a gradient
implies a truncated amplitude and hence an increased probability of extinction. From this reasoning we may infer that the number of species
present will decline somewhat towards a terminus
and that the number of additional species encountered will fall to zero before the terminus is
reached. Since it is likely that a few species will
persist near termini in spite of somewhat truncated amplitudes, the fauna inhabiting the immediate vicinity of a terminus might show a slightly
reduced mean amplitude. Species loss should con-

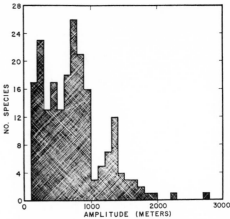

FIG. 5. Frequency distribution of elevational amplitudes of 207 species in the montane avifauna of the Cordillera Vilcabamba, Peru. Data are included only for
species whose upper and lower limits are adequately
known and which do not occur on the Apurimac Valley
floor. The mean amplitude is 741 m. Strong indications
of more than one peak in the frequency distribution suggest that two, three or possibly more separate mechanisms are involved in its determination.

tinue unabated up to any terminus for there is no
direct way in which a gradient-dependent population could be influenced by an extralimital boundary.

RESULTS AND DISCUSSION

Testing of the stated predictions of the models
requires two kinds of data: (i) quantitative measures of the abundances of individual species at
different points on a gradient and (ii) knowledge
of the entire fauna at several points so that congruity curves may be constructed. In dealing with
birds, observational data such as song censuses and
sight records may be used, although a more objective method is desirable for greater consistency
and reduced bias, especially in estimating abundances. For this reason, the results that follow
have been obtained largely through the use of
mist nets. These devices capture a random sample
(for any given species) of birds flying in the airspace between 0.2 and 2.0 m above the ground.
The four Vilcabamba expeditions have exploited
this technique to obtain large samples of birds at
15 levels on the transect (Table 2).

Since the ecotone model of species distribution
uniquely predicts discontinuities in faunal congruity curves, we may begin the inquiry by examining curves based on net samples taken at 11
elevations (Fig. 6). Each curve represents the
attenuation of the fauna contained in a sample of
219 to 300 birds as measured by samples captured

TABLE 2. Size and elevation of mist net samples

Median elevation of net line (meters)	Year and number of birds captured				
	1965	1966	1967	1968	Total
585	185	99			284
685				273	273
930		212	198	194	604
1,350		370			370
1,520		293	125		418
1,730			181	135	316
1,835			114	145	259
2,095			206		206
2,145		169	159	136	464
2,215			170		170
2,640			245	194	439
2,840			302	149	451
3,220			101	76	177
3,300			168	143	311
3,510				219	219
Total	185	1,143	1,969	1,664	4,961

above and/or below the sample in question. Extrapolation of the curves shows a faunal congruity of about 82% at the levels of the 11 samples against which the comparisons were made. Less than 100% congruity is obtained because a sample of 300 does not include all the potentially nettable species at any elevation on the transect. In only one set of net lanes (at 930 m) did we succeed in accumulating a sample of over 600 birds, enough to permit a check of the extrapolated values by a determination of the congruity in two independent samples of 300 from the same location. The agreement, as can be seen in Figure 6, is entirely satisfactory, indicating that the extrapolations provide fair approximations of the true shapes of the curves.

Inspection of the empirical curves reveals a consistent smoothness and symmetry, in accord with the predictions of models I and II. The only exception to constant or slightly concave slopes is found between 585 and 685 m across the lowland to montane rainforest ecotone. It is still unclear to us why the only resolvable ecotone effect should come at a point where the vegetational change affects mainly the upper canopy, rather than the understory from which the samples were captured. A second notable result is a trend towards reduced slopes on both the uphill and downhill sides of the curves representing the samples taken at higher elevations. An interpretation of this will be attempted later.

As most of the sampling stations were separated by vertical distances of 200–500 m, the possibility remained that ecotone effects were being blurred as an artifact of the low sampling density. Hence we undertook a thorough investigation of one ecotone; that at which montane rainforest and cloud forest meet. The bird faunas through and on both sides of the transition were sampled with a long net line that ran continuously between the elevations of 1,270 and 1,540 m. The catch of 663 birds was divided into four nearly equal samples, two above and two below the 1,380-m boundary. Congruity curves based on these data are as smooth as those based on more widely spaced samples, and confirm the lack of any faunal discontinuity coincident with the ecotone (Fig. 7).

A more sensitive test for faunal perturbations on a gradient can be devised from the information contained in congruity curves. One simply measures the slopes of a set of such curves on both

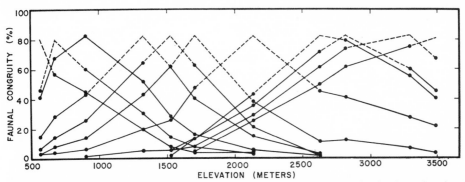

FIG. 6. Faunal congruity curves on the elevational gradient of the Cordillera Vilcabamba, Peru. Samples of approximately 300 birds were taken at each of eleven elevations. Each curve is constructed by comparing the faunal lists of all other samples to the list of a reference sample which represents the peak of the curve. To compute the points, the number of species each sample contains in common with the reference sample is divided by the total number of species in the reference sample and converted to a percentage. Duplicate samples at 930 m had 82% of species in common, so this is assumed to be the faunal congruity that would be found for duplicate samples at all elevations. Hence, all the curves peak at 82% for the reference elevation, but these extrapolated peaks are shown as dashed lines.

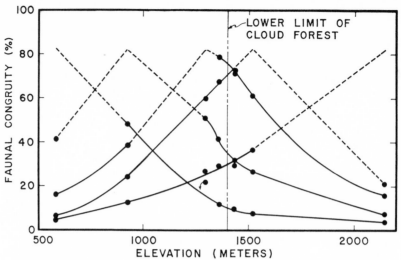

FIG. 7. Faunal congruity curves based on samples closely spaced about the lower limit of cloud forest in the Cordillera Vilcabamba, Peru.

the uphill and downhill sides at some given level of congruity (we used 50%), and notes the corresponding elevations by projecting the points onto the abscissa. A plot of these data is analogous to a derivative spectrum and serves to scan the entire gradient for irregularities in the faunal turnover rate (Fig. 8). Advancing upwards from the valley floor, the species gain (upper half of the figure) reaches 3% per 10 vertical meters in the vicinity of the lowermost ecotone and then quickly settles down to a nearly constant 1% per 10 m. Perhaps as a consequence of the abrupt appearance of a number of species at 600 m, the rate of species loss is noticeably reduced between 750 and 1,100 m (lower half of the figure). A small and possibly insignificant increase in the rate of species loss coincident with the lower limit of cloud forest (−1.3% vs. an average of ca. −1.0%) may reflect a slight ecotone effect. Above 2,000 m the loss and gain rates drop somewhat below 1.0% and appear to decrease slowly towards timberline. This last result will be taken up again after further relevant information has been introduced.

With the single exception of the few species involved in the high gain rate at the lowland-montane rainforest transition, it can be concluded that faunal turnover is continuous with elevation and hence that the ecotone model can account for the distributional limits of only a minor fraction of the bird species present in the Cordillera Vilcabamba.

We shall now turn to a consideration of evidence that pertains to the pattern prescribed by the competition model. Distributions which terminate because of competitive exclusion interactions ought to conform with the following three predictions: (i) mutual exclusion of congeners in linear series along the gradient, (ii) repulsion interactions, and (iii) amplitude compression in the larger series.

The first prediction is abundantly fulfilled by data of the kind shown in Figure 9. The forest-dwelling avifauna of the Cordillera Vilcabamba and Apurimac Valley floor (exclusive of all matorral species which do not enter into these results, see Terborgh and Weske 1969) amounts to somewhat more than 410 species. Of these, at least 180 (44%) are met without overlap, either above or below or on both sides by congeners. Exclusion interactions involve pairs (> 50), triplets (16) and quartets (6) of species. These series fall into 68 genera and 29 families, revealing a widespread evolutionary pattern in Andean birds. Elevational segregation of congeners has also been shown to occur on a large scale in the avifauna of the New Guinea highlands (Diamond 1969).

While in the midst of charting distributions in the field we became puzzled by apparent gaps between certain pairs of replacing species. A number of these are shown in the examples included in Figure 9. Some are surely due to experimental error while others were upheld by strenuous but unsuccessful efforts to find either species in the unoccupied zone. The large hiatus indicated between *Grallaria erythroleuca* and *G. rufula* is likely to be in error because Grallarias are notoriously difficult to observe and rarely enter nets. On the

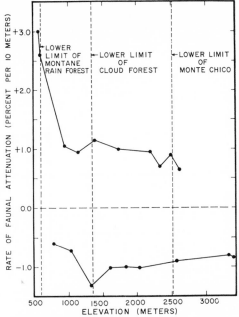

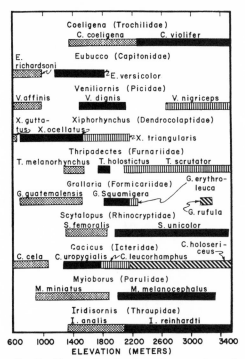

FIG. 8. Rates of faunal attenuation as a function of elevation in the Cordillera Vilcabamba, Peru. Points indicate the slopes of the congruity curves in Figure 6 at the 50% congruity level. Species gain with elevation is shown in the upper part of the figure (positive values) and species loss is shown in the lower part. These correspond, respectively, to the downhill and uphill halves of the congruity curves.

FIG. 9. Elevational replacement of congeneric species in 10 families of birds. Series of two, three and four replacing species are represented. Note the apparent hiatuses between some pairs of species and the consistency with which the uppermost species possesses the broadest amplitude.

other hand, the narrow gap between the two *Myioborus* warblers is probably real, or at least represents a zone of great scarcity, for both are abundant and conspicuous where present. We do not wish to delve into the possible interpretations of this phenomenon for the time being, yet it deserves mention for its novel implications. R. H. MacArthur has devised a theoretical scheme which predicts that competing species in a continuum will be separated by vacant zones under certain conditions (pers. commun.).

Especially favorable circumstances are required to demonstrate repulsion interactions between replacing congeners. The species must both be common and efficiently netted. Yet in a majority of the cases for which suitable data are available, the population density curves are conspicuously truncated on the side(s) towards an excluding congener (Figs. 10, 11 and 12). Generalizing from the examples for which sufficiently detailed information is at hand, it appears that repulsion interactions are a frequent but not universal corollary of mutual exclusion.

Arguments in support of the notion of compe-

tition between closely related species in nature are more often based on circumstantial fact than on tested hypotheses. Exclusion on a gradient, for example, is often put forth as confirming evidence, but it may be merely circumstantial and falls short of proving the case. Of the three explicit predictions of the competition model, that of amplitude compression is the one that most rigorously upholds the assumption of a state of tension between populations. It offers the only possibility for a direct comparison of sets of species that are presumed to be in competition with a set that is presumed not to be, or at least to a lesser degree.

If competition were the only factor to limit distributions then one could expect species which lacked competitors to occupy the entire gradient and those which formed excluding series to show reduced amplitudes in accordance with the length of the series. In such a case the amount of amplitude compression would be in strict proportion to the length of the excluding series. Obviously the real world is not so simple, for only five species span the entire gradient and fewer than two dozen

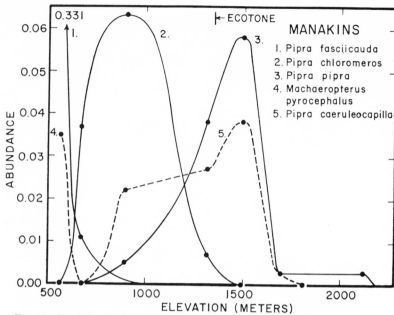

Fig. 10. Population density curves for species in the family *Pipridae* (Manakins). The ordinate represents the fractional abundance of the species in net samples. Species 1, 2 and 3 are of nearly equal size and distinctly larger than species 4 and 5. The large species show some overlap but this is due entirely to individuals in female or juvenile plumage. There is an hiatus between the two small species which taxonomists place in separate genera.

fill even half of it. The mean amplitude for all 207 species whose limits are well known and do not reach the lower terminus is 741 m (Table 3). These are the species whose amplitudes are portrayed in Figure 5. Species occurring on the valley floor are excluded because their amplitudes are arbitrarily truncated at 585 m; the real terminus is at sea level. Taking 741 m as a standard amplitude it could then be argued that competition alone should not lead to compression in series of fewer than four species, because nearly four of the standard amplitudes can be accommodated in a 3,000-m gradient. We may thus draw two conclusions from the data presented in Table 3: (i) Strong compression is apparent, even in series of two, but only in the lower-altitude members of the series and (ii) the members of excluding series are not able to spread at will along the gradient to achieve a compressionless accommodation.

T-tests of the significance of differences between the means of various groups of species yield the following results (Table 4):

1) Uppermost members of the three- and four-fold series possess significantly greater amplitudes than lower members, but the difference is not significant for twofold series. Evidence for a general trend towards expanded amplitudes in the upper half of the gradient has already been presented. By breaking the fauna into subsets it becomes clear that both monotypic forms and the uppermost members of replacing series contribute to the phenomenon detected in the congruity curves. (We use monotypic here in the restricted context of the Vilcabamba fauna for species having no congeners present on the gradient.) The mean amplitude of 16 monotypic species whose upper limits lie above 3,000 m is 1,348 m while that of 18 monotypic species whose upper limits lie between 2,000 and 3,000 m is only 815 m. The difference is significant at the 0.001 level. Consideration of the possible meaning of these results will be taken up at a later point.

2) Amplitudes of the lower members of series are all significantly reduced in comparison with the mean of monotypic species, but there are no significant differences between the set of monotypic species and the sets of uppermost species.

3) Lower members of two-fold series possess significantly greater amplitudes than the lower members of three- and four-fold series, but the means of lower members of the three- and fourfold series are not significantly different.

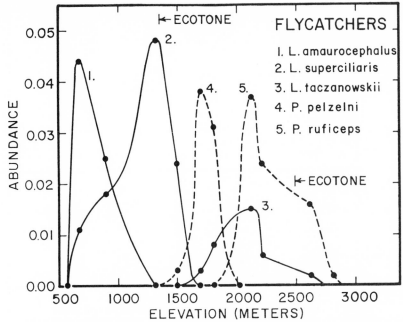

Fig. 11. Population density curves for species in family *Tyrannidae* (Flycatchers). Species 1 and 2 overlap broadly, but 2 reaches maximum abundance in the absence of 1. Both 1 and 2 appear abruptly above the lowland-montane rainforest ecotone. Truncation of the curves (repulsion interaction) is apparent in the zones of replacement of species 2 and 3 and of species 4 and 5. The genera are *Leptopogon* and *Pseudotriccus*.

The latter two findings substantiate the existence of amplitude compression in the lower members of series of replacing congeners. As the competition model predicts, the compression is progressive, but because few of the series have expanded to fill the entire gradient the degree of compression is less than proportional to the length of the series. It is of interest to note that the compression experienced by the lower members of fourfold series (56%) restricts their amplitude, and hence mean population size, to less than half of that of species having no obvious competitors in the fauna. If it is presumed that the probability of extinction is increased by reduced amplitude, then it is clear (i) why the number of series drops off rapidly with increasing length and (ii) why there are no series containing more than four members. (Parenthetically, it should be mentioned that a fivefold series in *Grallaria* is probable in other localities where *G. andicola* lives in scattered patches of brushwood at elevations around 4,000 m.)

Of the three models that were proposed to account for species distributions on environmental gradients, we have considered evidence pertaining to two: the ecotone and competition models. Un-

fortunately the gradient model offers no means of directly and unequivocally identifying species whose limits conform to its precepts. However, if the distributional limits that concur with the features of the other two models can be identified, the remaining limits, by default, can be assigned to the gradient model.

With these limitations of method in mind, it will be of interest to partition the upper and lower limits of netted bird species in accordance with the following procedure. (i) Distributions that extend to either the upper or lower terminus of the gradient cannot be assigned to any of the models and hence are set aside. (ii) Limits that, within 10 vertical meters, coincide with one of the three above-mentioned ecotones have been assigned to that model. Examples are the lower limits of *Leptopogon amaurocephalus* and *L. superciliaris* (Fig. 11) and the lower limit of *Basileuterus coronatus* (Fig. 12). (iii) Limits which lie at the boundary between two replacing congeners, whether or not they are separated by an hiatus, have been assigned to the competition model. (iv) In view of the general paucity of ecotone effects, congener pairs that happen to meet at ecotones have been presumed to do so coincidentally

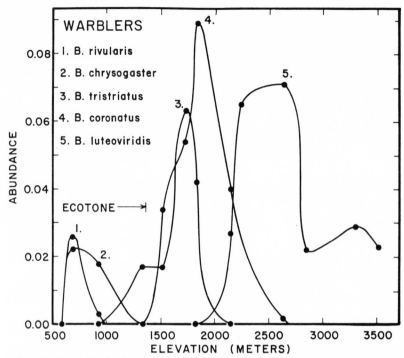

FIG. 12. Population density curves for species in the genus *Basileuterus, Parulidae* (Warblers). Species 1 and 2 overlap but are of different size, occupy different habitats and forage at different levels in the vegetation. Species 2, 3 and 5 replace each other without overlap, while species 4, which differs in size from both 3 and 5, reaches maximum abundance in the replacement zone. Notice that the lower limit of species 4 coincides with the ecotone.

and accordingly have been assigned to the competition model. (v) All remaining limits are arbitrarily assigned to the gradient model.

The outcome of this procedure reveals that the three presumed mechanisms of amplitude determination differ appreciably in their importance in the Vilcabamba avifauna (Table 5). Ecotones account for less than 20% of all distributional limits, a result that must come as a surprise to anyone familiar with the importance of habitat in predicting the census of North American bird species (MacArthur, MacArthur, and Preer 1962). Moreover, the stated figures are likely to represent overestimates since it can be expected that some of the included limits fell at ecotones coincidentally.

Approximately one-third of all limits are ascribed to competitive exclusion. In this case some underestimation is probable because of the omission of several replacing series of related species whose members have been placed by taxonomists in separate genera. (Examples are to be found among the guans, toucans, antbirds, etc.). A fur-

ther underestimation would result should the occurrence of any appreciable number of species be restricted by diffuse or other nonobvious kinds of competitive interactions. These considerations lead to the conclusion that, in the densely packed avifauna of the Andes, competition determines at least twice as many distributional limits as do ecotones. Where the pressure of high diversity is relaxed, as on islands or at temperate latitudes, the importance of competition is likely to be considerably reduced. Under such circumstances one could expect that ecotones would assume a more prominent role, as species amplitudes would in general be broader.

[*Editors' Note:* Material has been omitted at this point.]

LITERATURE CITED

Diamond, J. M. 1969. Preliminary results of an ornithological exploration of the North Coastal Range, New Guinea. Amer. Mus. Novitates **2362**: 1–57.

MacArthur, R. H., J. W. MacArthur, and J. Preer. 1962. On bird species diversity. II. Prediction of bird census from habitat measurements. Amer. Naturalist **96**: 167–174.

Terborgh, J., and J. M. Diamond. 1970. Niche overlap in feeding assemblages of New Guinea birds. Wilson Bull. **81**: 29–52.

Terborgh, J., and J. S. Weske. 1969. Colonization of secondary habitats by Peruvian birds. Ecology **50**: 765–782.

25

Reprinted from *Wilson Bull.*, **83**(3), 215–229, 235–236 (1971)

ORDINATIONS OF HABITAT RELATIONSHIPS AMONG BREEDING BIRDS

Frances C. James

I N an attempt to express habitat relationships in a new way, I have applied two methods of multivariate analysis to a large set of data pertaining to the habitats of 46 species of common breeding birds. The question asked was: How do these species distribute themselves with respect to the structure of the vegetation? This required (1) devising field techniques that would give quantitative measurements of the vegetation within the breeding territories of individual birds, (2) analyzing these by species in order to obtain a sample of the characteristic habitat dimensions of the species niche, (3) reconstructing the relationships among the species according to their relative habitat separation, and (4) considering the ability of the vegetational variables to describe differences among habitats mathematically.

Data were gathered in the spring and summer of 1967 in Arkansas. The vegetation was sampled in 0.1-acre circular plots, using singing male birds as the centers of the circles. The statistical procedures used which were principal component analysis and discriminant function analysis provided a tool for describing bird distribution objectively as ordinations of continuously-varying phenomena along gradients of vegetational structure. The relative positions of the species were located within multidimensional "habitat space." The relationship between this approach and studies involving ordinations of plant and animal communities is discussed.

FIELD METHODS

Estimates of the characteristics of the structure of the vegetation were obtained by means of sampling one 0.1-acre circular plot within the territory of each singing male bird. A 0.1-acre is a large enough area (radius 37 feet) that it should include an adequate sample of the vegetation. It is convenient to have a circular plot with its center at a singing perch selected by a territorial bird. This might give a biased view of habitat for species which occur in open areas and choose singing perches in places very different from their foraging areas, but this objection is minimized in the forest (including most of the species considered here).

The sampling technique was a modification of the range-finder circle method recommended by Lindsey, Barton, and Miles (1958) as a very accurate and efficient procedure. The range-finder itself was found to be unnecessary. Instead, I suspended a brightly colored yardstick at or below the spot where a territorial male bird was singing. This was sighted by holding at armslength a second yardstick having a mark equal to the length of the first when viewed from the perimeter of the circle. This proved to be an accurate and efficient way of determining whether I was within the area to be sampled. A total of 401 0.1-acre circles was measured in the territories of 46 species. No attempt was made to remain within a fairly uniform stand. In fact as many habitat types as

TABLE 1

FIFTEEN VARIABLES OF THE STRUCTURE OF THE VEGETATION CONSIDERED IN THE ANALYSIS
OF 0.1-ACRE PLOTS SHOWING THE CORRESPONDING SYMBOLS USED IN TABLES 2 AND 3

1	% GC	Per cent ground cover divided by 10
2	S/4	Number of shrub or tree stems less than 3 inches DBH per two armslength transects (0.02 acres) divided by 4
3	SPT	Number of species of trees
4	% CC	Per cent canopy cover divided by 10
5	CH	Canopy height divided by 10
6	T_{3-6}	Number of trees 3 to 6 inches DBH
7	T_{6-9}	Number of trees 6 to 9 inches DBH
8	T_{9-12}	Number of trees 9 to 12 inches DBH
9	T_{12-15}	Number of trees 12 to 15 inches DBH
10	$T_{>15}$	Number of trees greater than 15 inches DBH
11	$CH \times S$	Canopy height $\times$ shrubs (variable 2 $\times$ variable 5)
12	$CH \times T_{3-9}$	Canopy height $\times$ trees 3 to 9 inches DBH [variable 5 $\times$ variables (6 + 7)]
13	$CH \times T_{>9}$	Canopy height $\times$ trees greater than 9 inches DBH [variable 5 $\times$ variables (8 + 9 + 10)]
14	T^2_{3-9}	Number of trees 3 to 9 inches DBH squared [square of variables (6 + 7)]
15	$T^2_{>9}$	Number of trees greater than 9 inches DBH squared [square of variables (8 + 9 + 10)]

possible were sampled. Data were obtained in eighteen different counties in various parts of Arkansas. In the few cases in which two species were singing in the same 0.1-acre circle, data for that circle were used to describe one observation of each of the species. In the subsequent analysis data from the circles were organized by species of bird, regardless of where the data were obtained.

Each tree greater than three inches in diameter at breast height (DBH) within the circle was identified to species and the size class was recorded. The same sighting stick mentioned above was graded on the other side for three-inch size-class estimates of tree diameters. Calibrations on the stick were determined by using the formula $S = \sqrt{(aD^2)/(a + D)}$, where S is the graduation on the stick, a is the armlength of the observer, and D is the diameter at breast height (Forbes, 1955).

To estimate shrub density, two armlength transects together totalling 0.02 acres were made across the circle and the number of stems intersected that were less than three inches DHB was recorded. An estimate of ground cover was made by taking 20 plus-or-minus readings for the presence or absence of green vegetation sighted through a sighting tube 1.25 inches in diameter held at armslength. An estimate of canopy cover was made by taking 20 plus-or-minus readings for the presence or absence of green leaves sighted directly upwards on alternate steps of a transect of the circle. The average height of the canopy was measured with a clinometer. After some practice a level of efficiency was reached whereby the field data for one 0.1-acre circular plot could be obtained in 15 to 20 minutes of effort. A more detailed description of this sampling technique is given elsewhere (James and Shugart, 1970).

Measurements of 10 vegetational variables were made in each 0.1-acre circle (first 10 items in Table 1). To facilitate handling the data, percentage values for ground cover and canopy cover and the values for canopy height in feet were divided by ten. The number of shrub stems intersected in two transects was divided by four. The last five items in Table 1 are multiples of the first 10. These were used in the discriminant function analysis to determine whether variables were interacting in such a way that their combinations were more highly correlated with the specificity of bird habitats than were the originally measured variables.

THE NICHE-GESTALT

The assumptions underlying both the field methods and the analysis are somewhat different from those used in other recent studies of avian habitats. In the latter the experimental unit is generally the avian community. Analysis is of study plots large enough to support several coexisting species, and this permits interpretations concerning diversity, resource division, and the relative width of ecological niches (MacArthur and MacArthur, 1961; MacArthur and Pianka, 1966; MacArthur, Recher, and Cody, 1966; MacArthur and Levins, 1967; Cody, 1968; Wiens, 1969; and others). In the present study the advantages of community approach are sacrificed in favor of the opportunity to view habitat relationships among a large number of species occurring in a large geographic area as if each were dependent up a specific life form or configuration of vegetational structure. The experimental unit is the basic life form of the vegetation that characterizes the habitat of each particular species. Measurements from territories are organized by species without regard for which other species occurred nearby. This approach can be defended only if one assumes that predictable relationships exist between the occurrence of a bird and of its characteristic vegetational requirements. I have called this basic configuration of the ecological niche, the *niche-gestalt*.

It is not required that this configuration is directly meaningful to the bird, but this hypothesis could be tested by presenting it with different configurations to see whether it recognizes them as appropriate (see Klopfer, 1963, 1965; Wecker, 1963, 1964; Harris, 1952). Inherent in the term *gestalt* are the concepts that each species has a characteristic perceptual world (the Umwelt of von Uexküll, 1909), that it responds to its perceptual field as an organized whole (the Gestalt principle, see Köhler, 1947), and that it has a predetermined set of specific search images (Tinbergen, 1951). This is

BELL'S VIREO

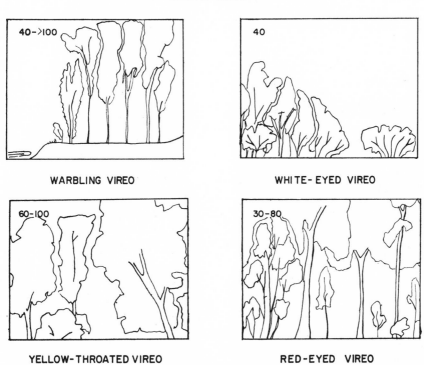

WARBLING VIREO   WHITE-EYED VIREO

YELLOW-THROATED VIREO   RED-EYED VIREO

FIG. 1. Outline drawings of the niche-gestalt for five species of vireos, representing the visual configuration of those elements of the structure of the vegetation that were consistently present in the habitat of each. Numbers give the vertical scale in feet.

FIG. 2. Outline drawings of the niche-gestalt for six species of warblers, representing the visual configuration of those elements of the structure of the vegetation that were consistently present in the habitat of each. Numbers give the vertical scale in feet.

FIG. 3. Marsh at the edge of Lake Sequoyah, five miles east of Fayetteville, Washington Co., Ark., where Bell's Vireos and Yellowthroats had breeding territories.

assumed to be at least partially genetically determined, but is surely also modifiable by experience and subject to ecological shift under varying circumstances. Whereas the community approach is sensitive to shifts in habitat due to such factors as competition for resources, the present approach is an attempt to define relationships among birds based upon the basic life forms of the vegetation which each species requires. Since the geographic range of every species is unique and since species are uniquely adapted to utilize certain aspects of their environment, I hope the reader will agree that this approach is justified.

The outline drawings (Figs. 1 and 2) are examples of visual descriptions of the life forms of the vegetation that were consistently present in the habitats of the species in question. These were made by comparing notes and photographs of each 0.1-acre circle where a species occurred and by selecting *only* the features in common. Conversely, if definable niche-gestalt units occur, it should be possible to discover as many of these units as there are pairs of breeding birds in any one place. For example the vegetational configuration in the drawings for the Bell's Vireo (Fig. 1) and the Yellowthroat (Fig. 2) can be identified in a photograph of a place where both occurred (Fig. 3). Likewise the configurations which characterize the habi-

Fig. 4. Vegetation along the Mulberry River, five miles east of Cass, Franklin Co., Ark., where pairs of White-eyed Vireos, Redstarts, and Parula Warblers were nesting.

tats of the White-eyed Vireo (Fig. 1), American Redstart, and Parula Warbler (Fig. 2) can be identified in Figure 4; a territorial male Red-eyed Vireo (Fig. 1), Hooded Warbler, and Ovenbird (Fig. 2) were each present where Figure 5 was photographed.

An attempt will be made to reconstruct relationships between species-specific niche-gestalt units from the quantitative data and to view them in multidimensional "habitat space." Of course this space also contains gradients in types of food, nest-sites, microclimate, etc. Although these variables are undefined in the present study, they would have to be included in a thorough analysis of the ecology of adaptation.

RESULTS

Correlations Among Vegetational Variables.—The vegetational variables are highly interrelated. In the correlation matrix (Table 2) all values of r greater than 0.39 are significant at $\alpha = 0.01$ (44 df). The first column, percentage of ground cover, is negatively correlated with all of the other variables. The second column, an estimate of shrub density, has a different pattern of variation from the last eight columns, which are all characteristics of trees. Shrub density varies concordantly with the number of small trees

Fig. 5.   Upland mesic forest at Cherry Bend, Franklin Co., Ark., in the Ozark National Forest, where Red-eyed Vireos, Ovenbirds, and Hooded Warblers had breeding territories.

and also with the number of species of trees and canopy cover. But shrub density varies independently of canopy height and trees greater than six inches DBH. Correlations between the number of species of trees per unit area, percentage of canopy cover and canopy height are particularly highly related to each other and to tree density by size classes (last five columns). This means that for a 10×46 data matrix of mean values of each vegetational variable for each species (see next section), a large amount of the variation is statistically attributable to these variables. Although there appears to be redundancy in the five interrelated variables for number of trees by size classes (last five items in Table 2), it will be shown in a later section that each contributes significantly to the statistical description of habitat differences among the species of birds.

PRINCIPAL COMPONENT ANALYSIS

Morrison (1967) defines principal components as those linear combinations of the responses which explain progressively smaller portions of the total sample variance. The components can be interpreted geometrically as the variates corresponding to the principal axes of the scatter of observations in space. If a sample of N trivariate observations had the ellipsoidal scatter

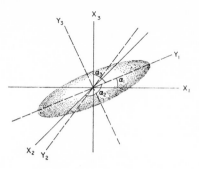

FIG. 6. Principal axes of trivariate observations (redrawn from Morrison, 1967).

plot shown in Figure 6, the swarm of points could be defined as having a major axis Y_1 and less well defined minor axes Y_2 and Y_3. If Y_1 passes through the sample mean point its position can be determined by its orientation with regard to the original response axes (angles a_1, a_2, a_3). The major axis passes through the direction of maximum variance in the points and represents a continuum of the first principal component of the system. The importance and usefulness of the component can be measured by the proportion of the total variance attributable to it. If this proportion is high, then it would be reasonable to express the variation in the data set along a single continuum rather than in N-dimensional space. The second principal component represents that linear combination of the responses that is orthogonal (perpendicular) to the first and has the maximum variance in this direction. The variances of successive components sum to the total variance of the responses. The advantage of the analysis is that it can take

TABLE 2

CORRELATION MATRIX (r) FOR 10 VEGETATIONAL VARIABLES

$N = 46$

	% GC	S/4	SPT	% CC	CH	T_{3-6}	T_{6-9}	T_{9-12}	T_{12-15}	$T_{>15}$
% GC										
S/4	−0.44**									
SPT	−0.67**	0.54**								
% CC	−0.76**	0.55**	0.80**							
CH	−0.51**	0.23	0.72**	0.77**						
T_{3-6}	−0.63**	0.54**	0.92**	0.76**	0.60**					
T_{6-9}	−0.58**	0.25	0.80**	0.79**	0.76**	0.81**				
T_{9-12}	−0.52**	0.06	0.61**	0.61**	0.63**	0.57**	0.77**			
T_{12-15}	−0.59**	0.15	0.69**	0.63**	0.65**	0.61**	0.68**	0.77**		
$T_{>15}$	−0.45**	0.16	0.66**	0.62**	0.81**	0.47**	0.55**	0.43**	0.48**	

** Significant at $\alpha = 0.01$.

TABLE 3

SUMMARY OF THE RESULTS OF THE PRINCIPAL COMPONENT ANALYSIS OF MEAN VALUES
OF EACH OF 10 VEGETATIONAL VARIABLES FOR 46 SPECIES OF BREEDING BIRDS

		Component		
	I	II	III	IV
Percentage of total variance accounted for	·64.8	12.5	7.7	4.9
Cumulative percentage of total variance accounted for	64.8	77.3	85.0	89.9
Correlations to original variables				
% GC	−0.77	0.21	0.15	0.53
S/4	0.46	−0.83	0.04	0.03
SPT	0.93	−0.16	0.03	0.17
% CC	0.91	−0.17	0.06	−0.12
CH	0.84	0.25	0.35	0.01
T_{3-6}	0.87	−0.25	−0.14	0.29
T_{6-9}	0.89	0.16	−0.12	0.25
T_{9-12}	0.76	0.41	−0.34	0.01
T_{12-15}	0.80	0.30	−0.27	−0.13
$T_{>15}$	0.71	0.22	0.62	−0.07

N-dimensional data and reduce it to a few new variables which account for known amounts of the variation in the original set.

In the present case, the basic ten vegetational variables (first 10 items in Table 1) are used as coordinates of a hypothetical ten-dimensional space. Each of the 46 species of birds has a position in this space according to the mean values of the variables for the 0.1-acre circles measured. This complex situation is analyzed so that a few new variables, the principal components are derived. The principal component analysis is summarized in Table 3.

The first or major component accounts for 64.8 per cent of the total variance and is highly correlated with all of the original variables. All values are positive except percentage of ground cover. The highest correlations are with number of species of trees per 0.1-acre, percentage of canopy cover, number of small trees, and canopy height. Species found where ground cover is high and where there are few shrubs and trees would be expected to have low values of the first component. Species found in mature forests, where ground cover is low and there are many trees of various species and sizes, would be expected to have high values of this component.

The second principal component accounts for an additional 12.5 per cent of the total variance (Table 3). Correlations between it and the original variables show that it represents an inverse interaction between medium-sized trees and shrub density. Species inhabiting dense shrubs would have low values of this component. Species found where there are medium-sized trees and few shrubs would have high values of the second component. The third component accounts for 7.7 per cent of the variance in addition to that already explained. It represents parkland, the presence of large trees with the absence of smaller ones. The fourth component, representing 4.9 per cent of the variance is most closely associated with ground cover. By means of these four newly-computed variables, it has been possible to account for 89.9 per cent of the variation in the original data set. The analysis has derived a parsimonious description of the dependence structure of the multivariate system.

Now it is possible to reconstruct the habitat relationships among these species using the components as coordinates. Figure 7 is a three-dimensional view of the position of each species listed in Table 4 along the axes of the first three principal components. The horizontal axis, representing the first component, has separated the species fairly regularly from open-country birds on the left found in places having high ground cover and few trees (Prairie Warbler, Bell's Vireo, Yellow-breasted Chat, Brown Thrasher) to birds on the right found in well-developed shaded forests (Ovenbird, Red-eyed Vireo, Wood Thrush). In the center along this axis falls a group of species that show remarkable latitude in their choice of habitat (Cardinal, Brown-headed Cowbird, Blue-gray Gnatcatcher). The axis of the second principal component extends backwards from species found in shrubs and low trees (Catbird, White-eyed Vireo, Kentucky Warbler) in the foreground toward species found where there is limited understory (Prothonotary Warbler, Robin, Red-headed Woodpecker). The axis of the third component extends vertically from species not dependent on large trees to those requiring large trees. The highest circles are for the Baltimore Oriole and Hooded Warbler.

Distances between species in Figure 7 represent ecological differences in "habitat space." Consider the positions of the five species of vireos. Their major separation is accomplished along the axis of the first principal component in the order Bell's, Warbling, White-eyed, Yellow-throated, and Red-eyed. This ordering corresponds to increases in the following: number of species of trees per unit area, percentage of canopy cover, number of small trees per unit area, and canopy height (see legend for Fig. 7). Along the axis of the second component (bases of the vertical lines) the same species fall in the order White-eyed, Red-eyed, Bell's, Warbling, and Yellow-throated. This axis is defined as increasing number of medium-sized trees and/or decreasing shrub density. Along the axis of the third component (height of circles) the

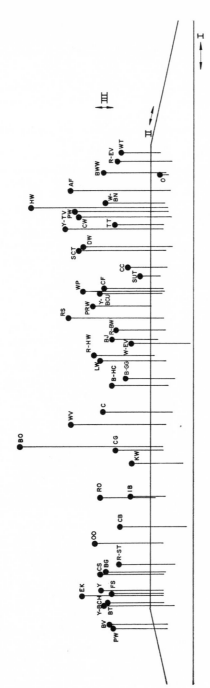

Fig. 7. Three-dimensional ordination of the distribution of 46 species of birds according to the first three principal components of their habitat relationships. The first component, extending from left to right and accounting for 64.8 per cent of the total variance, is highly correlated with all the vegetational variables measured but mainly with increasing number of species of trees, percentage of canopy cover, number of small trees, and canopy height. The second principal component, extending from front to back and accounting for an additional 12.5 per cent of the variance, represents an increasing number of medium-sized trees and/or decreasing shrub density. The third component, extending from low to high and accounting for 7.7 per cent of the variance, represents the presence of large isolated trees. The total variance explained by this ordination is 85 per cent. Symbols for the species are given in Table 4.

TABLE 4
LIST OF SPECIES IN ALPHABETICAL ORDER GIVING SYMBOLS USED IN FIGURES 7 AND 9

AF	Acadian Flycatcher	(*Empidonax virescens*)
BG	Blue Grosbeak	(*Guiraca caerulea*)
B-GG	Blue-gray Gnatcatcher	(*Polioptila caerulea*)
B-HC	Brown-headed Cowbird	(*Molothrus ater*)
BJ	Blue Jay	(*Cyanocitta cristata*)
BO	Baltimore Oriole	(*Icterus galbula*)
BT	Brown Thrasher	(*Toxostoma rufum*)
BV	Bell's Vireo	(*Vireo bellii*)
BWW	Black-and-White Warbler	(*Mniotilta varia*)
C	Cardinal	(*Richmondena cardinalis*)
CB	Catbird	(*Dumetella carolinensis*)
CC	Carolina Chickadee	(*Parus carolinensis*)
CF	Crested Flycatcher	(*Myiarchus crinitus*)
CG	Common Grackle	(*Quiscalus quiscula*)
CS	Chipping Sparrow	(*Spizella passerina*)
CW	Carolina Wren	(*Thryothorus ludovicianus*)
DW	Downy Woodpecker	(*Dendrocopos pubescens*)
EK	Eastern Kingbird	(*Tyrannus tyrannus*)
FS	Field Sparrow	(*Spizella pusilla*)
HW	Hooded Warbler	(*Wilsonia citrina*)
IB	Indigo Bunting	(*Passerina cyanea*)
KW	Kentucky Warbler	(*Oporornis formosus*)
LW	Louisiana Waterthrush	(*Seiurus motacilla*)
O	Ovenbird	(*Seiurus aurocapillus*)
OO	Orchard Oriole	(*Icterus spurius*)
PW	Prairie Warbler	(*Dendroica discolor*)
PAW	Parula Warbler	(*Parula americana*)
PRW	Prothonotary Warbler	(*Protonotaria citrea*)
RS	American Redstart	(*Setophaga ruticilla*)
RO	Robin	(*Turdus migratorius*)
R-BW	Red-bellied Woodpecker	(*Centurus carolinus*)
R-EV	Red-eyed Vireo	(*Vireo olivaceus*)
R-HW	Red-headed Woodpecker	(*Melanerpes erythrocephalus*)
R-ST	Rufous-sided Towhee	(*Pipilo erythrophthalmus*)
SCT	Scarlet Tanager	(*Piranga olivacea*)
SUT	Summer Tanager	(*Piranga rubra*)
TT	Tufted Titmouse	(*Parus bicolor*)
W-BN	White-breasted Nuthatch	(*Sitta carolinensis*)
W-EV	White-eyed Vireo	(*Vireo griseus*)
WP	Eastern Wood Peewee	(*Contopus virens*)
WT	Wood Thrush	(*Hylocichla mustelina*)
WV	Warbling Vireo	(*Vireo gilvus*)
Y	Yellowthroat	(*Geothlypis trichas*)
Y-BCH	Yellow-breasted Chat	(*Icteria virens*)
Y-BCU	Yellow-billed Cuckoo	(*Coccyzus americanus*)
Y-TV	Yellow-throated Vireo	(*Vireo flavifrons*)

Warbling and Yellow-throated Vireos have higher positions than the others, indicating that they require the presence of higher trees. These relationships can be checked by considering the drawings in Figure 1 in the order that the species fall along the respective axes. The same procedure can be applied to the six species of warblers for which the niche-gestalt is outlined in Figure 2.

Although the species in Figure 7 are fairly evenly distributed, several appear to be more isolated than the others, and these are birds that are not widely distributed in Arkansas in the breeding season. The Baltimore Oriole occurs in summer only in places having very large trees with clearings below. These are in towns and farmyards in the southern parts of the state and along river banks. Warbling Vireos are confined to cottonwoods (*Populus*) and willows (*Salix*) along major rivers or adjacent to them. Hooded Warblers occur in upland and lowland situations but only in the most mature mesic forests.

I do not want to exaggerate the validity of specific relationships. This analysis is based on mean values of the vegetational variables without regard for their variance. Sample sizes by species are small, and data pertain to a limited area of the breeding range of each. Nevertheless, a complex environmental situation has been reduced to a manageable mathematical and diagrammatic structure.

[*Editors' Note:* Material has been omitted at this point.]

REFERENCES

Cody, M. L. 1968. On the methods of resource division in grassland bird communities. Amer. Naturalist, 102:107–147.

Forbes, R. D. (ed.). 1955. Forestry handbook. The Ronald Press Company, New York.

Harris, V. T. 1952. An experimental study of habitat selection by prairie and forest races of the deermouse, *Peromyscus maniculatus.* Contrib. from the Laboratory of Vertebrate Biology, No. 56, Univ. Michigan Press, Ann Arbor.

James, F. C., and H. H. Shugart, Jr. 1970. A quantitative method of habitat description. Audubon Field Notes, 24:727–736.

Klopfer, P. 1963. Behavorial aspects of habitat selection: the role of early experience. Wilson Bull., 75:15–22.

Klopfer, P. 1965. Behavorial aspects of habitat selection: a preliminary report on stereotypy in foliage preferences of birds. Wilson Bull., 77:376–381.

Köhler, W. 1947. Gestalt psychology. Liveright Publ. Corp., New York.

Lindsey, A. A., J. D. Barton and S. R. Miles. 1958. Field efficiencies of forest sampling methods. Ecology, 39:428–444.

MacArthur, R. H., and R. Levins. 1967. The limiting similarity, convergence, and divergence of coexisting species. Amer. Naturalist, 101:377–385.

MacArthur, R. H., and J. W. MacArthur. 1961. On bird species diversity. Ecology, 42:594–598.

MacArthur, R. H., H. Recher, and M. Cody. 1966. On the relation between habitat selection and species diversity. Amer. Naturalist, 100:319–332.

MacArthur, R., and E. Pianka. 1966. On optimal use of a patchy envrionment. Amer. Naturalist, 100:603–609.

Morrison, D. F., 1967. Multivariate statistical methods. McGraw-Hill Book Co., New York.

Tinbergen, N. 1951. The study of instinct. Oxford Univ. Press, London.

Uexküll, J. von. 1909. Umwelt and Innenwelt der Tiere. Springer-Verlag, Berlin.

Wecker, S. C. 1963. The role of early experience in habitat selection by the prairie deermouse, *Peromyscus maniculatus bairdi.* Ecol. Monogr., 33: 307–325.

Wecker, S. C. 1964. Habitat selection. Sci. Amer., October, pp. 109–116.

Wiens, J. A. 1969. An approach to the study of ecological relationships among grassland birds. Ornithol. Monogr. 8.

26

Reprinted from *Proc. Natl. Acad. Sci. USA*, **69**(5), 1109–1113 (1972)

Niche Overlap as a Function of Environmental Variability

(food size/birds/ecology/exclusion/model)

ROBERT M. MAY* AND ROBERT H. MAC ARTHUR

Institute for Advanced Study, Princeton, New Jersey, and Department of Biology, Princeton University, Princeton, N.J. 08540

Contributed by Robert H. Mac Arthur, February 15, 1972

ABSTRACT The relationship between environmental variability and niche overlap is studied for a class of model biological communities in which several species compete on a one-dimensional continuum of resources, e.g., food size. In a strictly unvarying (deterministic) environment, there is in general no limit to the degree of overlap, short of complete congruence. However, in a fluctuating (stochastic) environment, the average food sizes for species adjacent on the resource spectrum must differ by an amount roughly equal to the standard deviation in the food size taken by either individual species. This mathematical result emerges in a nonobvious yet robust way for environmental fluctuations whose variance relative to their mean ranges from around 0.01% to around 30%. In short, there is an effective limit to niche overlap in the real world, and this limit is insensitive to the degree of environmental fluctuation, unless it be very severe. Recent field work, particularly on bird guilds, seems in harmony with the model's conclusion.

One of the central concepts in ecology is the competitive exclusion principle, which forbids the coexistence of two or more species making their livings in identical ways. Recently, an increasing amount of attention has been paid to the questions: How similar can competing species be if they are to remain in an equilibrium community? How identical is "identical"? How close can species be packed in a natural environment?

An answer to these questions may begin by noticing that in laboratory experiments, where the environment can be carefully kept unvarying, species whose ecology is well-nigh identical have coexisted for long periods (1). A conjecture (2, 3) is that in the real world, environmental fluctuations will put a limit to the closeness of species packing compatible with an enduring community, and that species will be packed closer or wider as the environmental variations are smaller or larger.

Motivated by these ideas, we consider a one-dimensional resource spectrum, sustaining a series of species, each of which has a preferred position in the spectrum, and a characteristic variance about this mean position, as given by some "utilization function" (see Fig. 1). For example, the resource spectrum may be food size, and the consumers may be bird species each having a utilization function that describes their mean food size and its variance. The dynamics of this situation may be plausibly modeled by a system of first-order differential equations, with competition coefficients that depend on how closely species are packed; that is, on the degree of niche overlap (on the ratio of d to w in Fig. 1).

In the stability analysis of such models, two qualitatively different circumstances need be distinguished. In the un-

realistic case when all the environmental parameters are strictly constant (deterministic), then in general the system remains stable even if an arbitrarily large number of species are packed in, arbitrarily close. On the other hand, when the relevant environmental parameters fluctuate (stochastic environment), there is a limit to the niche overlap consistent with long-term stability.

However, this limit to species packing depends on the environmental variance in a far-from-obvious and extremely interesting way (Fig. 3). If the fluctuations in the resource spectrum are severe, having variances comparable in magnitude with their mean values, the species packing is indeed roughly proportional to the environmental variance, as one would expect intuitively. But for fluctuations ranging from moderate to exceedingly small, the species packing attains an effective limiting value roughly equal to the width of the utilization functions. Thus, as the ratio between the variance and mean value in the resource spectrum, or other pertinent environmental parameter, falls from 0.3 to 0.0001, the closest species packing consistent with stability falls only from 2 to 1 times the utilization function variance. Moreover, our general result is a robust one, being rather insensitive to the details of the mathematical model.

Collecting these statements, we observe that the species packing parameter d indeed goes to zero when the environmental variance becomes strictly zero, but that for any finite environmental variance, d remains roughly equal to the utilization function width, w. This result, which at first glance seems odd, reflects the technical fact that the mathematics contains an essential singularity around $d = 0$ (Eq. [6] and Fig. 2), so that there is a qualitative difference between an environmental variance that is small but finite, and one that is zero.

Following Hutchinson's (4) initial observations, Mac Arthur (3) has recently reviewed a body of semiquantitative work

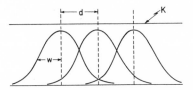

FIG. 1. The curve labeled K represents some resource continuum, say amount of food as a function of food size, that sustains various species whose utilization functions (characterized by a standard deviation w and a separation d) are as shown.

* On leave from the University of Sydney, Sydney, Australia.

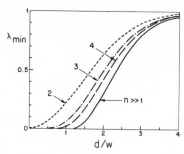

Fig. 2. The minimum eigenvalue of the stability matrix (Eq. [4]) as a function of niche overlap, d/w, for an n-species guild, where $n = 2,3,4$ and $n \gg 1$.

bearing on species packing and character displacement among competing species. These empirical data, which are discussed more fully in *section IV*, match the conclusions drawn from our model.

Two corollaries are worth mentioning.

First, most of the ideas advanced to account for the gradient in species diversity as one goes from the tropics to the poles may be summarized under three headings (1, 5): (*i*) as a matter of history, there has been more time for speciation in the tropics than most other places; (*ii*) *total* niche "volume" is greater in the tropics, which tend to be more productive, less seasonal, and more floristically complex, both in stratification and diversity; (*iii*) more niche overlap is permitted in the tropics by the unvarying environment. The potential number of species is the total volume [i.e., (*ii*)] divided by the effective niche volume per species [i.e., (*iii*)], and this potential will be realized if enough time is available [i.e., (*i*)]. The intuitive basis for the argument (*iii*) was set out in the second paragraph of this introduction, but the quantitative conclusion that niche overlap is only weakly dependent on the degree of environmental fluctuation (unless very severe) suggests that (*iii*) is a relatively unimportant factor in explaining the species diversity gradient, at least until one gets to extreme latitudes. It should be emphasized that our conclusion that species packing, d, is roughly proportional to utilization function width, w, implies only that *niche overlap* is largely independent of the environmental variance, σ^2. It remains true that the total number of species packed into an interval on the resource spectrum is greater if they are specialists (small w) than if they are generalists (large w); the question as to what ultimately determines w remains open.

Second, in this model, which explicitly treats only one trophic level, it is obvious that greater complexity (in the form of more species, more closely packed) makes for lesser stability. In a perfectly stable deterministic environment, arbitrarily close packing and rich speciation is possible, and to a certain limited extent the greater the environmental steadiness, the closer the packing, and the richer the consequent assembly of species. Insofar as this example adds a piece to the complexity–stability jigsaw puzzle, it is that complexity is a fragile thing, permitted in this instance by environmental steadiness: this is quite the opposite of the conventional "complexity begets stability" wisdom (6).

The details of the model are outlined in *section I*, and the results derived in *section II*. *Section III* contains a short account of work bearing on the insensitivity of the main results to the details of the model.

I. THE MODEL DEFINED

Suppose one has a one-dimensional continuum of resources, such as food size, or vertical habitat, or horizontal habitat, that may be schematically depicted as in Fig. 1, where the curve labeled K shows amount of food as a function of food size, or amount of habitat as a function of height, and, in general, amount of resource as a function of x. Suppose further that this resource sustains various species, each of which has a utilization function $f(x)$ as depicted in Fig. 1, which characterizes the species' use of the resource spectrum. In particular, we note the mean position and the standard deviation, w, about this mean for the various species; i.e., the mean and the variance of the food size, or of the habitat height, etc. The separation, d, between the mean positions of species that are adjacent on the resource continuum will clearly be a measure of how densely the species are packed.

Mac Arthur (3, 7) has established a criterion that ensures that the actual community utilization of the resource will provide the best least-squares fit to the available resource spectrum. This requires the populations of the n species, $N_i(t)$ [labeled sequentially $i = 1, 2, .., n$], to obey

$$\frac{dN_i(t)}{dt} = N_i(t) \left[k_i - \sum_{j=1}^{n} \alpha_{ij} N_j(t) \right] \qquad [1]$$

where the k_i are integrals with respect to x over the product of the resource spectrum and the utilization function of the ith species, and the competition coefficients α_{ij} are convolution integrals between the utilization functions of the ith and jth species. With this, we are assured both that the equilibrium populations (obtained by setting all $d/dt = 0$) minimize the squared difference between available and actual "production," and also that nonequilibrium initial populations will move in time towards this minimum configuration.

Eq. [1] is, of course, the Lotka–Volterra competition equation, but tied to the underlying model illustrated by Fig. 1, so that we have explicit recipes for the k_i and α_{ij} in terms of direct biological assumptions. Specifically, if we assume that all the species' utilization functions are the usual bell-shaped gaussian curves, with common width w, and that they are uniformly spaced along the resource continuum (common d), the competition coefficients are

$$\alpha_{ij} = (w^2\pi)^{-1/2} \int_{-\infty}^{\infty} dx \exp\left[-\frac{x^2}{2w^2} - \frac{(x - (i-j)d)^2}{2w^2} \right]$$
$$= [\alpha]^{(i-j)^2}, \qquad [2]$$

where we have for notational convenience defined

$$\alpha = \exp(-d^2/4w^2). \qquad [3]$$

Quite apart from the teleology implicit in the assumption that communities minimize anything, a choice of fit other than least-squares will lead to equations superficially different from [1]: however, their competition matrix characterizing small displacements from equilibrium will end up similar to that given below. As Lotka (8), and others since, have emphasized, Eq. [1] represents the first term in a Taylor expansion of a much wider class of equations, and thus should be useful in discussing the stability of potential equilibria.

In the stability analysis of equations such as [1], we first find the equilibrium populations, $N_i{}^*$, and then study small-amplitude perturbations by linearizing about this equilibrium. As a further simplification in our model, we rather arbitrarily choose the resource spectrum to be such that the community best-fit to it (i.e., the equilibrium community) has *equal* populations for all species; for a large number of species, $n \gg 1$; this means a flat resource spectrum. The conventional analysis then shows the stability of the system to be given simply by the eigenvalues of the $n \times n$ competition matrix A, which from [2] has the form

$$A = \begin{bmatrix} 1 & \alpha & \alpha^4 & \alpha^9 & . & . & \alpha^{n^2} \\ \alpha & 1 & \alpha & \alpha^4 & . & . & . \\ \alpha^4 & \alpha & 1 & \alpha & . & . & . \\ \alpha^9 & \alpha^4 & \alpha & 1 & . & . & . \\ . & . & . & . & . & & . \\ . & . & . & . & & . & . \\ \alpha^{n^2} & . & . & . & . & . & 1 \end{bmatrix}. \qquad [4]$$

In short, in this section we have made several particular assumptions, which have given a specific form for the competition matrix, namely [4]. Indeed, this is a form that can be plausibly justified from quite *ad hoc* considerations. In general, the system stability, and hence the permissible degree of niche overlap, hinges upon some such competition matrix. Other assumptions could give other (but similar) matrices, and the extent to which our conclusions are or are not tied up with the specific model outlined here is discussed in *section III*.

II. THE MODEL ANALYZED

Deterministic environment

As just described, all the parameters in our model system are unvarying constants. Consequently the equilibrium configuration is stable, with perturbations damping out, so long as all the eigenvalues, λ, of **A** are positive (notice that a minus sign was absorbed in the definition of the competition matrix). But **A** is a positive definite form for all $0 \leq \alpha < 1$ (i.e., for all d, see Eq. [3]), with the consequence that stability sets no limit to the species packing in a strictly deterministic environment. Moreover, in general the more species packed in, the better the least-squares fit to the resource spectrum.

Nevertheless it is interesting to see how the *smallest* eigenvalue of **A**, which sets the stability character, varies with niche overlap, as measured by d/w. For $n \gg 1$, we have (see *Appendix*)

$$\lambda_{\min} = 1 - 2\alpha + 2\alpha^4 - 2\alpha^9 + 2\alpha^{16} - \ldots. \qquad [5]$$

This series may be summed, by an elegant method, to get an approximation that is very accurate unless $d \gg w$ (see *Appendix*):

$$\lambda_{\min} = 4\pi^{1/2}(w/d) \exp\left[-\pi^2 w^2/d^2\right]. \qquad [6]$$

This is a remarkable result. For substantial niche overlap, i.e., d/w small, $\lambda_{\min}$ tends to zero faster than any finite power of d/w: there is an essential singularity at $d/w = 0$. Thus, although $\lambda_{\min}$ is indeed necessarily positive even for small d/w, it becomes exceedingly tiny, corresponding to extremely long damping times. This result foreshadows the results below.

Fig. 2 illustrates Eq. [6], along with numerical results for $n = 2, 3, 4$. Notice that for practical purposes, $n = 4$ is hard to distinguish from "$n = \infty$".

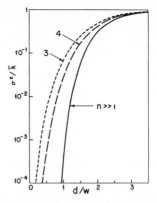

FIG. 3. The closest niche overlap, d/w, consistent with community stability in a randomly varying environment, whose fluctuations are characterized by a variance (relative to the mean) of $\sigma^2/\bar{k}$. The variance is plotted on a logarithmic scale to emphasize that, over a wide range, it has little influence on the species packing distance for $n > 2$.

Stochastic environment

More realistically, there will be random environmental fluctuations, so that the resource continuum will be noisy. This means the quantities k_i in Eq. [1] will not be constants, but rather will be random variables. We assume

$$k_i = \bar{k}_i + \gamma_i(t) \qquad [7]$$

where $\bar{k}_i$ is the constant mean value (having the common value $\bar{k}$ for large n), and $\gamma_i(t)$ is gaussian "white noise", with variance measured by σ^2.

In this stochastic problem, we may no longer talk of *the* species populations, but only of their joint probability distribution. To a good approximation, this is a multivariate normal distribution in the fluctuations about the means, and the probability of any species becoming extinct will be small (corresponding to the mechanical "stability" of the deterministic case) if the smallest eigenvalue of the competition matrix **A** roughly obeys

$$\lambda_{\min} > \sigma^2/\bar{k} \qquad [8]$$

This result (9) is commonsensical. In a randomly fluctuating environment, it is not enough that all the eigenvalues be positive, but rather they should be bounded away from zero by an amount roughly proportional to the environmental noise level.

Combining the qualitative equation, [8], with Fig. 2, we arrive at an estimate of the closest species packing, d/w, consistent with stability for a given environmental noise level, $\sigma^2/\bar{k}$. These results, illustrated in Fig. 3, are as discussed in the *Introduction*.

In particular, we see explicitly from Eq. [6] that for large n this closest degree of niche overlap depends on the environmental fluctuations only as $\sqrt{\ln \sigma^2}$, a very weak dependence. The results for $n = 3, 4$, although allowing a slightly closer limiting packing distance, display a similar insensitivity to the degree of random fluctuation, so long as it is not severe.

III. HOW ROBUST ARE THESE RESULTS?

The question arises, to what extent are these results peculiar to our particular model? We catalogue some answers.

(*i*) We chose gaussian utilization functions. Alternative $f(x)$ ranging from the opposite extremes of rectangles through to back-to-back exponentials or Lorentz lineshapes lead to **A** matrices different from [4], but the plot of λ_{min} as a function of d/w retains the essential features of Fig. 2 in all cases.

(*ii*) We chose the width and separation of the utilization functions to be constant. If the width w changes in some systematic way along the resource continuum, our results are preserved, as long as the separation d changes in the same proportion, keeping d/w roughly constant.

(*iii*) The resource spectrum of Fig. 1 was assumed to be such that, at equilibrium, all populations are equal. Extensive investigation of various resource spectrum shapes for $n = 2, 3,$ and 4 suggests that our results are not dependent on this feature, so long as all species are present in significant numbers in the equilibrium community.

(*iv*) The implications of use of Eq. [1] were discussed in *section I*.

(*v*) The stochasticity of the environment was taken to be gaussian "white noise," i.e., no correlation between the fluctuations at successive instants. In practice, this means only that fluctuations be correlated over times short compared to all other relevant time scales in the system (9).

(*vi*) Our model is for competition in one resource dimension. Cody's (10) classification of partitioning in the three-resource dimensions of horizontal habitat, vertical habitat, and food for 10 grassland bird communities around the world shows eight of them to be organized largely in one dimension (food selection), so that our model is not wholly unreasonable. Moreover, the model is directly relevant to niche overlap in two or more orthogonal resource dimensions, and may even be useful as a metaphor for more complicated circumstances.

IV. COMPARISON WITH REAL ECOSYSTEMS

In a classic paper, Hutchinson (4) observed that in various circumstances, including both vertebrate and invertebrate forms, character displacement among sympatric species leads to sequences in which each species is roughly twice as massive as the next; i.e., linear dimensions as measured by bills or mandibles in the ratio 1·2–1·4. Mac Arthur's more recent and quantitative reviews (3) of such data point to there being a limiting value to niche overlap in the natural world, corresponding to d/w in the range 1–2. Also pertinent is Simpson's (11) review of the factors making for latitudinal and altitudinal species diversity gradients among North American mammals; it concludes that degree of niche overlap is not an important contributing factor.

The work that seems to come closest to our one-dimensional model is that of Terborgh, Diamond, and Beaver on various guilds of birds in an assortment of habitats that have various degrees of environmental stability. Even so, such comparisons with the theory are necessarily approximate, partly because our α (Eq. [3]) comes from utilization functions that are not just percentage of time or of diet, but rather have weighting terms for resource renewal (3, 7); all available information from nature contains unweighted utilizations.

Terborgh (12) has shown five species of tropical antbird, segregating by foraging height, have mean heights separated by one standard deviation; i.e., $d/w \simeq 1$. Mac Arthur's analysis (3) of Storer's data (13) on the food weight distribution for three congeneric species of hawks also leads to $d/w \simeq 1$. Diamond's (14) extensive data on weights of tropical bird congeners that sort out largely (but not wholly, so that d/w should be smaller than our one-dimensional theory predicts) by size differences leads to weight ratios around 1·6–2·3; when Hespenheide's analysis (15) of the relation between weight ratio and α is used, Diamond's results become $\alpha \simeq 0·8$–$0·9$, i.e., $d/w \simeq 0·6$–$1·0$. In the Sierra Nevada, Beaver (personal communication) has shown that species packing in a brushland bird community appears equal to that in a forest foliage gleaning guild, although the microenvironment is thought to be significantly more unvarying in the forest.

In brief, the basic conclusion that emerges in a nonobvious but robust way from our mathematical model, namely that there is a limit to niche overlap in the natural world and that this limit is not significantly dependent on the degree of environmental fluctuation (unless it be severe, as in the arctic), seems to be in harmony with such facts as are known about real ecosystems.

APPENDIX

For large n, where "end effects" at the extremes of the resource spectrum are unimportant, we may pretend that the resource continuum is cyclic (so that the species labeled *1* adjoins that labeled n), whereupon the competition matrix **A** of Eq. [4] is slightly modified to become related to a class of matrices discussed by Berlin and Kac (16). Using their approach, one can obtain Eq. [5]. That this trick of imposing artificial cyclic boundary conditions does not alter the eigenvalues for $n \gg 1$ is a point made clear in the literature on the physicists' Ising model, from which comes Berlin and Kac's paper.

The series in Eq. [6] is identically equal to the contour integral

$$\frac{1}{2i} \oint_C \frac{\exp\,(z^2 \ln\,\alpha)\,dz}{\sin\,(\pi z)}$$

Here the contour C encloses all poles of the integrand up to $z = \pm n$, where the series has n terms. An $n \to \infty$, C is the circle at infinity in the complex plane. Using the standard Jordan contour, and ignoring correction terms of relative order $\exp(-4\pi^2 w^2/d^2)$, which are thoroughly negligible for $d/w < 3$ or so, we arrive neatly at Eq. [6].

At the other extreme, for $n = 2$ the eigenvalues of **A** are easily found directly. For other finite values, such as $n = 3, 4$, we take a meat axe and display λ_{min} as a numerical function of d/w.

This research was sponsored in part by the National Science Foundation, Grant GP-16147 A 1.

1. Miller, R. S. (1967) in *Advances in Ecological Research* (Academic Press, New York), Vol. 4, pp. 1–74 (see pp. 35–46).
2. Miller, R. S. (1967) in *Advances in Ecological Research* (Academic Press, New York), p. 67.
3. Mac Arthur, R. H. (1971) in *Avian Biology* (Academic Press, New York), Vol. I, pp. 189–221; (1972) *Geographical Ecology* (Harper and Row, New York), in press.
4. Hutchinson, G. E. (1959) *Amer. Natur.* **93**, 145–159.
5. Klopfer, P. H. (1962) in *Behavioral Aspects of Ecology* (Prentice-Hall, Englewood Cliffs), chap. 3; Pianka, E. R. (1966) *Amer. Natur.* **100**, 33–46; Mac Arthur, R. H. (1969) *Biol. J. Linn. Soc.* **1**, 19–30; (1969) *Diversity and Stability in Ecological Systems; Brookhaven Symposium in Biology No. 22* (Nat. Bur. Standards, Springfield, Va.).
6. May, R. M. (1971) *Math. Biosci.* **12**, 59–79.

7. Mac Arthur, R. H. (1969) *Proc. Nat. Acad. Sci. USA* **64**, 1369–1371; (1970) *Theor. Pop. Biol.* **1**, 1–11.

8. Lotka, A. J. (1925) *Elements of Physical Biology* (Williams and Wilkins, Baltimore), p. 62.

9. May, R. M. (1971) *Proc. Ecol. Soc. Aust.*, in press; Astrom, K. J. (1970) *Introduction to Stochastic Control Theory* (Academic Press, New York); Sykes, Z. M. (1969) *J. Amer. Stat. Ass.* **64**, 111–130.

10. Cody, M. (1968) *Amer. Natur.* **102**, 107–148.

11. Simpson, G. G. (1964) *Syst. Zool.* **13**, 57–73.

12. Terborgh, J. (1972), quoted in Mac Arthur, R. H. (1972) *Geographical Ecology* (Harper and Row, New York), Figure 6-4.

13. Storer, R. W. (1966) *Auk* **83**, 423–436.

14. Diamond, J. M. (1962) *The Avifauna of the Eastern Highlands of New Guinea* (Publ. Nuttall Ornithol. Club, Cambridge, Mass.), in press.

15. Hespenheide, H. A. (1971) *Ibis* **113**, 59–72.

16. Berlin, T. H. and Kac, M. (1952) *Phys. Rev.* **86**, 821–835.

27

Reprinted from *Proc. Natl. Acad. Sci. USA*, **71**(5), 2141–2145 (1974)

Niche Overlap and Diffuse Competition

(desert lizards/resource partitioning/community structure/species diversity)

ERIC R. PIANKA

Department of Zoology, University of Texas at Austin, Austin, Texas 78712

Communicated by Edward O. Wilson, March 11, 1974

ABSTRACT   Current theory predicts a distinct upper limit on the permissible degree of niche overlap; moreover, theory suggests that maximal tolerable overlap should be relatively insensitive to environmental variability. Data presented here demonstrate that, within the lizard subset of natural desert communities, niche overlap decreases both with increasing environmental variability and with increasing numbers of lizard species. The latter two factors are themselves positively correlated. A partial correlation analysis is interpreted as indicating that the extent of tolerable niche overlap does not necessarily decrease due to environmental variability, but rather that overlap is probably more closely related to the number of potential interspecific competitors in a community, or what has been termed "diffuse competition." This result lends support to the "niche overlap hypothesis," which asserts that maximal tolerable overlap should vary inversely with the intensity of competition. Moreover, this empirical discovery indicates that niche overlap theory could be profitably expanded to incorporate the number of competing species. Although the average amount of overlap between pairs of species decreases with the intensity of diffuse competition, the overall degree of competitive inhibition tolerated by individuals comprising an average species could nevertheless remain relatively constant, provided that extensive niche overlap with a few competitors is roughly equivalent to lower average overlap with a greater number of competitors.

NICHE OVERLAP THEORY

The ways in which species within ecological communities partition available resources among themselves is a major determinant of the diversity of coexisting species. All else being equal, a community with more resource sharing, or greater niche overlap, will clearly support more species than one with less niche overlap. In attempts to understand competition and determinants of species diversity, population biologists have reasoned that coexisting species must differ in their ecological requirements by at least some minimal amount to avoid competitive exclusion. Such thinking has led to the related concepts of "character displacement" (1), "limiting similarity" (2), "species packing" (3, 4), and "maximum tolerable niche overlap," which is simply the notion that there must be an upper limit on the permissible degree of niche overlap (5–8).

May and MacArthur (5) recently developed an elegant analytic model of niche overlap as a function of environmental variability. Their theory predicts an upper limit on the degree of tolerable overlap; moreover, the derivation suggests that maximal permissible overlap should be relatively insensitive to both number of species and environmental variability. The May–MacArthur niche overlap model assumes an equilibrium community in a fully saturated environment with all resources being used fully; as such, variation in the intensity of competition is not modelled (see also next paragraph). The model assumes a one-dimensional resource spectrum, but May (7) recently indicated that the argument can be expanded without qualitative change to a multidimensional niche space. In development of this theory, May and MacArthur express the inverse of niche overlap as a ratio of the distance between the centers of two "ultilization curves" (niche separation) over the standard deviation in utilization (niche breadth), with the latter assumed to be constant and identical for all species. Their model thus somewhat confounds niche overlap and niche breadth.

Estimates of overlap in resource utilization have often been equated with the "competition coefficients" or "alphas" of the much overworked Lotka–Volterra competition equations:

$$\frac{dN_i}{dt} = r_i N_i \left(\frac{K_i - N_i - \sum_{j \neq i}^{n} \alpha_{ij} N_j}{K_i} \right) \qquad [1]$$

where i and j subscript each of the n different species, N_i is the abundance of the ith species, r_i is its maximal intrinsic rate of increase per capita, K_i is the "carrying capacity" of species i, and α_{ij} represents the per capita competitive inhibition of species j on the population growth rate of species i. Alphas are extremely difficult to estimate directly except by population removal experiments, and ecologists have often equated estimates of overlap with competition coefficients (9). However, tempting though it may be, equating overlap with competition is an extremely dubious and misleading procedure (10, 11). Clearly niche overlap, in itself, need not necessitate competition; in fact, there may often be an inverse relationship between overlap and competition. If resources are not in short supply, two organisms can share them without detriment to one another. Thus, extensive niche overlap may actually be correlated with *reduced* competition. Similarly, disjunct niches may often indicate avoidance of competition in situations where it could potentially be severe. Such reasoning led me to propose that maximal tolerable niche overlap should be lower in intensely competitive situations than in environments with lower demand/supply ratios; I termed this the "niche overlap hypothesis" (8).

Diffuse competition

MacArthur (4) coined the term "diffuse competition" to describe the total competitive effects of a number of interspecific competitors. To illustrate the concept, consider Eq.

1. At equilibrium, all dN_i/dt must equal zero; that is

$$N_i{}^* = K_i - \sum_{j \ne i}^{n} \alpha_{ij} N_j \qquad [2]$$

where $N_i{}^*$ is the equilibrium abundance of species i. Eq. **2** must be true for all i at equilibrium. Note that the term, $-\sum_{j \ne i}^{n} \alpha_{ij} N_j$, increases with the number of competing species, n, and that the equilibrium abundance of species i, $N_i{}^*$, decreases as one sums over a greater number of competitors. Further, note that a little bit of competitive inhibition by a lot of other species (diffuse competition) can be equivalent to strong competitive inhibition by fewer competing species.

DESERT LIZARD COMMUNITIES

A series of 28 study areas at similar latitudes on three continents support from 4 to 40 sympatric species of desert lizards (12–14). Estimated species densities and lizard species diversities for these sites have been given elsewhere (14). My assistants and I recorded data on microhabitat, time of activity, and stomach contents of over 15,000 lizards of some 91 species on these desert study areas, which I use for the following analysis of niche overlap. Results presented rather briefly here are documented more fully elsewhere (14).

Environmental variability

In deserts, water is a master limiting factor, and long-term mean annual precipitation is very strongly correlated with average annual productivity. Moreover, standard deviation in annual precipitation can be considered an indicator of environmental variability since year-to-year variation in annual precipitation should generate temporal variability in food availability. I estimated both the long-term mean and standard deviation in annual precipitation from nearby weather stations for most study areas. Both precipitation statistics are significantly correlated with lizard species densities and diversities ($rs > 0.41$, $Ps < 0.05$ to 0.001).

Niche dimensionality

Although some pairs of sympatric competitors avoid competition primarily through differences in the use of a single resource gradient or niche dimension, it is far more prevalent for coexisting species to differ in their use of two or more niche dimensions simultaneously. Pairs with high overlap along one dimension often overlap relatively little along another, reducing overall effective niche overlap [see figure 6.7, page 198 in Pianka (11) and/or figure 1 in May (16)].

Like most animals, desert lizards subdivide resources in three major ways: they differ in what they eat, where they forage, and when they are active. Ecological differences in each of these three niche dimensions should reduce competition and thus facilitate coexistence of a variety of species. It is difficult or impossible to evaluate the degree of interdependence of these three niche dimensions for most lizard species because the animals move and are active over a period of time. However, the degree to which foods eaten depend upon microhabitat can be assessed in some relatively sedentary subterranean skinks (15); in these lizards, diet and microhabitat appear to be largely independent. Clear interactions among niche dimensions are apparent in other cases (13, 14). The vast majority of interspecific pairs of sympatric lizard species have substantial niche separation along one or more of

these three niche dimensions (trophic, spatial, and/or temporal), making it unnecessary to subdivide the three basic dimensions any further to analyze resource partitioning in these lizard communities.

Niche dimensionality has another important aspect: the number of potential neighbors in niche space increases more or less geometrically with the number of niche dimensions actually subdivided (4, 14). Hence a greater number of effective niche dimensions provides a greater potential for diffuse competition.

Niche overlap

Overlap has been quantified in numerous ways (2, 9, 10, 13). The particular overlap index used is somewhat arbitrary since similar qualitative results are obtained with a wide variety of indices. Here I use the following modification (13) of the equation first proposed by MacArthur and Levins (2) and Levins (12) for estimating competition coefficients, or alphas, from field data on resource utilization:

$$O_{jk} = O_{kj} = \frac{\sum_{i}^{n} p_{ij} p_{ik}}{\sqrt{\sum_{i}^{n} p_{ij}{}^2 \sum_{i}^{n} p_{ik}{}^2}}$$

where p_{ij} and p_{ik} represent the proportions of the ith resource used by the jth and kth species. May (16) discusses a mathematical rationale for the convenience of this symmetric measure over the original asymmetric form. I do not consider values obtained from this equation "competition coefficients," but rather merely measures of niche overlap (see above and refs. 10 and 11 for further discussion of the distinction between overlap and competition).

Thus calculated, the average extent of overlap along various dimensions differs among the three continental desert-lizard systems (Table 1). For example, overlap in microhabitat is high in North America where many lizards frequent the open spaces between plants, whereas dietary overlap is high in the Kalahari desert of southern Africa where termites dominate the diets of many species of lizards (13). Overlap is relatively low along all three niche dimensions in the most diverse lizard communities of Australia (13, 14).

Estimating overall niche overlap along three dimensions is difficult and can be quite treacherous (16). Ideally, a proper multidimensional analysis of resource utilization and niche separation along more than a single niche dimension should proceed through estimation of the simultaneous proportional utilization of all resources along each separate niche dimension. Thus, one would like to work with the proportion of prey type i captured in microhabitat j by species k, or the true multidimensional p_{ijk}s. However, in practice it is extremely difficult or even impossible to obtain such multidimensional utilization data, because animals usually integrate over both space and time (stomachs contain prey captured over a period of time and in a variety of microhabitats). Some progress toward understanding overall niche overlap along several dimensions can, however, be made using only the proportional utilizations along each of the component niche dimensions, as follows (for greater detail, see ref. 16).

Provided that niche dimensions are truly independent (orthogonal), with for example any given prey item being equally likely to be captured in any microhabitat, overall

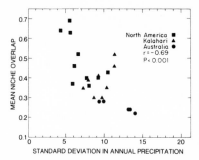

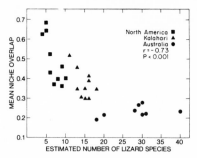

FIG. 1. Average overall summation niche overlap plotted against the standard deviation in annual precipitation. Continents coded by *shape*, as indicated. Although none of the correlations within continental desert systems is significant, the correlation coefficient for all areas is highly significant statistically. However, when the number of lizard species on various areas is held constant by partial correlation, this correlation disappears. Compare with Fig. 2.

FIG. 2. Average overall summation niche overlap plotted against the estimated number of lizard species. Continents coded by *shape*, as indicated. Similar inverse correlations exist with overall multiplicative overlap values and with three different estimates of maximal tolerable niche overlap (see *text*).

multidimensional utilization is simply the product of the separate unidimensional utilizations (16); that is, $p_{ijk} = p_{ik} \times p_{jk}$. In this case, overlaps along component niche dimensions can simply be multiplied to estimate overall multi-

dimensional niche overlap (16). However, should niche dimensions be entirely dependent upon one another (with for example, each prey type occurring in only a particular microhabitat), there is actually only a single resource dimension. Under such complete dependency, true "multidimensional" overlap is best estimated by the arithmetic mean of the overlaps along component dimensions; such "summation over-

TABLE 1. *Estimates of the number of lizard species and average niche overlap values for 28 desert study areas on three continents*

No. of lizard species	Average niche overlap			Estimates of average overall niche overlap				
				Multiplicative		Summation	Largest tenth	
	Food	Microhabitat	Time	(All)	(Nonzero)	(All)	(Multiplicative)	(Summation)
North America								
4	0.49	0.80	0.58	0.20	0.20	0.63	0.41	0.76
5	0.75	0.78	0.53	0.33	0.36	0.69	0.73	0.90
5	0.52	0.92	0.49	0.25	0.36	0.64	0.71	0.90
6	0.55	0.55	0.47	0.22	0.37	0.52	0.61	0.86
6	0.34	0.55	0.20	0.12	0.27	0.43	0.57	0.75
7	0.39	0.42	0.31	0.11	0.39	0.37	0.52	0.82
8	0.56	0.31	0.32	0.10	0.24	0.40	0.37	0.74
9	0.28	0.52	0.58	0.11	0.23	0.46	0.50	0.82
9	0.38	0.32	0.39	0.06	0.18	0.36	0.37	0.70
10	0.37	0.33	0.50	0.08	0.25	0.40	0.20	0.76
Kalahari								
11	0.92	0.35	0.28	0.18	0.41	0.52	0.69	0.89
13	0.36	0.39	0.30	0.08	0.26	0.35	0.50	0.80
13	0.56	0.47	0.34	0.13	0.36	0.46	0.61	0.85
14	0.56	0.21	0.15	0.04	0.27	0.31	0.19	0.61
15	0.45	0.23	0.21	0.04	0.22	0.30	0.31	0.70
15	0.56	0.25	0.24	0.06	0.23	0.35	0.44	0.76
16	0.72	0.22	0.23	0.09	0.35	0.39	0.64	0.85
16	0.44	0.22	0.24	0.05	0.22	0.30	0.34	0.73
16	0.71	0.28	0.26	0.11	0.36	0.42	0.69	0.88
18	0.51	0.26	0.27	0.07	0.24	0.35	0.44	0.77
Australia								
18	0.23	0.16	0.18	0.01	0.14	0.19	0.08	0.59
20	0.18	0.36	0.13	0.01	0.02	0.22	0.10	0.54
28	0.25	0.32	0.16	0.03	0.21	0.24	0.23	0.65
29	0.27	0.30	0.27	0.04	0.23	0.28	0.29	0.69
30	0.23	0.24	0.19	0.02	0.19	0.22	0.14	0.59
30	0.37	0.24	0.27	0.03	0.18	0.28	0.26	0.66
31	0.19	0.28	0.18	0.02	0.16	0.22	0.14	0.60
40	0.23	0.25	0.22	0.02	0.15	0.24	0.19	0.61

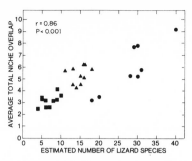

FIG. 3. Average total niche overlap (summation) plotted against the estimated number of lizard species, to show that total niche overlap increases with lizard species density even though overlap between average pairs of species decreases (compare with Fig. 2). Continents coded *by shape* as in previous figures.

TABLE 2. *Means and 95% confidence limits on ratios of niche separation over niche breadth*

Niche dimension	North America	Kalahari	Australia
Food	3.73(2.87–4.59)	2.51(2.21–2.81)	5.18(5.0–5.36)
Micro-habitat	5.79(4.83–6.74)	4.23(3.85–4.61)	4.60(4.48–4.72)
Time	2.78(2.0–3.56)	5.08(4.48–5.68)	5.11(4.93–5.29)
Overall	0.50(0.38–0.62)	1.54(1.16–1.92)	1.03(1.0–1.06)

laps" constitute upper bounds on the true multidimensional overlap (16). Since real niche dimensions are presumably seldom, if ever, either perfectly independent or perfectly dependent, neither the multiplication nor the summation technique is entirely satisfactory. Summation overlaps generally overestimate true niche overlap, whereas multiplicative overall overlaps often underestimate true multidimensional niche overlap (16).

In an attempt to overcome these very considerable difficulties, I computed estimates of overall overlap by both multiplication and summation of the overlaps along the three component niche dimensions (Table 1). When overlaps along the three dimensions are multiplied, the vast majority of interspecific pairs overlap very little or not at all (13, 14). The possible number of such nonoverlapping pairs increases markedly with the size of overall niche space, which is greater in more diverse saurofaunas (13, 14). Overlap between those pairs with some overlap is of greatest interest as it should reflect limiting similarity and/or maximal tolerable overlap. Average overlap values for each niche dimension in the various desert systems are listed in Table 1, along with means of all multiplicative overlaps and all summation overlaps (recall that the latter represent upper bounds on the true multidimensional overlap). Means of all nonzero multiplicative overlap pairs and the averages of the largest tenth of all multiplicative and summation overlaps are also given, as these are more likely to reflect maximal tolerable overlap. All five methods of estimating overall overlap produce strongly correlated values ($rs = 0.67$ to 0.97, $Ps < 0.01$ to 0.001). Estimates of overall niche overlap are strongly correlated with both the standard deviation in precipitation (Fig. 1) and with the number of lizard species (Fig. 2). At first glance, Fig. 1 seems somewhat at odds with May and MacArthur's prediction that maximal overlap should be insensitive to environmental variability. However, the overlap values used here are not entirely appropriate for testing the May–MacArthur theory, since this model is expressed in terms of the ratio of niche separation over niche breadth, effectively the inverse of niche overlap scaled by niche breadth. To approximate conditions of their model more closely, I estimated niche separation as one minus overlap for all interspecific pairs in each continental desert-lizard system, and expressed these values as ratios of separation over standardized niche breadths (Table

2). Such an analysis modified results presented in Figs. 1 and 2 only slightly. Distinct differences among the three continents are still apparent. Moreover, an area-by-area analysis also shows that niche separation over niche breadth ratios tend to increase with lizard species density. Clearly niche separation over niche breadth ratios are not constant between the three desert systems.

I used partial correlation analysis in an attempt to interpret factors influencing niche overlap. When the effects of lizard species density are held constant by partial correlation, average overall summation overlap and mean nonzero multiplicative overlap do not remain significantly correlated with the standard deviation in precipitation. However, the inverse correlations between lizard species density and both measures of overall niche overlap remain significant at the 0.01 level when standard deviation in precipitation is held constant by partial correlation. These results suggest that, as predicted, the extent of tolerable niche overlap is not necessarily a function of the degree of environmental variability, but rather that maximal overlap is more closely related to the number of competing species and the intensity of diffuse competition.

Evidently, stronger diffuse competition requires greater average niche separation among coexisting lizard species. Low overlap with lots of competitors may be similar to high overlap with fewer interspecific competitors. Rather than remaining constant, niche overlap seems to be adjusted to the number of competing species, perhaps resulting in a relatively constant level of interspecific competitive inhibition for an average species even in communities that differ widely in diversity. A first hypothesis might be that total overlap with sympatric species remains constant; Fig. 3 shows that total overlap actually increases with lizard species density, even though the average amount of overlap between pairs decreases.

In conclusion, empirical results presented here support the niche overlap hypothesis, which predicts that maximal tolerable niche overlap should decrease with increasing intensity of competition. Moreover, these data indicate that niche overlap theory needs to be modified to incorporate more fully the phenomenon of diffuse competition.

This research was supported by grants from the National Science Foundation (GB-5216, GB-8727, and GB-31006).

1. Brown, W. L. & Wilson, E. O. (1956) *Syst. Zool.* **5,** 49–64; Hutchinson, G. E. (1959) *Amer. Natur.* **93,** 145–159; Schoener, T. W. (1965) *Evolution* **19,** 189–213; Grant, P. R. (1972) *Biol. J. Linn. Soc.* **4,** 39–68.
2. MacArthur, R. H. & Levins, R. (1967) *Amer. Natur.* **101,** 377–385.
3. MacArthur, R. H. (1970) *Theor. Pop. Biol.* **1,** 1–11.

4. MacArthur, R. H. (1972) *Geographical Ecology* (Harper and Row, New York).
5. May, R. M. & MacArthur, R. H. (1972) *Proc. Nat. Acad. Sci. USA* **69**, 1109–1113.
6. May, R. M. (1973) *Stability and Complexity in Model Ecosystems* (Princeton Univ. Press, Princeton, N.J.).
7. May, R. M. (1974) *Theor. Pop. Biol.* **5**, in press.
8. Pianka, E. R. (1972) *Amer. Natur.* **106**, 581–588.
9. Pico, M. M., Maldonado, D. & Levins, R. (1965) *Carib. J. Sci.* **5**, 29–37; Schoener, T. W. (1968) *Ecology* **49**, 704–726; Orians, G. H. & Horn, H. S. (1969) *Ecology* **50**, 930–938; Pianka, E. R. (1969) *Ecology* **50**, 1012–1033; Culver, D. C. (1970) *Ecology* **51**, 949–958; Brown, J. H. & Lieberman, G. A. (1973) *Ecology* **54**, 788–797.
10. Colwell, R. K. & Futuyma, D. J. (1971) *Ecology* **52**, 567–576.
11. Pianka, E. R. (1974) *Evolutionary Ecology* (Harper and Row, New York).
12. Pianka, E. R. (1967) *Ecology* **48**, 333–351; Pianka, E. R. (1969) *Ecology* **50**, 498–502; Pianka, E. R. (1971) *Ecology* **52**, 1024–1029.
13. Pianka, E. R. (1973) *Annu. Rev. Ecol. Syst.* **4**, 53–74.
14. Pianka, E. R. (1975) in *The Ecology and Evolution of Communities* (Harvard Univ. Press, Cambridge, Mass.).
15. Huey, R. B., Pianka, E. R., Egan, M. E. & Coons, L. W. (1974) *Ecology* **55**, 304–316.
16. May, R. M. (1974) *Ecology* **55**, in press.

28

Reprinted from R. Levins, *Evolution in Changing Environments: Some Theoretical Explorations*, Princeton University Press, Princeton, N.J., 1968, pp. 14–20, 41–45, 50–55, by permission of Princeton University Press

EVOLUTION IN CHANGING ENVIRONMENTS: SOME THEORETICAL EXPLORATIONS

R. Levins

[*Editors' Note:* In the original, material precedes this excerpt.]

We will now introduce the method of fitness sets for the analysis of adaptive strategy. In Figure 2.1a we show the relation between a component of fitness and the environment for an organism which may be in one of two physiological states. The location of the peak gives the optimal environment for that phenotype, the height of the peak is the measure of the best performance in the optimal environment, and the breadth of the curve is a measure of the tolerance for non-optimal environments. This tolerance is a measure of homeostasis. If there were no restrictions on the curve in Figure 2.1a, the optimal curve would obviously be infinitely high and infinitely broad. In fact there are restrictions. The height at the peak is undoubtedly limited by the physico-chemical structures involved. But in addition we suggest that the breadth cannot be increased without lowering the height. Suppose that the phenotypes I and II in the figure refer to two enzymes with different pH or temperature optima but with curves of the same shape. If the same total amount of enzyme is divided equally between the two forms the combined system has a fitness curve shown by the broken line in Figure 2.1a. If the enzymes are mixed in any proportions the shape of the curve may alter but the total area under the curve remains constant. If the curves refer to different genotypes in the same population the same argument holds.

The researches of George Sacher (1966) allow us to extend this principle of allocation further. It was first observed that despite differences in the total life spans of different animals, the caloric life spans measured in energy expenditure are remarkably uniform. The major discrepancies from constant expenditure were associated with the homeostatic system—the bigger the brain the lower the rate of aging per calorie. Thus Sacher defines organizational entropy as a measure of the energy cost per unit of development (carrying the organism from one stage to another). He amassed a great deal of data on his $S_{org}(T)$ as a function of temperature. It has a minimum value at some optimal temperature and increases with the departure of the temperature from the optimum. This "entropy" curve is broader and flatter for insect eggs and pupae than for larvae, which being mobile can seek out preferred environments. Sacher's work therefore suggests that the cybernetic system which reduces the organizational entropy is itself costly. It can be extended to cover a wider range of environments, but only by reducing the efficiency at the optimum.

Thus we assert the principle of allocation: the fitness curve $W(s)$ for an environmental parameter s may vary in shape but is subject to the constraint

$$\int F\{W(s)\}\, ds = C. \tag{2.1}$$

We do not know the functional form of F in general, but where fitness is altered by mixing components such as enzymes, $F(W) = W$.

For the purposes of this study we can specify that for any phenotype y and environment s, $W(s - y)$ is a non-negative function with a maximum at $s = y$ and decreasing symmetrically toward zero as $|s - y|$ increases. The dual of the curve in Figure 2.1a for a fitness component over environ-

15

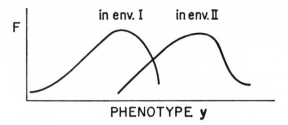

FIGURE 2.1b. Fitness as a function of phenotype in two environments. The peak occurs at the optimal phenotype for each environment.

ments is shown in Figure 2.1b, where for a fixed s we plot $W(s \doteq y)$ over a range of phenotypes y. For any two environments the curves overlap. If they are close enough so that their inflection points overlap, the average of the two curves (fitness in an environment which is half S_1, half S_2) will have a single peak in the middle. If S_1 and S_2 are farther apart the average curve has a minimum at the midpoint and two peaks near S_1 and S_2.

The fitness set representation presents the curves of Figure 2.1b in a different way. The two axes in the graph are now fitness components W_1 and W_2 in environments S_1 and S_2. The phenotype whose peak is at S_1 in Figure 2.1b gives the point farthest to the right in Figure 2.2. The phenotype corresponding to S_2 gives the uppermost point. All pheno-

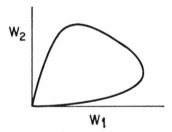

FIGURE 2.2. The fitness set. The coordinates of each point on the fitness set are the ordinates in Figure 2.1b, corresponding to the two environments for each phenotype.

16

types which have curves differing only in the location of their peaks will lie on the boundary of the fitness set shown in Figure 2.2. Those for which the area under the curve is less will lie inside the fitness set.

It can readily be seen that if S_1 and S_2 are sufficiently close (so that their inflection points overlap) the fitness set will be convex along the upper-right-hand boundary, while if S_1 and S_2 are farther apart the upper-right-hand boundary will be partly concave. This difference between the concavity and convexity of the fitness set will have important biological consequences. Therefore we define twice the distance from the peak to the inflection point as the tolerance of the phenotype and assert that the fitness set is convex or concave depending on whether the environmental range $|S_1 - S_2|$ is less than or greater than the tolerance of a single phenotype.

The importance of this distinction is that the fitness of a mixture of phenotypes in a population or of physiological states in an individual is represented by a point on the straight line joining their points on the fitness set. In a convex fitness set such mixed strategies will lie inside the set and therefore each one will be inferior to some single phenotype which lies above and to the right of it. But on a concave fitness set certain mixtures will lie outside (up and to the right) of the fitness set for single phenotypes so that mixed strategies may be optimal.

The fitness set alone does not define an optimum strategy. Over-all fitness in a heterogeneous environment depends on the fitnesses in the separate environments, but in a way which is determined by the pattern of environments. We therefore define the Adaptive Function $A(W_1, W_2)$, which measures fitness in the heterogeneous environment, to be a monotonic increasing function of its arguments. If the environment is sometimes S_1 and sometimes S_2, the individual must survive in both in order to survive. Let the probability of dying (or the loss in growth rate) in the interval Δt be $m(t) \, \Delta t$ where $m(t)$ takes on two values according to which environment S_1 or S_2 is currently present. Then the probability of survival up to time t is $P(t)$, and it satisfies

17

the relation

$$P(t + \Delta t) = P(t)[1 - m(t) \Delta t]$$

(assumes $m(t)$ unchanging for at least Δt). (2.2)

Where Δt is a whole generation,

$$P(t) = \Pi(1 - m(t) \Delta t) \qquad (2.3)$$

which is the product $W_1{}^p W_2{}^{1-p}$ for environments S_1 and S_2 occurring in the proportions $p : 1 - p$. But if Δt is very small the terms in $(\Delta t)^2$ and higher powers vanish, and

$$P(t) = P_0 e^{-\int m(t) dt}. \qquad (2.4)$$

This will be maximized when the integral $\int m(t) \, dt$ is smallest, which occurs when the linear average of the fitnesses $pW_1 + (1 - p)W_2$ is greatest. In the former case the environment is described as coarse-grained and presents itself to the individual as alternatives. Thus a coarse-grained environment is uncertain for the individual even if the proportions of the alternative environments remain fixed. In the latter case the environment is experienced as a succession of possibly different conditions. The differences present themselves to the organism as an average which is the same for all members of the population. There is no uncertainty. Of course there may be intermediate-grained environments, with the Adaptive Function intermediate between the linear and the multiplicative (or logarithmic).

The notion of grain comes of course from the size of patches of environment. If the patch is large enough so that the individual spends his whole life in a single patch, the grain is coarse, while if the patches are small enough so that the individual wanders among many patches the environment is fine-grained. But the concept can be made more general. For example, since most animals eat many times, food differences are fine-grained, whereas alternative hosts, when several are available for a parasite, are coarse-grained differences.

Thus for the situations we have described the Adaptive Function $A(W_1, W_2)$ may vary from the hyperbola-like $W_1{}^p W_2{}^{1-p}$ to the linear $pW_1 + (1 - p)W_2$. In any case the

18

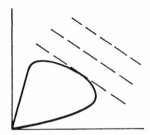

FIGURE 2.3a. Fine-grained environment on a convex fitness set.

optimal strategy is represented by the point on the fitness set (single phenotype or mixed strategy) which touches the curve $A(W_1, W_2) = K$ at the greatest K. In Figures 2.3a, 2.3b, and 2.3c we show several optimum strategies. In formal terms the results are the following:

1. On a convex fitness set (environmental range smaller than the tolerance) the optimum strategy is a single phenotype which is adapted to some intermediate value of S between S_1 and S_2, and does moderately well in both these environments.

2. On a concave fitness set (environmental range exceeds the tolerance) a fine-grained environment results in an optimum strategy of a single phenotype which is specialized to either environment S_1 or S_2 depending on p, the frequency of environment I. Therefore it does optimally in one environment and poorly in the other.

3. On a concave fitness set with a coarse-grained environment the optimum is a mixed strategy in which the two

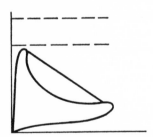

FIGURE 2.3b. Fine-grained environment on a concave fitness set.

19

246

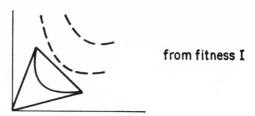

FIGURE 2.3c. Coarse-grained environment on a concave fitness set.

specialized phenotypes occur in proportions that depend on p.

When there are more than two environments new axes have to be added, and in a continuous environment we need infinitely many dimensions for the fitness set. However, as S_1 becomes arbitrarily close to S_2 any phenotype which does well in one will do well in the other, and the fitness set becomes a straight-line segment at 45° from the origin. Thus we only have to consider as distinct those environments different enough to change the ranking of fitnesses of the available phenotypes.

An alternative approach considers the environmentally determined optimum phenotype S to have a probability distribution $P(S)$, and attempts to maximize $\int A[W(S)]P(S)\,dS$ subject to the restriction $\int W(S)\,dS = C$. When $A(W)$ is linear, the optimum allocation is specialization to the most common S, while for a coarse-grained environment $A(W)$ is $W_1^p W_2^{1-p}$ or equivalently $p \log W_1 + (1 - p) \log W_2$, and the optimum is $W(S) = CP(S)$. Thus the fitness is assigned to each environment in proportion to its frequency. Qualitatively the results are the same—coarse-grained environments are uncertain and give rise to less specialized broad-niched populations, with the niche breadth increased by the uncertainty of the environment. The niche breadth interpretation is pursued further in Chapter 3.

On the basis of this formal theory we can now interpret biological situations by defining "phenotype" in various ways.

20

[*Editors' Note:* Material has been omitted at this point.]

NICHE BREADTH

Data on niche breadth come from three sources:

1. Survival experiments such as those of Tantawy's in Figure 3.1. Since fitness requires not only survival but also successful reproduction, this is clearly not a complete fitness measure but is an important component. In the figure it is seen that *D. simulans* has a narrower, higher, more specialized temperature niche than *D. melanogaster*. Any measure of spread could be used to quantify niche breadth.

2. Habitat or food selectivity in multiple choice experiments. For example, Martinez et al. (1965) set out several different kinds of bait (banana, tomato, potato, and oranges) in *Drosophila* traps less than ten feet apart. Thus any fly caught on any bait could have reached any of the others.

41

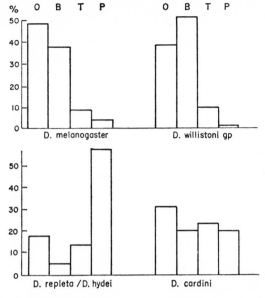

FIGURE 3.2. Food preference histograms for four species of *Drosophila*. O, orange; B, banana; T, tomato; P, potato. (From Martinez Pico et al., 1965.)

Some of their results are shown in Figure 3.2. A species such as *D. cardini*, which is attracted to each bait with almost equal frequency, would be said to have a broad niche for food as compared to *D. repleta*. Of course the baits offered did not span the whole range of *Drosophila* food, but it did include fruits differing in acidity, starchiness, and sugar content. All the flies were drawn from the same population, and since we did not observe evidence of aggressive exclusion of flies, the distributions are assumed to indicate true preferences and hence to correspond to the pre-competitive or potential niches.

3. Actual distributions of species over environments. (a) Environmental factor identifiable. We can use the frequency distribution of tidal organisms over levels across a beach, plant species vs. altitude, or flies against season of the year. (b) The environmental factor not identifiable but known to vary. Then the uniformity of a distribution over a presumed

42

249

patchy environment indicates a broad niche, and extreme clustering suggests a narrow niche provided we can exclude a clustering tendency per se and the persistence of progeny at the site of their birth. When the patches are small enough compared to the mobility of the species, the potential niche is measured.

Maldonado and Levins (in preparation) studied the microhabitat niche of *Drosophila* by setting out 20 traps with the same banana bait in a small area and classifying the species taken in each trap separately. Some of their results are shown in Table 3.2.

TABLE 3.2. Microhabitat niche breadths for *Drosophila*. $B = 1/\Sigma p_{ij}{}^2$

	Collection 5	Collection 6
Maximum	14	13
D. melanogaster	2.4	3.5
D. willistoni	5.3	7.5
D. latifasciaeformis	7.0	8.5
D. Dunni	5.4	5.2
D. ananassae	5.4	4.0
All flies	6.1	8.5

Two measures of niche breadth have been proposed:

$$\log B = -\Sigma p_i \log p_i \qquad (3.1)$$

and

$$1/B = \Sigma p_i{}^2. \qquad (3.2)$$

In both measures, p_i is the proportion of the species which is found in environment i, which selects environment i, or in the case of a viability measure

$$p_i = v_i \Big/ \sum_i v_i \qquad (3.3)$$

where v_i is the viability in environment i.

There is no very strong reason to prefer one measure over the other as yet. Both give niche breadths equal to N for N equally used resources or for uniform utilization over an interval of length N and no utilization outside. And both measures are similar, as is shown in Table 3.3. Finally, since the number of environmental classes is arbitrary, the meas-

43

TABLE 3.3. Two measures of seasonal niche breadth for *Drosophila*.*
Measure $1 = 1/\Sigma p_{ij}{}^2$; measure $2 = \exp(-p_{ij}\log p_{ij})$; maximum
value is 25. Data for Austin, Texas, from Patterson, 1943

Species	B1	B2
All	11.14	14.27
D. melanogaster + simulans	9.58	12.06
D. hydei	3.64	6.24
D. mulleri + aldrichi	4.23	6.92
D. macrospina	8.30	10.71
D. longicorni	4.39	6.95
D. affinis + algonquin	9.81	12.58
D. hematofila	3.51	5.16
D. putrida	3.52	6.16
D. pseudoobscura	4.65	6.51
D. melanica	10.35	13.07
D. busckii	3.21	4.51
D. meridiana	3.06	3.69
D. immigrans	1.70	1.84
D. robusta	2.53	3.44
D. tripunctata	6.72	9.35
Average	4.08	5.35
Correlation	.98	

* Species with fewer than 100 flies omitted.

ure B should be divided by the maximum number, which is
the number of classes, to give comparable measures of niche
breadth.

Once we have a measure of niche breadth we can ask
whether the abundant species tend to have broader niches
than the rare ones, whether climax species have narrower
niches than colonizing species. We can also compare niche
breadths for the biotae of different regions, zones, trophic
levels, or taxonomic groups. For the Puerto Rican *Drosophila*
we have found that the abundant species are usually the
ones which are broad-niched.

Immediately the question arises, how can we tell if we
are measuring the relevant factors so that the calculated
niche breadths have real meaning. We will show below that
the niche description leads to predictions of numbers of
species and other community properties which enable us to
check its completeness.

44

In the previous chapter we reached several conclusions about niche breadth. Qualitatively they all suggested that a broad niche is optimal in an environment which is uncertain. This uncertainty may derive from temporal variation in the environment from generation to generation, from a coarse-grained patchiness which is uncertain for each individual, or from a low density of usable resources or habitats (low productivity of the environment for the species in question). In a stable environment, fitness will be spread out only over environments which are so similar as to give a convex fitness set.

[*Editors' Note:* Material has been omitted at this point.]

NICHE OVERLAP

At issue here is the amount of competition or of ecological similarity among species. One measure would be some geometric distance, such as

$$d_{i,j} = \sum_h (p_{ih} - p_{jh})^2 \qquad (3.6)$$

where p_{ih} is the niche measure of species i in environment h. However not all biological differences reduce competition. Different birds may capture the same insects in flight or on a tree; fruit flies may use the same fruit at different hours. The p_{ih} must be limited to components of competition. In a coarse-grained environment, where individuals spend their whole lives in the same patch, viability in each patch type separately affects the competition, but in a fine-grained environment this will not be true.

The above measure has been used (Martinez et al., 1965)

50

TABLE 3.6. Maximum number of species for which the expected value of the community is positive

Average α	Covariance (α_{ij} with α_{ji})									
	.005	.01	.015	.02	.025	.03	.05	.075	.1	.3
.1	34	27	19	15	13	11	8	6	5	3
.2	24	22	16	13	11	10	7	6	4	
.3	21	18	13	11	9	8	6	5	4	
.4	17	14	11	9	7	7	5	4	4	
.5	14	11	8	7	6	5	4	3	3	
.6	12	8	—	5	5	5	3	3	3	

for food differences. However we prefer to measure niche overlap by the coefficient α_{ij} used by Gause. He started from the Volterra equations for increase of species x and y:

$$\mathrm{d}x/\mathrm{d}t = rx(K - x - \alpha y)/K \qquad (3.7)$$

and

$$\mathrm{d}y/\mathrm{d}t = ry(K - \beta x - y)/K. \qquad (3.8)$$

The coefficients α and β measure the reduction in the rate of increase of x caused by an individual of y compared to the effect of an individual of x, and vice versa. Since this competition depends on the probability that individuals of the two species meet (in the sense of seeking food in the same habitat, or pursuing the same kind of prey) a good approximation for α_{ij} is

$$\alpha_{ij} = \sum_h p_{ih}p_{jh} \Big/ \sum_h p_{ih}^2 \qquad (3.9)$$

which is

$$\alpha_{ij} = \sum_h p_{ih}p_{jh}(B_i), \qquad (3.10)$$

where B_i is the niche breadth. Thus α_{ij} is similar to a regression coefficient of one species' use of environment on that of the other. We see that because of the factor B_i, α_{ij} will not equal α_{ji} unless their niche breadths are equal. There is the unexpected result that α_{ij} can exceed one even without special mechanisms, such as environmental poisoning. A broad-niched species spread out over many environments will have a lower rate of encounter between members than would a more specialized species. Suppose for example that

51

TABLE 3.7. Coefficients of competition between species. First number is microhabitat α; number in parentheses is seasonal α. From unpublished studies at Mayagüez, Puerto Rico

Species	mel	lat	will	Dun	ana
D. melanogaster	1(1)	.30(.61)	.42(.76)	.61(.55)	.16(.75)
D. latifasciaeformis	.72(.70)	1(1)	.92(.79)	.72(.48)	.60(.50)
D. willistoni gp.	.88(.95)	.81(.85)	1(1)	.96(.57)	.47(.59)
D. Dunni	.90(.34)	.44(.26)	.67(.29)	1(1)	.38(.63)
D. ananassae	.18(.60)	.28(.35)	.25(.38)	.29(.81)	1(1)

one species uses environments A and B with frequencies 2/3 and 1/3, and that the second species is limited to the first environment. The probability of encounters between members of the first species is $(2/3)^2 + (1/3)^2$ or 5/9, the encounters of the second species with itself have a relative frequency of 1, and the encounters between species occur only in habitat A at the rate of $1 \times 2/3 = 6/9$. Therefore $\alpha_{12} = 6/5$ and $\alpha_{21} = 2/3$. This does not mean however that species 2 will exclude species 1, because the broad-niched species may have a greater K.

The coefficient of competition α can be measured with respect to any aspect of the environment separately, and the over-all α will then be a product of the individual ones. Once we have defined α we can raise questions about its statistical distribution. What is the average level of competition (how closely are niches packed), how variable is α, how does this differ in young and mature communities, etc.? In Table 3.7 we show some α's obtained from Puerto Rican *Drosophila*.

The logistic equations 3.7, 3.8 have been criticized from a number of points of view. However the use of the bracketed term in the equation to define the equilibrium conditions does not depend on the validity of the equations for describing the rate of change toward equilibrium. Nor does it matter that the α_{ij} may vary with population density.

THE COMMUNITY MATRIX

The simultaneous equations

$$\mathrm{d}x_i/\mathrm{d}t = r_i x_i (K_i - x_i - \Sigma \alpha_{ij} x_j)/K_i \qquad (3.11)$$

52

254

can give an equilibrium community when

$$K_i = x_i + \Sigma \alpha_{ij} x_j \qquad (3.12)$$

for all i. These equations can be expressed as a single matrix equation

$$AX = K, \qquad (3.12)$$

where X is the column vector $\begin{vmatrix} x_1 \\ x_2 \\ x_3 \\ \cdot \\ \cdot \\ \cdot \end{vmatrix}$,

K is the vector of the K_i, and A is the community matrix

$$A = \begin{vmatrix} 1 & \alpha_{12} & \alpha_{13} & \cdot & \cdot & \cdot \\ \alpha_{21} & 1 & \alpha_{23} & \cdot & \cdot & \cdot \\ \alpha_{31} & \alpha_{32} & 1 & \cdot & \cdot & \cdot \\ \cdot & & & & & 1 \end{vmatrix}$$

whose elements α_{ij} are the competition coefficients. Even without the solving of equations the matrix gives much information about the community. For competitors the α's are positive, while for predator-prey pairs α_{ij} and α_{ji} have opposite sign. For competitive α's, the average value indicates how closely species are packed. If the niche is one-dimensional, each species can have only a few positive α's and the others will be near zero. As the dimensionality of the niche increases the variance of the α's can be reduced. Thus the variance of the α's is an indicator of dimensionality. For competitors, $\alpha_{ij} = \alpha_{ji}$ when the niches are equal in breadth. Then the pairs of symmetrically arranged terms have a correlation of $+1$. As the niche breadths become more variable, this correlation decreases, and it is also reduced by predator-prey pairs.

With the descriptive parameters of the niche defined and the community matrix described, we can use them to study the original questions.

The simultaneous equations 3.12 could be solved to get the relative abundances if we knew the α_{ij} and the K_i. We

53

have suggested how to measure the α_{ij}, but K_i is less directly observable, since it is the maximum population attainable in the absence of competitors. Therefore we reversed the procedure, and used the known x_i and the calculated α's to find K_i. This was done for microhabitat separation in our Puerto Rican *Drosophila*. In Table 3.8 we show the frequencies of

TABLE 3.8. Carrying capacity K and relative niche breadth $B/\bar{B}$. From unpublished data, collections at Mayagüez, Puerto Rico, 12/26/65. $\bar{B}$ is the breadth for all flies combined.

Species	Fre-quency	K	$B/\bar{B}$	$K/(B/\bar{B})$
D. melanogaster	.03	.41	.42	.98
D. willistoni group	.54	.89	.87	1.02
D. latifasciaeformis	.32	.92	1.00	.92
D. Dunni	.01	.62	.61	1.02
D. ananassae	.08	.27	.47	.62
D. tripunctata species	.02	.35	.44	.79

species, their niche breadths B relative to the breadth for all flies, $\bar{B}$, the estimated K, and the ratio of K to $B/\bar{B}$. The uniformity of this ratio suggests the hypothesis for more general testing of whether the carrying capacity K is proportional to the niche breadth B.

There is no *a priori* reason to expect this result. It means that the species do not differ very much in the average efficiency with which they utilize their resources but only in how wide a range of microhabitats they can use. Another surprising result is the suggestion that in spite of seasonal changes in relative abundance of the species, at each moment the community is in equilibrium for the current environment but this equilibrium changes with the changes in the resources, keeping the community close to a moving equilibrium. Thus the community matrix approach helps to determine if we have in fact identified the important environmental factors and to determine if the community is near equilibrium.

At a more abstract level the community matrix leads to some results about the limit to the number of coexisting

54

species. In order for a community to be stable the determinant of its matrix must be positive. Furthermore, the symmetrized determinant whose element $\alpha_{ij}{}^* = \alpha_{ji}{}^* = (\alpha_{ij} + \alpha_{ji})/2$ must be positive, and so must each subdeterminant formed by striking out a row and the corresponding column. Therefore we can investigate the average value of these determinants in terms of the statistical distribution of the α. In Appendix I (at the end of this chapter) we derive the recurrence relation for the expected value D_n of a determinant of rank n (n rows and columns). The result is the pair of equations

$$D_n = D_{n-1} - (n-1)\bar{\alpha}^2 T_{n-1} - (n-1) \operatorname{cov}(\alpha_{ij},\alpha_{ji})D_{n-2}$$
(3.13)

$$T_n = D_{n-1} - (n-1)\bar{\alpha}T_{n-1}$$
(3.14)

where the initial values $D_0, D_1 = 1$. We see that the covariance term enters in such a way as to reduce D_n. Thus a community in which niches are equally broad can hold fewer species than one with high dimension (uniform α's, low variance) and non-uniform niche breadths. We suspect that as a community matures the variance of the α's decreases and more species could be accomodated. Thus a waif fauna of diverse origins should reach a demographic equilibrium with fewer species than old faunae hold. Recent work by Wilson and Taylor on Pacific ants (1967) seems to support this idea.

In the next section we will use the equations 3.13, 3.14 to estimate the number of species which can coexist in a community given their general similarity and heterogeneity.

[*Editors' Note:* Material has been omitted at this point.]

55

BIBLIOGRAPHY

Gause, G. F. 1934. *The Struggle for Existence.* Williams & Wilkins, Baltimore.

Martinez Pico, M., C. Maldonado, and R. Levins. 1965. Ecology and genetics of Puerto Rican *Drosophila*, I. Food preferences of sympatric species. *Carib. J. Sci.* 5: 29–38.

Patterson, J. T. 1943. The Drosophilidae of the southwest. University of Texas publ. 4313.

Sacher, G. 1966. The complementarity between development and aging—experimental and theoretical considerations. *N. Y. Acad. Sci. Conf. on Interdisciplinary Perspectives of Time,* Jan. 17–20, 1966.

Tantawy, A. O. and G. S. Mallah. 1961. Studies on natural populations of *Drosophila*, I. heat resistance and geographic variation in *Drosophila melanogaster* and *D. simulans. Evol.* 15: 1–14.

Wilson, E. O., and R. W. Taylor. 1967. The estimate of potential evolutionary increase in species density in the Polynesian ant fauna. *Evol.* 21: 1–10.

29

Reprinted from *Ecology*, **52**(4), 567–576 (1971), with permission of the publisher, Duke University Press, Durham, N. C.

ON THE MEASUREMENT OF NICHE BREADTH AND OVERLAP[1]

Robert K. Colwell

Department of Zoology, University of California, Berkeley

AND

Douglas J. Futuyma

Department of Ecology and Evolution, State University of New York, Stony Brook

Abstract. Measures of niche breadth and overlap that depend on the distribution of individuals among resource states (ecological categories) should be independent of the relative abundance of the species and of the number of resource states considered. Such measures should also take into account the degree of distinctness of the resource states from the point of view of the organisms concerned. An ecoassay of the distinctness of resource states may well be easier and more meaningful than measurements of physical and chemical factors. We propose that the species composition of communities utilizing different resource states may be used to develop weighting factors with which each state may be weighted in proportion to its degree of distinctness. The weighting factors are used in the development of indices of niche breadth and overlap that correct for variation in the range and distinctness of resource states and that suffer less from human subjectivity than do the measures used to date. The use of such indices and the relationship of niche overlap to competition are discussed.

The analysis of niche relationships in natural communities by painstaking observation of natural history (e.g., MacArthur 1958, Cody 1968) is both valuable and necessary, but methods capable of revealing greater generality at the expense of detail are also desirable. The methods of niche analysis we propose are designed to be used with numerical data arranged in ecological categories. The data will usually be animal or plant abundances, in number-of-individuals or equivalent units. The categories need not be orderable, and they may differ one from another in known ways (food types, habitat types, substrates, lures, climatic regimes, etc.) or in unspecified ways (randomly or systematically spaced quadrats, times of day or year, traps, etc.). We will refer to such categories as "resource states."

In this paper we shall deal only with niche breadth and overlap, although there are other niche metrics of interest (dimensionality, topological deformation, fitness density and center of mass, dispersion). In terms of the spatial model of the niche, as formalized by Hutchinson (1958) and expanded by Slobodkin (1962), Levins (1968), and MacArthur (1968), niche breadth is the "distance through" a niche along some particular line in niche space. Other terms have been used for niche breadth, including "niche width" (Van Valen 1965, McNaughton and Wolf 1970), "niche size" (Klopfer and MacArthur 1960, Willson 1969), and "versatility" (Maguire 1967). In all these cases, as in this paper, the property referred to is essentially the inverse of ecological specialization, a term which has been used in a quantitative sense by Kohn (1968). With respect to food preferences, for example, a koala is more specialized (has a smaller

trophic niche breadth) than a Virginia opossum. Less extreme cases are more debatable, of course, and generalizations about relative niche breadth among taxonomic groups or in different communities are not even worth discussing with only anecdotal evidence as data. Some method of quantifying niche breadth is required for any worthwhile study of the property.

Niche overlap is simply the joint use of a resource, or resources, by two or more species. In other words, it is the region of niche space (in the sense of Hutchinson 1958) shared by two or more contiguous niches. The difficult question of the relationship of niche overlap to interspecific competition will be treated in the final section. As with niche breadth, any legitimate comparison of niche overlap among the species of a community, and especially among communities, requires careful quantification.

THE RESOURCE MATRIX AND SOME SIMPLE MEASURES

The measures of niche breadth and overlap we will discuss are all based on the distribution of individual organisms, by species, within a set of resource states. The table formed by using species as rows, and resource states as columns will be called the "resource matrix." A heterogeneous habitat, for example, might be subdivided into sunny-dry, sunny-wet, shady-dry, and shady-wet resource states, or alternatively, into unnamed random quadrats, which would then be considered the resource states. The resource matrix for vascular plants in this habitat would consist of a table with the species of vascular plants as rows, and the named habitat subdivisions or the unnamed random quadrats as columns. The typical cell would contain the number of individuals (N_{ij})

[1] Received August 10, 1970; accepted February 20, 1971.

of species i found to be associated with resource state j. The total number of individuals of species i (the row total) will be called Y_i, and the total number of individuals of all species combined for resource state j (the column total) will be referred to as X_j. The total number of individuals in the matrix (the grand total) will be called Z. The number of species (rows) is s, and the number of resource states (columns) is r. We summarize this notation in Table 1.

TABLE 1. Notation for the resource matrix

	Resource states					
S	N_{11}	...	N_{1j}	...	N_{1r}	Y_1
p	.		.		.	.
e	.		.		.	.
c	.		.		.	.
i	N_{i1}	...	N_{ij}	...	N_{ir}	Y_i
e	.		.		.	.
s	.		.		.	.
	.		.		.	.
	N_{s1}	...	N_{sj}	...	N_{sr}	Y_s
	X_1	...	X_j	...	X_r	Z

The niche breadth of a species in the resource matrix can be estimated by measuring the uniformity of the distribution of individuals of that species among the resource states of the matrix. Levins (1968) suggests two simple measures of uniformity. In our notation, for the niche breadth of the ith species, these are

$$B_i = \frac{1}{\sum_j p_{ij}^2} = \frac{(\sum_j N_{ij})^2}{\sum_j N_{ij}^2} = \frac{Y_i^2}{\sum_j N_{ij}^2} \quad (1)$$

and

$$B'_i = -\sum_j p_{ij} \log p_{ij} \quad (2)$$

where $p_{ij} = N_{ij} / Y_i$, the proportion of the individuals of species i which is associated with resource state j. The measure B_i, equation (1), is closely related to the coefficient of variation, and is simply the inverse of Simpson's (1949) "measure of concentration." The measure B'_i, equation (2), is the Shannon-Wiener formula for information or uncertainty (Shannon and Weaver 1949). Both B_i and B'_i are maximized when an equal number of individuals of species i are associated with each resource state. This would imply that species i does not discriminate among the resource states (or discriminates, but requires or prefers them equally), and therefore has the broadest possible niche, with respect to those resource states. The measures are both minimized when all individuals of species i are associated with one resource state (minimum niche breadth, maximum specialization). The problem of standardizing the range of B_i and B'_i will be taken up in a later section.

Using the data of a resource matrix, niche overlap between two species can be estimated by comparing the distribution of the individuals of the two species among the resource states of the matrix. If the distributions are identical, the two niches overlap completely, with respect to the resource states in the matrix. If the two species share no resource states, their niches do not overlap at all, with respect to these states. Perhaps the simplest measure of proportional overlap is

$$C_{ih} = 1 - \frac{1}{2} \sum_j | p_{ij} - p_{hj} | \quad (3)$$

where p_{ij} is the same as before, and $p_{hj} = N_{hj} / Y_h$, where h is a second species in the matrix. The measure C takes its minimum value of 0 when species i and species h share no resource states, and its maximum value of 1 when the proportional distributions of the two species among the resource states are the same. Schoener (1970) uses this measure of overlap in the form of a percentage to measure ecological similarity.

WEAKNESSES OF THE SIMPLE MEASURES

There are two situations in which one might wish to compare niche metrics: among species within a community, and between species of different communities. In both cases, the paramount difficulty is the standardization of the procedure so that measurements are comparable for different species. The problem is related to the familiar sophistry that the koala is really a generalist on *Eucalyptus* leaves, while the Virginia opossum is specialized for eating garbage. The difficulty, of course, arises in our definition of categories, since "specialized" means "covering few categories" and "generalized" means "covering many categories."

The question of defining appropriate categories can be divided into three parts: (i) the problem of range, (ii) the problem of spacing, and (iii) the problem of nonlinearity. To illustrate these, suppose someone wished to compare niche breadth and overlap, with respect to soil moisture, among species of soil mites at several study sites. At each site, he samples from each of 10 quadrats spaced along a transect from stream bank to dry hillside. Now, for simplicity, suppose that in fact the same five species are present at each study site, and that their true niches can be represented as in Fig. 1 (top set of axes), with respect to soil moisture measured as percentage saturation. Let the 10 quadrats at each of four different sites be arranged with respect to soil moisture as are the dots along the lines in the center of Fig. 1. Finally, assume that the abundance of each species in a quadrat is proportional to its fitness in that quadrat, so that, for example, the abundance of Species 2 at Site I would be maximal in the fourth quadrat (counting from the dry end), much less in the third

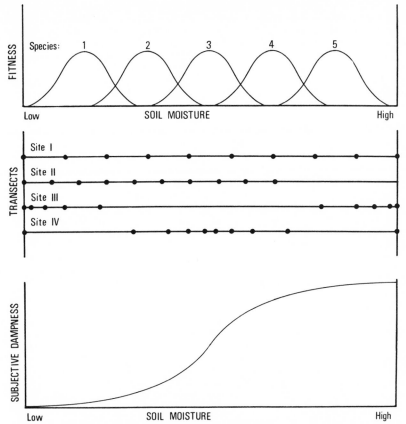

Fig. 1. A hypothetical example demonstrating the problems of range, spacing, and non-linearity in the measurement of niche breadth and overlap. The horizontal axis represents a gradient of soil moisture, from dry hillside (left) to moist stream bank (right). We consider 5 species of soil mites, each maximally fit at a different soil moisture (top graph). For simplicity we assume that, wherever they are found, the abundance of each mite species depends only on its fitness with respect to soil moisture. The same moisture gradient exists at each of four different study sites (Sites I, II, III, IV), but the sampling quadrats are placed in different patterns relative to soil moisture—shown by the large dots in the middle part of the figure. Niche breadth, being an intrinsic property of the species of mite, is the same wherever that species occurs. However, simple estimates of niche breadth will differ at different study sites, owing to differences in the patterns of spacing and total range of soil moisture covered by the quadrats. Estimates of niche overlap will be biased also. See text for a full explanation.

and fifth quadrats, and zero in all the other quadrats at that site.

The problem of range is demonstrated by a comparison of Site I with Site II. At Site II, Species 1 through 3 will be found in a larger number of quadrats than they are at Site I, so that the distribution of each of these species over the 10 quadrats (resource states) appears more uniform at Site II than at Site I. This uniformity would be reflected in higher estimates of niche breadth by equation (1) or (2) for the first three species at Site II than at Site I, even though they have the same tolerance for variation in

soil moisture at both sites. Differences in range, in themselves, will not affect proportional measures of niche overlap, such as equation (3), as long as the entire niche of each species is sampled (as is the case for Species 1 and 2 at Sites I and II).

The problem of range in the measurement of niche breadth is recognized by Ricklefs (1966), Cody (1968), and Maguire (1967), each of whom constructs an absolute scale by using as upper and lower bounds the extreme values found for all species combined, and then adjusts niche breadth estimates to that scale. McNaughton and Wolf (1970), on the other hand, make no attempt to standardize ranges in calculating niche breadths from data on vegetational succession and on the distribution of organisms along "intuitive gradients." Consequently, the negative correlation they find between the number of species on a transect and the mean niche breadth of those species may be entirely spurious. Consider again Fig. 1, Sites I and II. Site I includes all five species, while the quadrats at Site II reveal fewer species, and yield a higher average niche breadth, as shown earlier, thus giving a negative correlation between niche breadth and diversity—for the same set of species.

The problem of spacing can be seen by comparing the distribution of quadrats at Sites I, III, and IV in Fig. 1. All three sets of quadrats cover the same range of soil moisture but differ in spacing along the moisture gradient. If niche breadths were calculated for the five species at each of these sites by equation (1) or (2), the data for Site III would yield higher estimates of niche breadth for Species 1 and 5 than for Species 2, 3, and 4, while just the reverse would result from the data for Site IV, with Site I giving approximately equal estimates for all five species. Thus spacing alone can easily bias estimates of niche breadth, and the problem will often arise even when samples are taken at even intervals of elevation, geological time, surface distance, and so forth, since the resource may not vary linearly with physical parameters. Unless resource states have ecologically equivalent degrees of distinctness among them, comparisons between communities, and particularly within communities, are perilous.

The problem of spacing also affects estimates of niche overlap. If equation (3) were calculated for Species 1 and 2 in the figure, a considerably higher estimate of niche overlap would be given for Site II than for Site III. This is a result of the concentration of a disproportionate number of quadrats in the exclusive portion of the niche of Species 1 at Site III. There is probably a general tendency toward the underestimation of niche overlap from abundance data due to the abruptness, and sometimes even the inaccessibility, of many ecotonal areas.

The problem of nonlinearity arises when an ecological variable is measured on a physical or chemical scale. Within any given range of total variation, resource states should be equally distinct, or equally spaced, not in relation to some physical or chemical variable measured in ordinary units, but in relation to an ecological variable, ideally measured in units of "subjective" effect on the organisms in question. Returning to the example in Fig. 1, the quadrats of Site I, which are equally spaced with respect to soil moisture measured as percentage saturation, will correctly estimate the niche breadths and overlaps of the five species of soil mites only if the effect of soil

moisture on the mites is linearly related to percentage saturation within the range measured. In fact, for most ecological variables, this kind of linearity seems rather unlikely. Nevertheless, the methods of Cody (1968), Maguire (1967), McNaughton and Wolf (1970), and Ricklefs (1966) for measuring niche breadths all assume a linear relationship between physical parameters and ecological variables, although the assumption is not explicitly stated.

By analogy to such measures as "subjective loudness" in human psychophysics (see Stevens 1959), we might refer in the example to "subjective dampness." If subjective dampness is related to soil moisture (measured as percentage saturation) as on the bottom set of coordinates of Fig. 1, then Species 2, 3, and 4 actually have broader niches than Species 1 or 5, since the middle species are able to survive over a much broader range of subjective dampness. Thus Site IV would yield the best estimates of niche breadth and overlap, since its quadrats are placed at equal increments of subjective dampness. The problem of nonlinearity is a subtle one, partially because of the difficulty of avoiding circularity in measuring its effect. But the data of human psychometrics and the response curves of physiological ecology are compelling evidence of the need to take the problem seriously in the estimation of niche metrics.

RESOURCE STATE WEIGHTING FACTORS

The amount of error caused by nonlinearity and ecological inequality of spacing among resource states would be considerably reduced if we could discover a way of weighting each resource state by its degree of distinctness from the other resource states in the resource matrix. The problem of range could be alleviated by correcting for the total or "collective" heterogeneity of the set of resource states in each estimation of a niche metric.

Although physical and chemical parameters may certainly be used to provide information on the distinctness and heterogeneity of resource states, we contend that biological parameters provide the best estimate of the degree of ecologically important difference among resource states. In particular, the relative abundance among resource states of a sufficiently wide variety of other species in the community should provide adequate information on the ecological distinctness of resource states, in most cases. This might be called an "eco-assay" of the resource states.

Thus in the example of the moisture gradient, we would argue that a census of the plant and animal species in each quadrat contains more relevant information about the degree of distinction of that quadrat from other quadrats, from a mite's point of view, than does a series of physical and chemical measurements obtained with the same amount of effort. Not only are other species important to a mite directly—as competitors, predators, prey, and perhaps substrate and habitat—but the entire fauna and flora must reflect, indirectly, all relevant physical and chemical parameters.

We shall now develop a method for weighting resource states by their distinctness, and for estimating their collective heterogeneity. The method applies equally well, in theory, to chemical and physical parameters, although it is designed for biological weighting data.

Whether physical, chemical, or biological parameters are used, each resource state is scored for each parameter, as well as for the abundance of the species whose weighted niche breadth is to be calculated. The environmental parameters are cast into a resource matrix, as defined above and in Table 1, but one in which the rows may represent either species or physical-chemical parameters. In the latter case, the N_{ij}'s are scores or measurements, which need not be integers. (The possibility of weighting rows will be discussed later.) For simplicity, it will be assumed in what follows that the rows are all species, and that the abundance data for the species whose niche breadth is to be calculated is also in the matrix.

Now we define a new matrix in which each element π_{ij} is the proportion of the grand total represented by each N_{ij}, so that

$$\pi_{ij} = \frac{N_{ij}}{Z} . \tag{4a}$$

If we call the row (species) classification the "Y-classification," and the column (resource state) classification the "X-classification," then the π matrix is the joint probability distribution of the X and Y classifications, and π_{ij} is the probability (actually an estimate, of course) that a randomly chosen individual from the pooled species will be an individual of species i associated with resource state j. We next define the conditional probabilities

$$p_{ij} = \frac{N_{ij}}{Y_i} , \tag{4b}$$

$$P_j = \frac{X_j}{Z} , \tag{4c}$$

$$q_{ij} = \frac{N_{ij}}{X_j} , \tag{4d}$$

and

$$Q_i = \frac{Y_i}{Z} . \tag{4e}$$

Thus p_{ij} is the probability that an individual is associated with resource state j, given that it is of species i, and q_{ij} is the probability that an individual is of

species i, given that it is associated with resource state j. Finally, Q_i is the probability that an individual is of species i, disregarding resource states, and P_j is the probability that an individual is associated with resource state j, disregarding species identity. Therefore,

$$\sum_i \sum_j \pi_{ij} = \sum_j p_{ij} = \sum_i q_{ij} = \sum_j P_j = \sum_i Q_i = 1 . \tag{4f}$$

To obtain weighting factors for the resource states, we begin by defining the following information functions. The uncertainty with respect to resource state of one of the Z individuals in the resource matrix is

$$H(X) = - \sum_{j=1}^{r} P_j \log P_j , \tag{5}$$

and the uncertainty of an individual with respect to species is

$$H(Y) = - \sum_{i=1}^{s} Q_i \log Q_i . \tag{6}$$

Given an individual of species i, its uncertainty with respect to resource state is

$$H_i(X) = - \sum_j p_{ij} \log p_{ij} . \tag{7}$$

Given an individual associated with resource state j, its uncertainty with respect to species is

$$H_j(Y) = -\sum_i q_{ij} \log q_{ij} . \tag{8}$$

The uncertainty of an individual with respect to both species and resource state is

$$H(XY) = -\sum_i \sum_j \pi_{ij} \log \pi_{ij} . \tag{9}$$

We wish to determine the total heterogeneity of the matrix with respect to resource states, and the contribution of each resource state to this total, which will be the basis for a weighting factor for that state. If there is complete homogeneity of resource states (i.e., if $q_{ij} = Q_i$ for all i and j), then the X and Y classifications are independent, and $\pi_{ij} = Q_i P_j$. In this limiting case,

$$H(XY) = H(Y) + H(X)$$

(see Appendix A for proof). If, however, the X and Y classifications are not independent, the relation

$$H(XY) < H(Y) + H(X) \tag{10}$$

is always true, since the total uncertainty $H(XY)$ is reduced by any dependence between the two classifications [Khinchin (1957) gives a formal proof].

If we know the relative abundance of species (the Y_i's) then the total uncertainty $H(XY)$ is reduced by the amount $H(Y)$, equations (9) and (6). The remainder is defined as

$$H_Y(X) = H(XY) -- H(Y) , \tag{11}$$

which can be shown (see Appendix B) to be

$$H_Y(X) = -\Sigma\Sigma\; \pi_{ij} \log p_{ij}, \qquad (12)$$
$$\quad\; i\, j$$

or alternatively, the weighted average

$$H_Y(X) = \;\; \Sigma\; Q_i H_i(X) , \qquad (13)$$
$$\qquad\quad i$$

where $H_i(X)$ is defined by equation (7). [See Appendix B and Pielou (1969).]

It is clear that when the distribution of individuals over resource states differs among species, the X and Y classifications are not independent, and, from equations (10) and (11), $H_Y(X) < H(X)$. Thus a measure of the departure from independence is given by

$$m(X) = H(X) - H_Y(X) , \qquad (14)$$

which is sometimes called "mutual information" (Abramson 1963), since

$$m(X) = m(Y) = H(Y) - H_X(Y).$$

The limits of $m(X)$ are zero, for complete homogeneity of resource states, and $H(X)$, for complete heterogeneity of resource states [see Appendix A and Hays (1964)]. Thus a standardized measure of resource state heterogeneity is

$$M(X) = \frac{m(X)}{H(X)} = \frac{H(X) - H_Y(X)}{H(X)} = 1 - \frac{H_Y(X)}{H(X)} . \qquad (15)$$

$M(X)$ may be called "relative mutual information"; it has the range 0 to 1, and does not depend upon either the number of resource states or the magnitude and evenness of column (resource state) totals. Partitioning $M(X)$ by resource states, we obtain

$$M_j(X) = (\Sigma\; \pi_{ij} \log p_{ij} - P_j \log P_j)/H(X) , \qquad (16)$$
$$\qquad\quad i$$

or, for computation,

$$M_j(X) =$$
$$\frac{X_j(\log X_j - \log Z) - \sum_i N_{ij} \log (N_{ij}/Y_i)}{\sum_j X_j \log X_j - Z \log Z} .$$

The quantity $M_j(X)$ is the distinctness of the jth resource state in the resource matrix. The collective heterogeneity of the resource states is therefore $\Sigma\; M_j(X) = M(X)$, with range $(0, 1)$. We now define the "absolute" weighting factor for the jth resource state as

$$\delta_j = M_j(X) , \qquad (17)$$

and the "relative" weighting factor for the jth resource state as

$$d_j = \frac{M_j(X)}{\sum_j M_j(X)} = \frac{\delta_j}{M(X)} , \qquad (18)$$

so that $\Sigma\; d_j = 1$.
$\quad j$

WEIGHTED EXPANSION OF THE RESOURCE MATRIX AND WEIGHTED NICHE METRICS

The simple niche measures, equations (1–3), will be good estimators of the true niche breadth and overlap of species in a resource matrix only if the resource states are all equally distinct, one from another. Conceptually, this might be accomplished by adding new columns (resource states) to the matrix that are identical to the more-distinct original resource states, which would then be less distinct in the expanded matrix. For example, in the pattern "abaa," the letter "b" is more unusual than "a," but if we expand the pattern to "abbaa," and then to "abbbaa," the letter "b" becomes less and less unusual. Theoretically, such an expansion of the resource matrix could be carried out until the distinctness of all resource states, as measured by equation (17) or (18), was the same.

An equivalent procedure, which is much more tractable, is to choose a large number k (say 10.000), and "expand" the original matrix of r resource states to a new matrix having k resource states, with each of the original resource states represented $d_j k$ times in the expanded matrix $(\Sigma\; d_j k = k\Sigma d_j = k)$. Suppose, for example, that the original matrix has three resource states, and the abundance N_{ij} of species i on these states and the calculated values of d_j are as follows:

j :	1	2	3
N_{ij}:	2	5	1
d_j :	.5	.1	.4

Now if we set $k = 10$, we obtain the expanded abundance vector:

j :	1	2	3	4	5	6	7	8	9	10
N_{ij}:	2	2	2	2	2	5	1	1	1	1

in which each of the original resource states is represented $d_j k$ times. In practice, there would of course be more resource states, and a much larger k. Even though it is not possible to represent $d_j k$ resource states when that product is not an integer, the use of the weighting factors in the weighted niche metrics is legitimate mathematically, as can be seen from what follows.

The application of the simple formulas for niche breadth and overlap to such an expanded matrix will yield measures corrected for the original variation in distinctness and for any ecological non-linearity in the original resource states. To calculate the weighted niche measures directly from the original matrix and the vector of weighting factors, we first define Y^*_i as the total number of individuals of species i in the expanded matrix. Since the jth column of the original matrix is reproduced $d_j k$ times in the expanded matrix,

$$Y^*_i = \sum_j d_j k \, N_{ij} . \qquad (19a)$$

Now define

$$p^*_{ij} = \frac{N_{ij}}{Y^*_i} . \qquad (19b)$$

Then for niche breadth, by equation (1), we have

$$B_i = 1 / \sum_j (d_j k \, p^*_{ij}{}^2) .$$

This formulation has range $(1,k)$, so to standardize the range to $(0,1)$ we must calculate

$$\beta_i = \frac{B_{i(\text{observed})} - B_{(\text{min})}}{B_{i(\text{max})} - B_{(\text{min})}} ,$$

which is

$$\beta_i = \left[\frac{1}{\sum_j (d_j k \, p^*_{ij}{}^2)} - 1 \right] \left[\frac{1}{k-1} \right] \qquad (20)$$

Niche breadth by equation (2) for the expanded matrix becomes

$$B'_i = - \sum_j d_j k \, (p^*_{ij} \log p^*_{ij}) .$$

This expression has the range $(0, \log k)$, so to standardize the range to $(0, 1)$, we must calculate

$$\beta'_i = - \frac{k}{\log k} \sum_j d_j \, (p^*_{ij} \log p^*_{ij}) . \qquad (21)$$

For niche overlap, the modification for C_{ih} [equation (3)], for the expanded resource matrix is

$$\gamma_{ih} = 1 - \tfrac{1}{2} \sum_j d_j k \, | \, p^*_{ij} - p^*_{hj} \, | . \qquad (22)$$

A second measure of niche overlap follows directly from the definition of "relative mutual information" given in the section on weighting factors, equation (15). We will first define a new measure of simple overlap, and then add the resource state weighting factors. Now let $t_j = p_{ij} + p_{hj}$, where $p_{ij} = N_{ij}/Y_i$, as previously defined, and $p_{hj} = N_{hj}/Y_h$ for the hth species in the original resource matrix. To simplify notation, define the function $I(x) = x \log x$. Then an index of the difference in proportional distribution of species i and h over the r resource states of the original matrix is the relative mutual information of the vectors $(p_{i1} \ldots p_{ij} \ldots p_{ir})$ and $(p_{h1} \ldots p_{hj} \ldots p_{hr})$, which is

$$M_{ih} = 1 - \frac{\sum_j [(p_{ij}/2) \log (p_{ij}/t_j) + (p_{hj}/2) \log (p_{hj}/t_j)]}{2(\tfrac{1}{2} \log \tfrac{1}{2})} .$$

The complement of M_{ih} (that is, $1 - M_{ih}$), an index of niche overlap, reduces to

$$C'_{ih} = - \frac{1}{2 \log 2} \sum_j \left[I(p_{ij}) + I(p_{hj}) - I(t_j) \right] , \qquad (23)$$

which ranges from 0, when species i and h do not co-occur on any resource state, to 1, when the proportional distribution of the two species over resource states is the same. Equation 23 is formally similar to the "faunal overlap" index of Horn (1966), which is a two-sample formula for relative mutual information. The weighted form of C' is simply equation (23) calculated for the expanded matrix, or

$$\gamma'_{ih} = - \frac{1}{2 \log 2} \sum_j d_j k \left[I(p^*_{ij}) + I(p^*_{hj}) - I(t^*_j) \right] , \qquad (24)$$

where p^*_{ij} is as defined in equation (19b), $p^*_{hj} = N_{hj}/Y^*_h$, and $t^*_j = p^*_{ij} + p^*_{hj}$.

So far we have only considered weighted niche metrics which utilize the "relative" weighting factors (d_j's) defined by equation (18). All four of these measures—β, β', γ, and γ' [equations (20), (21), (22), and (24)]—have the range (0, 1), regardless of the total heterogeneity of the resource states in the original resource matrix, although they do give relative weight to each resource state proportional to its degree of distinctness. These measures are therefore appropriate only for comparisons among species in the same resource matrix, or among matrices of equal total heterogeneity.

To correct for differences in ecological range among matrices, we must take into account the total heterogeneity of resource states in each matrix by using the "absolute" weighting factors (δ_j's) defined by equation (17). The expressions for absolute niche breadth are the same as equations (20) and (21), but with δ_j substituted for d_j in the summations, as well as in the calculation of p^*_{ij} by equations (19a) and (19b). In the expressions for absolute niche overlap which correspond to equations (22) and (24), δ_j is substituted for d_j in the summations, but the values of p^*_{ij} and p^*_{hj} are calculated as in the original equations, using d_j. In addition, the summation must be subtracted from $\sum_j \delta_j$, instead of from 1, in the expression for absolute niche overlap corresponding to equation (22). The absolute measures will always be less than the corresponding relative measures, unless the resource matrix is maximally heterogeneous, in which case $\delta_j = d_j$ for all j.

The actual value chosen for the constant k is irrelevant for the niche overlap measures, but the measures of niche breadth depend upon the value of k used, so that the same value should be used consistently in any particular study. We have found 10,000 to be an adequate standard for entomological field data. In any case, k must be substantially larger than r to avoid positive logarithms in equation (21). The logarithms used in calculating the weighting factors and in equations (21) and (24) may be taken to any convenient base (as long as the same base is

TABLE 2. Hypothetical resource matrix and weighted niche metrics

A. Raw data, marginal totals, and resource state weighting factors

Species	Resource states												Y_i
1	1	1	1	1	1	1	1	1	1	1	1	1	12
2	1	1	1	1	0	0	0	0	1	1	1	1	8
3	0	0	0	0	1	1	1	1	1	1	1	1	8
4	0	0	1	1	1	1	0	0	1	1	1	1	8
5	0	0	0	0	2	2	2	2	2	2	2	2	16
6	7	0	0	0	0	0	0	0	0	0	0	0	7
7	1	6	0	0	0	0	0	0	0	0	0	0	7
8	0	1	6	0	0	0	0	0	0	0	0	0	7
9	0	0	1	6	0	0	0	0	0	0	0	0	7
10	0	0	0	7	0	0	0	0	0	0	0	0	7
X_j :	10	9	10	16	5	5	4	4	6	6	6	6	$Z = 87$
													Sum:
d_j :	.175	.150	.140	.226	.040	.040	.042	.042	.036	.036	.036	.036	.999[a]
δ_j :	.073	.062	.058	.094	.017	.017	.017	.017	.015	.015	.015	.015	.415

B. Niche breadth

Species	Unweighted	Relative	Absolute
1	1.000	1.000	.683
2	.824	.931	.577
3	.824	.490	.126
4	.824	.781	.425
5	.824	.481	.135

C. Niche overlap

Species		Unweighted	Relative	Absolute
1	2	.809	.919	.414
1	3	.809	.564	.255
1	4	.809	.770	.347
1	5	.809	.553	.250
2	3	.500	.308	.139
2	4	.750	.745	.336
2	5	.500	.310	.140
3	4	.750	.538	.243
3	5	1.000	1.000	.451
4	5	.750	.538	.243

[a] Differs from 1.000 by rounding error.

used consistently) without affecting the resulting values. To avoid circularity, the species for which niche breadth is being calculated, or the two species for which niche overlap is being calculated, should be excluded from the computation of weighting factors. This means there is actually a different set of weighting factors for each calculation of a niche metric.

To illustrate the properties of weighted niche metrics, unweighted, relative, and absolute niche breadth [by equation (21)] and niche overlap [by equation (24)] were computed for the first five species of the hypothetical resource matrix in Table 2. Although the resource state weighting factors given in the table were calculated for the entire set of 10 species, the weighting factors used to compute the weighted niche metrics were derived anew for each value by skipping the row (for niche breadth) or rows (for niche overlap) for which the metric was to be calculated.

It may be desirable to alter the weighting of the rows in the resource matrix, particularly if physical or chemical factors are included, since each row has an equal influence on the resource state weighting factors only if $Q_i = 1/s$ (where s is the number of rows or species) for all i. Thus if biological abundance data are used, commoner species have more effect than less common species on the values of the weighting factors. If some other relative weighting is desired (such as logarithm of relative abundance), define a vector of row weights $(w_1 \ldots w_i \ldots w_s)$, where w_i is the desired weight of the ith row in the resource matrix. Then the typical element in the row-weighted matrix is

$$N'_{ij} = w_i\, p_{ij} = \frac{w_i\, N_{ij}}{Y_i}.$$

For equal weighting of all rows, $N'_{ij} = p_{ij}$.

Finally, we wish to emphasize strongly that the methods we have developed will do nothing to improve a poor research design or meaningless data.

It is essential that ecological data to be used for comparing niche metrics within, and especially among, communities be taken in a standardized and meaningful way. The corrective measures we propose are simply the "fine adjustment"; the coarse adjustment must be present in the experimental design. It is still not possible to compare the food niche of the Virginia opossum and the koala.

Niche Overlap and Competition

Because of the widespread use of measures of overlap as estimates of competition for resources, we find it necessary to clarify the possible use of the measures we have proposed in studies of competition. To begin with, let us distinguish between the "actual" and "virtual" niches of a population. These terms are approximately equivalent to the "realized" and "fundamental" niches of Hutchinson (1958), but are operationally defined, and pertain specifically to the effects of competition on a local population level. Niche breadth and overlap, when measured under "natural" conditions, are "actual" metrics, while "virtual" niche breadth and overlap are the corresponding values measured in the absence of competition among species. Thus Levins (1968) refers to the virtual niche as "pre-competitive." Most commonly, competitive displacement or exclusion tends to reduce niche breadth and overlap among competing species, so that the actual niche is a proper subset of the virtual niche in environment space.

The realized niche of Hutchinson is defined for an entire species, and for species with a wide geographic range, or with clinal variation, it may be considerably larger than the actual niche of a local population. Likewise, Hutchinson's fundamental niche includes all regions of niche space in which a species has positive fitness, whether or not all such conditions would exist in nature, even if competitors were removed. Thus the fundamental niche is larger than the virtual niche for a local population, and usually even for an entire species.

We must emphasize that it is the conditions under which data are collected, rather than the method of calculation, that determine whether actual or virtual niches are measured. In general, natural history data (transects, quadrats, etc.) are not capable of yielding measurements of the virtual niche. Virtual niche measurements can be obtained either by actually removing competitors (e.g., Connell 1961, Culver 1970, Hespenheide 1971), by exploiting natural situations in which competitors are absent (e.g., Van Valen 1965, Culver 1970), or in some situations by creating an oversupply of a scarce resource in order to minimize competition (Levins 1968, Colwell 1969).

The distinction between actual and virtual niche measurements is particularly important when niche overlap is taken as a measure of competition. Paradoxically, simply demonstrating an overlap in resource use by two species in nature can be evidence either for or against the existence of competition between them. Competition may be operating, but exclusion or displacement may be incomplete, or even impossible (as with competition among plants for carbon dioxide in short supply). In this case, observed niche overlap is evidence of the existence of competition. However, overlap may be evidence of a lack of competition if the resource under consideration is in oversupply or is irrelevant to one or both species.

By similar arguments, it can be shown that lack of demonstrable overlap may also be evidence either for or against the existence of competition. The only way to demonstrate the existence of competition, and to measure its intensity, is by comparing actual to virtual niche overlap between suspected competitors: if actual and virtual overlap are both zero, or if they are equal and no change in the population of either species occurs in the absence of its putative competitor, then there is no evidence for competition at that time and place. If, on the other hand, virtual overlap is shown to exceed actual overlap, the existence of competition has been demonstrated.

Finally, we do not find it useful to speak of the niche of an individual, although that is conceptually possible. The dimensions of such an individual's niche might just as well be treated as characters of individual phenotype. A niche, then, like a gene pool, is a property of a set of individuals, usually a biological population. (In fact, it could be argued that the concept of niche should be replaced by some notion of "phene pool.") Consequently, the niche of a species is a statistical entity which changes whenever its constituents change.

Acknowledgments

We wish to thank the many people who have discussed the substance of this paper with us, or who have read and criticized one of its many drafts, especially M. Cody, D. Culver, R. Inger, M. Lloyd, R. Levins, and the students of Zoology 244 at Berkeley. The comments of a reviewer were very helpful. Colwell was supported by a Ford Postdoctoral Fellowship at the University of Chicago and by an N.I.H. Biomedical Sciences Support Grant award at the University of California during the writing of this paper.

Literature Cited

Abramson, N. 1963. Information theory and coding. McGraw-Hill, New York. 201 p.

Cody, M. L. 1968. On the methods of resource division in grassland bird communities. Amer. Natur. **102:** 107-147.

Colwell, R. K. 1969. Ecological specialization and species diversity of tropical and temperate arthropods. Doctoral dissertation, University of Michigan, Ann Arbor.

Connell, J. H. 1961. The influence of interspecific competition and other factors on the distribution of the barnacle *Chthalamus stellatus*. Ecology 42: 710–723.

Culver, D. H. 1970. Analysis of simple cave communities: niche separation and species packing. Ecology 51: 949–958.

Hays, W. L. 1964. Statistics for psychologists. Holt, Rinehart, and Winston, New York. 719 p.

Hespenheide, H. A. 1971. Flycatcher habitat selection in the eastern deciduous forest. Auk 88: 61–74.

Horn, H. S. 1966. Measurement of "overlap" in comparative ecological studies. Amer. Natur. 100: 419–424.

Hutchinson, G. E. 1958. Concluding remarks. Cold Spring Harbor Symp. Quant. Biol. 22: 415–427.

Khinchin, A. I. 1957. Mathematical foundations of information theory. Dover, New York. 120 p.

Klopfer, P. H., and R. H. MacArthur. 1960. Niche size and faunal diversity. Amer. Natur. 94: 293–300.

Kohn, A. J. 1968. Microhabitats, abundance, and food of *Conus* in the Maldive and Chagos Islands. Ecology 49: 1046–1061.

Levins, R. 1968. Evolution in changing environments. Princeton Univ. Press, Princeton, N. J. 120 p.

MacArthur, R. H. 1958. Population ecology of some warblers of northeastern coniferous forests. Ecology 39: 599–619.

———. 1968. The theory of the niche, p. 159–176. *In* R. C. Lewontin [ed.] Population biology and evolution. Syracuse Univ. Press, Syracuse, New York.

MacArthur, R. H., and R. Levins. 1967. The limiting similarity, convergence, and divergence of coexisting species. Amer. Natur. 102: 377–385.

Maguire, B. 1967. A partial analysis of the niche. Amer. Natur. 101: 515–523.

McNaughton, S. J., and L. L. Wolf. 1970. Dominance and the niche in ecological systems. Science 167: 131–139.

Pielou, E. C. 1969. An introduction to mathematical ecology. Wiley Interscience, New York. 286 p.

Ricklefs, R. E. 1966. The temporal component of diversity among species of birds. Evolution 20: 235–242.

Schoener, T. W. 1970. Non-synchronous spatial overlap of lizards in patchy habitats. Ecology 51: 408–418.

Shannon, C. E., and W. Weaver. 1949. The mathematical theory of communication. Univ. of Illinois Press, Urbana. 117 p.

Simpson, E. H. 1949. Measurement of diversity. Nature 163: 688.

Slobodkin, L. B. 1962. Growth and regulation of animal populations. Holt, Rinehart, and Winston, New York. 184 p.

Stevens, S. S. 1959. Measurement, psychophysics, and utility, p. 18–63. *In* C. W. Churchman and P. Ratoosh [ed.], Measurement: definitions and theories. Wiley, New York.

Van Valen, L. 1965. Morphological variation and width of ecological niche. Amer. Natur. 99: 377–390.

Willson, M. F. 1969. Avian niche size and morphological variation. Amer. Natur. 103: 531–542.

Appendix A

The limits of $H(XY)$ and $m(X)$

From equation (9), $H(XY) = -\Sigma\Sigma \, \pi_{ij} \log \pi_{ij}$. When the resource states are completely homogeneous, $q_{ij} = Q_i$, and thus $p_{ij} = P_j$ necessarily, for all i and j. Therefore

$$P_j \, Q_i = P_j \, q_{ij} \; = \frac{X_j}{Z} \frac{N_{ij}}{X_j} = \pi.$$

Substituting, we have

$$
\begin{aligned}
H(XY) &= -\Sigma\Sigma \, Q_i \, P_j \log Q_i \, P_j \\
&= -\Sigma\Sigma \, Q_i \, P_j \, (\log Q_i + \log P_j) \\
&= -\Sigma_j \, P_j \, (\Sigma_i \, Q_i \log Q_i) - \Sigma_i \, Q_i \, (\Sigma_j \, P_j \log P_j).
\end{aligned}
$$

Thus, from equations (5) and (6), $H(XY) = H(Y) + H(X)$. Since $H_Y(X) = H(XY) - H(Y)$, equation (11), we have $H_Y(X) = H(X)$, so that $m(X) = 0$, equation (14).

When the resource states are completely heterogeneous, with no species in common among them, each row in the resource matrix may have only one non-zero N_{ij} (this will not necessarily be true for the columns), call it $N_{ig(i)}$. Then $N_{ig(i)} = Y_i$, and

$$
\begin{aligned}
\pi_{ij} &= 0, \text{ for } j \neq g(i) \\
\pi_{ij} &= N_{ij}/Z = Y_i/Z = Q_i, \text{ for } j = g(i).
\end{aligned}
$$

Therefore $H(XY) = H(Y)$, where $H(Y)$ is given by equation (6). Likewise, since

$$p_{ij} = N_{ij}/Y_i = 0, \text{ for } j \neq g(i),$$

and

$$p_{ij} = 1, \text{ for } j = g(i),$$

for all i, we have $H_Y(X) = -\Sigma\Sigma \, \pi_{ij} \log p_{ij} = 0$, from equation (12), and thus $m(X) = H(X)$.

Note that $H(XY)$ has another "minimum" [i.e., $H(X)$] when there exists complete heterogeneity of species distribution (zero overlap for all pairs of species).

Appendix B

Derivation of expressions for $H_Y(X)$

From equation (11), we define

$$H_Y(X) = H(XY) - H(Y).$$

Substituting from equations (9) and (6), we have

$$
\begin{aligned}
H_Y(X) &= -\Sigma\Sigma \, \pi_{ij} \log \pi_{ij} + \Sigma_i \, Q_i \log Q_i \\
&= -\Sigma\Sigma \, \pi_{ij} \log \pi_{ij} + \Sigma_j \, p_{ij} \, \Sigma_i \, Q_i \log Q_i,
\end{aligned}
$$

since $\Sigma_j \, p_{ij} = 1$; thus

$$
\begin{aligned}
H_Y(X) &= -\Sigma\Sigma \left[\frac{N_{ij}}{Z} \log \frac{N_{ij}}{Z} - \frac{N_{ij}}{Y_i} \frac{Y_i}{Z} \log \frac{Y_i}{Z} \right] \\
&= -\Sigma\Sigma \left[\frac{N_{ij}}{Z} \left(\log \frac{N_{ij}}{Z} - \log \frac{Y_i}{Z} \right) \right] \\
&= -\Sigma\Sigma \, \frac{N_{ij}}{Z} \log \frac{N_{ij}}{Y_i}.
\end{aligned}
$$

Thus we obtain equation (12):

$$H_Y(X) = -\Sigma\Sigma \, \pi_{ij} \log p_{ij}.$$

The alternative form, equation (13), is derived from the above form:

$$
\begin{aligned}
H_Y(X) &= -\Sigma\Sigma \, \frac{Y_i}{Y_i} \, \frac{N_{ij}}{Z} \log \frac{N_{ij}}{Y_i} \\
&= -\Sigma_i \, Q_i \, \Sigma_j \, p_{ij} \log p_{ij} \\
H_Y(X) &= \Sigma_i \, Q_i \, H_i(X)
\end{aligned}
$$

where $H_i(X)$ is defined by equation (7). Precisely symmetrical derivations can be given for the expressions

$$H_X(Y) = -\Sigma\Sigma \, \pi_{ij} \log q_{ij} = \Sigma_j \, P_j \, H_j(Y),$$

where

$$H(XY) = H(X) + H_X(Y).$$

30

Reprinted from *Ecology*, 52(4), 587–596 (1971)

NICHE BREADTH AND DOMINANCE OF PARASITIC INSECTS SHARING THE SAME HOST SPECIES

PETER W. PRICE

Department of Entomology, Cornell University, Ithaca, New York
and
Forest Research Laboratory, Department of Fisheries and Forestry, Quebec, Canada

Abstract. A guild of six parasitic insects (Hymenoptera) which attack cocoons of the same sawfly populations was composed of four indigenous and two introduced species. The indigenous parasitoids included the ichneumonids, *Pleolophus indistinctus* (Prov.), *Mastrus aciculatus* (Prov.), *Gelis urbanus* (Brues), and *Endasys subclavatus* (Say), and the introduced species were the ichneumonid, *Pleolophus basizonus* (Grav.), and the eulophid, *Dahlbominus fuscipennis* (Zett.). Their distributions were sampled on five variable requirements, or resource sets. Litter moisture content and seasonal activity varied within plots, and host density, host species, and plant community varied between plots. Although the distributions of the indigenous species overlapped, in relative terms each species occupied one position in the niche space that was poorly exploited by all other guild members. Such an "enclave" permitted each of the first three species to develop a zone of dominance over the other parasitoids. The introduced species had no recognizable enclave. The most abundant guild member, *P. basizonus*, was dominant in the sites most favorable to litter-searching parasitoids. It had the broadest niche over the range of litter moisture content and length of seasonal activity, and it interacted more with all other species than any other guild member. It was a better competitor than the next most broadly adapted species, *P. indistinctus*. Competitive superiority is proposed as the driving force behind abundance, leading to dominance in favorable sites. The density-dependent interaction between individuals, which results in dispersal, appears to be responsible for the occupation of a broader niche.

Interest in natural insect populations has focused on mortality factors that are likely to contribute to population fluctuations. In some studies, mortality caused by parasitoids was the most potent single factor in the prediction of the subsequent host population (Morris 1959, Nielson and Morris 1964, Auer 1968). In these investigations and others, parasitoid numbers showed a positive response to an increasing host population (Klomp 1968, Varley and Gradwell 1968). In every study cited, this response reached a peak before the host population was completely exploited. The correlative method used in these large projects precludes an insight into the reasons for this truncated response. Because it is a prominent feature of natural host-parasitoid relationships, it is of interest to examine the interaction carefully. As one approach to the problem, I have studied the organization within the parasitoid complex as it is influenced by the host population and other environmental factors.

Field studies were made on six species of hymenopterous parasitoid which attack the same host, *Neodiprion swainei* Midd., in the cocoon stage. They constitute a guild of species, exploiting the same resource in a similar manner (Root 1967). Two species, *Pleolophus basizonus* (Grav.) (Ichneumonidae) and *Dahlbominus fuscipennis* (Zett.) (Eulophidae),

are introduced from Europe. The remainder are indigenous species of ichneumonids: *Pleolophus indistinctus* (Prov.), *Mastrus aciculatus* (Prov.) *Endasys subclavatus* (Say) and *Gelis urbanus* (Brues).

The parasitoids are active in June when the host cocoons are abundant (Fig. 1). The host emerges from the cocoon in late June, and only already parasitized cocoons remain as food supply for the active parasitoid population. In July, the most abundant parasitoids that attack larvae emerge, leaving only cocoons containing parasitoids that attack cocoons in the forest litter. Intense competition for oviposition sites results, and hyperparasitism becomes common until the supply of cocoons is replenished in September (Price and Tripp, *unpublished data*). At this time mature larvae, with and without larval parasitoids, fall to the ground and spin cocoons. In some years, a small proportion of hosts in the study area remain in prolonged diapause during the summer. Those that are not too deeply buried become parasitized early in the season.

Previous results (Price 1970a) indicated that *P. basizonus* was dominant at high host densities and acted as an organizer species in the parasitoid fauna. Two other species, *M. aciculatus* and *P. indistinctus*, became dominant in less favorable, peripheral zones. This paper examines more closely the qualities of *P. basizonus* that may contribute to its dominance in favorable sites. The causes and consequences of these qualities are suggested. The analyses also suggest how

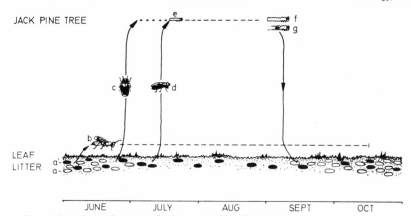

Fig. 1. Phenology of events affecting competition for hosts by cocoon parasitoids. The overwintered host-cocoon population is composed of unparasitized cocoons (a), and parasitized cocoons (a'). From a', cocoon parasitoids (b) and larval parasitoids (d) emerge. From a, the host sawfly (c) emerges to oviposit in the jack pine tree. Eggs hatch, the larvae (e) feed in the tree until maturity (f) and then drop to the ground to spin cocoons. Some host larvae contain larval parasitoids (g).

the indigenous species are able to coexist with this introduced competitor.

The study area was located on the west side of the St. Maurice River watershed in Quebec, Canada. The vegetation of this area has been described (Price 1970a). Field data were collected in 1967 and 1968 and laboratory observations were made in 1969.

METHODS

The description of the niche as an *n*-dimensional hypervolume, given by Hutchinson (1957), provides a conceptual basis for a method of estimating the niche size occupied by organisms. By studying the distribution of parasitoids on five variable requirements or resource sets, I have attempted to identify differences in niche occupation between parasitoid species. The resource sets were composed of units, in which sample points were located. When these units could be ranked, the set formed a gradient. Three resource sets represented gradients: litter moisture content (7 units), seasonal activity (12 units), and host density (3 units). Two sets could not be ranked: host species (2 units) and plant community (3 units). The first two resource sets varied within sample plots. They represented microenvironmental differences, and the remaining sets varied between plots, since the units in them were large in comparison to the sample plots. The space that each unit occupied in the study area and the distance between units on a gradient were not measured. They were treated as if equal.

The within-plot factors are important in determining which species can coexist in the same small locality. Between-plot factors refer to major environ-

mental requisites that species may utilize as enclaves from competition. Because these effects represent different scales of influence on parasitoid distributions, they are treated separately in this study.

The distributions of parasitoids in the microenvironments were evaluated by locating sample points in different types of soil cover. These units represented a spectrum of moisture contents and, when arranged in order, gave an estimate of the activity pattern of parasitoids on a moisture gradient. The gradient was composed of seven soil-cover types, listed from dry to wet: sand, pine litter, a mixture of pine litter and lichens (*Cladonia* spp.), lichens alone, a mixture of lichens and moss [*Pleurozium schreberi* (Brid.) Mitt.], moss alone, and the hardwood litter of small shrubs and trees in forest stands of jack pine, *Pinus banksiana* Lamb. Sampling was continued from June through August so that the seasonal activity pattern of each species could be studied within a jack pine stand.

Sample points were also located in different jack pine stands and other plant communities, to detect the macroenvironmental influences on parasitoid distribution. Three levels of host (*N. swainei*) density were sampled: high (about 3–5 cocoons per 30 by 30 cm sample), moderate (about 0.5–1 host cocoons per sample), and low (about 0.01 host cocoons per sample). High populations of two host species, *N. swainei* and *Neodiprion pratti banksianae* Roh., that occur in jack pine stands were also sampled. The latter species was sampled only in 1968. These data were used as an estimate of the host species set in calculations for niche overlap in 1967. Three plant communities were sampled: jack pine; black spruce,

Picea mariana (Mill.) BSP.; and hardwood. The hardwood stands included paper birch, *Betula papyrifera* Marsh.; trembling aspen, *Populus tremuloides* Michx.; and maples, *Acer rubrum* L. and *A. spicatum* Lam.

Each sample point consisted of a 15 by 15 by 5 cm cage of galvanized wire cloth, with wires 6.5 mm apart, containing humus, level with that outside the cage. Five cocoons were placed on the humus and covered with the same type of soil cover as was around the cage. The cage was closed and left for a week, after which time the cocoons were removed and immediately replaced by another five cocoons. Female parasitoids searching in the litter could enter the cage, oviposit, and leave unimpeded. Therefore, the sampling procedure did not appreciably influence the parasitoid population outside the cage. The removed cocoons were reared to maturity. From the rearings, the species of female that laid the eggs could be determined. The cages were necessary to protect the cocoons from small mammal predation. As the sample points were minute compared to the area outside, they sampled without interference the natural parasitoid population outside the cages which was influenced by small mammal predation of cocoons.

Sampling lasted 12 weeks each season. A total of 146 and 140 cages, and 8,760 and 8,400 cocoons, were used in 1967 and 1968, respectively. They yielded 688 and 1697 parasitoid individuals on which the analyses are based. The percentage parasitism of the planted cocoons was therefore 8% and 20% in the 2 years.

The extent to which each species of parasitoid exploited the different portions of the resource sets available to it was calculated by Levins' (1968) niche breadth formula,

$$B = 1 / \sum_{i=1}^{n} pi^2$$

where p_i is the proportion of a species found in the ith unit of the resource set and n is the number of units in the set. For the range of values obtained in this study, B was closely correlated with the niche width measure, W, proposed by McNaughton and Wolf (1970) (e.g., on the moisture gradient the correlation coefficient, $r = .99$, $df = 11$, $P < .01$; on the time gradient, $r = .97$, $df = 11$, $P < .01$).

Interaction between species was estimated by Levins' (1968) formula for niche overlap,

$$\alpha_{ij} = \sum_{h} p_{ih} p_{jh} (B_i)$$

where α_{ij} is the niche overlap of species i over species j, p_{ih} and p_{jh} are the proportions of each species in the hth unit of the resource set, and B_i is the niche breadth of species i.

The similarity in distribution of species was calculated using the estimate of proportional similarity,

$$PS = \sum_{i=1}^{n} p_i$$

where p_i is the proportion of the less-abundant species of a pair in the ith unit of a resource set with n units. This formula (using percentages) has been employed frequently in comparative studies of communities and species (e.g., Whittaker 1952, Whittaker and Fairbanks 1958, Southwood 1966).

As the parasitoids operate in a highly competitive situation (Price and Tripp, *unpublished data*), factors which may influence their direct interaction were examined. These observations were restricted to the two most closely related species, *P. basizonus* and *P. indistinctus*. The length and width of the eggs of each species were measured to give an estimate of the relatives sizes of the first instar larvae. When the larvae hatch they attack and kill any other eggs or larvae present in the host cocoon (Price 1970b) so that the larger larvae should be the stronger competitors in direct aggressive encounters. To test this directly, one egg of each species was introduced into a host cocoon following the method developed by Green (1938). As only one larva can survive on a host, rearings to maturity indicated which species had the superior competitive ability.

The longevity and fecundity of *P. basizonus* and *P. indistinctus* were measured using the techniques already described (Price 1970b, c).

RESULTS

Distribution within plots

The three most abundant species of parasitoid, *P. basizonus*, *P. indistinctus*, and *M. aciculatus*, exhibited different patterns of distribution on the moisture gradient (Fig. 2). The pattern of each species was similar in 1967 and 1968, although *P. basizonus* was much more abundant in 1968. The broad and flat distribution of *P. basizonus* contrasted with the peaked distributions of *P. indistinctus* and *M. aciculatus*. The abundance of *P. indistinctus* reached a peak in the moist units of the gradient, while *M. aciculatus* was concentrated in the drier units of the gradient. The minor species, *E. subclavatus*, *G. urbanus*, and *D. fuscipennis*, showed narrow occupation of the moisture gradient in 1967 and broad occupation in 1968. The small sample sizes for these latter species were probably responsible for this large variation.

May temperatures were higher in 1968 than in 1967, so sampling was started earlier in the warmer spring (Fig. 3). *Pleolophus basizonus* was active for longer than any other species in 1967, although its abundance was less evenly distributed throughout the season than that of *P. indistinctus*. In 1968 the activity period of *P. basizonus* must have been longer

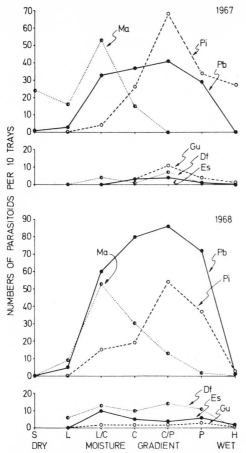

FIG. 2. Distribution of parasitoids on a moisture gradient. The six species are divided into the three major and the three minor guild members for each year. S, sand; L, pine litter; L/C, litter plus lichen; C, lichen; C/P, lichen plus moss; P, moss; H, hardwood litter. Pb, *P. basizonus;* Pi, *P. indistinctus;* Ma, *M. aciculatus;* Es, *E. subclavatus;* Gu, *G. urbanus;* Df, *D. fuscipennis.*

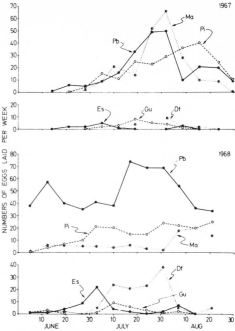

FIG. 3. Seasonal activity of parasitoids, estimated by the number of eggs laid in all cocoons exposed in that week. Division of species and symbols as in Fig. 2.

than any other species, judging by the level of activity at the start of sampling. Also, its proportional distribution of activity was more even throughout the season than other species, indicating that *P. basizonus* tended to exploit the time available in a season more fully than other species.

Distribution between plots

In the set of host densities sampled, *P. indistinctus* was the most evenly distributed species (Fig. 4). In contrast, the numbers of *P. basizonus, M. aciculatus,* and *D. fuscipennis* were concentrated in the moderate and high host-density units.

Only *E. subclavatus* was evenly distributed be-

tween the two host species *N. swainei* and *N. pratti banksianae* (Fig. 4). *Pleolophus basizonus* was dominant in jack pine stands where *N. swainei* was abundant and *P. indistinctus* was even more dominant in *N. pratti banksianae* populations.

Of the three plant communities studied, jack pine stands supported the only appreciable parasitoid population (Fig. 4). The host species required by the parasitoids were clearly concentrated in jack pine stands. Only *G. urbanus* was evenly distributed between plant communities, indicating a different spectrum of host utilization.

The difference in distribution of species on the moisture gradient within a plot, and host densities between plots were combined to show how the species divide up the area available to them (Fig. 5). There was a strong interaction between these two influences on parasitoid distribution. *Mastrus aciculatus* was dominant in dry sites having a high host density. Conversely, in a high host density on moist sites, *P. basizonus* was dominant, while in a low host density on moist sites, *P. indistinctus* was dominant. No species exploited low host populations in dry sites. In each case, dominance was associated with reduced species diversity (H'), leaving the moderate host densities, in mesic sites, with the highest species diversity of parasitoids. The most abundant species,

272

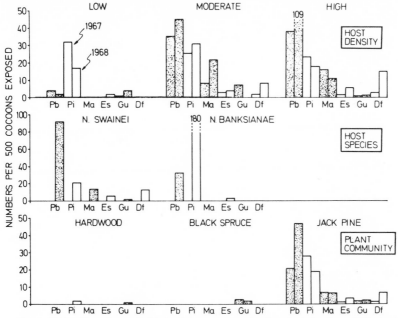

Fig. 4. Distribution of parasitoids in three resource sets: top, host density; middle, host species; bottom, plant community. Symbols as in Fig. 2. In each unit of the resource set each species is represented by two bars, one for each year, except in the host species set which was sampled only in 1968. Alternate species are stippled for clarity.

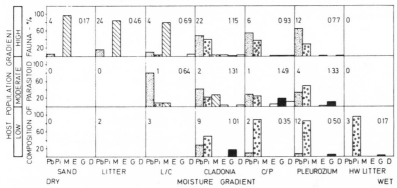

Fig. 5. Percentage of the parasitoid fauna represented by each species in each unit of area on two gradients, moisture and host density, 1967. In each unit, the species are arranged in order as in Fig. 4, and the units are as in Fig. 2. The number of sampling trays per unit is shown on the left of each unit and the species diversity index, H', on the right.

P. basizonus, was dominant in the most favorable site, where hosts were abundant and where the moist litter conditions resulted in little physiological stress.

Niche breadth

The proportion of the available resource that was utilized by each species on the five resource sets was compared by calculating their niche breadths, B (Fig.

6). In both within-plot variables, moisture and time, *P. basizonus* had the largest average niche breadth. Of the three major species, *P. indistinctus* had the next broadest niche and *M. aciculatus* the narrowest niche. The values obtained for the minor species in different years on the moisture gradient were highly variable.

Pleolophus basizonus did not have the broadest

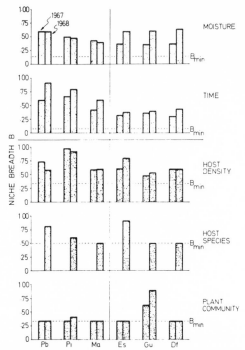

FIG. 6. Niche breadths, B, of parasitoid species within plots, above transverse line, and between plots, below. Major species are on left of vertical line, minor species on the right. Symbols as in Fig. 2. B_{min} is the reciprocal of the number of units in the habitat set.

For each species five values were obtained, there being five other guild members any one species could interact with. The mean of these five values gave an estimate of the relative interaction each species experienced with the remainder of the parasitoid complex. To obtain absolute estimates of mean overlap each unit of a resource set should be calibrated equal to other units, and the same scale should be utilized for each resource set. These conditions are not met in the present study, and the values calculated are therefore used only as relative estimates of overlap.

The mean niche overlap within plots was greatest for *P. basizonus* (Fig. 7). Its closest relative, *P. in-*

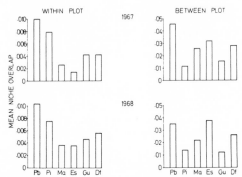

FIG 7. Mean niche overlap for each parasitoid species, derived from the products of niche overlap values in within-plot and between-plot habitat sets. Symbols as in Fig. 2.

niche on resource sets that varied between plots. However, it was broadly distributed on two sets, host density and host species. *Pleolophus indistinctus* had a broader niche in the range of host densities, *E. subclavatus* in the two host species, and *G. urbanus* in the plant communities.

Niche overlap

The niche overlap, α, of one species over another was calculated for every combination of species in each resource set. This value gave an estimate of the impact one species was likely to have on another. As Levins (1968) pointed out, the overlap of activity, or impact, will be lessened by the influence of more than one resource set operating together, provided that each set influences each species in a different way. Therefore the overlaps occurring on the resource sets, when multiplied together, gave a relative value to the amount each species overlapped another when factors were considered together. These values were calculated for each pair of species and the products obtained for within-plot effects were kept separate from those for between-plot effects. The reasons for this have been given in the methods.

distinctus, had the next broadest mean niche overlap. *Dahlbominus fuscipennis,* one of the introduced species, showed moderate overlap over all other species. The mean niche overlap between plots was again greatest for *P. basizonus* in 1967. The lower value obtained for this species in 1968 was due to its great increase in numbers in the most favorable sites, which resulted in less-even distribution on the host-density gradient. The outcome was reduced niche breadth, in contrast to the increased niche breadth of the minor species, *E. subclavatus.* The latter species had the broadest mean niche overlap in 1968 although, as a minor species, it could never greatly influence the parasitoid fauna. Between plots, *P. basizonus* therefore had the greatest influence on the parasitoid complex.

Of the four indigenous parasitoid species, two, *M. aciculatus* and *E. subclavatus,* had effective means of partitioning the resources so that they interacted little with the other species. This was reflected in the low values for mean niche overlap within plots. Conversely, the remaining species, *P. indistinctus* and *G. urbanus,* had low values for mean niche overlap in between-plot resource sets. By considering the habitat sets together, an enclave for each of the indigenous species has been identified where they can exist without severe interference from the other species. This amelioration of interaction over parts of their distributions has not been observed for the two introduced species.

Once the relative value for overlap had been calculated, the species that were the most similar in their distributions were determined. This indicated the species that experienced the greatest interference from the most abundant species, *P. basizonus*. The proportional similarities, *PS,* were calculated using the same approach as for niche overlap, only the means were not used, leaving five values of similarity for each species. Within plots, when litter moisture and seasonal activity were considered together, *P. basizonus* and *P. indistinctus,* the two most closely related species, showed the greatest similarity (Fig. 8). With such broadly overlapping distributions as

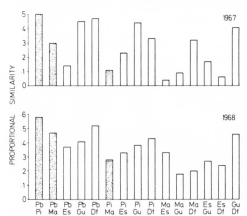

FIG. 8. Within-plot proportional similarity of parasitoid distribution. Columns for similarities between major species are stippled. Symbols as in Fig. 2.

50% in 1967 and 58% in 1968, competitive interaction was likely to be severe. Many of the other high values of similarity involved one or both of the introduced species.

Between-plot similarities showed that the introduced species had high values when compared to the indigenous species (Fig. 9). In five of the six high values of similarity, the introduced species were involved. The indigenous species generally showed low values of similarity, indicating an effective partitioning of the area between species with respect to the three habitat sets.

The value of between-plot similarity for *P. basizonus* and *P. indistinctus* was low each year. Therefore, the existence of the latter species was not threatened since it had an effective refuge from severe interaction with *P. basizonus*.

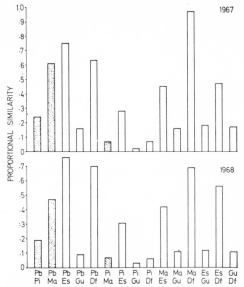

FIG. 9. Between-plot proportional similarity of parasitoid distribution. Columns for similarities between major species are stippled. Symbols as in Fig. 2.

Interactions between P. basizonus and P. indistinctus

Since *P. basizonus* and *P. indistinctus* showed the greatest similarity in their distributions on the two within-plot sets, these species were likely to show the strongest competitive interaction. The fecundity and longevity of each species were not significantly different (mean fecundity: *P. basizonus* 1.68 eggs per day, *P. indistinctus* 2.18 eggs per day, $P_{Pb=Pi} > .40$, $n = 6$ and 4, respectively; mean longevity: *P. basizonus* 23.1 days, *P. indistinctus* 28.7 days, $P_{Pb=Pi} > .10$, $n = 24$ and 17, respectively). *Pleolophus basizonus* was active earlier in the spring than *P. indistinctus* (Fig. 3) at a time when host cocoons were most abundant and competition for them was least severe. *Pleolophus basizonus* therefore had an early advantage.

Pleolophus basizonus had a larger second generation than *P. indistinctus*. Of the eggs laid in 1967, 51% of *P. basizonus* females emerged the same year and 49% emerged in 1968. Similarly, of the eggs laid in 1968, 58% emerged in that year, and 42% emerged in 1969. There were two well-defined, more or less equally abundant generations each year. In *P. indistinctus* populations, the second generation in a year was very poorly developed. Eggs laid in 1967 produced 2% of the females in the same year and 98% in 1968. Eggs laid in 1968 produced 12% of the females in the same year and 88% in 1969. The response of this native species to high host numbers must be slower than that of its European relative.

Pleolophus basizonus eggs were 7.5% longer and 10.0% wider than those of *P. indistinctus* (*P. basizonus* mean egg size, 1.14 by 0.33 mm, *P. indistinctus* mean egg size, 1.06 by 0.30 mm. $P_{P_b, P_i} < .005$ and .001, $n = 20$). Therefore, *P. basizonus* larvae have an advantage when confronted with *P. indistinctus* larvae within the cocoon. This advantage is reflected in the results of the experiment when an egg of each species was introduced within the cocoon. Of the 100 replicates, *P. basizonus* emerged from 68% and *P. indistinctus* from 32%.

Discussion

Usually, in competition for a nonrenewable resource, time of utilization does not provide an effective means of avoiding competition. However, although host cocoons were not renewed naturally during the 12-week period of study each year, the course of time presented the parasitoids with a complex set of conflicting selective pressures. Although hyperparasitism was frequent during the summer, at each successive attack there was progressively less food in a cocoon. Thus there was a partial advantage to early oviposition in a cocoon that had the maximum food supply available. Also, the act of searching and ovipositing helped to reduce immediate competition among the progeny, since other females would avoid the trail left by the previous female (Price 1970d). The earliest egg laid in a cocoon had the greatest chance of hatching first and consuming other eggs laid subsequently. However, when the parasitoid larva was larger, it became subject to parasitism itself, and the parasitoid acting later in the season had a partial advantage. Differences in seasonal activity also resulted in the different exploitation of the two hosts, *N. swainei* and *N. pratti banksianae* (Fig. 4). *Pleolophus basizonus*, with its early first generation, exploited *N. swainei* while this host was abundant in the spring. The life cycle of *N. pratti banksianae* was quite different: the eggs overwintered and cocoons were spun in late July. Therefore, the cocoons became available just as *P. indistinctus* was reaching its peak abundance for that season. These factors, acting on the temporal differences in parasitoid activity, probably exerted a strong selective pressure on each guild member for an efficient strategy to survive the difficult period through the summer. It appears that the combination of early seasonal activity, two equal generations per season, and superior competitive ability within the cocoon has enabled *P. basizonus* to exploit most successfully the most abundant host, *N. swainei*.

I found that *P. indistinctus* was dominant in plots with a low host density, but this species was competitively displaced from high host populations by *P. basizonus* (Price 1970a). Also, *M. aciculatus* prevailed in dry plots, and *G. urbanus* had an enclave

in black spruce and hardwood stands. Similar zones of dominance for these species were evident in the present study. However, the distributions of *M. aciculatus* and *E. subclavatus* included enclaves of reduced interaction with the other species when litter moisture and time dimensions were considered as microenvironmental variables. The remaining indigenous species, *P. indistinctus* and *G. urbanus*, experienced little interaction with other guild members in between-plot influences, just as they were identified in the previous study.

It is significant that no clearly defined enclaves from interference were observed for the two introduced species, *P. basizonus* and *D. fuscipennis*. These parasitoids were selected for introduction as biological control agents because of their abundance and wide distribution in Europe (Morris and Cameron 1935; Morris, Cameron, and Jepson 1937). On introduction to Canada they might therefore be expected to show broad niche exploitation and competitive superiority. There is evidence that *D. fuscipennis* becomes more abundant in the decline phase of a host population (Bobb 1965, Nielson and Morris 1964). Its narrow niche observed on some dimensions in this study may result from incomplete exploitation of the available resources since the host population was still increasing.

Another possible manifestation of the dominance of *P. basizonus* is its broad natural distribution, not only in Europe but throughout the Palearctic region (Townes, Momoi, and Townes 1965). Its range includes the center of distribution for the tribe Echthrini, to which the genus *Pleolophus* belongs (Townes 1969). Therefore, this species may have the potency expected of species existing in the center of distribution for the taxon (Darlington 1959). In this species, potency is manifested as competitive ability which acts as a general adaptation (Brown 1958) not influenced greatly by local conditions.

The characteristics of the most abundant species, *P. basizonus*, in the parasitoid guild were dominance in the most favorable sites (Fig. 5), broadest niche on microenvironmental dimensions (Fig. 6), greatest mean niche overlap (Fig. 7), competitive superiority and lack of an enclave from interference (Fig. 7). *Pleolophus basizonus* was the only species to respond constantly and positively to increasing host density in the field, and may displace *P. indistinctus* into peripheral zones of the host population (Price 1970a).

There are probably several factors that permit high relative abundance. The first species to reach a new resource can increase rapidly, free from competition, and become abundant. In the presence of others, one species can specialize so that it becomes competitively superior over a portion of a resource set. If the conditions in this portion are widely distributed, then the species can become abundant (McNaughton and Wolf 1970). A third situation can result when a generally adapted species, which exploits a broad spectrum of resources, becomes abundant in the presence of other species. This broad distribution of activity must depend on competitive ability of a very general nature, different from the ability derived from specialization (Brown 1958). Specialization, with the consequent narrowing of the exploitation curve, tends to reduce competition. Broad exploitation increases competition. Thus what may be termed "competitive dominance" involves a broadly adapted organism that becomes the most abundant species through competitive ability. It results in the dominant occupying potential niche space of other species. This prevents the subordinates from attaining an abundance that could be realized in the absence of the dominant. Since food is frequently in limited supply, the influence of the dominant species will usually involve coexisting members of the same guild and trophic level. Therefore the characteristics of *P. basizonus* listed above are probably typical of many competitive dominants.

One difficulty with this concept is that dominance cannot easily be demonstrated, since abundance and distribution data usually measure only realized niche occupation, or the outcome of competition. Ideally, the *process* of competition should be analyzed to determine potential niche occupation by the subordinate species. Dominance is more readily observed experimentally (e.g., Connell 1961) or when a species is introduced into an area, as in the present study.

In this concept competitive superiority is seen as the driving force behind dominance. Four of the characters of the dominant species in this study that may contribute to its competitive prowess are: (i) it is the earliest to emerge in the spring, (ii) there are two equally abundant generations per year, (iii) it is the largest parasitoid in the guild (Price, *unpublished data*), and (iv) it lays the largest eggs (*unpublished data*). These characters are specific adaptations of *P. basizonus* which may be important only within the context of the guild and resources under study. The adaptations that confer competitive ability will differ between systems and can be understood only within the context of a particular system. Thus, an insight into community organization may depend upon an understanding of the detailed strategy of the dominants.

Competitive ability enables *P. basizonus* to become abundant in the face of competition, and ultimately dominant in the guild. Abundance permits high genetic diversity within a population, and this, in turn, may result in broad exploitation of a resource (McNaughton and Wolf 1970). The colonization of new niche space through genetic diversity is probably a slow process, taking several generations, with a relatively permanent result.

In contrast to the genetic diversity effect, I suggest that density-dependent dispersal can result in the rapid occupation of uncolonized niche space. This dispersal, from centers of abundance in favorable sites, would be driven by increasing interaction between individuals. Female parasitoids recognize and avoid the trails of others, both intra- and interspecifically, and tend to walk faster and to fly, with increased exposure to these trails (Price 1970c, d). This behavior provides the density-sensitive mechanism for dispersal.

The suggested forces leading to the exploitation of a broad niche are summarized on the diagram. The inputs to competitive superiority will be specific to each dominant species.

The historical details of parasitoid distribution in the study area lend support to the idea that niche breadth is a density-dependent phenomenon for these parasitoids. In 1965 host populations were high throughout the study area. In August of that year some areas were sprayed with insecticide, which resulted in the areas of low host density sampled in 1967 and 1968. The moderate and high host populations remained in areas left as controls. Thus in early 1965, parasitoid populations were similar in all locations (Price 1971). If genetic diversity had generated broad niche occupation by *P. basizonus*, assuming that insecticidal action was equal on each phenotype, then niche occupation should not have changed appreciably within 2 years. In fact the niche breadth of *P. basizonus* on the moisture gradient was 0.59 in 1967 in high host densities and only 0.30 in low host densities (cf. Fig. 5). The same trend was evident in 1968 (0.56 and 0.41, respectively) indicating a strong density-dependent response in niche occupation.

The ability of a female to leave viable progeny depends on the compromise between remaining in a host population where exploitation by parasitoids is reaching completion, and emigration into areas of unknown host resources. This emigration stage should be reached before the host resource is fully utilized because hyperparasitism becomes less rewarding, hosts become harder to find, and the chances of progeny being hyperparasitized are greater. Thus the density-dependent emigration caused by mutual interaction can be adaptive. Insect herbivores can reach high population densities (e.g., 10 cocoons per 900 cm² of litter by *N. swainei*) and interaction between parasitoid females is likely to become severe at 1 per 900 cm². Therefore the capacity of the habitat to support a parasitoid population is more likely to be dictated by the level of interaction between individuals than by the supply of host cocoons. This would result in the inability of parasitoid populations to fully exploit hosts at high population levels.

ACKNOWLEDGMENTS

I acknowledge the profound influence of Richard B. Root in every stage of this work. Discussions with his other graduate students at Cornell University, Kathleen Eickwort, J. Tahvanainen, and W. Tostowaryk have been of great value in developing this paper. William L. Brown, Jr., made valuable criticisms in his review of the manuscript. J. Boulerice, G. Castonguay, L. Gerczuk, F. Millette, and M. Ruzayk gave able assistance in the field and laboratory.

LITERATURE CITED

Auer, C. 1968. Erste Ergebnisse einfacher stochasticher Modelluntersuchungen uber die Ursachen der Populationsbewegung des grauen Larchenwicklers, Zeiraphera diniana Gn. (= Z. griseana Hb.) im Oberengadin, 1949/66. Z. Angew. Entomol. **62**: 202–235.

Bobb, M. L. 1965. Insect parasite and predator studies in a declining sawfly population. J. Econ. Entomol. **58**: 925–926.

Brown, W. L., Jr. 1958. General adaptation and evolution. Syst. Zool. **7**: 157–168.

Connell, J. H. 1961. The influence of interspecific competition and other factors on the distribution of the barnacle Chthamalus stellatus. Ecology **42**: 710–723.

Darlington, P. J., Jr. 1959. Area climate and evolution. Evolution **13**: 488–510.

Green, T. U. 1938. A laboratory method for the propagation of Microcryptus basizonus Grav. Entomol. Soc. Ontario Annu. Rep. **69**: 32–34.

Hutchinson, G. E. 1957. Concluding remarks. Cold Spring Harbor Symp. Quant. Biol. **22**: 415–427.

Klomp, H. 1968. A seventeen-year study of the abundance of the pine looper, Bupalus piniarius L. (Lepidoptera: Geometridae), p. 98–108. In T. R. E. Southwood [ed.] Insect abundance. Symp. Roy. Entomol. Soc. London **4**.

Levins, R. 1968. Evolution in changing environments: some theoretical explorations. Princeton Univ. Press, Princeton, N.J. ix + 120 p.

McNaughton, S. J., and L. L. Wolf. 1970. Dominance and the niche in ecological systems. Science **167**: 131–139.

Morris, K. R. S., and E. Cameron. 1935. The biology of Microplectron fuscipennis Zett. (Chalcid.), a parasite of the pine sawfly, Diprion sertifer Geoffr. Bull. Entomol. Res. **26**: 407–418.

Morris, K. R. S., E. Cameron, and W. F. Jepson. 1937. The insect parasites of the spruce sawfly (Diprion polytomum Htg.) in Europe. Bull. Entomol. Res. **28**: 341–393.

Morris, R. F. 1959. Single-factor analysis in population dynamics. Ecology **40**: 580–588.

Nielson, M. M., and R. F. Morris. 1964. The regulation of European spruce sawfly numbers in the Maritime Provinces of Canada from 1937–1963. Can. Entomol. **96**: 773–784.

Price, P. W. 1970a. Characteristics permitting coexistence among parasitoids of a sawfly in Quebec. Ecology **51**: 445–454.

———. 1970b. Biology of and host exploitation by Pleolophus indistinctus (Provancher) (Hymenoptera: Ichneumonidae). Ann. Amer. Entomol. Soc. **63**: 1502–1509.

———. 1970c. Dispersal and establishment of Pleolophus basizonus (Gravenhorst) (Hymenoptera: Ichneumonidae). Can. Entomol. **102**: 1102–1111.

———. 1970d. Trail odors: recognition by insects parasitic on cocoons. Science **170**: 546–547.

———. 1971. Immediate and long-term effects of insecticide application on parasitoids in jack pine stands in Quebec. Can. Entomol., in press.

Root, R. B. 1967. The niche exploitation pattern of the blue-gray gnatcatcher. Ecol. Monogr. **37**: 317–350.

Southwood, T. R. E. 1966. Ecological methods with particular reference to the study of insect populations. Methuen and Co., London. xviii + 391 p.

Townes, H. 1969. The genera of Ichneumonidae, part 2: Gelinae. Mem. Amer. Entomol. Inst. **12**: 1–537.

Townes, H., S. Momoi, and M. Townes. 1965. A catalogue and reclassification of the eastern Palearctic Ichneumonidae. Mem. Amer. Entomol. Inst. **5**: 1–661.

Varley, G. C., and G. R. Gradwell. 1968. Population models for the winter moth, p. 132–142. In T. R. E. Southwood [ed.] Insect abundance. Symp. Roy. Entomol. Soc. London **4**.

Whittaker, R. H. 1952. A study of summer foliage insect communities in the Great Smoky Mountains. Ecol. Monogr. **22**: 1–44.

Whittaker, R. H., and C. W. Fairbanks. 1958. A study of plankton copepod communities in the Columbia Basin, south-eastern Washington. Ecology **39**: 46–65.

Part V

VARIATION IN SPACE
AND TIME

Editors' Comments
on Papers 31 Through 34

Some of the measures discussed in Part IV can be used not only to compare species in a community, but to observe changes in the niche of a given species. We may comment on the perspective of such research. The niche is a characteristic of a species population, its way of relating to environment and other species within a given community. The dimensions of the niche are determined in part by characteristics of the species derived from its longer–term evolution, in part by interplay in the present with other species populations—as resources, competitors, and predators or control mechanisms—in the community in which it occurs. Two parasitic wasps, *a,* and *b,* feed upon sawfly pupae of species *C* as hosts. Wasp species *b* is capable also of feeding on sawfly species *D.* If a higher proportion of the individuals of *b* feeding on host *D* survive and reproduce, because of the effect of competition with species *a* in feeding on host *C,* this difference in survival (or fitness) will be "information return" that will modify the genetic composition of *b*'s population. The implication of this feedback, given many generations, is divergence of niche so that *b* becomes a specialist on host *D,* leaving *a* a specialist on host *C.*

The niche characteristics of a species are thus to a degree labile; they can change with difference in environmental and community context. The change can be either phenotypic, involving direct response by individuals, or genotypic, involving change in the population's genetic inheritance by feedback. The genetic changes are the more interesting ones, for the student of evolution at least. We often know little or nothing about the genetics of natural populations, but can reasonably assume that many differences in morphology have a genetic basis. Differences in bill lengths of birds, for example, we assume are determined primarily by heredity. Bill length also expresses range of food-size adaptation, hence a major niche characteristic. We can consequently compare means and ranges of bill lengths in different bird populations and assume both that differences in these express niche differences, and that these differences are inherited characteristics of the populations. Bill lengths can then be used as an index for our present question: How may niches change with change in the context—difference in environments, faunas, or communities—affecting different populations of a species?

Consider first, as a kind of natural experiment, two species that are congeners and rather close competitors. Geographically they occur both separately and together; what happens to their characteristics where their areas overlap and they are in competition in some of the same communities? In principle we should expect them to diverge in characteristics where they compete, whereas they might be more alike where they occur in different geographic areas, without competitive influence on one another. It is most significant that such divergences in the area of overlap do occur; and the phenomenon, called *character displacement,* is described in the Paper 31 by Brown and Wilson. A theoretical introduction to the phenomenon was provided in the preceding section; a review of observations on it is given by Grant (1972); Cody (1973) reexamines character displacement and the related phenomenon of character convergence.

In Paper 32 several aspects of niche evolution are approached through bill lengths of birds. Schoener focuses concern on sympatric congeneric species and on the effects of different contexts, competitors, and food habits on differences in bill length. The critical index is then the ratio of the bill lengths of two competing species populations, first discussed by Hutchinson (1959); Schoener terms such a ratio a *character difference.* He observes that these ratios are higher in some combinations of species than others and that the reasons they are higher are interpretable.

From Schoener's examination of bill-length difference as an expression of niche difference (and, by implication, relative overlap), we go to a study of the dispersion of bill lengths as an expression of niche width. Van Valen's article (Paper 33) examines the genetic and evolutionary meaning of the range of bill lengths in a population, while comparing such ranges for island and mainland species. The island species in general occur with fewer competitors. One might then expect that, under reduced competitive pressure, the ranges of bill lengths and of

food sizes taken might expand; one would expect the island populations to have broader niches. Such expansion of niche breadth with reduced interspecies competition may be termed *character release.* Van
Valen's article, which follows, has been sharply criticized by Soulé and
Stewart (1970).

Islands clearly provide us with many unplanned experiments on
change in species' niche (and habitat) with change in environmental and
community context. Some of the best experiments occur when species
are newly introduced into areas, and we can observe the result. Many of
these species must fail because they cannot find adequate niches—offer-
ing sufficient resources without excessive competition or predation
pressure—in the new communities. Some of these species, relieved of
the natural enemies that control their populations in their native areas,
may show aggressive expansion in niche and area after introduction; the
species is then a "pest." Such expansions can appear in species introduced to new continents as well as those new to islands. Niche contraction is also a possibility. The biology of introduced species has been extensively discussed by Elton (1958) and a symposium edited by Baker
and Stebbins (1965). One notes a less-studied reverse process: species
extinctions, or removals from areas as in the experiments of Paine
(Paper 12), change the niches of the remaining species. The experimental manipulation and defaunation of islands provides a powerful tool for
the study of niche development and the question of ecological expansion and contraction is well summarized in Paper 34 by MacArthur and
Wilson.

We can only mention a further evolutionary question: the extent to
which niche relationships that evolve among species in different geographic areas may converge. Convergence in morphology and some
broad niche relationships is well known to occur as a consequence of
adaptive radiation, as in the evolution of types of marsupial mammals in
Australia that are convergent with types of placental mammals in the
Northern Hemisphere. A classic case of adaptive radiation, with implications for niche convergence, among birds is that of the finches of the
Galápagos Islands described by Lack (1947). For continental birds some
convergence is indicated for niche relationships in the studies of grassland birds by Cody (1968), and for the consequent species diversities of
bird communities on different continents as compared by Recher
(1969). But the convergence is surely imperfect. Lack (1969) comments
on the occurrence of six species of tits with extensive sympatry in
Europe, versus only two in North America. The difference in niche division among members of this group on the two continents results, prob-
ly, from difference in length of time the group has evolved in the two
areas. Cody (1970) found that bird communities in Chile changed less

in composition along habitat gradients than in California, but that individual bird communities in Chile had on the average higher species diversity and equitability than those in California. These observations suggest that the Chilean birds have evolved toward broader habitats, but narrower niches, than the Californian birds.

An additional question may thus be asked. Suppose certain, different species of a group of organisms reached two continents in the distant evolutionary past, and in each area they evolved and radiated independently. Is it not likely that there would be chance differences, or a kind of founder effect, in the manner in which the species would evolve their relationships to resources and niche axes as the numbers of species in the two areas increased? Evolution might then lead to intensification of some of the initial differences in manner of niche differentiation in the two areas, rather than convergence. Such effects are suggested by the studies of lizard communities of deserts on three continents by Pianka (1971), which showed that niche relationships among lizard species, and niche and diversity relations between lizards as a group, and birds and small mammals, differed in the three areas. These aspects of community evolution lead soon into the topic of species diversity, however, and that topic is dealt with in Ruth Patrick's volume *Diversity* in this Benchmark series.

REFERENCES

Baker, H. G., and G. L. Stebbins. 1965. *The Genetics of Colonizing Species.* Academic Press, New York. 599 pp.

Cody, M. L. 1968. On the methods of resource division in grassland bird communities. *Amer. Naturalist* **102**: 107–147.

Cody, M. L. 1970. Chilean bird distribution. *Ecology* **51**: 455–464.

Cody, M. L. 1973. Character convergence. *Ann. Rev. Ecol. Syst.* **4**: 189–211.

Elton, C. 1958. *The Ecology of Invasions by Animals and Plants.* Methuen, London. 181 pp.

Grant, P. R. 1972. Convergent and divergent character displacement. *Biol. J. Linn. Soc.* **4**: 39–68.

Hutchinson, G. E. 1959. Homage to Santa Rosalia, or why are there so many kinds of animals? *Amer. Naturalist* **93**: 145–159.

Lack, D. 1947 *Darwin's finches.* Cambridge University Press, Cambridge, England. 208 pp.

Lack, D. 1969. Tit niches in two worlds; or homage to Evelyn Hutchinson. *Amer. Naturalist* **103**: 43–49.

Pianka, E. R. 1971. Lizard species density in the Kalahari desert. *Ecology* **52**: 1024–1029.

Recher, H. F. 1969. Bird species diversity and habitat diversity in Australia and North America. *Amer. Naturalist* **103**: 75–80.

Soulé, M., and B. R. Stewart. 1970. The "niche variation" hypothesis: a test and alternatives. *Amer. Naturalist* **104**: 85–97.

31

Reprinted from *Systematic Zool.*, **5**(2), 49–64 (1956), with permission of the Society of Systematic Zoology

Character Displacement

W. L. BROWN, JR. and E. O. WILSON

IT IS the purpose of the present paper to discuss a seldom-recognized and poorly known speciation phenomenon that we consider to be of potential major significance in animal systematics. This condition, which we have come to call "character displacement," may be roughly described as follows. Two closely related species have overlapping ranges. In the parts of the ranges where one species occurs alone, the populations of that species are similar to the other species and may even be very difficult to distinguish from it. In the area of overlap, where the two species occur together, the populations are more divergent and easily distinguished, i.e., they "displace" one another in one or more characters. The characters involved can be morphological, ecological, behavioral, or physiological; they are assumed to be genetically based.

The same pattern may be stated equally well in the opposite way, as follows. Two closely related species are distinct where they occur together, but where one member of the pair occurs alone it converges toward the second, even to the extent of being nearly identical with it in some characters. Experience has shown that it is from this latter point of view that character displacement is most easily detected in routine taxonomic analysis.

By stating the situation in two ways, we have called attention to the dual nature of the pattern: species populations show displacement where they occur together, and convergence where they do not. Character displacement just might in some cases represent no more than a peculiar and in a limited sense a fortuitous pattern of variation. But in our opinion it is generally much more than this; we believe that it is a common aspect of geographical speciation, arising most often as a product of the genetic and ecological interaction of two (or more) newly evolved, cognate species during their period of first contact. This thesis will be discussed in more detail in a later section.

Character displacement is not a new concept. A number of authors have described it more or less in detail, and a few have commented on its evolutionary significance. We should like in the present paper to bring some of this material together, to illustrate the various aspects the pattern may assume in nature, and to discuss the possible consequences in taxonomic theory and practice which may follow from a wider appreciation of the phenomenon.

Two Illustrations

An example of character displacement outstanding for its simplicity and clarity has been reviewed most recently by Vaurie (1950, 1951). This involves the closely related rock nuthatches *Sitta neumayer* Michahelles and *S. tephronota* Sharpe. *S. neumayer* ranges from the Balkans eastward through the western half of Iran, while *S. tephronota* extends from the Tien Shan in Turkestan westward to Armenia. Thus, the two species come to overlap very broadly in several sectors of Iran (Fig. 1). Outside the zone of overlap, the two species are extremely similar, and at best can be told apart only after careful examination by a taxonomist with some experience in the complex (Vaurie, personal communication). Both species show some geographical variation, and it seems clear from Vaurie's account (1950, Table 5, pp. 25–26) that such races as bear names have been raised for character discordances in various combinations. It therefore appears safe to ignore the subspecies analysis as such and to concentrate on the variation of the independent characters themselves.

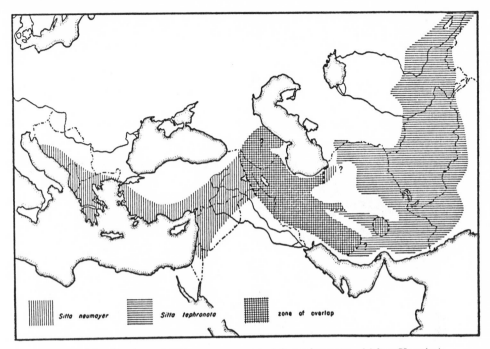

Fig. 1. Distribution of *Sitta neumayer* and *S. tephronota*. (After Vaurie.)

These show quite remarkable displacement phenomena in the Iranian region of overlap between the species, where the two species apparently usually occur in more or less equal numbers (see Fig. 2). In this region, *S. neumayer* shows distinct reductions in overall size and bill length, as well as in width, size, and distinctness of the facial stripe. *S. tephronota,* on the other hand, shows striking positive augmentation of all the same characters in the overlap zone, so that it is distinguishable from sympatric *neumayer* at a glance. Vaurie concludes, we think quite correctly, that the differences within the zone of overlap constitute one basis upon which the two species can avoid competition where they are sympatric. The case of these two nuthatches has already received considerable attention both in the literature and elsewhere, and it bids fair to become the classic illustration of character displacement.

A more complicated case involving multiple character displacement is seen in the ant genus *Lasius* (Wilson, 1955). Where they occur together, in forested eastern North America, the related species *L. flavus* (Fabr.) and *L. nearcticus* Wheeler show differences in the following seven characters: antennal length, ommatidium number, head shape, degree of worker polymorphism, relative lengths of palpal segments, cephalic pubescence, and queen size. In western North America and the Palaearctic Region, where *nearcticus* is absent, *flavus* is convergent to it in all seven characters. In this shift, each character behaves in an independent fashion; e.g., scape length becomes exactly intermediate between that of the two eastern populations, ommatidium number increases in variability and overlaps the range of the two, and queen size changes to that of *nearcticus*. In North Dakota, at the western fringe of the *nearcticus* distribution, the *flavus* population is at an intermediate level of convergence (Fig. 3).

There is some evidence that this dual displacement-convergence pattern is associated with competition and ecological

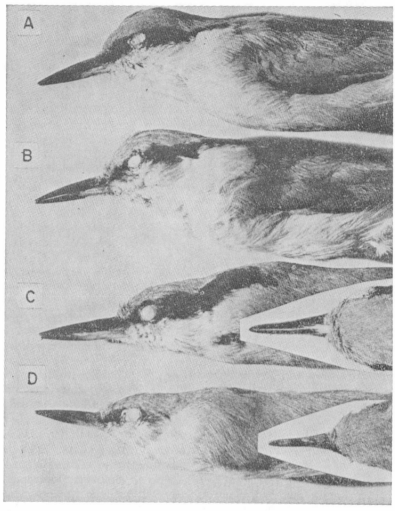

Fig. 2. Size and shape of the bill and facial stripe in *Sitta neumayer* and *S. tephronota*: *A, S. neumayer* from Dalmatia; *B, S. tephronota* from Ferghana; *C, S. tephronota* and *D, S. neumayer*, both from Durud, Luristan, in western Iran. (After Vaurie.)

displacement between the two species. So far as is known, they have similar food requirements. But in eastern North America, where they occur together, *flavus* is mainly limited to open, dry forest with moderate to thin leaf-litter, while *nearcticus* is found primarily in moist, dense forest with thick leaf-litter. There is little information available on the western North American and Asian *flavus* populations, but in northern Europe this species is known to be highly adaptable, preferring open situations, but also occurring commonly in moist forests.

Some Additional Examples

In the following paragraphs we wish to present a number of cases selected from the literature (with two additional unpublished examples) which we have interpreted as showing character displacement. In so doing we are trying to document the thesis that character displacement occurs widely in many groups of animals and in a range of particular patterns. But at the same time we are obliged to give warning, perhaps unnecessarily for the critical reader, that most of these cases in-

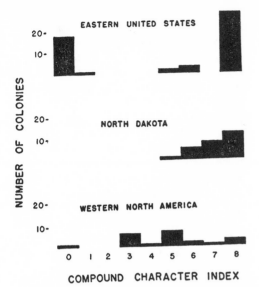

FIG. 3. Frequency histograms of the compound character index of the ants *Lasius nearcticus* (0–1) and *L. flavus* (3–8) in three broad geographic samples. For each colony typical *nearcticus* characters are given a score of 0, typical eastern *flavus* characters a score of 2, and intermediate characters a score of 1. The four characters most clearcut in the eastern United States are used: maxillary palp proportions, antennal scape index, compound eye ommatidium number and head shape. Thus, completely typical *nearcticus* colonies score a total of 0 and completely typical eastern *flavus* 8, with the various ranks of intermediates falling in between (after Wilson, 1955).

volve discontinuously distributed populations, that as a result the species status of these populations with respect to one another has not been ascertained with complete certainty, and that explanations alternative to character displacement are therefore assuredly possible. We ask only that the reader bear through and consider our interpretation in each case.

Birds of the Genus Geospiza. A striking case of character displacement has been described by David Lack in his classic, *Darwin's Finches* (1947). Lack has shown that in the Galapagos certain species of *Geospiza* are often absent on smaller islands, in which case their food niche is filled by other species of the genus. The populations of the latter tend to converge in body size and beak form to the absent species, so much so as to make placement of these populations to species difficult. Lack has demonstrated that body size and beak form are generally important in *Geospiza* in both food getting and species recognition. The dual displacement-convergence pattern we are interested in occurs, at least once, in the following situation. The larger ground-finch *Geospiza fortis* Gould and the smaller *G. fuliginosa* Gould differ from each other principally in size and beak proportion. On most of the islands, where they occur together, the two species can be separated easily by a simple measurement of beak depth, i.e., a random sample of ground-finches (excluding from consideration the largest ground-finch *G. magnirostris* Gould) gives two completely separate distribution curves in this single character. But on the small islands of Daphne and Crossman a sample of ground-finches gives a single unimodal curve exactly intermediate between those of *fortis* and *fuliginosa* from the larger islands. Analysis of beak-wing proportions has shown that the Daphne population is *fortis* and the Crossman population is *fuliginosa;* according to Lack's interpretation each has converged toward the other species, filling the ecological vacuum its absence has created.

Birds of the Genus Myzantha. Among the Australian honey-eaters of the genus *Myzantha*, a light-colored species, *M. flavigula*, occupies the greater part of the arid inland. Toward the wet southwestern corner of the continent, *flavigula* blends gradually into a darker population, usually referred to as "subspecies *obscura.*" In southeastern Australia, in higher-rainfall country, *flavigula* is replaced by two forms—*M. melanocephala*, mostly in the wettest districts, and *M. melanotis* of the subarid Victorian-South Australian mallee district. The southwestern (*obscura*) and one of the southeastern populations (*melanotis*) are extremely similar, differing by what are described as trifling characters of plumage shading, so that some authors consider them conspecific.

The members of an ornithological camp-out in the Victorian mallee, however, have found that *melanotis* there nests sympatrically with both *melanocephala* and *flavigula,* and that at this place the three behave as distinct species without intergradation. Thus we find the two morphologically very similar forms, *obscura* and *melanotis,* flanking the much more widely distributed and differently colored species, *flavigula,* but showing exactly opposite interbreeding reactions with *flavigula. Obscura* appears to represent merely the terminus of a cline for melanism produced by *flavigula* in the southwest, where, it may be noted, there is no other competing dark form of the same species group (Fig. 4).

Judging by the findings of the mallee observers, *melanotis* is clearly to be regarded as a species distinct from *flavigula,* including the southwestern *obscura* population. In this we follow Condon (1951), and not Serventy (1953), though

the latter has furnished the most comprehensive analysis of the situation.

Serventy's dilemma is keyed by his statement that ". . . it would be unreal to treat *melanotis,* obviously so akin to south-western *obscura,* as a separate species from it. . . ." Here one plainly sees the conflict between two species criteria: one based on morphological similarity, and one on interbreeding reaction in the zone of sympatry.

From the data presented, we interpret the *Myzantha* situation as a case of character displacement. *M. flavigula* tends to produce, in the less arid extremities of its range, populations with darker plumage. In the southwest, it has done just this; presumably, melanism is connected adaptively in some way, directly or indirectly, with increased moisture ("Gloger's Rule"), or plant cover, or both. In the southeastern mallee, however, the melanistic tendencies presumed to be latent or potential in *flavigula* toward the wetter extremes of its range are suppressed in the presence of the darker species *melanotis* (and possibly also *melanocephala*). It would be interesting to know more about

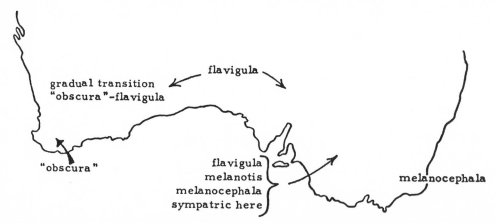

FIG. 4. Map showing the geographical relationships of three species of the bird genus *Myzantha* in southern Australia, based on the discussion of Serventy, 1953. *M. flavigula,* the light-colored bird of arid central Australia, grades into a darker population ("race *obscura*") in southwestern higher-rainfall districts. In southeastern Australia, in the Victorian mallee belt, transitional and mixed ecological conditions allow three non-intergrading species to breed side by side: *M. flavigula; M. melanotis,* a species characteristic of the mallee scrub; and *M. melanocephala,* a southeastern bird of the higher-rainfall districts. *M. melanotis* and the "*obscura*" population are extremely similar, and have been considered synonymous or at least conspecific in the past.

the ecological distribution, food, and habits of the three *Myzantha* species within the region where they occur together.

Parrots of the Genus Platycercus. Serventy (1953) also reviews, among other cases that may involve character displacement, the situation in the rosellas of southeastern Australia (Fig. 5). The crimson rosella (*Platycercus elegans*) is a species of the wooded eastern areas— mostly those with higher rainfall nearest the coast. On Kangaroo Island, off the coast of South Australia, occurs a crimson population that appears to be *elegans* from a strictly morphological viewpoint. Beginning on the mainland opposite Kangaroo Island is a cline connecting the crimson form to an inland, arid-country yellow form (*P. flaveolus*) inhabiting the red gums of the rivers and dry creeks in

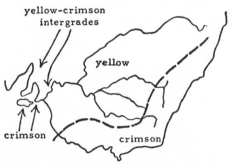

FIG. 5. Map showing the approximate distribution of color forms of the rosellas (parrots) of the *Platycercus elegans* complex in southeastern Australia. The heavy pecked line indicates roughly the inland margin of the southeastern highlands and the higher-rainfall districts, and also the inland limit of the range of the crimson-trimmed *P. elegans.* Inside this line, along the upper reaches of the Murray-Darling river systems, the closely related *P. flaveolus,* a yellow-trimmed form, approaches and may even meet the range of *P. elegans* at some points without producing intergrades. Downstream, *P. flaveolus* grades through a series of intermediately-colored populations culminating in the crimson-trimmed flocks of Kangaroo Island, which are apparently outwardly indistinguishable from those of the true eastern *elegans.* (Adapted from Cain, 1955.)

the Murray-Darling Basins. However, in the Albury district of the upper Murray River and elsewhere up the other rivers, *flaveolus* overlaps or closely approaches the true southeastern *elegans* along a wide front without interbreeding (for a recent detailed account, see Cain, 1955).

It is interesting to note that the cline from yellow to crimson in South Australia follows broadly the regional increase in moisture and luxuriance of forest vegetation; both rise to peaks in the ravines at the western end of Kangaroo Island. We suggest that the South Australian clinal population on the mainland, and probably even the crimson populations of Kangaroo Island, are referable to *flaveolus,* which can here produce a wet-adapted crimson form free of displacement pressure from *elegans.*

Birds of the Cape Verde Islands. Bourne (1955) in his review of the birds of the Cape Verde Islands, has presented several cases of character displacement so concisely and pointedly that we can quote him directly:

The two shearwaters [breeding in the Cape Verde Islands], Cory's shearwater *Procellaria diomedea* and the Little Shearwater *Procellaria baroli,* take similar foods (fish and cephalopods) differing only in size; competition for food between the two species is reduced by the development of different breeding seasons. Elsewhere in its range *Procellaria diomedea* breeds at the same stations as the medium Manx Shearwater *Procellaria puffinus,* which takes similar foods but breeds slightly earlier. There is a dramatic difference in size, and particularly the size of bill, between those races of *Procellaria diomedea* which breed with *Procellaria puffinus* and the form [*P. diomedea*] *edwardsi* which breeds alone at the Cape Verde Islands, the latter having a bill exactly intermediate in size between that of the northern races and that of *Procellaria puffinus.* It seems likely that *edwardsi* takes the food that is divided between both species elsewhere. It may be remarked that one race of *Procellaria puffinus, mauretanicus* of the Balearic Islands, avoids competition with *Procellaria diomedea* by breeding unusually early and leaving the area when the larger species prepares to nest; it is significant that this is the only race of the species which has a large bill resembling that of *P. d. edwardsi.* It would appear that the bill-size and the breeding seasons of these shearwaters vary with the amount of competition occurring between different species breeding at the same site. . . .

Where the two kites *Milvus milvus* and *Milvus migrans* occur together the latter is the species which commonly feeds over water. The race of *Milvus milvus* found in the Cape Verde Islands closely resembles *Milvus migrans* in the field, and very commonly feeds along the shore and over the sea. It may replace *Milvus migrans,* but it seems likely that with the Raven *Corvus corax,* which also abounds along the shore, it replaces the gulls *Larus* spp. which usually scavenge along the shore elsewhere but have failed to colonize the barren coast of the islands.

Bourne's opinion concerning which species are replaced is a little confusing in this case, since elsewhere *Milvus,* notably *M. migrans* in India, often tends to replace or at least dominate the gulls in scavenger-feeding situations around seaports (Brown, personal observation). The absence of *migrans* seems to us the probable chief reason for the convergence characteristics in the Cape Verde Islands populations of *milvus.*

Bourne cites one additional case:

The Cane Warbler *Acrocephalus brevipennis* [a species precinctive to the Cape Verde Islands] is closely related to large and small sibling species *Acrocephalus rufescens* and *A. gracilirostris* which occur together in the same habitats on the [African] mainland. Where the ranges of these two species overlap they are sharply distinct in size and voice; where they occur apart these distinctions are less marked (Chapin, 1949). *A. brevipennis* is probably related to the larger species, *A. rufescens,* but in the absence of the smaller species it is exactly intermediate in all its characters except the bill, which is large, resembling that of *A. rufescens.* The large bill may be part of the general trend seen on islands, or a consequence of competition for food with the smaller *Sylvia* warblers.

Birds of the Genus Monarcha. Mayr (1955 and personal communication) has described a case of displacement in the monarch flycatchers of the Bismarck Archipelago. *Monarcha alecto* and *M. hebetior eichhorni* occur together through the main chain of the Bismarcks, from New Britain north onto New Hanover, but beyond, on isolated St. Matthias, *M. hebetior hebetior* occurs alone; this last is an ambiguous variant combining several features of *alecto* and *eichhorni.* Mayr suggests the following evolutionary scheme: *hebetior* differentiated from *alecto* as an isolate on St. Matthias and later reinvaded the range of *alecto* on New Britain and New Ireland, where it diverged further under displacement pressure from the latter until it became the present *eichhorni.*

It seems to us that this situation can be more simply explained by assuming that the Bismarcks were first populated by a stock which evolved within the Archipelago and became the species *hebetior.* The later entry of *alecto* into the chain was followed by the displacement of *hebetior* as far as the sympatry extended, leaving the St. Matthias isolate to represent the undisplaced relict of the original *hebetior.*

Fishes of the Genus Micropterus. The two basses *Micropterus punctulatus* and *M. dolomieu* have ranges which include a large part of the eastern United States and are mostly coextensive (Hubbs and Bailey, 1940). Of the two, however, only *punctulatus* is known to occur in Kansas, western Oklahoma, and the Gulf States south of the Tennessee River drainage system. In the Wichita Mountains of western Oklahoma there is a population, described as *M. punctulatus wichitae,* which is intermediate between typical *punctulatus* and *dolomieu.* Its affinity to *punctulatus* is shown by the fact that in a number of characters it grades without a break into *punctulatus,* so that some specimens are indistinguishable from typical *punctulatus,* and in its agreement with *punctulatus* in the critical character of scale-row counts. Hubbs and Bailey seem to favor the theory of a hybrid origin for *wichitae,* but they consider this "no more plausible than the view that the similarities between *wichitae* and *dolomieu* are caused by parallel development, or the view that *wichitae* is a relict of a generally extinct transitional stage between *punctulatus* and *dolomieu*." We, of course, are inclined to favor parallel development, resulting specifically from the absence of the displacing influence of *dolomieu*, as the simplest and most plausible explanation.

Away to the south, many of the Texas populations of *punctulatus* are peculiar in showing converging trends toward *dolomieu*, but less strongly, so that Hubbs and Bailey consider them as possible intermediates between *punctulatus* and *wichitae*. In northern Alabama and Georgia there is a form described as a distinct species (*M. coosae*), which combines some of the characters of *punctulatus* and *dolomieu*, besides showing some peculiar to itself. *Coosae* is completely allopatric to *dolomieu*, and there is some evidence that it may hybridize extensively with the sympatric *punctulatus*. We should like to suggest the possibility here that *coosae* is conspecific with *punctulatus* and represents a section of the *punctulatus* population tending to converge toward *dolomieu* where that species is absent.

In summary, it appears to us likely that *wichitae*, the Texas populations, and possibly even *coosae*, each of which shows intermediate characters, are not products of introgressive hybridization, but may instead represent true *punctulatus* stocks that have tended to converge toward *dolomieu* in the absence of displacing influence from that species.

Frogs of the Genus Microhyla. W. F. Blair (1955) concludes from his study of two North American frogs of the genus *Microhyla:*

The evidence now available shows that there are geographic gradients in body size in both *Microhyla olivacea* and *M. carolinensis*. The former species shows a west to east decrease in body length, while the latter shows an east to west increase. The clines are such, therefore, that the largest *carolinensis* and the smallest *olivacea*, on the average, occur in the overlap zone of the two species. This pattern of geographic variation in body size parallels the pattern of geographic variation in mating call reported by W. F. Blair (1955) [in press] in which the greatest call differences in frequency and in length occur in the overlap zone. One of these call characteristics, frequency, probably is directly related to body size, for smaller anurans of any given group tend to have a higher pitched call than larger ones of the same group. The other, length of call, appears unrelated to size.

The differences in body size, like those in mating call, belong to a complex of isolation mechanisms (W. F. Blair, 1955) which tends to restrict interspecific mating in the overlap zone of the two species. The existence of the greatest size differences as well as the greatest call differences where the two species are exposed to possible hybridization supports the argument (*op. cit.*) that these potential isolation mechanisms are being reinforced through natural selection.

Frogs of the Genus Crinia. A most interesting case in the Australian genus *Crinia* has recently been called to our attention by A. R. Main (*in litt.*). Where they occur together in Western Australia, as around Perth, the two species *C. glauerti* and *C. insignifera* have markedly different calls. *C. glauerti* has a rattling call resembling "a pea falling into a can and bouncing"; oscilloscope analysis shows this to consist of evenly spaced single impulses at the rate of about 16 per second. *C. insignifera* produces a call "similar to a wet finger being drawn over an inflated rubber balloon . . . we refer to this call as a 'squelch.'" Oscilloscope analysis shows the squelch to have a duration of about 0.25 second and to consist of impulses crowded together. Around Perth and in other localities where it is sympatric with *insignifera*, *glauerti* individuals are occasionally heard to produce the beginnings of the "squelch" by running 12–15 impulses together, but this occurrence is extremely rare. Along the south coast of Western Australia, however, where *glauerti* occurs alone, the call is commonly modified by running 30 or more single impulses together to produce a squelch almost identical to the ear with that of *insignifera*. Thus, in effect, where this species occurs alone it has extended the variability of its call to include the sounds typical of both species. According to Main, the two species show color differences in the breeding males and different ecological preferences; laboratory crosses show reduced F_1 viability. It seems evident to us (Brown and Wilson) that displacement in this case is associated with the reinforcement of reproductive barriers, the breakdown of which would result in inferior hybrids. This aspect will be discussed more fully in a later section.

Ants of the Genus Rhytidoponera. The ants of the Australian *Rhytidoponera metallica* group (revised by Brown, ms.) are widespread and often among the dominant insects of given localities. The common greenhead (*R. metallica*) is the most successful species—a metallescent green or purple ant adapted to a variety of habitats ranging from desert to warm, open woodland, and the only species of the group at all abundant across the dry interior of Australia. In the southeastern and southwestern ("Bassian") corners of

the continent, where the rainfall is higher and luxuriant forests occur, *metallica* is replaced by similar species of the same group that nearly or quite completely lack metallic coloration (Fig. 6).

In the east, two such species make the replacement, *R. tasmaniensis* Emery and *R. victoriae* André. *R. tasmaniensis* is the larger of the two, has the fine gastric sculpture of *metallica*, and is usually reddish brown, with bronzy-brown gaster. It is virtually identical with *metallica*, except for color. *R. victoriae* is smaller, more blackish, and has relatively coarser gastric striation. *R. tasmaniensis* is found in a variety of woodland situations, but apparently is excluded from the very wettest forests, which are occupied by *victoriae*. Nevertheless, the two species exist in abundance side by side over large parts of southeastern Australia without a sign of interbreeding. At some points, such as on the moist temperate grasslands west of Melbourne, both species occur together with *metallica*, but maintain their distinctness.

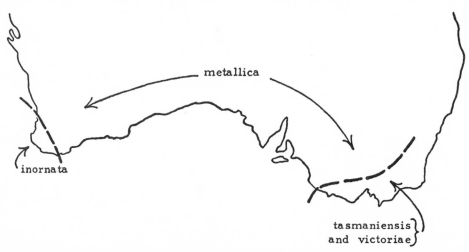

FIG. 6. Map showing the approximate distribution in southern Australia of four closely related common species of ants of the *Rhytidoponera metallica* group. *R. metallica* is nearly or quite the only representative of its group in the more arid central regions, and occurs in open situations in the southeast and southwest as well. In the moister forests of the southeast, *metallica* is replaced by the small, dark *R. victoriae* and the larger, more reddish *R. tasmaniensis*, which frequently occur side by side in the same localities. In the mesic wooded areas of the extreme southwestern corner of Australia, *metallica* is replaced by *R. inornata*, a distinct species which in size and color resembles closely, and broadly overlaps in variation, the two eastern forest species.

In the southwestern corner of Australia, *metallica* is replaced in the wetter parts of the region by non-metallescent *R. inornata* Crawley (though the two species overlap in the Darling Range and undoubtedly elsewhere). The interesting feature here is that *R. inornata* varies in size and color so as to cover the variation in these attributes of both southeastern non-metallic species, *tasmaniensis* and *victoriae*. In fact, one might speak of the two southeastern forms as mutually-displacing equivalents of the southwestern *inornata*, the latter being nearest the generalized type of the group because it has never suffered close competitive pressure and the character displacement that helps to relieve that pressure. This example illustrates the existence of a dual character-displacement pattern where the convergent population is clearly at, or above, the species level.

Slave-making Formica *Ants.* A simpler case in the ants involves the famous Holarctic slavemakers of the *Formica sanguinea* group. In a recent revision (Wilson and Brown, 1955) only three really distinct species are recognized in the group: *F. sanguinea* Latreille, widely distributed through temperate and northern Eurasia, where it is the only species; *F. subnuda* Emery, of boreal and subboreal North America; and *F. subintegra* Emery, ranging through temperate North America and overlapping the range of *subnuda* in the northern United States and along the Rocky Mountain chain.

The two most different forms are *subnuda* and *subintegra*, which can be separated on several external characters. *F. sanguinea* is closely related to *subnuda* in form and habits and is treated as a separate species only arbitrarily, on the basis of slight morphological discontinuities. At the same time, *sanguinea* has pilosity intermediate between that of the two American species, and its clypeal notch, a second important diagnostic character, is more like that of *subintegra* than like that of *subnuda*. We have interpreted this

pattern to represent a displacement of *subnuda* away from *subintegra* where these two species meet and interact, while the Palaearctic equivalent of *subnuda* (i.e., *sanguinea*) has tended to converge toward *subintegra* as a consequence of its filling the "adaptive vacuum" which a companion species might otherwise occupy. Of course in this case, as in all others under present consideration, there is no way of determining how much "displacement" has occurred as a process in the sympatric populations as opposed to "convergence" in the unispecific one. The final pattern observed may in fact be the result of one of these two processes alone.

American Scarabaeid Beetles. Howden (1955, p. 207) discusses the status of two geotrupine beetles considered by him to represent subspecies of the species *Eucanthus lazarus* (Fabricius). His *E. l. lazarus* is stated to range widely over the United States, but records from the Gulf States, excepting Florida, are scanty. *E. l. subtropicus* Howden, on the other hand, is restricted to the southeastern states, and is best represented in Florida, Georgia, Alabama and neighboring states.

Howden is puzzled by the apparent fact that "intermediates" between the two forms came from areas "not bordering the Gulf of Mexico," despite the circumstance that it is in this region that the main overlap falls. Intergrades came from areas "on the East Coast," and from Miami, Florida, and, "Occasional northern specimens appear to exhibit most of the characters of *subtropicus*." However, in particular limited localities, presumably near or in the zone of overlap, Howden was able to name the populations one way or the other with little difficulty.

Although the situation in the Florida Peninsula is not clear from Howden's account, the "intermediate" and more typical-appearing *E. l. lazarus* occurring together with *subtropicus* in the Miami area may really represent undisplaced populations of *subtropicus*. If this is the case, then we would favor Howden's alternative interpretation, and consider *lazarus* and *subtropicus* as closely related but distinct species.

Crabs of the Genus Uca. Jocelyn Crane (in Allee *et al.*, 1950, p. 620) notes that in fiddler crabs of the genus *Uca* differentiation in behavior and often in the coloration of the male is greater if the species are found together than if they are found in different habitats or regions.

The Evolution of Character Displacement

Divergence between two species where they occur together, coupled with convergence where they do not, is a pattern that strongly suggests some form of interaction in the evolutionary history of the pair. The usual case may be one in which the members of the pair are cognate (derived from the same immediate parental population) and have recently made secondary contact following the geographical isolation that has mediated their divergence to species level. In such cases, the "terminal" populations, to which overlap does not yet extend, are not affected by the contact and remain closely similar to each other. But where contact has been made, there are two important ways in which the sympatric populations can interact to augment their initial divergence.

The first type of interaction might best be termed *reinforcement*[1] of the reproductive barriers. It may happen that the species continue to interbreed to some extent, and either the resulting inseminations are ineffectual, or the hybrids produced are inviable or sterile, resulting in what geneticists have termed "gamete wastage." Consequently, any further ethological or genetic divergence reducing this wastage will be strongly favored by natural selection (Dobzhansky, 1951; Koopman, 1950; Kawamura, 1953).

Of conceivably equal or greater importance is the process of *ecological displacement*. It seems clear from an *a priori*

[1] *Reinforcement* is a familiar term in psychology that has been applied to speciation processes (Blair, 1955).

basis that any further ecological divergence lessening competition between the overlapping populations will be favored by natural selection if it has a genetic basis (Mayr, 1949). That such a process actually occurs is suggested by abundant indirect evidence from ornithology (Lack, 1944), as well as the cases already cited above.

It seems unnecessary to go into a detailed discussion of these previously elaborated concepts, except to point out that secondary divergence of this nature inevitably entails phenotypic "characters" of the type employed in ordinary taxonomic work. Character displacement therefore may be considered as merely the aspects of such divergence that are recognizable to the taxonomist and some other favored organisms. It is interesting to note that the tendency toward displacement of characters is opposed by the pressure for mimicry. One can imagine some elaborate interactions between the two tendencies, particularly in the evolutionarily fertile tropics.

Competition

The concept of competition has been the focus of much important disagreement among ecologists and other biologists, and it deserves close and persistent investigation. However, were it not that Andrewartha and Birch (1954) criticize the use of the concept by Lack and others to explain distribution and variation of birds and other animals, we might well have avoided discussing it here altogether. Andrewartha and Birch (p. 25) seem to consider that competition is an idea of lesser, perhaps even negligible, importance in biology. They think that the tendency for closely related species to inhabit different areas or exploit different ecological niches (as reported, for instance, by Lack) may conceivably have originated from causes "quite different" from competition. They do not offer alternatives that seem to us anything like as satisfactory as Lack's hypotheses.

Andrewartha and Birch make a point

when they ask for more direct evidence for the action of competition, but it is clear that they have failed to appreciate the amount of evidence that does exist in the literature. However, interspecific competition of the direct, conspicuous, unequivocal kind is apparently a relatively evanescent stage in the relationship of animal individuals or species, and therefore it is difficult to catch and record (just as is the often parallel crisis in the rise of reproductive barriers between two newly diverging species). What we usually see is the result of an actually or potentially competitive contact, in which one competitor has been suppressed or is being forced by some form of aggressive behavior to take second choice, or in which an equilibrium has been established when the potential competitors are specialized to split up the exploitable requisites in their environment. A third possible result is the dispersion of potential competitors in space (Lack, 1954). Surely the cases of character displacement we have considered above, especially those for which we have some ecological data, are pertinent examples of correlation between sympatry (with the possibility of competition) and genetic fixation of specializations resulting in the avoidance of competition. The respective convergent unispecific populations outside the sympatric zones are the "controls" for these observations.

The case in which Lack (1944) cites the distribution of the chaffinches (*Fringilla*) in the Canary Islands is held up to special criticism by Andrewartha and Birch. Lack demonstrates that *F. teydea*, endemic to the islands of Gran Canaria and Tenerife, occupies only the coniferous forests at middle altitudes. On the same islands there also occurs a form of the widespread *F. coelebs*, presumably a relatively recent arrival from the Palaearctic mainland, but this bird occurs only in the tree-heath zone above, and in the broadleaf forests below, the coniferous belt. On the island of Palma, however, *F. teydea* is absent, and there a form of *F.*

coelebs occupies the coniferous forest as well as the broadleaf zone. Andrewartha and Birch conclude that, "So far as the case is stated, there is no direct evidence that the two species could not live together if they were put together." It is obvious from this that Lack's critics are not going to be satisfied by any ordinary kind of evidence.

What emerges starkly from contemporary discussion of "competition" is the great variation in the meanings with which different authors freight the word. Andrewartha and Birch, while differing with Nicholson (1954) on most important points, do manage to agree with him that the correct kernel of meaning of competition is contained in the expression "together seek." We would adopt the part of their definition that deals with the common striving for some life requisite, such as food, space or shelter, by two or more individuals, populations or species, etc. This seems to us to be close to the definitions preferred by the larger dictionaries we have consulted.

But Andrewartha and Birch, following many other writers, allow their competition concept to include another idea—that expressing direct interference of one animal or species with the life processes of another, as by fighting. On the surface, this inclusion of aggression as an element of competition may seem to some familiar and reasonable, but we wonder whether the concept of competition could not be more useful in biology if it were more strictly limited to "seeking, or endeavoring to gain, what another is endeavoring to gain at the same time," the first meaning given in *Webster's New International Dictionary, Second Edition, Unabridged*. It is noteworthy that competition as defined by this dictionary fails to include the idea of aggression in any direct and unequivocal way.

It may therefore be more logical in the long run to regard the various kinds of aggression between potential competitors (the outcome of which is so often predictable) as another method, parallel with character displacement and dispersion—and genetically conditioned in a similar fashion—by which organisms seek to lessen or avoid competition. Surely it is significant that aggressive behavior often seems most highly developed in cases where a conspecific, or closely related, potential competitor occurs with the aggressor, yet shows little or no displacement in behavior or form. In contrast are the many cases of complete mutual tolerance shown by closely related organisms that live side by side and are differentially specialized in behavior or form.

Character Displacement versus Hybridization

Since both divergent and "intermediate" populations are involved in the displacement patterns we have been describing, it is clear that the convergent populations might easily be mistaken as representing products of interspecific hybridization between the two species displacing each other. This is especially true if the convergent populations are small and isolated, or if only a single one is developed. Lack, for instance, in an early paper (1940) interpreted the Daphne and Crossman populations of *Geospiza* as being of hybrid origin, changing his mind only after he had begun to consider more fully the influence of competition on speciation (in *Darwin's Finches,* 1947).

To take another possible example, Miller (1955) describes what he calls a "hybrid" between the woodpeckers *Dendrocopos scalaris* and *D. villosus.* This specimen, a female, was shot in the Sierra del Carmen, Coahuila, Mexico, at about 7000 feet altitude, near the lower limits of the coniferous belt capping the Sierra. Up to, or near, this altitude, Miller found the Sierra to support a population of *scalaris,* but despite intensive collecting, he found no sign of occupancy by the other putative parent species, *villosus.* *D. scalaris* reaches a higher point in these mountains than it usually does in the neighboring regions of desert scrub and bottomland—its habitat wherever it has been studied—

in Mexico, Arizona, New Mexico and parts of Texas. In general, the *villosus* populations of this part of North America are restricted to the higher coniferous belts, but *villosus* and *scalaris* are in contact at some stations where pinyon-oak-juniper meets coniferous forest. Presumably *scalaris* extends farther vertically in the Sierra del Carmen because *villosus* is not present to limit its upward expansion. According to Miller, *villosus* probably does not occur within 200 miles of the Sierra at the present time.

The specimen, thoroughly described and figured by Miller, is indeed intermediate in many respects between the *scalaris* and *villosus* of northern Mexico. However, there seems to be nothing in the information presented to prevent one's interpreting this as a large, unusually dark specimen of *scalaris,* instead of as a hybrid. There is no good reason to deny the possibility that *scalaris* can produce somewhat *villosus*-like variants at the upper limits of its range when *villosus* is absent.

Other examples we have already cited in the present paper show the difficulty in deciding between displacement and hybridization where the species involved are incompletely known. This situation adds considerable complication to the analysis of interspecific hybridization in nature, for it is clear that the alternative explanation of displacement should at least be taken into account.

One thing seems certain; the "hybrid index," better called "compound character index," can by itself be no sound indication that the situation plotted really involves hybridization. This leads us to ask whether even such elaborate and beautifully documented studies of "hybrid" situations as that made by Sibley (1950, 1954) on the towhees of southern Mexico (*Pipilo erythrophthalmus s. lat.* and *P. ocai*) are not really just illustrations of character displacement. In some of the higher mountains of the southeast (Orizaba, Oaxaca), the two very differently colored forms (species) meet but remain distinct. Farther west are found various populations that apparently grade between the extreme *erythrophthalmus* form and the *ocai* form to various degrees of intermediacy, as expressed by Sibley in his "hybrid index."

Some of the *ocai*-form populations at the western end of the range (*P. ocai alticola*) are stated to be distinct from the other races of *ocai* by a characteristic melanization of the head region, which Sibley thinks is due to introgression from *erythrophthalmus* populations found to the north in the Sierra Madre Occidental. Despite this indication of introgression, the western populations at the *ocai* end of the gradients studied are indexed at, or extremely close to, zero, the figure indicating a population of "pure" *ocai*. Aside from what seems to be a variation in "purity" standards for *ocai* here, it is interesting to note that the western populations and those others among the apparent intermediates of the southern Plateau Region can all conceivably, on present evidence, be interpreted as *erythrophthalmus* that have converged toward *ocai* in the absence of the "true" *ocai* form represented by the upland, sympatric southeastern samples.

It seems possible that some strong selective pressure may be acting in the southern Plateau region to produce an *ocai* coloration-type in finchlike birds, and that *erythrophthalmus* may yield to this pressure wherever the true *ocai* is absent in this area. A very *ocai*-like bird of a related genus, *Atlapetes brunneinucha*, reaches the northern limit of its range in the southern Plateau area, and it is possible that the striking similarity marks some adaptive relationship to which both it and the *Pipilo* stock respond. It might even be that mimicry is involved between the sympatric *Atlapetes* and *Pipilo* stocks, although this is nothing more than the sheerest speculation in view of our very incomplete knowledge of the relative distribution of the two forms and other aspects of their biology and their environment, including their predators. At any

rate, character displacement must for the time being be considered a reasonable alternative explanation of the variation of southern Mexican *Pipilo* in this group.

It may perhaps be argued that the "hybrid" populations of *Pipilo* are more variable than the presumed parental populations, and that this in itself is a strong indication of hybridization. We do not believe, however, that the case should be decided on this kind of evidence. To start with, tailspot length, the one character used in Sibley's study that has also been analyzed at length in other populations of *P. erythrophthalmus*, shows very considerable variation in areas far removed from the likely influence of *ocai*. According to the data of Dickinson (1953), the Florida population ("race *alleni*") has a coefficient of variation in this character of about 22 in the male; the range of variation is from 6.1 to 27.5 mm. The northeastern (nominate) race shows a corresponding coefficient of about 12, with a range of variation of from 24.0 to 55.0 mm. Furthermore, the chestnut-tinted pileum characteristic of *ocai-erythrophthalmus* "hybrids" occasionally crops up in the eastern North American samples of *erythrophthalmus*. But even if it were true that variation in the direction of *ocai* could be demonstrated only in the *ocai* "area of influence," this could not be taken as proof of hybridization, because an increase in variation is also a common quality of the "convergent" populations in character displacement patterns.

Character Displacement and Taxonomic Judgment of Allopatric Populations

Foremost among the problems of taxonomic theory today is the tantalizing conundrum concerning the status of the allopatric (isolated) population. Few authors hesitate to assign such populations either subspecific or specific rank, and most, it is hoped, appreciate the fact that their decisions are essentially arbitrary. As Mayr (1942) says, "The decision as to whether to call such forms species or subspecies is often entirely arbitrary and subjective. This is only natural, since we cannot accurately measure to what extent reproductive isolation has already evolved." There does not seem to be any definable threshold between polytypic species composed of such subspecific "units" and the superspecies composed of allopatric sister species. However, it is entirely possible that by the time an isolated population attains an ascertainable level of character concordance, it has already passed the species line; i.e., the more sharply defined an isolated subspecific population is by conventional standards, the less likely it is to be infraspecific in reality.

The phenomenon of character displacement should be borne heavily in mind in considering this matter of allopatric populations. If the present conception is correct, related sympatric species will generally show more morphological differences than similarly related allopatric ones. Hence the degree of observed difference between sympatric species cannot be considered a reliable yardstick for measuring the real status of related allopatric populations, nor can the differences among the latter be taken too seriously as indications of their relationships. In fact, the morphological standards set for determining which completely allopatric populations have reached species level may be much too strict in current practice. Despite impressions that might be gained from recent literature, many systematists have realized that in different allopatric populations (of the same species-group or genus), the degree of morphological divergence may be poorly correlated with the amount of reproductive isolation holding between them (Moore, 1954; Kawamura, 1953). In other words, where there is any question whatsoever about the objective species status of two closely related but geographically separated populations, morphology alone cannot be expected to answer it definitely.

Unfortunately, allopatric species or "subspecies" designated as such on a purely morphological basis frequently enter into theoretical discussions as though they were objectively established realities, when in fact they are usually no more than arbitrary units drawn for curatorial convenience.

Summary

Character displacement is the situation in which, when two species of animals overlap geographically, the differences between them are accentuated in the zone of sympatry and weakened or lost entirely in the parts of their ranges outside this zone. The characters involved in this dual divergence-convergence pattern may be morphological, ecological, behavioral, or physiological. Character displacement probably results most commonly from the first post-isolation contact of two newly evolved cognate species.. Upon meeting, the two populations interact through genetic reinforcement of species barriers and/or ecological displacement in such a way as to diverge further from one another where they occur together. Examples of the phenomenon, both verified and probable, are cited for diverse animal groups, illustrating the various aspects that may be assumed by the pattern.

Character displacement is easily confused with a different phenomenon: interspecific hybridization. It is likely that many situations thought to involve hybridization are really only character displacement examples, and in cases of suspected hybridization, this alternative should always be considered. Displacement must also be taken into account in judging the status (specific *vs.* infraspecific) of completely allopatric populations. It is clear that, in the case where the species are closely related, sympatric species will tend to be more different from one another than allopatric ones. Thus, degrees of difference among related sympatric populations cannot be used as trustworthy yardsticks to decide the status of apparently close, allopatric populations.

Acknowledgements

We are grateful for information, advice and other aid received from numerous colleagues in the course of preparing this contribution. Especially to be thanked are J. C. Bequaert, W. J. Bock, W. J. Clench, P. J. Darlington, A. Loveridge, A. R. Main, E. Mayr, A. J. Meyerriecks, K. C. Parkes, R. A. Paynter, and E. E. Williams. Dr. C. Vaurie kindly offered the use of his figures to illustrate the *Sitta* case and gave us the benefit of some unpublished observations. Our acknowledgement is not meant to imply that any of those listed necessarily support the arguments we advance.

REFERENCES

ALLEE, W. C., EMERSON, A. E. and others. 1950. Principles of animal ecology. W. B. Saunders Co.

ANDREWARTHA, H. G., and BIRCH, L. C. 1954. The distribution and abundance of animals. Univ. Chicago Press.

BLAIR, W. F. 1955. Size differences as a possible isolating mechanism in *Microhyla*. *Amer. Naturalist*, 89:297–301.

BOURNE, W. R. P. 1955. The birds of the Cape Verde Islands. *Ibis*, 97:508–556, cf. 520–524.

CAIN, A. J. 1955. A revision of *Trichoglossus haematodus* and of the Australian platycercine parrots. *Ibis*, 97:432–479, cf. 457–461, 479.

CONDON, H. T. 1951. Notes on the birds of South Australia: occurrence, distribution and taxonomy. *S. Aust. Ornith.*, 20:26–68.

DICKINSON, J. C. 1952. Geographical variation in the red-eyed towhee of the eastern United States. *Bull. Mus. Comp. Zool. Harv.*, 107:273–352.

DOBZHANSKY, TH. 1951. Genetics and the origin of species. 3rd Ed. Columbia Univ. Press.

HOWDEN, H. F. 1955. Biology and taxonomy of the North American beetles of the subfamily Geotrupinae . . . *Proc. U. S. Nat. Mus.*, 104:159–319, 18 pls.

HUBBS, C. L., and BAILEY, R. M. 1940. A revision of the black basses (*Micropterus* and *Huro*) with descriptions of four new forms. *Misc. Publ. Zool. Univ. Mich.*, No. 48, 51 pp.

KAWAMURA, T. 1953. Studies on hybridization in amphibians. V. Physiological isolation among four *Hynobius* species. *J. Sci. Hiroshima Univ. (B, 1)* 14:73–116.

KOOPMAN, K. F. 1950. Natural selection for reproductive isolation between *Drosophila pseudoobscura* and *Drosophila persimilis*. *Evolution*, 4:135–148.

LACK, D. 1940. Evolution of the Galapagos finches. *Nature, 146*:324–327.

—— 1944. Ecological aspects of species formation in passerine birds. *Ibis*, 86:260–286.

—— 1947. Darwin's finches. Cambridge Univ. Press.

—— 1954. The natural regulation of animal numbers. Oxford Univ. Press.

MAYR, E. 1942. Systematics and the origin of species. Columbia Univ. Press.

—— 1949. Speciation and selection. *Proc. Amer. Phil. Soc.*, 93:514–519.

—— 1955. Notes on the birds of northern Melanesia. *Amer. Mus. Novitates*, No. 1707: 1–46, cf. p. 29.

MOORE, J. A. 1954. Geographic and genetic isolation in Australian amphibia. *Amer. Naturalist*, 88:65–74.

MILLER, A. H. 1955. A hybrid woodpecker and its significance in speciation in the genus *Dendrocopos*. *Evolution*, 9:317–321.

NICHOLSON, A. J. 1954. An outline of the dynamics of animal populations. *Australian J. Zool.*, 2:9–65.

SERVENTY, D. L. 1953. Some speciation problems in Australian birds . . . *Emu*, 53:131–145, with further references.

SIBLEY, C. G. 1950. Species formation in the red-eyed towhees of Mexico. *Univ. Calif. Publ. Zool.*, 50:109–194.

—— 1954. Hybridization in the red-eyed towhees of Mexico. *Evolution*, 8:252–290.

VAURIE, C. 1950. Notes on Asiatic nuthatches and creepers. *Amer. Mus. Novitates*, No. 1472:1–39.

—— 1951. Adaptive differences between two sympatric species of nuthatches. *Proc. Xth Internat. Ornith. Congr., Uppsala, June 1950*:163–166, 3 figs.

WILSON, E. O. 1955. A monographic revision of the ant genus *Lasius*. *Bull. Mus. Comp. Zool. Harv.*, 113:1–205, ill.

WILSON, E. O., and BROWN, W. L., JR. 1955. Revisionary notes on the *sanguinea* and *neogagates* groups of the ant genus *Formica*. *Psyche*, 62:108–129.

WILLIAM L. BROWN, JR. is Associate Curator of Insects at the Museum of Comparative Zoology, Harvard University. EDWARD O. WILSON is a Junior Fellow of the Society of Fellows of Harvard University.

32

Reprinted from *Evolution*, **19**(2), 189–191, 199–203, 212–213 (1965)

THE EVOLUTION OF BILL SIZE DIFFERENCES AMONG SYMPATRIC CONGENERIC SPECIES OF BIRDS

Thomas W. Schoener

Biological Laboratories, Harvard University, Cambridge, Massachusetts

Recently, Hutchinson (1959), Klopfer and MacArthur (1961), and Klopfer (1962) have introduced the use of the quantitative comparison of bill size differences among groups of sympatric congeneric species of birds in the study of the evolution of niche overlap and size. In his analysis of character displacement, Hutchinson (1959) has shown that the ratio of the size of the larger to smaller trophic appendages of congeneric species generally falls between 1.2 and 1.4 where they are sympatric, but is less where they are allopatric. Klopfer and MacArthur (1961), using Ridgway's data, found a much smaller ratio for certain associations of sympatric tropical birds. This paper presents the results of an analysis of bill length, in which representatives of 46 bird families inhabiting temperate, subtropical, and tropical zones are compared. Several models are proposed to explain interfamilial, regional, and intrafamilial differences. Finally, the implications of this study for the concepts of niche overlap and behavioral stereotypy are discussed.

Character Displacement and Character Difference

It is perhaps best to limit the term *character displacement*, as Brown and Wilson (1956) have proposed, to "the situation in which, when two species of animals overlap geographically, the differences between them are accentuated in the zone of sympatry and weakened or lost entirely in the parts of their ranges outside this zone." So defined, character displacement is a quantity or number. Such studies as that of Klopfer and MacArthur (1961) and the one presented here are simply attempts to measure the amount of difference in size of trophic appendages among sympatric congeneric species, and the broader term *char-acter difference* will be used in referring to these latter cases.

Methods

In order to determine the food dimensions of the fundamental niche of a species (in Hutchinson's 1957 sense) by a study of bill size, one must accept the postulate, suggested first by Huxley (1942) and Lack (1944), that differences in bill size among species reflect, in some significant way, differences in the nature of their food. It is reasonable to suppose that such culmen differences will often be correlated with differences in the size of food taken. Betts (1955) has shown this to be true for certain sympatric species of the genus *Parus* in Great Britain. For many birds of prey, body size is probably a better indicator of the size of food preferred than bill size, although the two dimensions are often directly proportional. For this reason, wing lengths as a measure of comparative body size are included in the tables for the Accipitridae, Falconidae, and Strigidae. Lack (1946) has demonstrated that certain sympatric congeneric birds of prey, differing greatly in body size, feed on animals of different sizes. An especially striking example of this is the Palearctic sympatric pair, *Accipiter nisus* and *A. gentilis* (wings of females 230–240 and 340–375 mm., respectively), which do not have a single important prey species in common, a prey species being considered important when it provides at least five per cent of the diet.

Size of food alone may not be the only factor which determines the adaptive optimum for bill size. Bowman (1961) concluded from his analysis that differences of culmen size and shape among certain species of *Geospiza* are correlated with differences in ability to break open hard seeds, and in

Camarhynchus are correlated with a differential ability to cut deeply into woody tissues. The bill sizes and shapes of certain other birds, such as woodpeckers, would be expected to be correlated more with properties of the immediate food environment, rather than of the food itself. A related factor, the method of obtaining food, also influences dimensional properties of the bill. It is assumed in this paper that length of the bill is generally a good indicator of the nature of food taken, although its width, depth, and shape are also obviously important.

Bill size differences may sometimes be used for species recognition. Lack (1947) has proven this for *Geospiza* of the Galápagos Islands, but believes this factor secondary to that of food separation.

In order to determine how the niche of a species is limited by partitioning of the available food supply, it would be most desirable to examine the effects of all potential and actual food competitors of the species concerned. The extreme difficulty in carrying out such an analysis with present methods makes it more practical to concentrate on the most closely related competitors, *i.e.*, sympatric congeneric species. The justification of such a restriction depends heavily on the ability of the taxonomist to classify closely related species into genera, and differences of opinion as to what comprises a genus may add an unwanted variable to a study of this kind.

In Table 1, the culmen lengths of sympatric congeneric species are listed, along with the ratios of the culmens of large to small species. Data have been gathered from Ridgway and Friedmann (1901–1950) for most of the New World species listed, Rand and Rabor (1960) for Philippine Island birds, and Betts (1955), Brown and Wilson (1956), Dilger (1956), Hamilton (1959), MacArthur (1958), Ripley (1959), and Wetmore and Swales (1931) for individual genera, or have been obtained from measurements I have made of specimens in the Museum of Comparative Zoology, Harvard University. This survey covers most

families of New World land birds and certain Old World families as well. Some of the freshwater bird families are included, but others, such as the Anatidae, are excluded, partly for lack of enough data, and partly because their bill sizes may not be sufficiently correlated with preferred food size. Oceanic and most shorebirds are also excluded from this analysis.

Mayr's (1963) definition of sympatry as "the existence of a population in breeding condition within the cruising range of individuals of another population" is used for this study. One feature of this cruising potential is of special importance: species whose geographic breeding ranges overlap but which show differences in habitat preference are considered geographically sympatric, even though breeding individuals may never leave their preferred habitats. This is because one of the purposes of this study is to consider the relationship of varying habitat preferences to morphological differences. Associations in which one or more members prefer a different major habitat (*e.g.*, vegetation zone or proximity to bodies of water) are marked "h" in the tables. This is not to say that such species show any greater habitat separation than unmarked birds, because species which occupy different microhabitats within a major vegetation zone may be equally exclusive in their feeding habitats. Whenever possible, measurements used were taken from specimens collected within the area of overlap. If this was not possible, the mean culmen length throughout a larger area including the area of overlap is listed, followed by a "W." Species whose geographic breeding ranges overlap only marginally are not considered in this study, nor are species separated by altitudinal gaps.

General Results of This Study

The results of this study indicate that large ratios of character difference usually are found in three general situations:

1. *Large ratios occur among the members of certain families which appear to feed on food of relatively low abundance.—*

Families which commonly contain sympatric congeneric species associations possessing large ratios of character difference are the woodpeckers (Picidae: Table 1, 19), kingfishers (Alcedinidae: 15), sunbirds (Nectariniidae: 38), sandpipers (Scolopacidae: 5), rails (Rallidae: 6), parrots (Psittacidae: 8), and nuthatches (Sittidae: 30). Each of these families as a group is either specialized to take food in relatively restricted immediate food environments, *e.g.*, on or within bark, or to feed on rather restricted food types, *e.g.*, fish. Such restrictions imply that the total biomass of food available to these families will be relatively small in proportion to the body sizes of the feeding birds.

In contrast, those birds which appear to feed on more abundant food, in proportion to their body sizes, as do most small passerines and some non-passerines, show small ratios of character difference within their sympatric congeneric species associations. These birds include the small-billed representatives of the Phasanidae (4), Columbidae (7), Caprimulgidae (11), Apodidae (12), Trochilidae (13), Trogonidae (14), Todidae (16), Furnaridae (21), Formicariidae (22), Cotingidae (23), Pipridae (24), Tyrannidae (25), Pittidae (26), Paridae (29), Pycnonotidae (32), Troglodytidae (33), Mimidae (34), Turdidae (35), Sylviidae (36), Muscicapidae (37), Dicaeidae (39), Vireonidae (40), Parulidae (41), Icteridae (42), Thraupidae (43), Coerebidae (44), Ploceidae (45), and Fringillidae (46). Each of these families as a whole probably partitions or helps partition a larger biomass of food, relative to the body sizes of the feeding birds, than those in the former list, because the members of each of these families as a whole consume abundant small invertebrate and/or plant food, often in less restricted immediate food environments.

2. *Large ratios of character difference are found in sympatric congeneric species associations on islands, especially small islands.* — Lack (1947) and Hutchinson (1959) have noted the large differences in culmen size in *Geospiza* and *Camarhynchus* of the Galápagos Islands, Amadon (1947) has cited even more extreme cases for certain sympatric congeneric birds of the Hawaiian Drepaniidae, and Lack (1944, 1947) noted other instances of insular forms differing greatly in bill and body size. In this study, a systematic comparison of associations of the West Indies with related associations of North and Middle America was made. All Greater and Lesser Antillean sympatric species groups belonging to the families studied are listed at the beginning of each table. Of the eight pairs of sympatric species occurring in the Lesser Antilles (areas 62–687 square miles), seven have a ratio greater than 1.14 and four greater than 1.24. For the Greater Antilles (areas 3,423–44,218 square miles), 56.4 per cent of the thirty-nine ratios are greater than 1.14, and 43.6 per cent greater than 1.24; the figures for the same families on mainland North and Middle America (193 ratios) are 31.6 per cent and 17.6 per cent. Within families, the West Indian genera of the Mimidae (34), Accipitridae (1), Strigidae (3), Icteridae (42), Tyrannidae (25), Turdidae (35), Vireonidae (40), and Columbidae (7—*Geotrygon, Zenaida*) show a greater character difference than the same or related genera on the mainland, and associations of the latter four families possess a greater range of bill size than comparable (often more diverse) mainland associations. Certain genera within the Psittacidae (8), Rallidae (6), Corvidae (28), Trochilidae (13), Columbidae (*Columba*), Parulidae (13), and Fringillidae (46) do not vary much in the magnitude of character difference from mainland forms, and all inhabit the Greater Antilles except two species of *Columba*, found together on some of the northern Lesser Antilles. Of the families which have sympatric congeneric associations on both the Greater and Lesser Antilles, the Tyrannidae, Columbidae, Psittacidae, and Mimidae of the latter area display a greater character difference, whereas that of the Turdidae is less.

[*Editors' Note:* Material has been omitted at this point.]

RELATIVE FEASIBILITY OF FOOD PARTI-
TIONING BY SIZE AND HABITAT

A taxon which is undergoing diversification in a certain large area to the extent that sympatric species associations are being formed will have to partition the available food resources in order for its species to avoid or reduce competition. If a census were taken for this area to determine the relative abundance in terms of biomass of certain size classes of the food which is acceptable to the given taxon, whether because of the major habitat preferences, gross morphological limitations, foraging behavior, or preferred type of food which all its members have in common, the results could be plotted on a graph such as those illustrated in Fig. 1, where the horizontal axis represents size of food and the vertical axis represents food abundance in terms of biomass. The resulting curve, called the "food abundance curve," forms a boundary of the food space with respect to size of the given taxon, the other three boundaries being X_S and X_L, the lower and upper limits of food size consumed by this group, and the horizontal axis. X_S is defined as the small size limit beyond which a bird would have to spend an overly great amount of time feeding in order to support its metabolism and body size. X_L, where it exists, would be determined by structural and behavioral limitations of the taxon under consideration.

Such a food space may be used as the basis for models which predict the relative feasibility of food partitioning by size among species of families whose food abundance curves are of different heights. Assuming for the models in Fig. 1 that the distribution of food biomass is uniform with respect to food particle size, and species require approximately equal amounts of biomass (F) to support a stable population, it can be seen from a comparison of Figs. 1a and 1b that families possessing a low food abundance curve will at best contain a small number of species, whereas it is possible (although not necessary) for families feeding on more abundant food to show greater species diversity. If species occupied non-overlapping, contiguous niches with respect to food size, the food space would be partitioned as indicated by the dotted lines in Fig. 1. In actual fact, there is probably some minimum span of food size consumed by a species below which it is more efficient for the species to develop other means of reducing competition with sympatric species, such as partitioning food by microhabitat, rather than evolve the more precise discriminatory perceptual mechanisms necessary to further reduce the range in food size responded to. This minimum span, labeled "r" in the models, might be expected to vary somewhat among different groups of birds, but for simplicity is assumed constant.

Species feeding on relatively less abundant food, with few potential competitors, will find it possible to partition food by size, due to the small or zero overlap in "r's" possible (Fig. 1a). It will also be possible in some cases for such species to partition food by differing values of certain dimensional properties of the immediate food environment, *e.g.*, depth within bark or below the surface of water. In contrast, when the "r's" of sympatric species overlap greatly, food partitioning by size as a prin-

303

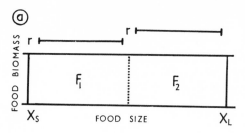

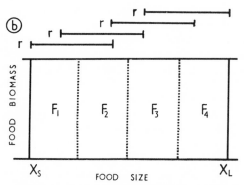

FIG. 1. Model of relative feasibility of food partitioning by size for abundant and rare food. Supporting biomasses of food for each species population represented by equal areas, F_1 to F_n. *a*. Species feeding on relatively rare food. *b*. Species feeding on relatively abundant food. F's of (a) equal F's of (b); *r* is the minimum span of food size consumed below which it is more efficient for species to partition food in ways other than by size, and where the *r*'s overlap considerably, food partitioning by size as the principal means of avoiding competition will not occur. Further explanation in text.

Such families contain individuals which generally feed on large packets of food, and if the total number of individuals in their species populations is not very much less than that of species feeding on abundant food, will possess larger "F's" than those assumed in Fig. 1b and hence will be less diverse.

Most families partitioning food on small islands are in a situation similar to that of the rare food families on the mainland in that the relatively small total biomass of food, itself limited by the area of the island, is one important factor which holds their diversity at a lower level than that of mainland groups which feed on abundant food. Hence the species of many families which on the mainland have too many competitors for a given range in food size (X_L–X_S) to make food partitioning by size feasible are able to efficiently partition food by size on islands.

Another important factor which must be considered when examining the relative feasibility of food partitioning by size is the relative feasibility of its most important alternative, food partitioning by habitat. Families whose members feed on extremely abundant food and are of small size will be able not only to occupy different major habitats, but also to become specialists at partitioning food by differences in microhabitat and foraging behavior within a major habitat, as do certain *Dendroica* (MacArthur, 1958) and *Parus* (Dixon, 1961). In contrast, those birds which feed on relatively rare food will not be able to find enough of it in a small area to maintain themselves individually or collectively as a stable population, and hence will often be more wide ranging and less restricted in habitat. Likewise, as McNab (1963) has pointed out for mammals, large species often have a larger home range than smaller ones, because they consume more energy and need a greater area in which to find this energy. Again, very large species will probably need to be less habitat specific, or maintain populations with large geographic ranges, or both, thus decreasing

cipal means of avoiding competition will not occur. This situation will exist if the number of potential competitors, both congeneric and extrageneric, is very high for a given range in food size (X_L–X_S) or dimensional property of the food environment, as is most likely to be the case for species feeding on relatively abundant food (Fig. 1b). Hence, such species will partition food by evolving significant differences in habitat preference or foraging behavior patterns.

Families whose food abundance curves are as high or higher than those of the latter group will still be able to partition food by size, if they are composed of large birds.

the likelihood that food separation by habitat or at least microhabitat will be utilized. When separation by habitat occurs in such groups it often consists of the occupation of different major habitats. For example, the owls *Asio otus* and *A. flammeus*, of similar bill and body size, occupy areas of fairly dense tree growth, especially conifers, and open plains, marshes and dunes respectively (Bent, 1938). Another example is the pair of rails *Rallus elegans* and *R. longirostris* (6), which occupy fresh- and salt-water marshes respectively (Peterson, 1947).

EVOLUTION OF CHARACTER DIFFERENCE

It has been argued in the last section that it is not feasible for species with many potential competitors to partition food by size, and that such species will likely feed in different habitats. However, the models do not predict that where there is a small number of competing species, food partitioning by size will necessarily occur, but only that it can occur. Another conceivable method of partitioning food in such situations is for species of similar size to occupy different major habitats or combinations of microhabitats, yet this second possibility is less often realized in nature.

A principal reason why this is so may be related to the pattern of invasion of an area or a major habitat and is perhaps best applicable to islands. If a hypothetical species A invades a small island on which there are no potentially competing species, it may eventually come to occupy a rather wide range of habitats, expanding into those which are occupied by closely related competitors on mainland areas. It will do this largely because of the absence of competitors, but in some cases also because the lesser abundance of its accustomed mainland habitat on the island forces it to expand into other habitats to find enough food biomass to maintain a stable population. However, even though a large range of food size (X_L-X_S) is available to it, due to structural limitations and for reasons of efficiency, it will at best be able to feed on

a certain subdivision of that range. In other words, each species, like the families considered in the last section, will also have an upper limit, X_L', determined by the morphology of its trophic structures, and an X_S', defined as the small size limit below which the amount of time spent feeding is not repayed sufficiently in biomass of food consumed.

If this island is now invaded by species B, and B differs greatly in size from A, it will be able to coexist in all habitats with A by taking differently sized foods. If B is somewhat different in size, there may be considerable selection for the individuals which are the most different in size from those of A, resulting in character displacement. There may also be selection operating on individuals of A, but since presumably A occupies large areas in which B is not present, gene flow from these areas will tend to mitigate B's effect on A. Such an argument can also be applied to species which partition food by different dimensional properties of the immediate food environment, such as woodpeckers.

If B is very similar in size to A, and takes similarly-sized foods, it must force A out of a considerable part of its habitat, especially on a very small island, if it is to coexist with A and maintain a population of a stable size. Unless B is very much superior to A in certain habitats it will not be able to do this, due to the much smaller number of individuals which it has on the island, all of which are in direct competition with those of A, as compared to the large supply of non-competing, already established individuals available to A. Although extinction of A or the forcing of A out of a considerable part of its habitat are possible outcomes, it is more likely that B will become extinct. Of course, if A does not have a wide habitat preference, because of recency of invasion or for some other reason, at the time of invasion by B, then it will be more likely that B will have nonoverlapping habitat preferences with A, and thus immediately be able to coexist, whether size is different or not.

A comparison of the two possible modes of coexistence discussed reveals another important fact. Two similarly-sized species feeding on the same range of food size in different habitats will be able to consume much less food between them than two species which coexist in all habitats but feed on different ranges of food size or food with differing dimensional properties of its immediate environment.

Such considerations lead one to expect more cases of partitioning by food size or environmental dimension than by habitat for insular species, resulting in many large ratios of character difference on small islands, a lesser number on larger islands, and still less on the mainland. As pointed out above, this is the case for West Indian sympatric congeneric species associations. Although it could be argued that larger ratios of character difference on islands might be produced mainly by random sampling from all the mainland species, it is not likely that random sampling alone of an area where the great majority of species are similarly-sized small birds could be responsible for the much greater proportion of large ratios of character difference on islands. The higher ratios of character difference in the West Indies could not have been produced simply by the dropping out of certain species of mainland genera because the number of species in the most comparable mainland sympatric congeneric association is usually the same or sometimes less than that of the West Indian association. Furthermore, the total ranges in bill size of West Indian congeneric associations are often greater, even if the number of species these associations contain is less, probably due to the absence of certain extrageneric competitors allowing expansion into food size categories additional to those utilized by the same or related genus on the mainland. It might be expected that the factors causing large ratios of character difference in the West Indies would also effect character difference in the Philippine Island associations listed in Table 1, although the nearness of these islands to one

another would partly counteract any tendency for much larger ratios than those found on mainland areas.

The above considerations of invasion patterns and efficiency of food utilization, while perhaps most appropriate for small islands, may apply to many situations in which food with a small total biomass relative to the size of the feeding birds is being partitioned, although on mainland areas other complicating factors, such as the existence of individuals of both species which do not occur in the area of overlap, may be in operation. However, species which partition rare food or those which feed on large packets of food are similar to many insular species in that they would tend to occupy a wide range of microhabitats as mentioned above, and hence would have to be displaced from a considerable area by invading competitors, unless these competitors differed greatly in size.

From these considerations, the following general pattern of the evolution of species partitioning a common food space can be postulated which probably applies to many associations on small islands and to certain mainland associations as well. A first stage in the evolution of sympatric species associations is often food partitioning by size or dimensional properties of the immediate food environment, with considerable overlap in feeding habitat. In situations where species diversity is limited, either by the small biomass of food which is being partitioned or for some other reason, such as the inability of invading species to establish a foothold without possessing sufficient differences, the final result will be a small number of species with large ratios of character difference. If, however, diversification of the partitioning species continues, as is possible for small birds which feed on extremely abundant food, food partitioning by size becomes impossible as more and more species are evolved, and must be abandoned in favor of separation by feeding habitat or foraging behavior pattern. This is not to say that the latter species all take food in the same size range, but rather that such ranges are so overlapping as to make other, more exclusive, subdivisions of the food space necessary in order to avoid competition.

Another reason why food partitioning by habitat is more likely to be found among small birds feeding on abundant food, even in the absence of a large number of competitors, as on some islands, is that the invading species will have to drive the established species out of a much smaller range of habitats in order to coexist, because such birds can maintain stable species populations in more restricted areas than larger species or those which feed on less dense food. It may also be that such birds, being less wide ranging, fail to expand their habitats as quickly or as completely as do more wide-ranging forms, even in the absence of all competition. The presence of certain associations of the Columbidae (7), Parulidae (43), Cuculidae (8), and Fringillidae (46) in the West Indies which do not have greater ratios of character difference than related mainland groups might be explained in such ways. However, there are associations of the former two families which do show larger ratios on the West Indies.

[*Editors' Note:* Material has been omitted at this point.]

LITERATURE CITED

AMADON, D. 1947. Ecology and the evolution of some Hawaiian birds. Evolution, **1**: 63–68.

BENT, A. C. 1919–1933. Life histories of North American birds. Bulletins U. S. Nat. Mus.

BETTS, M. M. 1955. The food of titmice in oak woodland. J. Anim. Ecol., **24**: 283–323.

BLAKE, E. R. 1953. Birds of Mexico: A guide for field identification. Univ. Chicago Press.

BOND, J. 1961. Birds of the West Indies. Houghton Mifflin.

BOWMAN, R. I. 1961. Morphological differentiation and adaptation in the Galápagos finches. Univ. California Publ. Zool., **58**: 1–302.

BROWN, W. L., AND E. O. WILSON. 1956. Character displacement. Syst. Zool., **5**: 49–64.

CARRIKER, M. A., JR. 1910. An annotated list of the birds of Costa Rica, including Cocos Island. Ann. Carnegie Mus., **6**: 314–915.

DELACOUR, J., AND E. MAYR. 1946. Birds of the Philippines. Macmillan.

DICKEY, D. R., AND A. J. VAN ROSSEM. 1938. The birds of El Salvador. Field Mus. Nat. Hist., Zool. Ser., Publ. **406**: 1–609.

DILGER, W. C. 1956. Adaptive modifications and ecological isolating mechanisms in the thrush genera *Catharus* and *Hylocichla*. Wilson Bull., **68**: 171–199.

DIXON, K. L. 1961. Habitat distribution and niche relationships in North American species of *Parus*. In W. F. Blair, ed., Vertebrate speciation. Univ. Texas Press, pp. 179–216.

EISENMANN, E. 1952. Annotated list of birds of Barro Colorado Island, Panama Canal Zone. Smithsonian Misc. Coll., **117**(5): 1–62.

——. 1955. The species of Middle American birds. Trans. Linnean Soc. New York, No. 7.

ENGELS, W. L. 1940. Structural adaptations in thrashers (Mimidae: Genus *Toxostoma*) with comments on interspecific relationships. Univ. California Publ. Zool., **42**: 341–400.

FRIEDMANN, H., L. GRISCOM, AND R. T. MOORE. 1950 and 1957. Distributional check-list of the birds of Mexico. Pts. 1 and 2. Pacific Coast Avifauna nos. 29 and 33, Cooper Ornithological Club.

GRISCOM, L. 1932. The distribution of bird-life in Guatemala. Bull. Amer. Mus. Nat. Hist., **64**: 1–439.

——. 1935. The ornithology of the Republic of Panama. Bull. Mus. Comp. Zool., **78**: 261–382.

HAMILTON, T. H. 1959. Adaptive variation in the genus *Vireo*. Wilson Bull., **70**: 307–346.

HAMILTON, T. H., AND R. H. BARTH, JR. 1962. The biological significance of season change in male plumage appearance in some New World migratory bird species. Amer. Nat., **96**: 129–144.

HUTCHINSON, G. E. 1957. Concluding remarks. Cold Spring Harbor Symposia on Quantitative Biology, **22**: 415–427.

——. 1959. Homage to Santa Rosalia, or Why are there so many kinds of animals? Amer. Nat., **93**: 145–159.

HUXLEY, J. 1942. Evolution: The modern synthesis. Allen and Unwin.

KLOPFER, P. H. 1962. Behavioral aspects of ecology. Prentice Hall.

KLOPFER, P. H., AND R. H. MACARTHUR. 1960. Niche size and faunal diversity. Amer. Nat., **94**: 293–300.

——. 1961. On the causes of tropical species diversity: Niche overlap. Amer. Nat., **95**: 223–226.

LACK, D. 1944. Ecological aspects of species formation in passerine birds. Ibis, **86**: 260–286.

——. 1946. Competition for food by birds of prey. J. Anim. Ecol., **15**: 123–129.

——. 1947. Darwin's finches. Cambridge University Press.

MACARTHUR, R. H. 1958. Population ecology of some warblers of northeastern coniferous forests. Ecol., **39**: 599–619.

MAYR, E. 1963. Animal species and evolution. Belknap Press of Harvard Univ. Press.

McNAB, B. K. 1963. Bioenergetics and the determination of home range size. Amer. Nat., **97**: 133–140.

PETERSON, R. T. 1947. A field guide to the birds. 2nd rev. ed. Houghton Mifflin.

——. 1961. A field guide to Western birds. Houghton Mifflin.

PETERSON, R. T., G. MOUNTFORT, AND P. A. D. HOLLOM. 1954. A field guide to the birds of Britain and Europe. Houghton Mifflin.

POUGH, R. H. 1949. Audubon bird guide. Doubleday.

——. 1956. Audubon water bird guide. Doubleday.

RAND, A. L., AND D. S. RABOR. 1960. Birds of the Philippine Islands: Siquijor, Mount Malindang, Bohol, and Samar. Fieldiana: Zool., 35, No. 7.

RIDGWAY, R., AND H. FRIEDMANN. 1901–1950. Birds of North and Middle America. Bull.

U. S. Nat. Mus., **50**: parts 1–8, 1901–1919 by Ridgway; parts 9–10, 1941 and 1946, by Ridgway and Friedmann; part 11, 1950, by Friedmann.

RIPLEY, S. D. 1959. Character displacement in Indian nuthatches (*Sitta*). Postilla, Yale Peabody Mus. Nat. Hist., No. 42.

RUSSELL, S. M. 1964. A distributional study of the birds of British Honduras. Ornithological Monogr. no. 1, Amer. Ornith. Union.

SALT, G. W. 1957. An analysis of avifaunas in the Teton Mountains and Jackson Hole, Wyoming. Condor, **59**: 373–393.

SKUTCH, A. F. 1946. Life history of the Costa Rican Tityra. Auk, **63**: 327–362.

——. 1951. Congeneric species of birds nesting together in Central America. Condor, **53**: 3–15.

——. 1954a. Life history of the White-winged Becard. Auk, **71**: 113–129.

——. 1954b, 1960. Life histories of Central American birds. Pacific Coast Avifauna, 31 and 34, Cooper Ornithological Club.

SLUD, P. 1960. The birds of Finca "La Selva," Costa Rica: A tropical wet forest locality. Bull. Amer. Mus. Nat. Hist., **121**: 49–148.

VAN TYNE, J., AND A. H. BERGER. 1959. Fundamentals of ornithology. John Wiley and Sons.

WAGNER, H. O. 1946. Food and feeding habits of Mexican hummingbirds. Wilson Bull., **58**: 69–93.

WETMORE, A., AND B. H. SWALES. 1931. The birds of Haiti and the Dominican Republic. Bull. U. S. Nat. Mus., **155**: 1–483.

33

Reprinted from *Amer. Naturalist,* **99**(908), 377–390 (1965)

MORPHOLOGICAL VARIATION AND WIDTH OF ECOLOGICAL NICHE

LEIGH VAN VALEN

Department of Vertebrate Paleontology, American Museum of
Natural History, New York

How does a species maintain its genetic variability in a form suitable for adaptive evolution without an unbearable genetic load? One of the answers to this question has been that different individuals in a population may be adapted to somewhat different environments. This hypothesis normally requires the existence of all the following three conditions, where the population may be divided into sets a, b, ... n and the niche inhabited by the population into more or less separable units (although complete gradation is by no means excluded for either phenotype or environment), A, B, ... M, at least at the time of year when population size is limited:

1) The individuals of some set a survive or reproduce better than those of some set b in some environment A, while the reverse is true in some environment B.

2) The above difference between a and b is in part genetic.

3) Some appropriate mechanism of distributional or mating preference exists. The third condition can be met by individuals of set a choosing environment A with a greater than random frequency, or by homogamic mating (at the limit of which a and b are two reproductively isolated species) if A and B differ spatially and there is a positive correlation between the locations of birth and mating.

It is important to note that continuous variation as well as polymorphism can be maintained by this model, the advantage of one or the other depending in part (not as an absolute condition in either direction) on the degree of regularity and continuity of the environment. Stabilizing selection is weaker with greater intra-niche diversity, and may even be absent or negative (destabilizing selection) when strong discontinuities are present within the niche.

Intrapopulation variation in niche occupied permits a greater population size than would otherwise be possible, and is advantageous even when a polymorphism is maintained by some other means. Genetic control need not be present in such cases. As an example of this advantage, if a and b are the sexes (however determined), equally abundant and of equal size, and eat different foods of the same abundance, and if population size is limited by the availability of food, then the population size will be twice what it would be if the sexes ate the same food.

Coexistence of morphologically or physiologically different phenotypes, as well as the behavioral or microgeographic diversity mentioned above, would be useful on both the individual and the population levels for exploitation of a diversified niche, and it is therefore probable that under the above conditions the population would be more variable than if it occupied a narrower niche.

Various aspects of the above theory have been formalized by Ludwig (1950), Levene (1953), Li (1955a, b), Lewontin (1955), Maynard Smith (1962), and Levins (1962, 1963); see also Dempster (1955). It was perhaps first explicitly proposed by Mayr (1945), although others approached it earlier. Mayr (1963) has summarized some of the evidence for the subsidiary hypotheses of the model.

The niche-variation model is compatible with a dependence of fitness on relative frequency of genotypes or phenotypes, but neither situation implies the other. For example, learning by predators can produce a frequency-dependent polymorphism in their prey, and if one part of the niche cannot be tolerated or reached by some individuals, reduction in the frequency of individuals occupying that part will not change the relative fitness of those excluded from it. Even in the more usual case, where a reduction in frequency of one form leads to an increased relative fitness in its part of the niche, this occurs only when that part of the niche remains habitable and would not apply when an environmental change reduced the frequency by reducing the parts of the niche differentially.

In the present paper I will compare the variation within populations of some birds that differ regionally as to width of niche. In one region, usually on the zoogeographic mainland, the niches of the species are relatively tightly packed together by the action of stabilizing selection imposed by ecologically adjacent species. In the other region, usually on islands, the environment available for the birds is partitioned into wider niches, which implies weaker stabilizing selection by adjacent species. The wider niches would permit greater phenotypic variation if this variation is controlled to a significant extent by the adaptive diversity of the niche. The results suggest that this is in fact the case.

By "niche width" I mean the proportion of the total multidimensional space of limiting resources used by a species or segment of a community. "Niche" is used here in an observational sense, not a theoretical one. The relative sizes of the niches in the regions compared have been studied for the species used and found to be different in the directions stated (see next section). The relative number of species in the total fauna is a related question, but one that I will not consider because it seems relevant only to the average relative width of the niches in each fauna, and the actual relative width for each species is known from direct observation.

Let us consider a segment of a community called a "trophic-A level" dependent upon some set, usually a relatively diverse set, of limiting resources. There are three extreme ways in which such a segment of a community can exploit the available supply of the limiting resources:

1) There can be relatively few species, each individual of each of which is adapted to occupy a relatively broad segment of the resource space.

Each individual may occupy this entire segment, or there may be a behavioral partitioning of the segment.

2) There can be relatively few species, but these are relatively variable and different individuals within a species are fitted to, and do in fact, occupy on the average different narrow niches (or, equivalently, different parts of a broad niche).

3) There can be a relatively large number of species, each of which is restricted to a relatively narrow segment of the resource space, more or less uniform for each individual.

These alternatives are not mutually exclusive on the community level, and most trophic-A levels of most communities probably contain more than one kind of species. Species intermediate between the extremes given are presumably common. Genetically controlled interindividual behavioral and morphological variation related to the occupation of the niche should on the average be large in the second kind of species and small in the first and third, although environmentally controlled behavioral (and in a few taxa morphological) variation may sometimes also be large in the first kind. In small and unstable communities (for example, on islets), and in stable trophic-A levels where the limiting resource changes in quality from time to time (for example, changes in composition of prey), I would expect the first kind of species to predominate. In large and exceptionally stable communities I would expect a preponderance of the third kind, which should also predominate among the rare species of most communities, excluding species maintained by immigration. The set of moderately to very stable communities should contain the greatest proportion of the second kind of species, which perhaps may never comprise a majority of the species in a community. This list is obviously suggestive rather than exhaustive, and may be compared with an analogous list, with supporting theory and involving mainly other criteria, given by Levins (1963).

I regard the fact that the present comparisons are between island and mainland populations as logically (not necessarily causally) irrelevant insofar as the relative number of species in the avifaunas is concerned. It is unknown to what extent niches are vacant or incompletely occupied on the islands, and to what extent the relative impoverishment of the island biotas is cybernetic by, in itself, restricting the number of potential niches for any trophic-A level. Because of these considerations, no *a priori* prediction of greater average niche width on the islands can be given with assurance. Island and mainland populations of most of the species on which comparisons have been made of niche width are, at the levels of resolution used, indistinguishable in niche width. I have ignored these species, although they are relevant to any consideration of relative niche width on the islands. Some species clearly have broader niches on the islands, and these species have been used here. In one case the island niche is clearly narrower than that on the mainland, and it is interesting that in this case the mainland populations, rather than the island ones, seem to be more variable.

MATERIAL

Aside from cases due to human influence or to differences in the availability of some resources (testable cases if the difference is of long standing), the only instances I know in which intraspecific differences in width of ecological niche have been demonstrated involve some of the birds inhabiting Bermuda, Curaçao, the Azores, and the Canaries. Lack and Southern (1949) observed that several species on the Canaries occurred in a different (usually wider) variety of local habitats than did the same species on the mainland, and Marler and Boatman (1951) and Vaurie (1957) made similar observations for other species on the Azores. Of these species five are present in American collections in sufficient numbers for statistical tests to be made, and these five species are therefore used in the present study. They are *Phylloscopus collybita* (the chiffchaff), *Parus caeruleus* (the blue tit), and *Fringilla coelebs* (the chaffinch) from the Canaries, and *Fringilla coelebs*, *Motacilla cinerea* (the gray wagtail), and *Regulus regulus* (the goldcrest) from the Azores. C. Vaurie (verbal communication) has informed me that *F. coelebs* has a wider niche in natural habitats in the Azores than on the mainland. Marler and Boatman (1951) merely note that it has expanded its habitat into towns, a change I would otherwise ignore.

Because the above observers (except Vaurie) are British, comparison has been made with British (not Irish) specimens wherever possible. In the cases of Phylloscopus and Parus, series from Bonn, Germany, have also been included, and for Motacilla, Britain and the entire European mainland west of Russia (excluding Scandinavia and Iberia) were used. All the insular populations are placed in different subspecies from those of Europe (including Britain), and in some cases different subspecies are recognized between individual islands. Each island was treated as the locality of a single population, and on the mainland (including Britain) the maximum area used for one population was the county of Devon. The only exception was Motacilla, a very recent invader of Europe, in which all European specimens were treated as a single population because of lack of any series from a single locality. The sexes were treated separately; unsexed specimens, damaged specimens, and those with the bill clearly constricted by string were ignored.

A fortunate circumstance permits some control over factors other than width of niche. In the case of *Fringilla coelebs* of the Canaries, the niche is not wider but narrower on two islands (Gran Canaria and Teneriffe) than on the mainland. This difference is correlated with the presence on only these islands of a second species (*F. teydea*), which occupies the pine-forest habitat that is part of the range of *F. coelebs* elsewhere on the Canaries and on the mainland. Therefore only specimens from Gran Canaria, Teneriffe, and the mainland were used for one comparison of *F. coelebs*. In the Azores the niche is, as with the other species used, wider than on the mainland.

The measurement used was the width of the culmen at the anterior border of the left nostril. The bill is not necessarily a constant character (no morphological character in birds has been shown to be invariant within an individual both seasonally and throughout the span of adult life), but the dates of collection were recorded for most of the specimens measured and the seasonal variation in collection time is closely comparable between island and mainland. Sample sizes and usually a scarcity of specimens collected in the middle third of the year did not permit an accurate determination of possible seasonal change within a population; any change present is not great. Davis (1954, 1961) has shown that the length of the bill of some birds varies seasonally because of differential attrition, although seasonal differences in its growth were not excluded as a contributory factor; the cycle demonstrated would presumably have less effect, if any, on bill width.

It is possible that migration may have caused some mixture of populations in mainland samples, but, as will be seen, any such effect does not eliminate the differences present.

On Curaçao and nearby islands two species have been reported by Voous (1957, p. 39) to have unusually wide niches in comparison with the rest of their range: *Mimus gilvus* (the southern mockingbird) and *Crotophaga sulcirostris* (the groove-billed ani). Voous remarked on a possible causal relation of this difference in niche in Mimus to a difference in its variation, but gave no comparative data. Crotophaga apparently reached Curaçao only in the last century, and would not necessarily be expected to have established a stable pattern of variation. Mimus is subspecifically distinct from the mainland forms. Voous (1957) gives measurements for both species from the islands, and I have made similar measurements on mainland samples of Mimus from single localities. Because there are only six degrees of freedom for each sex in Voous's data for Crotophaga (eight specimens, two localities), I did not measure mainland specimens of it. The measurement used for Mimus is length of bill. My measurement may not be exactly the same as that of Voous because of possible differences in techniques, but this should not affect the variation.

STATISTICAL METHODS

The hypothesis to be tested is that the intrapopulation coefficient of variation is larger (except Fringilla of the Canaries) on the islands than on the mainland, including Britain. As tests are available for the variance but not for the coefficient of variation as such, the variance is used and a scaling correction made for any difference in mean.

Aside from the biological impossibility of making any other kind of correction for size in the present case, I believe that the use of coefficients of variation or their equivalent is at least as justified as assuming, with Fisher (1937) and some others, that large and small groups are biologically equally variable and then calculating an empirical relationship among the variances of groups with different means. A change of scale produces a

change in the variance and in any transformation of the variance not equivalent to the coefficient of variation; I believe that this fact is adequate
justification for use of the coefficient of variation under any circumstances
I can imagine.

Homogeneity of intrapopulation variance within a region (that is, among
the islands or mainland populations) was tested by Bartlett's test. Only
two comparisons were significant at the 5 per cent (but not at the 1 per
cent) level; such a frequency of spuriously significant results is to be expected when 26 comparisons (within region and sex) are made, and they
will be ignored. The test is probably unnecessary in any event (Box,
1953).

The intrapopulation variance for a region was obtained by the equation

$$s^2 = \frac{\displaystyle\sum_{j=1}^{m} \sum_{i=1}^{n} (x_{ij} - \overline{x}_j)^2}{\left(\displaystyle\sum n\right) - m}$$

where there are m populations and n individuals in one population.

Homogeneity of means within a region and species was tested approximately by a two-way hierarchical analysis of variance, with population and
sex as the variables. The probabilities for each region were then combined to give a joint probability for the region, by both Fisher's method
(1958) and Barton's modification of mine (Barton, 1964; Van Valen, 1964).
The hypothesis tested here is the presence of interpopulation variation in
one or more species on the mainland and also on the islands. The populations of Mimus from the American mainland were not used in this test; their
means are obviously heterogeneous and the distance between populations
is much greater than for the European populations.

Whether or not there was a significant difference in mean between the island and mainland populations, the mainland variance was multiplied by
the square of the ratio of the island mean to the mainland mean. This correction makes relatively little difference in the significance of the results
(it usually slightly lowers the probability of significance), and is done to
remove the usual proportionality, from scaling effects, between the mean
and standard deviation. The only other statistical effect of the correction
is a slight reduction in the power of distinguishing populations with different variation, because the means are not uniquely known. The resulting
variances were compared by the F test. Comparison of the uncorrected
variation by Levene's test (1960) gives similar results; the latter test (on
which see also Van Valen, in press) could be refined by multiplying each
mainland measurement, before testing, by the ratio of the means, but this
was not done.

Measurement error was estimated by 21 replicate measurements per species, and was assumed to be homogeneous within a species. The esti-

mated variance is the sum of the measurement variance and the true variance of the population. Because it is the latter that is of interest, I subtracted the measurement variance from each estimated variance and corrected as above for the difference in means. A final correction on the variance ratio, to remove a small effect of different sample size, is to divide by the appropriate F value at 50 per cent. The resulting ratio is an unbiased estimate of the ratio of the intrapopulation variation on the islands to that on the mainland (except Fringilla of the Canaries), although significance tests should be performed on the ratio given in the preceding paragraph.

RESULTS AND DISCUSSION

From the results given in table 1 there seems to be no real room for doubt that the species studied are on the average more variable in the region with a wider niche. In none of the 12 independent tests are the populations with a narrower niche more variable than the others; the probability of this happening by chance is less than 0.001. Combining the 12 probabilities also gives a joint one-tailed probability of less than 0.001.

The unweighted mean of the 14 corrected F ratios is 2.20, and the weighted mean is 2.00. In other words, the birds with the more variable niche are about twice as variable in bill width or length as the others. This suggests that adaptation to different aspects of the environment is a major cause of variation, at least in passerine birds.

There is significant interpopulation heterogeneity in mean bill width on both the islands and the mainland ($P < 0.01$ in each case). This heterogeneity did not suppress the island versus mainland difference in *Motacilla cinerea*, if heterogeneity was present on the mainland in this species, but may well have contributed to the failure of Crowell (1962) to find a similarly consistent difference in comparisons of three species between Bermuda and North America. Because of scarcity of material he was forced to use the territory east of Indiana from Pennsylvania to Florida as a single population; any interpopulation variation within this region would increase the mainland variance and so reduce the difference with Bermuda. Crowell also found little difference between island and mainland in diversity of feeding habits in the species he studied, although the total niche was apparently broader on Bermuda.

There is no indication of a greater difference between the sexes on the islands than on the mainland.

The greater variation found on the islands was predicted by the hypothesis stated in the introduction, but clearly does not prove that the hypothesis is correct. There may be correlated factors that would also lead to a greater variation. It is conceivable, for instance, that the food of the young is more heterogeneous from nest to nest on the islands and that the greater phenotypic variance found there is not related to a greater genetic variance. The case of Fringilla of the Canaries, with a prediction and ostensible result opposite to those for the other species and the same species in

TABLE 1

Comparative within-population statistics of bill measurements
(in millimeters) for several species of birds (see text)

	Population with wider niche	$\overline{X}$	s^2	C.V.	Corrected F	$P(s_1^2 \leq s_2^2)$	Degrees of freedom
Phylloscopus collybita							
males							
Canaries	x	2.28	0.0254	6.99			8
mainland		2.14	0.0141	5.55	1.86	0.10–0.20	23
females							
Canaries	x	2.30	0.0088	4.07			10
mainland		2.31	0.0047	2.96	2.11	0.20–0.25	6
Parus caeruleus							
males							
Canaries	x	3.47	0.0381	5.62			21
mainland		3.55	0.0141	3.35	3.10	0.001–0.005	35
females							
Canaries	x	3.37	0.0143	3.54			20
mainland		3.46	0.0128	3.27	1.19	0.40–0.50	29
Fringilla coelebs							
males							
Azores	x	6.49	0.0553	3.62	1.78	0.01–0.025	87
Canaries		6.10	0.0229	2.55			15
mainland	x	5.67	0.0242	2.74	1.11	0.40–0.50	64
females							
Azores	x	6.05	0.0625	4.13	2.08	0.025–0.05	36
Canaries		6.01	0.0171	2.28			7
mainland	x	5.37	0.0241	2.89	1.65	0.20–0.25	24
Motacilla cinerea							
males							
Azores	x	3.41	0.0306	5.13			24
mainland		3.00	0.0203	4.75	1.21	0.25–0.40	31
females							
Azores	x	3.33	0.0414	6.11			15
mainland		3.06	0.0131	3.75	2.88	0.01–0.025	19
Regulus regulus							
males							
Azores	x	2.18	0.0417	9.37			32
mainland		1.70	0.0123	6.53	2.33	0.025–0.05	31
females							
Azores	x	2.24	0.0360	8.47			17
mainland		1.77	0.0070	4.73	4.12	0.025–0.05	10
Mimus gilvus							
males							
Curaçao	x	26.92	2.231	5.55			15
mainland		21.42	0.362	2.81	3.94	0.005–0.01	16
females							
Curaçao	x	26.13	0.727	3.26			12
mainland		22.10	0.362	2.72	1.47	0.20–0.25	17

the Azores, suggests that width of niche is involved in some way. Comparison with species that have similar food habits as on the mainland would be useful in testing other possibilities, but although no island-mainland differences in food or habitat are known for several species of the Canaries and the Azores, none of these species have as yet been sufficiently well studied to be sure that important differences do not exist.

It is unknown from field observations whether the greater diversity of food presumably eaten on the islands is done by largely different individuals, or whether each bird partakes of the entire array of food of its species. In the latter case no clear-cut prediction could be made. On the one hand the environmental component of the phenotypic variance is probably larger in a more diverse environment, and greater heterozygosity (related to a greater genetic variance) may be useful in coping with a more diverse environment, but on the other hand the heterozygotes themselves may be less variable. Heterozygosity may, however, be greater on the mainland because of migration.

It is possible that most of the island populations are more homozygous than those on the mainland. The effect of this possible difference on the phenotypic variance is uncertain, although it would probably not produce much increase if any. In the only direct study of this relationship Bader (1956) found almost as much phenotypic variance in laboratory-raised inbred lines of three rodents as in wild populations. The studies of the relationship between abundance and variation, cited below, show only a small average effect with large differences in abundance. *Fringilla coelebs* is a very common species in the Azores and rare in the Canaries; its pattern of variation is the reverse of that predicted by the hypothesis that canalization (insensitivity to environmental variation) is reduced in rare species.

The distributions of the measurements for the island populations give no suggestion of bimodality. The greater variation is therefore probably not maintained by any system of polymorphism but rather by the various mechanisms available for continuous variation.

In other animals there is some evidence for a relationship of variation with width of niche. In several species of Drosophila (*D. robusta, D. nebulosa, D. acutilabella, D. americana texana, D. nigromelanica, D. immigrans, D. willistoni*, parts of *D. pseudoobscura* and *D. subobscura*, and perhaps *D. euronotus*) and in *Chironomus tentans* there is a decrease of inversion polymorphism in geographically peripheral and ecologically marginal areas (da Cunha, Burla, and Dobzhansky, 1950; Townsend, 1952; da Cunha, Brncic, and Salzano, 1953; da Cunha and Dobzhansky, 1954; Brncic, 1955; Carson, 1955, 1956, 1958; da Cunha et al., 1959; Acton, 1959; Dobzhansky et al., 1963; Stalker, 1964a, b; Carson and Heed, 1964; Sperlich, 1964; Prevosti, 1964), although this is not true for all species or populations. A possibly similar situation occurs in *D. funebris* between urban and rural environments (Dubinin and Tiniakov, 1946, 1947); however, both urban and rural environments have been interpreted in the literature as more variable for Drosophila. The alternative or supplementary hypothesis of Carson (for example, 1955), involving a greater potentiality for recombination in marginal areas, does not consider the compensating increase of recombination elsewhere in the genome caused by a heterozygous inversion, and so is perhaps of less than full applicability. How to reconcile it with the apparently superior homeostasis of inversion heterozygotes and of polymorphic populations is also obscure. Genetic drift in one or more of its forms could also produce a similar pattern of distribution. In the *D. cardini* species group (Heed, 1963) there is a particularly interesting situation in that color polymorphism is less frequent in peripheral areas; the polymorphism is differently determined in different species and the two morphs even belong to different species in some areas, which strongly suggests an ecological control. Other studies are summarized by Mayr (1963); see also Sieburth (1964).

There is also some evidence (Fisher and Ford, 1928; Fisher, 1937; Dobzhansky, Burla, and da Cunha, 1950; da Cunha, Brncic, and Salzano, 1953; Salzano, 1955; data discussed but not presented by Darwin, 1859) that numerically abundant species tend to be more variable than their rarer relatives, although for the data used by Fisher (1937) and in part those used by Darwin (1859) the variation discussed is over quite wide geographic areas. This relationship is probably best explained by the variation-niche hypothesis; the evidence is not as clear-cut as for comparisons within species, and more studies are needed.

The few known examples in nature of destabilizing (centrifugal, disruptive, diversifying, fractionating) selection are presumably examples of the same phenomenon (maintenance of variation by adaptation to different parts of a niche), although destabilizing selection need not be involved in it. Different intensities of stabilizing selection could produce the same result. These cases (demonstrated or probable) of destabilizing selection are in a grasshopper (White, Lewontin, and Andrew, 1963, p. 159), butterflies (Brower, 1959; Clarke and Sheppard, 1962; Creed et al., 1959, 1962; Ford, 1964), a gull (Smith, 1964), and a mouse (Van Valen, 1963), in addition to cases such as that studied by Bradshaw (1963) and Jowett (1964) where sharp ecological boundaries cross an interbreeding population. The latter cases grade into the usual adaptive geographic ("ecotypic") variation. Variation produced by spatially varying use of pesticides and antibiotics is similar in principle. The frequently nongenetic stunted nature of plants grown on poor soil or under other adverse conditions is clearly adaptive in permitting any to survive and reproduce (see Harrison, 1959).

On the basis of the new and old evidence discussed in the present paper, it seems probable that adaptation to different aspects of the environment is a major cause of variation in at least higher animals and plants. In other words, much variation is probably adaptive in itself and is not part of the genetic or phenotypic load.

SUMMARY

In six bird species convenient for study, niches are known to be broader on some islands than on the mainland. In every case there is also greater variation in a bill measurement on the islands, except for one species in which the niche on the Canary Islands is narrower than that on the mainland. The adjusted variances average about twice as great in the broader niche. These and other results suggest that continuous variation within local populations is often adaptive in itself and is not part of the genetic or phenotypic load.

ACKNOWLEDGMENTS

I am particularly indebted in Dr. Charles Vaurie for introducing me to birds. He and Drs. Robert H. MacArthur, George E. Watson, Richard C. Lewontin, Edward O. Wilson, and Ernst Mayr also made various useful suggestions. Dr. Vaurie, Dr. Philip S. Humphrey of the United States National Museum, and Dr. Raymond A. Paynter of the Museum of Comparative Zoology permitted measurements of specimens in their care.

LITERATURE CITED

Acton, A. B., 1959, A study of the differences between widely separated populations of *Chironomus* (=*Tendipes*) *tentans* (Diptera). Proc. Roy. Soc. London (B) 151: 271–296.

Bader, R. S., 1956, Variability in wild and inbred mammalian populations. Quart. J. Florida Acad. Sci. 19: 14–34.

Barton, D. E., 1964, Combining the probabilities from significance tests. Nature 202: 731.

Box, G. E. P., 1953, Non-normality and tests on variances. Biometrika 40: 318–335.

Bradshaw, A. D., 1963, The analysis of evolutionary processes involved in the divergence of plant populations. Proc. XI Int. Cong. Genet. 1: 143.

Brncic, D., 1955, Chromosomal variation in Chilean populations of *Drosophila immigrans*. J. Hered. 46: 59–63.

Brower, J. vZ., 1959 (discussion after paper by Thoday). Proc. XV Cong. Zool.: 130.

Carson, H. L., 1955, The genetic characteristics of marginal populations of *Drosophila*. Cold Spring Harbor Symp. Quant. Biol. 20: 276–287.

   1956, Marginal homozygosity for gene arrangement in *Drosophila robusta*. Science 123: 630–631.

   1958, The population genetics of *Drosophila robusta*. Adv. Genet. 9: 1–40.

Carson, H. L., and W. B. Heed, 1964, Structural homozygosity in marginal populations of nearctic and neotropical species of Drosophila in Florida. Proc. Nat. Acad. Sci. 52: 427–430.

Clarke, C. A., and P. M. Sheppard, 1962, Disruptive selection and its effect on a metrical character in the butterfly *Papilio dardanus*. Evolution 16: 214–226.

Creed, E. R., W. H. Dowdeswell, E. B. Ford, and K. G. McWhirter, 1959, Evolutionary studies on *Maniola jurtina*: the English mainland, 1956–57. Heredity 13: 363–391.

   1962, Evolutionary studies on *Maniola jurtina*: the English mainland, 1958–60. Heredity 17: 237–265.

Crowell, K. L., 1962, Reduced interspecific competition among the birds of Bermuda. Ecology 43: 75–88.

da Cunha, A. B., D. Brncic, and F. M. Salzano, 1953, A comparative study of chromosomal polymorphism in certain South American species of Drosophila. Heredity 7: 193–202.

da Cunha, A. B., H. Burla, and Th. Dobzhansky, 1950, Adaptive chromosomal polymorphism in *Drosophila willistoni*. Evolution 4: 212–235.

da Cunha, A. B., and Th. Dobzhansky, 1954, A further study of chromosomal polymorphism in *Drosophila willistoni* in its relation to the environment. Evolution 8: 119–134.

da Cunha, A. B., Th. Dobzhansky, O. Pavlovsky, and B. Spassky, 1959, Genetics of natural populations. XXVIII. Supplementary data on the chromosomal polymorphism in *Drosophila willistoni* in its relation to the environment. Evolution 13: 389–404.

Darwin, C., 1859, On the origin of species by means of natural selection. J. Murray, London.

Davis, J., 1954, Seasonal changes in bill length of certain passerine birds. Condor 56: 142–149.

1961, Some seasonal changes in morphology of the rufous-sided towhee. Condor 63: 313–321.

Dempster, E. R., 1955, Maintenance of genetic heterogeneity. Cold Spring Harbor Symp. Quant. Biol. 20: 25–32.

Dobzhansky, Th., H. Burla, and A. B. da Cunha, 1950, A comparative study of chromosomal polymorphism in sibling species of the *willistoni* group of *Drosophila*. Amer. Natur. 84: 229–246.

Dobzhansky, Th., A. Hunter, O. Pavlovsky, B. Spassky, and B. Wallace, 1963, Genetics of natural populations. XXXI. Genetics of an isolated marginal population of *Drosophila pseudoobscura*. Genetics 48: 91–103.

Dubinin, N. P., and G. G. Tiniakov, 1946, Structural chromosome variability in urban and rural populations of *Drosophila funebris*. Amer. Natur. 80: 393–396.

1947, Inversion gradients and selection in ecological races of *Drosophila funebris*. Amer. Natur. 81: 148–153.

Fisher, R. A., 1937, The relation between variability and abundance shown by the measurements of the eggs of British nesting birds. Proc. Roy. Soc. London (B) 122: 1–26.

1958, Statistical methods for research workers. 13th ed. Oliver and Boyd, Edinburgh.

Fisher, R. A., and E. B. Ford, 1928, The variability of species in the Lepidoptera, with reference to abundance and sex. Trans. Entomol. Soc. London 76: 367–384.

Ford, E. B., 1964, Ecological genetics. Methuen, London.

Harrison, G. A., 1959, Environmental determination of the phenotype, p. 81–86. *In* A. J. Cain (ed.), Function and taxonomic importance. Systematics Assoc. Pub. No. 3.

Heed, W. B., 1963, Density and distribution of *Drosophila polymorpha* and its color alleles in South America. Evolution 17: 502–518.

Jowett, D., 1964, Population studies on lead-tolerant *Agrostis tenuis*. Evolution 18: 70-80.

Lack, D., and H. N. Southern, 1949, Birds on Tenerife.   Ibis 91: 607-626.

Levene, H., 1953, Genetic equilibrium when more than one ecological niche is available.   Amer. Natur. 87: 331-333.

   1960, Robust tests for equality of variances, p. 278-292. *In* Olkin, Ghurye, Hoeffding, Madow, and Mann (eds.), Contributions to probability and statistics. Stanford University Press.

Levins, R., 1962, Theory of fitness in a heterogeneous environment. I. The fitness set and adaptive function.   Amer. Natur. 96: 361-373.

   1963, Theory of fitness in a heterogeneous environment. II. Developmental flexibility and niche selection.   Amer. Natur. 97: 75-90.

Lewontin, R. C., 1955, The effects of population density and competition on viability in *Drosophila melanogaster*.   Evolution 9: 27-41.

Li, C. C., 1955a, The stability of an equilibrium and the average fitness of a population.   Amer. Natur. 89: 281-296.

   1955b, Population genetics.   Univ. Chicago Press, Chicago.

Ludwig, W., 1950, Zur Theorie der Konkurrenz.   Neue Ergeb. Prob. Zool., Klatt-Festschrift: 516-537.

Marler, P., and D. J. Boatman, 1951, Observations on the birds of Pico, Azores.   Ibis 91: 607-626.

Maynard Smith, J., 1962, Disruptive selection, polymorphism, and sympatric speciation.   Nature 195: 60-62.

Mayr, E., 1945, Some evidence in favor of a recent date: symposium on the age of the distribution pattern of the gene arrangements in *Drosophila pseudoobscura*.   Lloydia 8: 70-83.

   1963, Animal species and evolution.   Harvard Univ. Press, Cambridge, Mass.

Prevosti, A., 1964, Chromosomal polymorphism in *Drosophila subobscura* populations from Barcelona (Spain).   Genet. Res. 5: 27-38.

Salzano, F. M., 1955, Chromosomal polymorphism in two species of the *guarani* group of *Drosophila*.   Chromosoma 7: 39-50.

Sieburth, J. M., 1964, Polymorphism of a marine bacterium (*Arthrobacter*) as a function of multiple temperature optima and nutrition.   Narragansett Marine Lab., Univ. Rhode Island, Occ. Pub. 2: 11-16.

Smith, N. G., 1964, Evolution of some arctic gulls (*Larus*): a study of isolating mechanisms.   Dissert. Abst. 24: 3901.

Sperlich, D., 1964, Chromosomale Strukturanalyse und Fertilitätsprüfung an einer Marginalpopulation von *Drosophila subobscura*.   Z. Vererb. 95: 73-81.

Stalker, H. D., 1964a, Chromosomal polymorphism in *Drosophila euronotus*. Genetics 49: 669-687.

   1964b, The salivary gland chromosomes of *Drosophila nigromelanica*. Genetics 49: 883-893.

Townsend, J. I., 1952, Genetics of marginal populations of *Drosophila willistoni*.   Evolution 6: 428-442.

Van Valen, L., 1963, Intensities of selection in natural populations.   Proc. XI Int. Cong. Genet. 1: 153.

1964, Combining the probabilities from significance tests. Nature 201: 642.

1965, Selection in natural populations. III. Measurement and estimation. Evolution. (In press)

Vaurie, C., 1957, Systematic notes on Palearctic birds. No. 25. Motacillidae: the genus *Motacilla*. Amer. Mus. Novitates 1832: 1-16.

Voous, K. H., 1957, The birds of Aruba, Curaçao, and Bonaire, p. 1-260. *In* P. H. Hummelinck (ed.), Studies on the fauna of Curaçao and other Caribbean islands, Vol. 7. Martinus Nijhoff, The Hague.

White, M. J. D., R. C. Lewontin, and L. E. Andrew, 1963, Cytogenetics of the grasshopper *Moraba scurra*. VII. Geographic variation of adaptive properties of inversions. Evolution 17: 147-162.

34

Reprinted from R. H. MacArthur, and E. O. Wilson,
The Theory of Island Biogeography, Princeton University Press, Princeton, N. J.
1967, pp. 105–108, by permission of Princeton University Press

THE THEORY OF ISLAND BIOGEOGRAPHY

Robert H. MacArthur and E. O. Wilson

[*Editors' Note:* In the original, material precedes this excerpt.]

Ecological Expansion and Contraction

Compressibility of an elementary kind is illustrated in the comparison between the Puerto Rico and Panama bird censuses given in Tables 9 and 10. Clearly the Puerto Rican birds are found in many habitats, while the Panamanian ones are much more restricted. A diversity of similar effects is shown by tropical ant faunas (Wilson and Taylor, 1967; Wilson and Hunt, 1967). The ants of the Wallis Islands, which are located between Fiji and Samoa, exemplify the complex changes that accompany reduction in species diversity. The native fauna of Uvéa, the principal island of the group, is a small subset of the native Fijian fauna. Its species occur in different abundances from the conspecific populations on Fiji. Surrounding Uvéa are 20 islets, which in turn are populated by small subsets of the Uvéa species. Despite the fact that the islets are strung close together on a coral reef and are similar in appearance, their faunulae show marked differences in the composition, relative abundances, and even nest sites of the resident species.

Entry into a smaller fauna is often accompanied by ecological release. On New Guinea some of the most widespread of the Indo-Australian ant species, notably *Rhytidoponera araneoides, Odontomachus simillimus, Pheidole oceanica, P. sexspinosa, P. umbonata, Iridomyrmex cordatus,* and *Oecophylla smaragdina,* are mostly or entirely limited to species-poor "marginal" habitats, such as grassland and gallery forest. But in the Solomon Islands, which has a smaller native fauna, these same species also penetrate the rain forests, where they are among the most abundant species. In the New Hebrides, which has a truly impoverished ant fauna, the species of *Odontomachus* and *Pheidole* just listed almost wholly dominate the rain forests as well as the marginal habitats. Ecological release in the opposite direction, from central to marginal habitats, has also occurred. In Queensland and New Guinea, *Turneria* is a genus of rare species mostly confined to rain forests. It is also the only genus of the subfamily Dolichoderinae to have reached the northern New Hebrides. On Espiritu Santo, one of the latter islands, two species of *Turneria* are among the most abundant arboreal insects in both marginal habitats and virgin rain forest.

The degree of compression or release in new environments varies among species and is difficult to predict in advance. A case in point is the marked difference in behavior between two of the thirteen ant species that have succeeded in colonizing the Dry Tortugas, the outermost of the Florida Keys. In the presence of such a sparse fauna, *Paratrechina longicornis* has undergone extreme expansion. In most other parts of its range it nests primarily under and in sheltering objects on the ground in open environments. On the Dry Tortugas it is an overwhelmingly abundant ant and has taken over nest sites that are normally occupied by other species in the rest of southern Florida: tree-boles, usually occupied by species of *Camponotus* and *Crematogaster*, which are absent from the Dry Tortugas; and open soil, normally occupied by the crater nests of *Conomyrma* and *Iridomyrmex*, which genera are also absent from the Dry Tortugas. In striking contrast is the behavior of *Pseudomyrmex elongatus*. This ant is one of ten species that commonly nest in hollow twigs of red mangrove in southern Florida. It tends to occupy the thinnest twigs near the top of the canopy and is only moderately abundant. *P. elongatus* is also the only member of the arboreal assemblage that has colonized the Dry Tortugas, where it has a red mangrove swamp on Bush Key virtually all to itself. Yet it is still limited primarily to thinner twigs in the canopy and, unlike *Paratrechina longicornis*, has not perceptibly increased in abundance.

Cameron (1958) has described an interesting case of expansion followed by compression in the arctic hare (*Lepus arcticus*), which was initially the only hare on Newfoundland, where it expanded to occupy both its mainland tundra habitat and also forest land. When the varying hare (*Lepus*

106

americanus) was introduced, the habitat of the arctic hare contracted back to the tundra.

Once compressibility of any degree has been demonstrated, the next general question to ask is whether diet or habitat or both should contract when competitors are present. The converse question is whether diet or habitat or both should expand where interspecific competition is relaxed. The theoretical answer is quite interesting: when competition is increased, the variety of occupied habitats (or more correctly, the space searched) should shrink, or at least be altered, but the range of foods within the occupied habitats should not. MacArthur and Pianka (1966) give a formal proof of this hypothesis, but the following verbal argument is adequate. (1) On hunting for food within a fine-grained patch of habitat an efficient species should clearly accept or reject each item encountered on its own merits, which are nearly independent of how rare other such items of the same species are. That is, the choice is made item by item as the items of food are found; and the finding within a searched area is not controlled by preference of one food or another. Hence if a new competitor enters the patch and reduces the density of a particular food species, it will not greatly affect whether a previous species includes that food in its diet, although it may reduce the *proportion* of the food in the diet. (2) However, each habitat may be thought of as a mosaic of fine-grained patches. Which of these patches a species hunts depends very much on competition, because this choice is made ahead of time, and, by avoiding a patch that is heavily foraged by a competitor, a species can increase its harvest of food. Actually, competition may not only restrict the types of patches in which the species feeds; it may also cause new, previously unsuitable patch types to be included in the itinerary. This is quite consistent with the hypothesis and has in fact been demonstrated in ant thrushes by Willis (1966). In sum, a species can, without marked waste of time, forage only in the more profitable kinds of patches. But within each patch it cannot avoid coming upon the full spectrum of food, since the patches are fine-grained, and thus be forced to decide independently

107

upon each item. Consequently, on being freed from competition on an island, a species can be expected to alter and usually to enlarge its habitat, but not its range of diet—at least initially—although the variance of items in the diet may be enlarged. Looked at another way, as a new island is occupied by progressively more and more species, both the preferred food and habitat should be initially as varied as the species' morphology allows, but the habitat primarily should change with the entry of new species. There are limits to this process. As the species rejects more and more kinds of patch, in favor of the most productive kind, it eventually begins to lose an appreciable amount of time in travelling. Also, as it restricts the patches in which it feeds, it may automatically eliminate some kinds of food and thus alter its diet. At this point, further habitat restriction is unlikely. The relationships can be simply visualized in terms of species packing and unpacking, as shown in Figure 33.

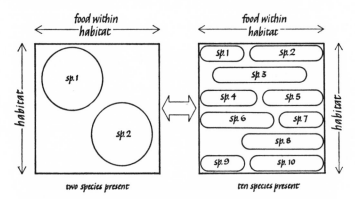

FIGURE 33. The compression hypothesis. As more species invade and are packed in, the occupied habitat shrinks (although some marginal patches may be added), but not the range of acceptable food items within the occupied habitat. The actual diet, reflecting both acceptability and abundance of items, may become more concentrated, but the range of items should not greatly change. Conversely, as species invade a species-poor island from a species-packed source, only the occupied habitat expands. The hypothesis applies only to short-term, non-evolutionary changes.

108

[*Editors' Note:* Material has been omitted at this point.]

REFERENCES

Cameron, W. A., 1958, Mammals of the islands in the Gulf of St. Lawrence. *Natl. Museum Canada, Bull.* 154.

MacArthur, R. H. and E. Pianka, 1966, On optimal use of a patchy environment. *Am. Naturalist,* 100:603–609.

Willis, E. O., 1966, Interspecific competition and the foraging behavior of plain-brown woodcreepers. *Ecology,* 47:667–672.

Wilson, E. O. and G. L. Hunt, Jr., 1967, The ant fauna of Futuna and the Wallis Islands, stepping stones to Polynesia. *Pacific Insects* (in press).

Wilson, E. O. and R. W. Taylor, 1967. An estimate of the potential evolutionary increase in species density in the Polynesian ant fauna. *Evolution* 21: 1–10.

Part VI

CASE HISTORIES

Editors' Comments
on Papers 35 Through 38

By now, it should be clear that our attitude regarding niche is a
functional one, tightly interwoven with the question of population regulation. The niche of the population is determined by and with those
factors which regulate the population. As such, we turn now to case
studies of species coexistence patterns, beginning with experimental
studies of community structure (Paper 35) and extending to field studies of natural populations (Papers 36, 37, and 38).

Vandermeer (Paper 35) and Wilbur (1972) use experimental manipulations to derive insights regarding the underlying interspecies interactions and hence the niche definitions. As did Ayala (1969), Ayala et
al. (1973), and Gilpin and Ayala (1973), they conclude by challenging
the gospel of the oversimplified Lotka–Volterra equations, arguing for
the importance of higher-order interactions. Wilbur argues that natural
species groupings are organized by interspecific interactions that can be
discovered and evaluated by experimentally dissecting the community.
The Vandermeer article is reprinted as Paper 35, and we also recom-
mend the reader's attention to the longer article of Wilbur (1972) that
we could not reprint here.

Robert MacArthur's many contributions to theory of niche and diversity include a classic study of niche difference in what seems a tick-
lish case—the several species of warblers feeding on insects in the same
trees in a spruce forest (Paper 36). Niche differences among these spec-

ies do appear. Two species—the baybreasted warbler and the Cape May warbler—are fluctuant or relatively opportunistic, compared with the others. The other three, more stable, species differ in their behavior and use of the feeding space in the crowns of the trees. The study thus illustrates the subtlety of the ways the principle of competitive exclusion, or of Volterra and Gause, may apply to real species in communities. A point of note is that the niches do overlap, and yet differ in their frequency distributions in time or in space within the spruce trees. The niches are thus not characterized by boundaries of exclusion; the differences between species are statistical (p. 188). Further studies of foraging relationships include Morse (1971) on island warblers and Stallcup (1968) on nuthatches and woodpeckers, and Power (1971) on measurement of warbler species relationships.

We owe to Broadhead and Wapshere an intensive and incisive study of an insect guild—a group of species of closely related niche, in this case species of psocids. Psocids are bark lice, members of one of the less familiar insect orders (Corrodentia or Psocoptera) that vaguely resemble aphids or plant lice, but have chewing mouth parts. The species studied in Paper 37 feed on the film of algae and other microorganisms that develops on the bark of trees. Broadhead and Wapshere first consider niche differences—primarily time of development, and concentration on living versus dead branches—for the several species of the guild. They then report a most detailed analysis of niche similarities and differences, and possible control mechanisms, for two species. There are differences in relative preference for living and dead branches, in relative effects of parasites, and in habitat distributions; but the key difference appears to be one of choice of twig size for oviposition. The authors give an effective discussion of the Gause hypothesis and different possibilities of coexistence for species. From their long article we give their comments on general biological features of the two species (p. 328–329), and their discussion of the Gause hypothesis as it bears on these species (pp. 376–381).

Harry Recher discusses in Paper 38 a group of interacting bird species that seems larger than a guild—the full assemblage of shorebirds of an estuary along the California Coast. An estuary is a complex ecosystem, for not only the mixing of salt and fresh water and consequent salinity gradients, but for differences in substrate, in tide levels and duration of exposure, and in kinds of plant and animal communities related to these and other factors. Part of the differentiation among shorebirds can be based on use of different parts of estuaries; but Recher observes a number of directions of difference among these birds—in their times of arrival and use of the area, their patterns of use of the estuaries and other coastal environments, food size and kind, depth of food in water

331

and mud, and relation to tide levels. He suggests that competition between young and adults is reduced by difference in time of migration. Recher is thus describing many-factored differentiation among these birds in not only niche but habitat. He is characterizing the birds of this group as occupants of different ecotopes, as we term the combination of niche and habitat factors in the last article (Part VII). For a recent, detailed study of six shorebird species, see also Baker and Baker (1973).

REFERENCES

Ayala, F. J. 1969. Experimental invalidation of the principle of competitive exclusion. *Nature* 224: 1076–1079.

Ayala, F. J., M. E. Gilpin, and J. G. Ehrenfeld. 1973. Competition between species: theoretical models and experimental tests. *Theoret. Pop. Biol.* 4: 331–356.

Baker, M. C., and A. E. M. Baker. 1973. Niche relationships among six species of shorebirds on their wintering and breeding ranges. *Ecol. Monogr.* 43: 193–212.

Gilpin, M. E., and F. J. Ayala. 1973. Global models of growth and competiiton. *Proc. Natl. Acad. Sci. USA* 70: 3590–3593.

Morse, D. H. 1971. The foraging of warblers isolated on small islands. *Ecology* 52: 216–228.

Power, D. M. 1971. Warbler ecology: diversity, similarity, and seasonal differences in habitat segregation. *Ecology* 52: 434–443.

Stallcup, P. L. 1968. Spatio-temporal relationships of nuthatches and woodpeckers in ponderosa pine forests of Colorado. *Ecology* 49: 831–843.

Wilbur, H. M. 1972. Competition, predation, and the structure of the *Amblystoma-Rana sylvatica* community. *Ecology* 53: 3–21.

35

Reprinted from *Ecology*, **50**(3), 362–371 (1969), with permission of the publisher, Duke University Press, Durham, N. C.

THE COMPETITIVE STRUCTURE OF COMMUNITIES: AN EXPERIMENTAL APPROACH WITH PROTOZOA

John H. Vandermeer[1]

Department of Zoology, University of Michigan, Ann Arbor, Michigan

(Received November 25, 1968; accepted for publication January 31, 1969)

Abstract. An empirical test of the existence of higher order interactions was carried out using four ciliate protozoans, *Paramecium caudatum*, *P. bursaria*, *P. aurelia*, and *Blepharisma* sp.

All four ciliates were cultured individually, and their population histories were described quite well by the simple logistic equation.

Attempts to explain minor deviations of the data from the logistic by use of the one or two time lag logistic failed. A more complicated time lag phenomenon must be operative.

Every possible pair of the four ciliates was cultured and a trial and error procedure was used to estimate α and β of the Gause equations. In all cases the simple Gause equations seemed adequately to describe the data.

All four ciliates were cultured together and compared to predictions made by use of the competition coefficients estimated from pair-wise competition and population parameters estimated from single species population growth. The correspondence between prediction and data suggests that the higher order interactions have slight or no effect on the dynamics of this artificial community.

The central goal of community ecology is to understand the dynamics of community organization. Most likely that goal would be best approached from the point of view of mechanism, beginning with basic principles and deducing how communities should behave. However, such attempts have in the past been stifled by certain methodological difficulties, diverting the attention of community ecologists toward more empirical observations. These empirical observations have usually taken the form of measuring the relative abundance of species (number of species with a given number of individuals).

Most of the work concerned with the latter may be dichotomized as follows. One series of papers (Fisher, Corbet, and Williams 1943; Preston 1948, 1962a, b; Hairston and Beyers 1954; Hairston 1959) emphasizes underlying mathematical distributions and causes of deviance from these distributions. Another series, stimulated by Margalef (1957), and rigorously formalized by Pielou (1966a, b), is concerned with measuring the relative abundance of species, usually to compare communities to each other.

[1] Present address: Department of Biology, University of Chicago, 1103 E. 57th Street, Chicago, Illinois 60637.

There have also been some attempts at building either conceptual or mathematical models of communities using supposedly basic components. Kendall (1948) has shown how certain patterns of population growth will lead to Fisher's logarithmic series. MacArthur (1955) discussed several aspects of community stability as deduced from considerations of energy transfer among populations, and Watt (1964) later tested some of these ideas. MacArthur (1960) suggested mechanisms whereby lognormal distributions should arise. Hairston, Smith, and Slobodkin (1960) presented a general explanation for the structure of terrestrial communities which stimulated a recent controversy (Murdoch 1966, Ehrlich and Birch 1967, Slobodkin, Smith, and Hairston 1967). Garfinkel (1967) studied stability properties of theoretical systems using the Lotka-Volterra predator-prey equations.

These approaches to community ecology have established the basic attitude for the following empirical tests. Simple assumptions are made,

namely that populations grow according to the logistic equation, and populations on the same trophic level obey the Gause equations of competition. I shall be concerned with communities which derive their structure through inter- and intraspecies competition.

The basic Gause competition equations may be extended to include m species. The differential equation for the ith species is,

$$\frac{dN_i}{dt} = \frac{r_i N_i}{K_i} \left\{ K_i - N_i - \sum_{j=1}^{m} \alpha_{ij} N_j \right\} \quad j \neq i \tag{1}$$

(MacArthur and Levins 1967), where there are m species in the community, α_{ij} is the effect of the jth species on the ith species, r is the intrinsic rate of natural increase, N is number of individuals and K is saturation density. Assumed in the above extension is that there are no higher order interaction terms. That is, we need not write,

$$\frac{dN_i}{dt} = \frac{r_i N_i}{K_i} \left\{ K_i - N_i - \sum \alpha_{ij} N_j - \sum_{jk} \beta_{ijk} N_j N_k - \ldots - \sum_{jk\ldots m} \omega_{ijk\ldots m} N_j N_k \ldots N_m \right\} \tag{2}$$

where $j \neq i, k \neq i, \ldots, m \neq i,$ and the Greek letters are interaction coefficients of increasingly higher order, e.g. β_{ijk} is the joint effect of species j and k on the ith species. Communities represented by equations (1) will be called noninteractive communities, and communities represented by equations (2) will be called interactive communities.

Equations (1) (noninteractive communities) are fairly tractable and have already led to some rather interesting conclusions about community structure (Levins 1968, Vandermeer 1968). On the other hand, equations (2) (interactive communities) are difficult to work with analytically.

The qualitative effects of higher order interactions are made clear in the following discussion. Consider the competitive effect, on one particular species (say species A), of adding further species, holding the average α constant and assuming the higher order terms to be negligible. The mean α is held constant in such a way that each new species overlaps with 25% of species A. If another species is added, species A only occupies 75% of its former niche without competitive pressure. If another species is added only 50% of its niche is free of competition. In general, as competitors accumulate in the system, if the average competition coefficient remains constant, the weight of competition eventually gets so strong that other species cannot make their way into the community.

Now suppose the three way interactions (the β's in equation (2)) are important. Considering again species A, after the addition of another species, as before only 75% of the former niche is occupied without competition. If an additional species is added, part of the niche overlap will be contained in a three way overlap. Thus, instead of having the niche reduced to 50% of its former size, something greater than 50% but less than 75% of the niche remains free of competition (since a portion of the potential competition is absorbed in the three way interaction).

For instance, consider a hypothetical organism living in a stream. Suppose species A occupies the entire length of the stream. Suppose further that experiments are undertaken in which species B and C are added one at a time. If B and C each have X effect on species A, we would expect that the combination of B and C together would have effect 2X. But suppose that the result of B and C competing with A is that in either case A is eliminated only from the upstream section of the stream. It does not matter to species A which of B or C does the eliminating; the fact is, it is being eliminated. Thus, the predicted effect of B and C together on A would be 2X but the observed effect would be X.

The type of higher order interaction discussed above implies $\beta, \gamma, \ldots, \omega$ of equation 4 are positive. If, on the other hand, the higher order terms were negative, it would imply a coalition formed by two or more species against some other species. The higher order interaction reported by Hairston, et al. (1969) is an example of a coalition type interaction.

Thus, it is not possible at the present time to predict the general effect of higher order interactions except qualitatively. However, it is clear that their effects may be highly significant. It, therefore, would be quite important to discover whether or not they commonly occur in nature.

The motivation for undertaking the experiments reported below derives from the foregoing discussion of interactive communities. Since several interesting consequences are deducible from the premises of noninteractive communities, it would be desirable to know either the importance of higher order interactions in real situations, or the theoretical effect which higher order interactions would have on the consequences derived from the noninteractive assumptions. As remarked earlier, it is virtually certain that higher order interactions will have an effect on the outcomes of the model, but the nature and extent of this effect are difficult to assess owing to the intractable nature of the model when applied to interactive communities. Thus, the simplest approach is to verify the presence or absence of higher order interactions in nature, and this is the approach of this paper.

Induction in biology—the implied approach suggested in the above discussion—has often been eased by first studying extremes. If the existence of higher order interactions can be demonstrated for a group of species which would be expected on an *a priori* basis not to exhibit them, it may be safely assumed that such interactions are more or less of universal occurrence. If that group of species does not exhibit higher order interactions, the broad implications for which the experiment was designed are lost.

Obtaining an extreme example and satisfying the basic postulates of the model have been the two guidelines in choosing experimental subjects and situations. Different species of free-living ciliates from different localities seem to satisfy both of these guidelines. Ciliates are simple organisms from the population point of view, and in several cases are known to conform to simple population mathematics, making it likely that most of the basic premises of the model are satisfied.

Therefore, four species of ciliates from four different localities were studied in a laboratory situation. Parameters of population growth were estimated for each species separately. Competition coefficients were estimated for each possible pairwise combination of species. Then all four species were cultured together and compared to the prediction made on the basis of the simple growth

parameters and competition coefficients measured earlier.

METHODS AND MATERIALS

Three of the four ciliates, *Paramecium bursaria* (PB), *Paramecium aurelia* (PA), and *Paramecium caudatum* (PC) were obtained from three different localities in the vicinity of Ann Arbor, Michigan. The fourth species *Blepharisma* sp. (BL), was obtained from a biological supply house. The Varieties are unknown except for PA which is Variety 3.

Culture techniques are those of Sonneborn (1950), with various modifications after Hairston and Kellerman (1965). Cultures were maintained at 15°C and all experiments were done at 25°C. All cultures and experiments were kept in darkness—i.e., PB was without effective symbiotic algae.

Experiments were done in 10-ml test tubes, filled with 5 ml of culture. Each day a sample of 0.5 ml was rapidly removed with a coarse pipette after the test tube had been vigorously shaken. The tubes were shaken after a section of rubber glove had been sterilized in hot distilled water and positioned as a covering for the test tube. Verification of the test tube sampling procedure was accomplished by repeated sampling and subsequent replacement of samples in several groups of test tubes. The sampling procedure was extensively experimented with to obtain a reasonably close correspondence between the mean and variance of a set of repeated samples; i.e., to assure a random sampling technique.

Bacterized culture medium was added to each culture to replace exactly the 0.5 ml which had been removed as a sample. The culture medium had been aged for various lengths of time and was used in an orderly sequence based on this age. This was originally done to introduce a forced environmental oscillation on the experiments. Since absolutely no relationship between food age (food quality) and population growth or competition was observed anywhere in the study, the food source is considered as a constant in this paper.

POPULATION GROWTH

The mathematical model used for the simple population growth experiments was the familiar logistic equation, $dN/dt = (rN/K)(K - N)$, where N is the number of individuals, K is saturation population density, and r is the intrinsic rate of natural increase. The parameters K, r, and N_0 were estimated exactly as in Gause (1934) excepting the case of PC in which several points were discarded as wild points for purposes of estimating the parameters—not in plotting the figures. In all cases the intercept of the equation $\ln[(K-N)/N] = a + rt$ was used to estimate N_o, which was needed to generate the expected curve. Expected curves were obtained from the recurrence relation given by Leslie (1957),

$$N_{t+1} = \lambda N_t/(1 - bN_t)$$

where $\lambda = e^r$ and $b = (\lambda - 1)/K$.

Statistical testing of the fit of the data to the model is at best difficult (Smith 1952). Leslie's (1957) procedure depends on the assumption of independence of observations from one time to the next. It was obvious by inspection that there was a rather high correlation between times within replicates—i.e., that replicate which had the highest value at time t was most likely to be the replicate with the highest value at time $t + 1$. Because of this statistical difficulty, and the difficulties raised by Smith (1952), the statistical testing of Leslie (1957) was not used.

Figure 1 presents the expected and observed values for population growth of BL. The fit is

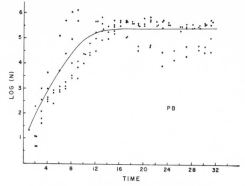

Fig. 2. Population growth of PB.

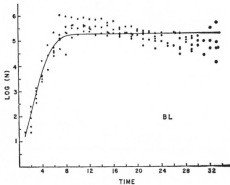

Fig. 1. Population growth of BL.

generally good, except that a systematic deviation of observed from expected is notable. This deviation is biased above the curve at early times, and below the curve at later times. The pattern is reminiscent of that obtained when time lags are added to the logistic equation (Cunningham 1954).

Using the equation of Wangersky and Cunningham (1956), the value of the reproductive time lag τ_1, and density time lag τ_2 were estimated. Systematic variation in the value of these two parameters provided no improvement of fit. If the deviations of fit are due to a time delay, the relationship is more complicated than that described by Wangersky and Cunningham's equation.

In Figure 2 are shown expected and observed values for PB. Of the four species this is the

most highly variable in its population growth. The relatively large variance in observed values is understood by noting that two of the replicates are suggestive of a very large time lag and three are suggestive of a smaller, if at all existent, time lag. The curve predicted by the logistic (Fig. 2) seems to be an adequate representation of the basic trend of population growth in this species.

The expected and observed values for PC and PA are shown in Figures 3 and 4. The fits are obviously excellent and need no further discussion here.

The parameters for population growth for all four species are summarized in Table 1.

TWO-WAY COMPETITION

The mathematical methods used here are even more inexact than those used in single species population growth. A smooth curve was drawn by eye through the observed data points, and estimates of α and β were obtained for each time from 2 to 32, with the equations given by Gause (1934),

$$\alpha = \frac{1}{N_2}\left[K_1 - \frac{(dN_1/dt)K_1}{r_1N_1} - N_1\right]$$

$$\beta = \frac{1}{N_1}\left[K_2 - \frac{(dN_2/dt)K_2}{r_2N_2} - N_2\right]$$

The K's and r's were obtained from the single species experiments. The N's were taken off of the smooth curve, and the derivatives were estimated as

$$\frac{dN_t}{dt} = \frac{N_{t+1} - N_{t-1}}{2}$$

The resultant values of α and β—31 values of each for each of the six competition experiments—were highly variable within experiments. The values were exceptionally extreme at very early times and very late times. Intermediate values

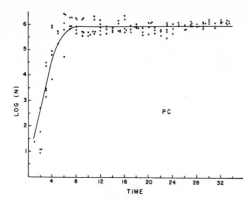

FIG. 3. Population growth of PC.

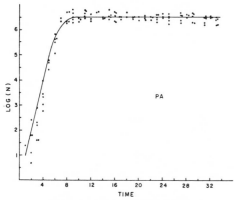

FIG. 4. Population growth of PA.

TABLE 1. Parameters of the logistic equation

Species	r	K	N_o
PA	1.05	671	2.5
PC	1.07	366	5.0
PB	0.47	230	5.0
BL	0.91	194	3.0

were somewhat less variable, but the variation seemed always to form some sort of trend. However, under the hypothesis of this paper, α and β are defined as constants so the objective is to obtain those values of α and β which produce the curve which best fits the observed data.

To obtain a reasonable first approximation to α and β I took the average of the intermediate values as described above—i.e., the average of those values that tended to form a reasonably invariant clump. With these values, integral curves were computed using the predictor corrector method of Hamming (Ralston and Wilf 1960; see also System/360 Scientific Subroutine Pack-

age 1968). Usually, these first approximations gave curves that rather poorly represented the observed data. A trial and error procedure was then pursued in which α and β were systematically varied and the behavior of the integral curves investigated under this variation. It eventually became apparent that the curves obtained were about the best obtainable under the present model. In all cases the boundary conditions were considered as random variables, and were thus relatively free to vary in the trial and error procedure.

The comments about statistical testing made in the previous section apply equally well here. However, a relative measure of the goodness of fit is needed for a later section, and is introduced here to facilitate easy comparison of the goodness of fit of one experiment to that of another. The

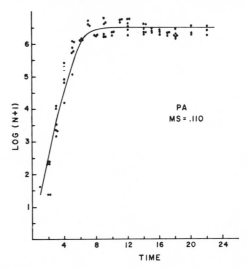

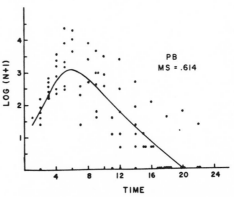

FIG. 5. Results of competition between PA and PB.

arbitrary criterion used is mean square deviation of logs of observed from log of expected.

The results of the six pairwise competition experiments are presented in Figures 5–10, in order of decreasing mean square deviation. The aforementioned variability of PB is reflected in Figure 5, but appears to be somewhat dampened by the presence of BL (Fig. 9). Figures 6 and 7 both show some indication of a time lag, but since the population growth data for these two species (PA and PC) alone did not suggest any frictional components, one must presume that the presence of a competitor is the factor which initiates a nonlinear response. However, the difference between the linear and nonlinear response in this case appears to be negligible.

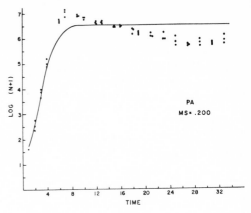

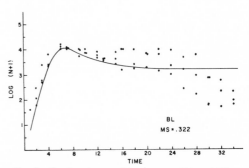

FIG. 7. Results of competition between PA and BL.

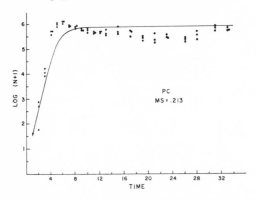

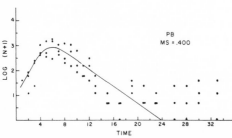

FIG. 6. Results of competition between PC and PB.

Thus, the six cases of pairwise competition are represented quite well by the basic theory, minor deviations being interpretable as a result of unknown but insignificant frictional components or excessive inherent variability. Since at least the general trend of the data is in agreement with that of the model, these small deviations most likely will not interfere with the basic hypothesis to be tested.

In Table 2 is presented the community matrix for this artificial community. The α_{ij}th entry refers to the effect of the jth species on the ith species.

FOUR-WAY COMPETITION

Having estimates of the intrinsic rate of natural increase and saturation density for each species, and estimates of all interaction coefficients on a pairwise basis, it is now necessary to substitute into equations (1)—four-way competition without higher order interactions—solve the equations, and see if the solution agrees with the data obtained from the laboratory experiments. It must be emphasized that up to this point all curves in this paper have been fit. That is, the expected curves are based on the data which they are describing and therefore cannot deviate too greatly. In the case of four-way competition, however, the expected curves are based on independent estimates of the parameters, estimates which were made from different experiments. As a consequence a reasonable expectation, if higher order

TABLE 2. Community matrix for four species of protozoa in test tubes

ith Species	jth Species			
	PA	PC	PB	BL
PA	1.00	1.75	-2.00	-0.65
PC	0.30	1.00	0.50	0.60
PB	0.50	0.85	1.00	0.50
BL	0.25	0.60	-0.50	1.00

interactions are not important, is that the fit to the four-way situation is, at best, as good as the fits to the pair-wise competitions. Though it cannot be expected that the four-way competition experiments will fit the model as well as the two-way competition, it is not clear exactly how "good" a fit is necessary to judge a model as being valid. Such a judgment, of course, depends on the purposes for which the model is to be used. In the present case it seems that there are three vaguely distinquishable levels of purpose which might be used in helping to judge the validity of this model. First, if only the equilibrium conditions of the model are needed—say in predicting the maximum number of species in an equilibrium community— we might only require that the model predict accurately which species will persist and which will

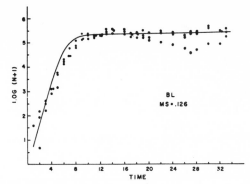

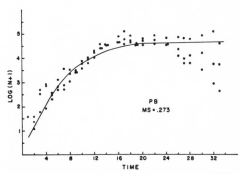

FIG. 9. Results of competition between BL and PB.

go to extinction. Second, the relative numbers of individuals at any particular point in time may be desired—say, in a theoretical study of the relative abundance of species—or, more or less equivalently, the general behavior of the populations within the framework of the larger system must be known—i.e., does species i oscillate, does it tend to be peaked and then drop off fast, and similar general questions. Under this requirement, we accept the model as valid if the general shape of the curves reflects the general trend of the data. Third, we may wish to make very precise predictions with the model such as exactly how much pesticide must be applied at time x to produce y per cent change in species i by time $x + a$. At this level we require the fit to four-way competition to be almost as good as the fit to the two-way competitions. Certainly these three levels are not discrete, but are merely useful points in a continuum, defined so as to be able to judge the relative validity of the model as applied to a simple community.

In Figures 11–14 are shown the results of simulating four-way competition with the community

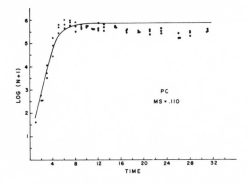

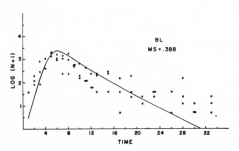

FIG. 8. Results of competition between PC and BL.

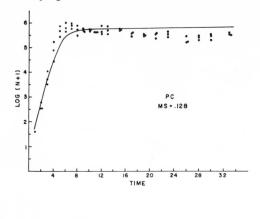

PC
MS = .128

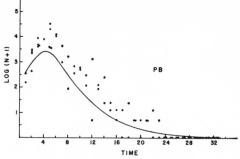

FIG. 12. PB in four-way competition.

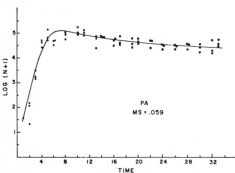

PA
MS = .059

FIG. 10. Results of competition between PC and PA.

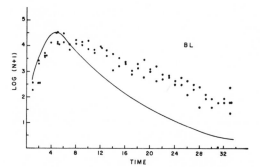

FIG. 13. PC in four-way competition.

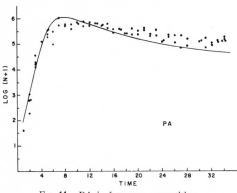

PA

FIG. 11. PA in four-way competition.

FIG. 14. BL in four-way competition.

matrix of the last section and the r's and K's from the single species population growth experiments. Also shown are the observed data from the four-way competition experiments.

Under the first criterion of validation—only the final outcome is desired—the data are described perfectly by the model. The model predicts that PB will be extinct; BL will be at a very low popu-

lation density; PA and PC will have higher densities. As can be easily seen from the figures, the data correspond exactly to these predictions.

Likewise under the second criterion the model is validated. The general trends are predicted in every case, and the rank order of abundance is as predicted, at virtually all points in time. The general trends of growth, peaking, and decay are predicted fairly accurately. The early peak for PC reflected in the experiments is essentially the same as that species exhibits in two-way competition (see Fig. 6, 8, and 10).

Finally, consider the third level of validation. Somewhere in between the second and third level the validation becomes questionable. The third level requires that precise predictions be made. The three species PC, PA and PB conform quite well, but BL seems to deviate excessively from the expected. However, as discussed previously, it is not sufficient to examine the data and expecttation alone, but they must be considered in the light of the pair-wise experiments, since the correspondence between expected and observed can only be as good as that of the basic model. It is for this reason that the mean square deviations were computed earlier (Fig. 5–10). In Table 3 are listed the pooled mean squares for each species—pooled over all two-way competitions in which that species was a part—and the mean

TABLE 3. Mean squares for two-way and four-way competition experiments, and ratio between the two (taking largest mean square as numerator). Numbers in italics are degrees of freedom

statistic	Species			
	PA	PC	PB	BL
Pooled mean square for 2-way competition	0.11 *212*	0.15 *197*	0.43 *230*	0.26 *209*
Mean square for 4-way competition	0.10 *74*	0.22 *74*	0.36 *74*	1.28 *74*
Ratio	1.1	1.5	1.2	4.9

squares for the same species in four-way competition. The first three species show very similar mean squares between two-way and four-way competition; indeed PA and PB are even better predicted in the case of four-way than in the case of two-way competition. The case of BL is somewhat less encouraging. The mean square is roughly five times greater in four-way than in two-way. The fact that the experimental points are always greater than predicted, implies that whatever higher order interaction there may be, they act solely to decrease the competitive effect felt by BL. They are noncoalition type higher order interactions.

However, it might be argued that if one is interested in this model from the standpoint of the community, it is not really valid to compare predictions on a specific basis. We should instead be interested in the general performance of the model in predicting community dynamics. To this end the sum of squares pooled over all two-way competition cases was computed as 0.246 and the

same pooled over all four-way cases was 0.487, only a two-fold difference. Thus, the four-way model predicted about half as well as the two-way model, implying that, from the point of view of the community as a whole, the higher order interactions are rather unimportant.

The above discussion is similar to the problem of multiple comparisons tests in statistics. Given a group of means, we wish to test for differences. In setting an error rate we are faced with deciding on the relative importance of pair-wise or experiment-wise rates. That is, is it more important to minimize the probability of making any mistake at all, or the probability that a given pair of means may be judged different when they are truly the same. Similarly in the present context, we are faced with deciding whether we want the model to make predictions about the community as a whole, or about the components of the community. The former criterion provides a single judgment of the model. The latter provides as many judgments as there are components.

Thus, if we are concerned with judging the present model under the third criterion—absolutely precise predictions—and if we are interested in the individual components, the noninteractive model is "good" three times and "bad" once. On the other hand, it appears to be simply "good" on the level of the whole community.

CONCLUSION

The conclusions of these experiments were basically stated in the previous section. However, it might be well at this point to recall statements made earlier about the choice of the experimental system. It was the author's prejudice that higher order interactions are very important in nature. Thus, the experimental system was chosen to maximize the chance of not finding higher order interactions, hoping, of course, that even then such interactions would be significant. Then, because of the way the experimental system was put together, one could make a fairly powerful statement about the universal occurrence of higher order interactions in nature. Unfortunately, the basic models proved to be excellent predictors (i.e., higher order interactions did not seem important), and one can only cite this study as a single example of the insignificance of higher order interactions.

ACKNOWLEDGMENTS

I am deeply indebted to Nelson G. Hairston who provided guidance and encouragement throughout the course of this work. I wish to thank James T. MacFadden, Lawrence B. Slobodkin, and Donald W. Tinkle for constructive comments and criticisms. I also am thankful to Douglas J. Futuyma for much valuable advice, and to my wife Jean for assistance in tabulating data and typing the manuscript. Computer time was granted by the University of Michigan Computation Center. This research was supported by an NSF summer fellowship for graduate teaching assistants, and an NIH predoctoral fellowship.

Literature Cited

Cunningham, W. J. 1954. A nonlinear differential-difference equation of growth. Proc. Nat. Acad. Sci. **40**: 708–713.

Ehrlich, P. R. and L. C. Birch. 1967. The "balance of nature" and "population control." Amer. Naturalist **101**: 97–107.

Fisher, R. A., A. S. Corbet, and C. B. Williams. 1943. The relation between the number of species and the number of individuals in a random sample from an animal population. J. Animal Ecol. **12**: 42–58.

Garfinkel, D. 1967. A simulation study of the effect on simple ecological systems of making rate of increase of population density-dependent. J. Theoret. Biol. **14**: 46–58.

Gause, G. F. 1934. The struggle for existence. Williams and Wilkins, Baltimore.

Hairston, N. G. 1959. Species abundance and community organization. Ecology **40**: 404–416.

Hairston, N. G. and G. W. Byers. 1954. The soil arthropods of a field in southern Michigan. A study in community ecology. Contrib. Lab. Vert. Biol., Univ. Mich. **64**: 1–37.

Hairston, N. G., F. E. Smith, and L. B. Slobodkin. 1960. Community structure, population control, and competition. Amer. Naturalist **94**: 421–425.

Hairston, N. G. and S. L. Kellermann. 1965. Competition between varieties 2 and 3 of *Paramecium aurelia:* the influence of temperature in a food limited system. Ecology **46**: 134–139.

Hairston, N. G., J. D. Allan, R. K. Colwell, D. J. Futuyma, J. Howell, J. D. Mathias, and J. H. Vandermeer. 1969. The relationship between species diversity and stability: an experimental approach with protozoa and bacteria. Ecology **49**: 1091–1101.

International Business Machines, 1968; System/360 scientific subroutine package (360A-CM-03X) Version III Programmer's manual. IBM Technical Publications Department, White Plains, N. Y.

Kendall, D. G. 1948. On some modes of population growth leading to R. A. Fisher's logarithmic series distribution. Biometrika **35**: 6–15.

Leslie, P. H. 1957. An analysis of the data for some experiments carried out by Gause with populations of the Protozoa, *Paramecium aurelia* and *Paramecium caudatum.* Biometrika **44**: 314–327.

Levins, R. 1968. Evolution in changing environments. Some theoretical explorations. Monographs in population biology, Princeton Univ. Press.

MacArthur, R. H. 1955. Fluctuations of animal populations, and a measure of community stability. Ecology **36**: 533–536.

——— 1960. On the relative abundance of species. Amer. Naturalist **94**: 25–36.

MacArthur, R. H. and R. Levins. 1967. The limiting similarity, convergence, and divergence of coexisting species. Amer. Naturalist **101**: 377–385.

Margalef, R. 1957. Information theory in ecology. (English trans. by Hall, W.) Gen. Systems **3**: 36–71.

Murdoch, W. W. 1966. Community structure, population control, and competition—a critique. Amer. Naturalist **100**: 219–226.

Pielou, E. C. 1966a. Species-diversity and pattern-diversity in the study of ecological succession. J. Theor. Biol. **10**: 370–383.

———. 1966b. The measurement of diversity in different types of biological collections. J. Theor. Biol. **13**: 131–144.

Preston, F. W. 1948. The commonness, and rarity, of species. Ecology **29**: 254–283.

———. 1962a. The canonical distribution of commonness and rarity. Ecology **43**: 185–215.

———. 1962b. The canonical distribution of commonness and rarity. Part II. Ecology **43**: 410–432.

Ralston, A. and H. S. Wilf. 1960. Mathematical methods for digital computers. Wiley, New York.

Slobodkin, L. B., F. E. Smith, and N. G. Hairston. 1967. Regulation in terrestrial ecosystems, and the implied balance of nature. Amer. Naturalist **101**: 109–124.

Smith, F. E. 1952. Experimental methods in population dynamics: a critique. Ecology **33**: 441–450.

Sonneborn, T. M. 1950. Methods in the general biology and genetics of *Paramecium aurelia.* J. Exper. Zool. **113**: 87–147.

Vandermeer, J. H. 1968. The structure of communities as determined by competitive interactions: a theoretical and experimental approach. Ph.D. thesis, Univ. Mich., Ann Arbor.

Wangersky, P. J. and W. J. Cunningham. 1956. On time lags in equations of growth. Proc. Nat. Acad. Sci. Wash. **42**: 699–702.

Watt, K. E. F. 1964. Comments on fluctuations of animal populations and measures of community stability. Can. Entomol. **96**: 1434–1442.

36

Reprinted from *Ecology*, 39(4), 599–612, 617–619 (1958)

POPULATION ECOLOGY OF SOME WARBLERS OF NORTHEASTERN CONIFEROUS FORESTS[1]

ROBERT H. MACARTHUR

Department of Zoology, University of Pennsylvania

INTRODUCTION

Five species of warbler, Cape May (*Dendroica tigrina*), myrtle (*D. coronata*), black-throated green (*D. virens*), blackburnian (*D. fusca*), and bay-breasted (*D. castanea*), are sometimes found together in the breeding season in relatively homogeneous mature boreal forests. These species are congeneric, have roughly similar sizes and shapes, and all are mainly insectivorous. They are so similar in general ecological preference, at least during years of abundant food supply, that ecologists studying them have concluded that any differences in the species' requirements must be quite obscure (Kendeigh, 1947; Stewart and Aldrich, 1952). Thus it appeared that these species might provide an interesting exception to the general rule that species either are limited by different factors or differ in habitat or range (Lack, 1954). Accordingly, this study was undertaken with the aim of determining the factors controlling the species' bundances and preventing all but one from being exterminated by competition.

LOGICAL NATURE OF POPULATION CONTROL

Animal populations may be regulated by two types of events. The first type occurs (but need not exert its effect) independently of the density of the population. Examples are catastrophes

[1] A Dissertation Presented to the Faculty of the Graduate School of Yale University in Candidacy for the Degree of Doctor of Philosophy, 1957.

such as storms, severe winters, some predation, and some disease. The second type of event depends upon the density of the population for both its occurrence and strength. Examples are shortages of food and nesting holes. Both types seem to be important for all well-studied species. The first kind will be called density independent and the second density dependent. This is slightly different from the usual definitions of these terms which require the effects upon the population and not the occurrence to be density independent or dependent (Andrewartha and Birch 1954).

When density dependent events play a major role in regulating abundance, interspecific relations are also important, for the presence of an individual of another species may have some of the effects of an individual of the original density dependent species. This is clearly illustrated by the generalized habitats of the few species of passerine birds of Bermuda contrasted with their specialized habitats in continental North America where many additional species are also present (Bourne 1957).

If the species' requirements are sufficiently similar, the proposition of Volterra (1926) and Gause (1934), first enunciated by Grinnell (1922), suggests that only one will be able to persist, so that the existence of one species may even control the presence or absence of another. Because of this proposition it has become customary for ecologists to look for differences in food or habitat of related species; such differences, if found, are then cited as the reason competition is not eliminating all but one of the species. Unfortunately, however, differences in food and space requirements are neither always necessary nor always sufficient to prevent competition and permit coexistence. Actually, to permit coexistence it seems necessary that each species, when very abundant, should inhibit its own further increase more than it inhibits the other's. This is illustrated in Figure 1. In this figure, the populations of the two species form the coordinates so that any point in the plane represents a population for each species. Each shaded area covers the points (*i.e.*, the sets of combined populations of the species) in which the species corresponding to the shading can increase, within a given environment. Thus, in the doubly-shaded area both species increase and in the unshaded area both species decrease. The arrows, representing the direction of population change, must then be as shown in the figure for these regions. In order that a stable equilibrium of the two species should exist, the arrows in the singly shaded regions obviously must also be as in the figure; an interchange of the species represented by the shading would reverse the directions of these arrows resulting in a situation in which only one species could persist. Thus, for stability, the boundaries of the shaded zones of increase must have the relative slope illustrated in the figure with each species inhibiting its own further increase more than the other's. The easiest way for this to happen would be to have each species' population limited by a slightly different factor. It is these different limiting factors which are the principal problem in an investigation of multispecific animal populations regulated by density dependent events.

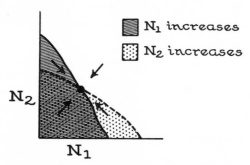

FIG. 1. The necessary conditions for a stable equilibrium of two species. The coordinate axes represent the populations of the species.

An example which has not received sufficient attention is competition in a heterogeneous environment. As has been often pointed out (Kluijver and Tinbergen 1953, Lack 1955, Hinde 1956) birds may emigrate or disperse from the most suitable areas where reproduction is successful into marginal habitats. Consider such a species which will be called A. Let B be a species that lives only in the area that is marginal for species A. Now, even if in an unlimited environment of this type, species B would eliminate species A by competition, in the heterogeneous environment species B may be eliminated from its own preferred habitat. For, if there is sufficient dispersal by species A, it may maintain, partly by immigration, such a high population in the marginal habitat that species B is forced to decrease. This process is probably very important in considering the environmental distributions of birds and implies that small areas of habitat typical for one species may not contain that species.

The study of limiting factors in nature is very difficult because ideally it requires changing the amount of the factor alone and observing whether this change affects the size of the population. Theoretically, if more than one factor changes, the analysis can still be performed, but in practice, if more changes of known nature occur, more of an unknown nature usually also occur. Limiting factors have been studied in two ways. The best way is artificially to modify single factors in the environment, observing the effect upon the birds. MacKenzie (1946) reviews some experiments of this type. The most notable was the increase from zero to abundant of pied flycatchers (*Muscicapa hypoleuca*) when nest boxes were introduced in the Forest of Dean. This showed conclusively that lack of nesting sites had limited the population. Such simple modifications are not always feasible. For instance, changing the food supply of an insectivorous bird is nearly impossible. The most feasible approach in such a case is to compare the bird populations in two regions which differ in the abundance of the factor being considered. Ideally, the two regions should differ only in this respect, but this is very improbable. A good example of this method of study is the work of Breckenridge (1956) which showed that the least flycatchers (*Empidonax minimus*) were more abundant in a given wood wherever the wood was more open.

The present study of the factors limiting warblers was conducted by the second approach. This is slightly less accurate than the first method, but permits studying more factors and requires less time. There are actually four parts to the study. First, it is shown that density dependent events play a large role in controlling the populations of the species. Second, a discussion of the general ecology of the species (food, feeding zones, feeding behavior, territoriality, predators, and mortality) is presented. The observations were made in the summers of 1956 and 1957. Third, the habits of the different species in different seasons are compared to see what aspects of the general ecology are invariant and hence characteristic of the species. Some observations on the species' morphology are discussed in the light of these characteristics. This was the project of the fall and winter of 1956 and the spring of 1957. Finally, a wood-to-wood comparison of species abundances, relative to the important constituents of their niches as determined in the earlier stages, is presented. This work was done in the summers of 1956 and 1957.

Density Dependence

It is the aim of this section to demonstrate that the five species of warbler are primarily regulated by density dependent events, that is, that they increase when rare and decrease when common (relative to the supply of a limiting factor). The strongest argument for this is the correlation of abundances with limiting factors discussed later. However, to avoid any risk of circularity, an independent partial demonstration will now be given.

If density independent events do not occur randomly but have a periodic recurrence, then a population controlled by these events could undergo a regular oscillation nearly indistinguishable in form from that of a population regulated by density dependent events. The distinction can be made, however, by observing the effect of the presence of an ecologically similar species. Here it will first be shown that increases and decreases are not random; then an argument will be given which renders the density independent explanation improbable.

If increases, I, and decreases, D, occur randomly, the sequence of observed I's and D's would have random order. A run of I's (or D's) is a sequence (perhaps consisting of one element) of adjacent I's (or D's) which cannot be lengthened; *i.e.,* the total number of runs is always one greater than the number of changes from I to D or D to I. If more runs of I's or D's are observed than would be expected in a random sequence, then an increase makes the following change more likely to be a decrease and conversely. This is what would be expected on the hypothesis of density dependent events.

There have been very few extensive censuses of any of the five species of warbler that are studied here. The longest are reproduced below

from the data of Smith (1953) in Vermont, Williams (1950) in Ohio, and Cruickshank (1956) in Maine. The populations of myrtle, black-throated green, and blackburnian are listed; only those species are mentioned that are consistently present.

Myrtle
Maine   7 5 7 7 6 7 8 10 9 8 10 10 10
           8 10 7 10

Black-throated green
Ohio   3 3 5 3 4 3 2 4 0 1 – 3 3
           3 4 3 3 4 3
Vermont 7 7 7 6 8 5 6 6 7 8 8 6 8
           8 3 6 3
Maine   8 9 11 10 10 10 11 11 11 10 11 9 8
           9 8 9

Blackburnian
Vermont 11 11 10 7 8 7 7 10 10 8 5 6 3
           3 6 6 5
Maine   2 4 3 3 5 5 7 5 6 5 7 5 5
           7 6 6 3

The increases, I, and decreases, D, of these censuses, in order are as follows:

Myrtle
D I D I I I D D I D I D I

Black-throated green
I D I D D I D I I I D I D
D I D I I I D I D I D
I I D I D I D D I D I

Blackburnian
D D I D I D D I D I D
I D I I D I D I D I D D

In this form the data from different censuses are perfectly compatible and, since censuses end or begin with I or D in no particular pattern, all the censuses for a given species may be attached:

Myrtle
D I D I I I D D I D I D I   (10)

Black-throated green
I D I D D I D I I I D I D D I D

I I I D I D I D I I D I   (27)

D I D D I D I

Blackburnian
D D I D I D D I D I D I D I I D I

D I D I D D   (19)

Here the groups of letters underlined are runs and the number of runs is totalled in parentheses. From the tables of Swed and Eisenhart (1943), testing the one-sided hypothesis that there are no more runs than would be expected by chance, each of these shows a significantly large number of runs (the first less than 5% significance, the others less than 2.5%). That is, each species tends to decrease following an increase and to increase following a decrease, proving population control nonrandom. The mean periods of these fluctuations can easily be computed. For a run of increases followed by a run of decreases constitutes one oscillation. Thus, the periods of the oscillations of the three species are $13/5 = 2.6$, $70/27 = 2.6$, and $46/19 = 2.4$ years respectively. These fluctuations would require an unknown environmental cycle of period approximately 2.5 years if a regularly recurring density independent event were controlling the populations. Thus, from these data alone, it seems very probable that the three species (myrtle, black-throated green, blackburnian) are primarily regulated by density dependent events.

A species may be regulated by density dependent events and yet undergo dramatic changes in populations due to changes in the limiting factor itself. In this case tests by the theory of runs, used above, are likely to be useless. However, if a correlation can be made of the population with the environmental factor undergoing change, then not only can density dependence, *i. e.* existence of a limiting factor, be established, but also the nature of the limiting factor. For, if an increase in one environmental variable can be established, an experiment of the first type described above has been performed. That is, the habitat has been modified in one factor and a resulting change in bird population has been observed. Therefore, because the population changes, that one factor has been limiting.

This is apparently what happens in populations of Cape May and bay-breasted warblers. Kendeigh (1947), examining older material, established the fact that these species are abundant when there is an outbreak of *Choristoneura fumiferana* (Clem.), the spruce budworm. More recent information confirms this. To correlate with the fact (Greenbank 1956) that there have been continuously high budworm populations since 1909, there is the statement of Forbush (1929) that the Cape May warbler became more common about 1909, and the statement of Bond (pers. comm.) that the winter range of the species has been increasing in size. An outbreak of spruce budworms started in northern Maine in the late 1940's, and Stewart and Aldrich (1951) and Hensley and Cope (1951) studied the birds during 1950 and 1951. Cape May and bay-breasted warblers were among the commonest birds present, as in the earlier outbreaks, although both species were formerly not common in Maine (Knight 1908). The outbreak has continued through New Brunswick, where current bird studies (Cheshire 1954) indicate that bay-breasted is again the commonest bird, although

for unknown reasons the Cape May has not been observed.

In conclusion, it appears that all five species are primarily regulated by density dependent events, and that a limiting factor is food supply for bay-breasted and Cape May warblers.

GENERAL ECOLOGY

The density dependence tentatively concluded above implies that the presence of individuals of a species makes the environment less suitable for other individuals of that species. It would also be expected then that the presence of individuals of one species may make the environment less suitable for individuals of a different species. This is called interspecific competition. As mentioned above, this seems to mean that two sympatric species will have their populations limited by different factors so that each species inhibits its own population growth more than it inhibits that of the others. The factors inseparably bound to a species' persistence in a region are, then, its relation to other species and the presence of food, proper feeding zone, shelter from weather, and nesting sites (Andrewartha and Birch 1954, Grinnell 1914). In this section these factors as observed during the breeding season of the five species of warbler will be discussed.

The summer of 1956 was devoted to observations upon the four species, myrtle, black-throated green, blackburnian, and bay-breasted warblers, on their nesting grounds. The principal area studied was a 9.4 acre plot of mature white spruce (*Picea glauca*) on Bass Harbor Head, Mt. Desert Island, Hancock County, Maine. On 7 July 1956 the site of observations was changed to the town of Marlboro in Windham County, Vermont, where a red spruce (*P. rubens*) woodlot of comparable structure was studied. In the summer of 1957 more plots were studied. From 30 May until 5 June, eighteen plots of balsam fir (*Abies balsamea*), black spruce (*P. mariana*), and white spruce near Cross Lake, Long Lake, and Mud Lake in the vicinity of Guerette in Aroostook County, Maine, were studied. The remainder of the breeding season was spent on Mt. Desert Island, Maine, where five plots were censused. These will be described later.

Feeding Habits

Although food might be the factor for which birds compete, evidence presented later shows that differences in type of food between these closely related species result from differences in feeding behavior and position and that each species eats what food is obtainable within the characteristic feeding zone and by the characteristic manner of feeding. For this reason, differences between the species' feeding positions and behavior have been observed in detail.

For the purpose of describing the birds' feeding zone, the number of seconds each observed bird spent in each of 16 zones was recorded. (In the summer of 1956 the seconds were counted by saying "thousand and one, thousand and two, . . ." all subsequent timing was done by stop watch. When the stop watch became available, an attempt was made to calibrate the counted seconds. It was found that each counted second was approximately 1.25 true seconds.) The zones varied with height and position on branch as shown in Figure 2. The height zones were ten foot units measured from the top of the tree. Each branch could be divided into three zones, one of bare or lichen-covered base (B), a middle zone of old needles (M), and a terminal zone of new (less than 1.5 years old) needles or buds (T). Thus a measurement in zone T3 was an observation between 20 and 30 feet from the top of the tree and in the terminal part of the branch. Since most of the trees were 50 to 60 feet tall, a rough idea of the height above the ground can also be obtained from the measurements.

There are certain difficulties concerning these measurements. Since the forest was very dense, certain types of behavior rendered birds invisible. This resulted in all species being observed slightly disproportionately in the open zones of the trees. To combat this difficulty each bird was observed for as long as possible so that a brief excursion into an open but not often-frequented zone would be compensated for by the remaining part of the observation. I believe there is no serious error in this respect. Furthermore, the comparative aspect is independent of this error. A different difficulty arises from measurements of time spent in each zone. The error due to counting should not affect results which are comparative in nature. If a bird sits very still or sings, it might spend a large amount of time in one zone without actually requiring that zone for feeding. To alleviate this trouble, a record of activity, when not feeding, was kept. Because of these difficulties, non-parametric statistics have been used throughout the analysis of the study to avoid any *a priori* assumptions about distributions. One difficulty is of a different nature; because of the density of the vegetation and the activity of the warblers a large number of hours of watching result in disappointingly few seconds of worthwhile observations.

The results of these observations are illustrated in Figures 2-6 in which the species' feeding zones are indicated on diagrammatic spruce trees. While

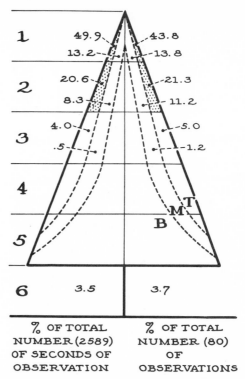

% OF TOTAL   % OF TOTAL
NUMBER (2589)   NUMBER (80)
OF SECONDS OF   OF
OBSERVATION   OBSERVATIONS

Fig. 2. Cape May warbler feeding position. The zones of most concentrated activity are shaded until at least 50% of the activity is in the stippled zones.

the base zone is always proximal to the trunk of the tree, as shown, the T zone surrounds the M, and is exterior to it but not always distal. For each species observed, the feeding zone is illustrated. The left side of each illustration is the percentage of the number of seconds of observations of the species in each zone. On the right hand side the percentage of the total number of times the species was observed in each zone is entered. The stippled area gives roughly the area in which the species is most likely to be found. More specifically, the zone with the highest percentage is stippled, then the zone with the second highest percentage, and so on until at least fifty percent of the observations or time lie within the stippled zone.

Early in the investigation it became apparent that there were differences between the species' feeding habits other than those of feeding zones. Subjectively, the black-throated green appeared "nervous," the bay-breasted slow and "deliberate." In an attempt to make these observations objective, the following measurements were taken on feeding birds. When a bird landed after a flight, a count

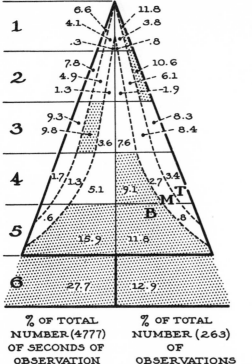

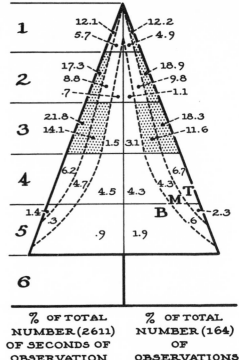

% OF TOTAL NUMBER (4777) OF SECONDS OF OBSERVATION   % OF TOTAL NUMBER (263) OF OBSERVATIONS

FIG. 3. Myrtle warbler feeding position. The zones of most concentrated activity are shaded until at least 50% of the activity is in the stippled zones.

% OF TOTAL NUMBER (2611) OF SECONDS OF OBSERVATION   % OF TOTAL NUMBER (164) OF OBSERVATIONS

FIG. 4. Black-throated green warbler feeding position. The zones of most concentrated activity are shaded until at least 50% of the activity is in the stippled zones.

of seconds was begun and continued until the bird was lost from sight. The total number of flights (visible uses of the wing) during this period was recorded so that the mean interval between uses of the wing could be computed.

The results for 1956 are shown in Table I. The results for 1957 are shown in Table II. Except for the Cape May fewer observations were taken than in 1956.

By means of the sign test (Wilson, 1952), treating each observation irrespective of the number of flights as a single estimate of mean interval between flights, a test of the difference in activity can be performed. These data are summarized in the following inequality, where $\genfrac{}{}{0pt}{}{95}{<}$ is interpreted to mean "has smaller mean interval between flights, with 95% certainty."

their time searching in the foliage for food, some appear to crawl along branches and others to hop across branches. To measure this the following procedure was adopted. All motions of a bird from place to place in a tree were resolved into components in three independent directions. The natural directions to use were vertical, radial, and tangential. When an observation was made in which all the motion was visible, the number of feet the bird moved in each of the three directions was noted. A surprising degree of diversity was discovered in this way as is shown in Figure 7. Here, making use of the fact that the sum of the three perpendicular distances from an interior point to the sides of an equilateral triangle is independent of the position of the point, the proportion of motion in each direction is recorded within a triangle. Thus the Cape May

$$\text{Black-throated green} \quad \genfrac{}{}{0pt}{}{95}{<} \left\{\begin{matrix}\text{Blackburnian}\\ \text{Myrtle}\end{matrix}\right\} \quad \genfrac{}{}{0pt}{}{99}{<} \left\{\begin{matrix}\text{Cape May}\\ \text{Bay-breasted}\end{matrix}\right\}$$

The differences in feeding behavior of the warblers can be studied in another way. For, while all the species spend a substantial part of

moves predominantly in a vertical direction, black-throated green and myrtle in a tangential direction, bay-breasted and blackburnian in a radial direc-

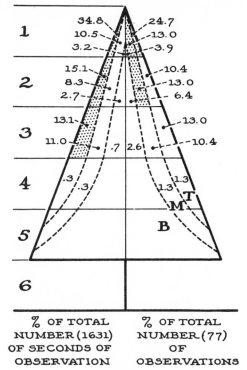

% OF TOTAL NUMBER (1631) OF SECONDS OF OBSERVATION

% OF TOTAL NUMBER (77) OF OBSERVATIONS

FIG. 5. Blackburnian warbler feeding position. The zones of most concentrated activity are shaded until at least 50% of the activity is in the stippled zones.

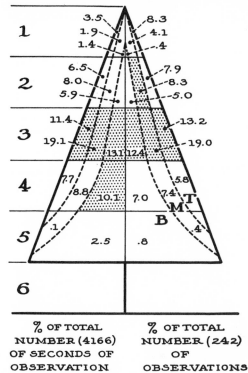

% OF TOTAL NUMBER (4166) OF SECONDS OF OBSERVATION

% OF TOTAL NUMBER (242) OF OBSERVATIONS

FIG. 6. Bay-breasted warbler feeding position. The zones of most concentrated activity are shaded until at least 50% of the activity is in the stippled zones.

tion. To give a nonparametric test of the significance of these differences Table III is required.

Each motion was classified according to the direction in which the bird moved farthest. Thus, in 47 bay-breasted warbler observations of this type, the bird moved predominantly in a radial direction 32 times. Applying a χ^2 test to these, bay-breasted and blackburnian are not different but all others are significantly (P<.01) different from one another and from bay-breasted and blackburnian.

There is one further quantitative comparison which can be made between species, providing additional evidence that during normal feeding behavior the species could become exposed to different types of food. During those observations of 1957 in which the bird was never lost from sight, occurrence of long flights, hawking, or hovering was recorded. A flight was called long if it went between different trees and was greater than an estimated 25 feet. Hawking is distinguished from hovering by the fact that in hawking a moving prey individual is sought in the air, while in hovering a nearly stationary prey indi-

vidual is sought amid the foliage. This information is summarized in Table IV.

Both Cape May and myrtle hawk and undertake long flights significantly more often than any of the other species. Black-throated green hovers significantly more often than the others.

At this point it is possible to summarize differences in the species' feeding behavior in the breeding season. Unfortunately, there are very few original descriptions in the literature for comparison. The widely known writings of William Brewster (Griscom 1938), Ora Knight (1908), and S. C. Kendeigh (1947) include the best observations that have been published. Based upon the observations reported by these authors, the other scattered published observations, and the observations made during this study, the following comparison of the species' feeding behavior seems warranted.

Cape May Warbler. The foregoing data show that this species feeds more consistently near the top of the tree than any species expect blackburnian, from which it differs principally in type

TABLE I. The number of intervals between flights (I) recorded in 1956 and the total number of seconds (S) of observation counted

MYRTLE		BLACK-THROATED GREEN		BLACK-BURNIAN		BAY-BREASTED	
I	S	I	S	I	S	I	S
1	40	4	45	1	5	5	55
4	32	8	20	13	77	3	22
4	13	6	60	2	11	2	33
4	17	5	35	5	18	1	7
2	10	5	23	5	24	3	37
1	10	3	7	3	16	2	20
5	25	9	35	3	18	1	7
5	10	8	25	3	15	4	10
5	11	1	12	4	11	11	60
6	30	7	20	3	12	5	41
3	10	7	39	4	46	3	50
1	5	13	25	2	26	1	17
13	68	5	10			1	3
1	5	5	12			1	49
1	7	2	37			4	42
4	26					5	60
						5	35
						5	26
						3	14
						4	38
						3	14
						1	11
						3	22
						2	29
Total 60	319	88	405	48	279	78	702
Total Adjusted to True seconds 60	399	88	506	48	349	78	876

TABLE II. Intervals between flights and seconds of observation in 1957

Cape May		Myrtle		Black-throated green		Blackburnian		Bay-breasted	
I	S	I	S	I	S	I	S	I	S
3	47	4	18	3	18	1	22	7	110
12	35	1	62	5	22	3	22	2	115
5	50	5	47	3	15	9	110	1	18
2	15	3	17	13	89	7	26	1	22
1	15	1	5	23	40	3	8	2	19
1	45	12	86					2	47
1	20							3	27
4	29							8	112
11	129							7	46
6	47								
1	15								
4	50								
3	20								
21	122								
1	12								
1	34								
4	20								
10	79								
Total 91	782	26	235	47	184	24	188	33	576
Adjusted 1956 Total		60	399	88	506	48	349	78	876
Grand Total 91	782	86	644	125	690	72	537	111	1392
Mean Interval Between Flights 8.59		7.48		5.52		7.47		12.53	

TABLE III. Number of times each species was observed to move predominantly in a particular direction. (Numbers ending in .5 result from ties)

Species	Radial	Tangential	Vertical	Total
Cape May	5.5	1.5	25	32
Myrtle	4.5	11.5	9	25
Black-throated green	4	21	5	30
Blackburnian	11	1	3	15
Bay-breasted	32	7	8	47

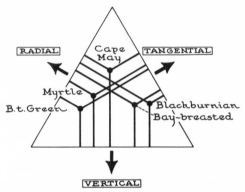

FIG. 7. Components of Motion. From the dot representing a species, lines are drawn to the sides of the triangle. The lengths of these lines are proportional to the total distance which the species moved in radial, tangential, and vertical directions, respectively.

TABLE IV. Classification of the flights observed for each species

Species	Long Flights	Hawking	Hovering	No. of Observ.
Cape May	35	12	0	53
Myrtle	25	9	0	62
Black-throated green	1	0	7	42
Blackburnian	0	1	0	35
Bay-breasted	4	2	2	57

Applying χ^2 tests to pairs of species, the following conclusions emerge.

of feeding action. It not only hawks far more often than the blackburnian, but also moves vertically rather than radially in the tree, causing its feeding zone to be more restricted to the outer shell of the tree. Myrtle warblers when feeding in the tips of the trees nearly duplicate the feeding behavior of the Cape May. During rainy, windy, and cold weather Cape Mays were not found in the tree tops, but were instead foraging in the low willows (*Salix* sp.) and pin cherries (*Prunus pensylvanica*). Here they often fed among the flowers, for which their semitubular tongue (Gardner 1925) may be advantageous.

Because of this species' irregular breeding distribution, both in space and time, and its former rarity, there are very few published descriptions of its feeding behavior. Knight (1908), although he lived in Maine, had never seen one. Brewster (Griscom 1938) wrote that:

"It keeps invariably near the tops of the highest trees whence it occasionally darts out after passing insects. . . . In rainy or dark weather they came in numbers from the woods to feed among the thickets of low firs and spruces in the pastures. Here they spent much of their time hanging head downward at the extremity of the branches, often continuing in this position for nearly a minute at a time. They seemed to be picking minute insects from under the surface of the fir needles. They also resorted to a thicket of blossoming plum trees directly under the window, where we were always sure of finding several of them."

He also said that it was more active than the bay-breasted. Bond (1937) stated that all feeding was done more than twenty feet above the ground. Kendeigh (1947) said males tend to sing about seven feet from the tops of the trees, and that feeding is done at the same level. He also mentions that the birds sometimes hawk after passing insects. The rainy weather observations indicate behavior very much like that of winter and migration to be discussed later.

Myrtle Warbler. This species seems to have the most varied feeding habits of any species. Although it moves slightly more in a tangential direction than any except black-throated green, it is probably more correct to think of the myrtle as having the most nearly equal components (radial, tangential, and vertical) of any species. This is shown by its most nearly central location in Figure 7. It is also seen to have the most widely distributed feeding zone, although the ground feeding was nearly, but not completely, restricted to the gathering of emerging Tipulids for newly hatched young. Sometimes a substantial amount of this is hawking for flying insects; at other times it is largely by rapid peering (Grinnell 1921) amid the thick foliage near the tree tops. Myrtle, along with Cape May, makes a much higher proportion of flights to other trees than do the other species, often flying from one side of its territory to the other with no apparent provocation. The other three species tend to search one tree rather thoroughly before moving on. Further evidence of the plasticity of the myrtle warbler's feeding habits will be presented when the other seasons are considered. Grinnell and Storer (1924) stated that the Audubon's warbler (which often hybridizes with the myrtle and with it froms a superspecies (Mayr 1950)) also feeds in peripheral foliage and does a greater amount of hawking than other species. Kendeigh (1947) said that

birds fed from ground to tops of trees, and also that two males covered two and four acres respectively in only a few minutes. Knight (1908) said "Many of the adult insects are taken on the wing, the warblers taking short springs and flights into the air for this purpose. The young for the first few days are fed on the softer sorts of insects secured by the parents, and later their fare is like that of the parents in every way."

Black-throated Green Warbler. Compared to the myrtle warbler this species is quite restricted in feeding habits. As seen in Figure 4, it tends to frequent the dense parts of the branches and the new buds, especially at mid elevations in the tree. Most of its motion is in a tangential direction, keeping the bird in foliage of a nearly constant type. It has the shortest interval between flights of any of the five species and thus appears the most active. Almost all feeding seems to be by the method of rapid peering, necessitating the frequent use of the wings which the observations indicate. The foliage on a white spruce is a thick, dense mat at the end of the branch, changing to bare branch rather more sharply than in the red spruce. The black-throated greens characteristically hop about very actively upon these mats, often, like the other species, looking down among the needles, and just as often, unlike the myrtle and bay-breasted, peering up into the next mat of foliage above. When food is located above, the bird springs into the air and hovers under the branch with its bill at the point whence the food is being extracted. While other species occasionally feed in this fashion, it is typical only of the black-throated green. After searching one branch, the black-throated green generally flies tangentially to an adjacent branch in the same tree or a neighboring one and continues the search. Only rarely, during feeding, does it make long flights. While it occasionally hawks for flying insects (missing a substantial proportion), this is not a typical behavior and the birds seldom sit motionless watching for flying insects in true hawking behavior. During its feeding, this species is very noisy, chipping almost incessantly, and, if it is a male and if it is early in the season, singing frequently. The other species are very quiet. A portion of this behavior can be confirmed from the literature. Knight (1908) said "Only rarely do they take their prey in the air, preferring to diligently seek it out among the branches and foliage" and Stanwood (Bent 1953) said "The bird is quick in its movements, but often spends periods of some length on one tree." Like the myrtle, this species enlarges its feeding zone while gathering food for its young. This is similar to the results of Betts (1955) indicating that the young tits eat different food from the adults.

Blackburnian Warbler. This species generally feeds high in the trees but is otherwise more or less intermediate between black-throated green and bay-breasted in its feeding behavior. This is true both of its flight frequency and its preferred feeding position on the limb (Figure 5). It is also intermediate in its method of hunting, usually moving out from the base to the tip of the branches looking down in the fashion of the bay-breasted and occasionally hopping about rapidly upon the mat of foliage at the branch tips looking both up and down for insects and even hovering occasionally. They seem to use the method of rapid peering, only occasionally hawking after a flying insect. As further evidence, Knight (1908) wrote "As a rule they feed by passing from limb to limb and examining the foliage and limbs of trees, more seldom catching anything in the air." Kendeigh (1945) said "It belongs to the treetops, singing and feeding at heights of 35 to 75 feet from the ground."

Bay-breasted Warbler. The usual feeding habits of this species are the most restricted of any of the species studied. All of the observations in the T1 zone and most in the T2 zone refer to singing males. This species uses its wings considerably less often than the other species, although it still appears to use the method of rapid peering in its hunting since it moves nearly continuously. These motions are, however, predominantly radial and seldom require the use of wings. The bird regularly works from the licheny base of the branches well out to the tip, although the largest part of the time is spent in the shady interior of the tree. It frequently stays in the same tree for long periods of time. This species very rarely hovers in the black-throated green fashion, and appears much less nimble in its actions about the tips of the branches, usually staying away from those buds which are at the edge of the mat of foliage. When it does feed at the edge of the mat, it is nearly always by hanging down rather than peering up. Other observers have emphasized the slowness. Brewster (Griscom 1938) called this warbler "slow and sluggish," and Kendeigh (1947) said "The birds do not move around much, but may sing and feed for long periods in the same tree." Forbush (1929) stated that it spends most of its time "moving about deliberately, after the manner of vireos."

Food

Two species may eat different food for only three reasons: 1. They may feed in different places

or different times of day; 2. They may feed in such a manner as to find different foods; 3. They may accept different kinds of food from among those to which they are exposed. (Of course, a combination of these reasons is also possible.)

In the previous section it was shown that the warblers feed in different places and in a different manner, thus probably being exposed to different foods for the first and second reasons mentioned above. It is the aim of this section to show that the five warbler species have only small differences of the third kind. Theoretically, such differences, unaccompanied by morphological adaptations, would be disadvantageous, for, lacking the adaptations required to give greater efficiency in food collecting, and suffering a reduction in the number of acceptable food species, a bird would obtain food at a lesser rate. When the necessary adaptations are present, they usually consist of quite marked differences in bill structure such as those reported by Huxley (1942), Lack (1947), and Amadon (1950). As Table V shows, the mean bill measurements in millimeters of the five species of warbler considered in this study are quite similar. Twelve specimens of each species from the Peabody Museum of Natural History at Yale University were measured for each of the means given.

Table V. Mean dimensions of the bills of 12 specimens of each species

Species	Bill Length	Height at Nares	Width at Nares	Width 2.5mm from tip
Cape May	12.82	2.85	2.93	0.96
Myrtle	12.47	3.26	2.12	1.33
Black-throated green	12.58	3.38	3.15	1.34
Blackburnian	12.97	3.24	3.36	1.17
Bay-breasted	13.04	3.69	3.58	1.43

The Cape May alone has a noticeably different bill, it being more slender, especially at the tip. This bill houses a semi-tubular tongue as mentioned above, which is unique in the genus. These may be useful adaptations for their rainy weather flower feeding, but would seem ill-adapted for the characteristic flycatching of the breeding season (Gardner 1925). It is doubtless useful in other seasons, as will be discussed later. Aside from the Cape May, all other species differ in bill measurement by only a small fraction of a millimeter. Thus, for theoretical reasons, no pronounced differences of the third kind would be expected. Empirically, there is evidence to support this belief.

McAtee (1932) reported upon the analysis of eighty thousand bird stomachs, in an effort to disprove mimicry. Although his results were not conclusive, he claimed that insects appeared in bird stomachs about proportionately to their availability. Kendeigh (1945) agreed with this conclusion. Although McAtee (1926) said that no detailed studies of warbler food habits had been made, and no general ones seem to have appeared since, two very suggestive sets of analyses covering the five species have been published. Kendeigh (1947) reported upon the stomach contents of a collection made near Lake Nipigon, and Mitchell (1952) analyzed the stomachs of many birds taken during a budworm infestation in Maine. These data show, first, that most species of warbler eat all major orders of local arboreal arthropods. Furthermore, although there are differences in proportion of types of foods eaten by various species, these differences are most easily explained in terms of feeding zone. Thus, black-throated green and blackburnian which are morphologically the most similar of the five species have quite different foods. Kendeigh's table shows that black-throated green eats 4% Coleoptera, 31% Araneida, and 20% Homoptera, which blackburnian eats 22% Coleoptera, 2% Araneida, and 3% Homoptera. Dr. W. R. Henson has pointed out (pers. comm.) that Coleoptera can reasonably be assumed to come from inner parts of the tree where blackburnian has been shown to feed, whereas the Homoptera and most of the Araneida would be caught in the current year's growth where the black-throated green feeds more often, thus explaining the observed difference. Black-throated green and bay-breasted, with the most vireo-like bills (high at the nares), seem to eat more Lepidoptera larvae which are typical vireo food, but Mitchell's table shows that the other species too can eat predominantly Lepidoptera larvae when these are abundant. Otherwise, the food of the bay-breasted is more like that of the blackburnian, the feeding habits of which are similar. There are not sufficient data to analyze Cape May in this fashion. The myrtle warbler's feeding behavior is so flexible that no correlation between insects caught and a specific feeding zone or behavior is expected. This is shown by its having the most even distribution of food of any of the species considered. Thus, these correlations show that the differences in warbler food can be readily explained by morphological and zonal characteristics of the species.

Nest Location

The position of the nest is quite characteristic in warbler species. Nearly all species of the genus *Dendroica* nest off the ground. Figure 8 shows heights of the nests of the five species of warbler studied here. These data result from a combination of the records of Cruickshank (1956), the information in the egg collection of the American Museum of Natural History in New York, and that gathered in this study. Since the distributions are skewed and irregular, the median and confidence intervals for the median (Banerjee and Nair 1940) are appropriate measurements. As the figure shows, the Cape May, with 95% confidence interval for the median of 40-50 feet, and the blackburnian, whose interval is 30-50 feet, have quite similar nest heights, probably reflecting their tendency to feed at high elevations. The Cape May's nest is virtually always near the trunk in the uppermost dense cluster of branches in a spruce or occasionally a fir. The blackburnian may nest in a similar location or may nest farther out toward the branch tips. Myrtle and black-throated green have similar nesting heights, both species having 95% confidence interval for the median nest height of 15-20 feet. The black-throated green seems to prefer smaller trees for its nest, and is thus more likely to place its nest near the runk, but, in keeping with its other characteristics, the myrtle seems quite varied in this respect. Finally, the bay-breasted, which has the lowest feeding zone, has the lowest nest position, the median height being between 10 and 15 feet (95% confidence). Thus, the nest positions of the five species of warbler reflect their preferred feeding zones.

Territoriality

Defining territory as any defended area, warbler territories in the breeding season are of what Hinde (1956) called type A ("Large breeding area within which nesting, courtship, and mating and most food-seeking usually occur"). He pointed out that, since the behavioral mechanisms involved in defending a territory against others of the same species are the same as those involved in defending it against other species, this distinction need not be specified in the definition. From the ecological point of view, the distinction is of very great importance, however, for, as G. E. Hutchinson pointed out in conversation, if each species has its density (even locally) limited by a territorial behavior which ignores the other species, then there need be no further differences between the species to permit them to persist together. A weaker form of the same process, in which territories were compressible but only under pressure of a large population, would still be effective, along with small niche differences, in making each species inhibit its own population growth more than the others'—the

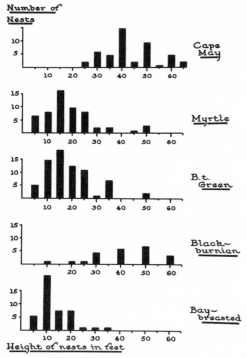

Fig. 8. Nesting heights of warblers.

necessary condition for the persistence of sympatric species.

Two further conditions make territory important for regulating populations. First, to have density dependent regulation, a species' regulating mechanism must have information of its own population density. Second, a predator ideally should keep its prey at that population level which permits the greatest rate of production. This means that the prey would not normally be particularly scarce. This, combined with the varied prey of the birds and the varied predators of the insects, would make food density a poorer criterion of a given bird species' density than size of territory. Thus, competition for food would be reduced from a "scramble" to a "contest" (Haldane 1955).

While the true nature of birds' territories has proved very elusive (Lack 1954, Hinde 1956), two separate lines of evidence suggest strongly that territories contribute to the regulation of local densities in warbler populations. Stewart and Aldrich (1951) and Hensley and Cope (1951) removed adult birds from their territories in 1949 and 1950 respectively, in a 40 acre plot in a budworm-infested area of Maine. The vacated territories were always filled by new pairs, the males

singing the vigorous song of a bird setting up a territory. It seems nearly inescapable that these were part of a large floating population of birds only prevented from breeding by the absence of unoccupied territories. Since this was in a budworm outbreak, there seems little doubt that there would have been adequate food for a larger breeding population.

In a study over a series of years on the birds in New Brunswick, Cheshire (1954) recorded the populations and territory sizes of the various species as a budworm outbreak began and progressed. He showed that while the bay-breasted warbler (the commonest bird during the outbreak) underwent a five- to seven-fold increase as the outbreak began, their mean territory sizes remained constant instead of decreasing correspondingly. That is, there had been unoccupied interstices between territories initially; these were filled in by the incoming birds but territory sizes were left unchanged.

The facts suggest that the territory size is more or less fixed in this region (although, of course, it may vary from region to region) and that if territorial compression occurs during high population densities, it only does so during higher population densities than those observed. Of course, if high population densities persisted, natural selection might be expected to reduce territory size, but this is a different situation.

As for interspecific territoriality, there is no exclusion of the kind found in intraspecific territoriality, as is clearly shown in Kendeigh's (1947) territory maps. It is very difficult to distinguish a mild repulsion of other species by territorialism from a preference for slightly different habitats. Adequate information does not exist to make the distinction at present. However, it seems quite certain that interspecific territoriality is weaker than intraspecific and, therefore, that the effect of a large density of one species is greater on that species than on the others. It is thus probable that, in the warblers, territoriality helps reduce competition and acts as a stabilizing factor (as well as performing the well-known functions of pair formation and maintenance).

Natality and Mortality

In a population which has reached an equilibrium size, abundance is independent of birth and death rates. For species in equilibrium, then, a study of birth and death rates is not necessary to understand the control of the equilibrium abundance. However, as Darwin (1859) said, "A large number of eggs is of some importance to those species which depend upon a fluctuating amount of food, for it allows them rapidly to increase in numbers."

The five species of warbler studied here are very interesting in this respect. Table VI is a summary of the nesting data of the Museum of Comparative Zoology at Harvard, the American Museum of Natural History in New York, and the data of Harlow published by Street (1956).

TABLE VI. Mean clutch sizes for the 5 species

Species	No. of Nests	CLUTCH SIZE					Mean	St. Dev.
		3	4	5	6	7		
Cape May........	48		4	11	24	9	5.792	.850
Myrtle............	24	1	19	4			4.125	.449
Black-throated green.........	45	2	39	4			4.044	.366
Blackburnian......	44	7	32	5			3.955	.526
Bay-breasted......	49		5	21	20	3	5.429	.752

Cape May and bay-breasted warblers' nests were enough of a prize that it is quite certain that all found were kept and that the collections do not reflect any bias. There is a possibility of slight bias, collectors perhaps prizing larger clutches, in the other three species in the museum collections, but their clutch sizes are so constant that this seems improbable. The data of Harlow are not subject to his criticism, since he recorded all nests found.

While the sources of these collections vary in latitude from that of the Poconos of Pennsylvania to that of northern New Brunswick, there appears to be very little change in clutch size in this range of latitude. Thus the mean clutch size of 16 nests of the black-throated green in the Poconos is 4.06, while for the combined collections from Nova Scotia and New Brunswick (12 nests) the mean clutch is 4.17. The nests of the other species are from a narrow range of latitude and would not be expected to vary. Thus it was felt permissible to combine the data from different latitudes.

It is immediately apparent that Cape May and bay-breasted, the species which capitalize upon the periodic spruce budworm outbreaks, have considerably larger clutches than the other species, as Darwin would have predicted. It is of interest that the only other warbler regularly laying such large clutches is the Tennessee warbler (*Vermivora peregrina*) which is the other species regularly fluctuating with the budworms (Kendeigh 1947). Thus it seems that Darwin's statement provides an appropriate explanation for the larger clutches. It is also interesting that the standard deviation of the Cape May and bay-breasted warblers' clutch sizes is greater. This suggests a certain plasticity which can be verified, for the

bay-breasted at least, as follows. If the time of the budworm outbreak in New Brunswick is taken as 1911-1920 (Swaine and and Craighead 1924), and other years from 1903 until 1938 are called non-budworm years, the bay-breasted warbler clutches from northeastern New Brunswick can be summarized as follows:

	Clutch Size			
	4	5	6	7
Budworm Years	1	5	15	3
Non-budworm Years	4	8	5	0

The U test (Hoel 1954) shows this to be significant at the .0024 level; that is, bay-breasted warblers lay significantly larger clutches during years of budworm outbreaks. There are not sufficient data to make a corresponding comparison for Cape May warblers. It is known (Wangersky and Cunningham 1956) that an increase in birth rate is likely to lead to instability. The easiest way to increase the stability, while still maintaining the large clutch which is desirable for the fluctuating food supply, is to have the clutch especially large when food is abundant. This is apparently the solution which the bay-breasted warbler, at least, has taken.

Mortality during the breeding season is more difficult to analyze. Disease is not normally important as a mortality factor in passerines (Lack, 1954) and this appeared to be the case for the warblers under observation. Predation may be important, however. Saw-whet owls (*Aegolius acadica*), Cooper's hawks (*Accipiter cooperii*), goshawks (*A. gentilis*), ravens (*Corvus corax*), crows (*C. brachyrhynchos*), and herring gulls (*Larus argentatus*) all occasionally were noted in the Maine woods, but no evidence was obtained of their preying upon the warblers. In fact, none of the established pairs of birds were broken up by predation of this type. Red squirrels (*Tamiasciurus hudsonicus*) were continually present in all plots and were frequently observed searching for nests. They certainly destroyed the nest of a black-throated green and of a brown creeper (*Certhia familiaris*) and were quite probably responsible for plundering one myrtle warbler nest which was robbed soon after eggs were laid. The most common evidence of mortality, however, was the frequent observation of parents feeding only one or two newly fledged young. Thus two pairs of myrtle warblers in 1956 and one in 1957 were observed the day the young left the nest feeding four young. One of the 1956 pairs succeeded in keeping all four young alive for at least three days, at which time they could no longer be followed. The remaining two pairs were only feeding two young on the day following the departure of the young from the nest. Similarly, of two black-throated green pairs (one in 1956, one in 1957) where young could be followed, one kept all four young alive and the second only raised two of the fledged four. It was difficult to determine the number of young the parents were feeding. It was also difficult to be at the nest site when the young left the nest to determine the number of fledged young. Consequently, no more observations suitable to report were made. When the young leave the nest, they fly to nearby trees quite independently of one another and apparently never return to the nest. The result is that within a few hours the young are widely scattered. In this condition they are very susceptible to predators and exposure, and should one fly when its parents were not nearby, it would rapidly starve. Normally, the young only fly or chatter loudly when a parent with food is calling nearby, and the parents seem remarkably good at remembering where the young have gone. At best, however, this is a very dangerous period. It is of some interest to note that adult warblers will feed not only young of other birds of their own species but also of other species. Skutch (1954) reviewed several published cases of this. Hence, when a wood is densely settled with warblers, the members of a large clutch might have a better chance of surviving, the straying young being fed by neighbors. This high density is, of course, the situation which obtains during a budworm outbreak when bay-breasted and Cape May warblers are so successful.

Time of Activities

So far, the nature and position of the species' activities during the breeding season have been compared. The time of these activities would also be a potential source of diversity. There could either be differences in the time of day in which feeding took place, or there could be differences in the dates during which eggs were laid and the young fed. The first type of difference (time of day) seems inherently improbable since, at least while feeding the fledged young, the parents are kept busy throughout the daylight hours gathering food. Record was taken of the time at which the various warbler species began singing in the morning of 19 June 1956. The results (Eastern Daylight Time) are: 0352, first warbler (magnolia, *D. magnolia*); 0357, first myrtle; 0400, both myrtle and magnolia singing regularly; 0401, first black-throated green; 0402, first parula warbler (*Parula americana*); 0403, first bay-breasted;

0405, all warblers singing regularly. Thus, within 13 minutes after the first warbler sang all species were singing regularly. The sequence of rising corresponds to the degree of exposure of the usual feeding zones for that date (see Figures 2-6), and therefore probably depends only upon the time at which the light reaches a certain intensity.

As for the breeding season, there is good evidence of differences in time of completion of clutches. Since date of completion can be expected to change from place to place, comparisons must be made at one fixed locality. Of 15 nests of black-throated green warbler found by Harlow in the Poconos of Pennsylvania, the mean date of clutch completion was June 3, and of 21 blackburnian the mean date of clutch completion was June 1. Thus, for this region, and there is no reason to think that the relative dates are different in other regions, there is little difference in time of nesting between blackburnian and black-throated green warblers. From the extensive collections of P. B. Philipp near Tabusintac, N. B., now in the American Museum of Natural History, and from a smaller number collected in the same region by R. C. Harlow, now in the Museum of Comparative Zoology, bay-breasted and Cape May warbler nest dates can be compared (Figure 9). It is quite clear that the bay-breasted with the median date of nest discovery of 25 June (95% confidence interval for the median 23-27 June) nest substantially later than the Cape May whose median date is 17 June (95% confidence interval for the median of 16 June-20 June). As the figure shows, the small number of nests of black-throated green and myrtle from the same region show a fairly wide spread but strongly suggest median dates intermediate between Cape May and bay-breasted. (The dates recorded by Palmer (1949) for Maine give a roughly similar sequence; myrtle, 30 May-6 June; black-throated green, 26 May-20 June; bay-breasted, after 7 June.)

It might be expected that the insects caught by the species which feed in the T zones and near the tree tops would reach peak abundance sooner thus making it desirable for those species to nest earlier. The sequence of nesting dates just presented seems to be consistent with this hypothesis.

[*Editors' Note:* Material has been omitted at this point.]

Discussion and Conclusions

In this study competition has been viewed in the light of the statement that species can coexist only if each inhibits its own population more than the others'. This is probably equivalent to saying that species divide up the resources of a community in such a way that each species is limited by a different factor. If this is taken as a statement of the Volterra-Gause principle, there can be no exceptions to it. Ecological investigations of closely-related species then are looked upon as enumerations of the divers ways in which the resources of a community can be partitioned.

For the five species of warbler considered here, there are three quite distinct categories of "different factors" which could regulate populations. "Different factors" can mean different resources, the same resources at different places, or the same resources at different times. All three of these seem important for the warblers, especially if different places and times mean very different—different habitats and different years.

First, the observations show that there is every reason to believe that the birds behave in such a way as to be exposed to different kinds of food. They feed in different positions, indulge in hawking and hovering to different extents, move in different directions through the trees, vary from active to sluggish, and probably have the greatest need for food at different times corresponding to the different nesting dates. All of these differences are statistical, however; any two species show some overlapping in all of these activities.

The species of food organisms which were widespread in the forest and had high dispersal rates would be preyed upon by all the warblers. Thus, competition for food is possible. The actual food eaten does indicate that the species have certain foods in common. The slight difference in habitat preference resulting from the species' different feeding zones is probably more important. This could permit each species to have its own center of dispersal to regions occupied by all species. Coexistence in one habitat, then, may be the result of each species being limited by the availability of a resource in different habitats. Even although the insects fed upon may be basically of the same type in the different habitats, it is improbable that the same individual insects should fly back and forth between distant woods; consequently, there would be no chance for competition. The habitat differences and, equivalently, the feeding zone differences, between blackburnian, black-throated green, and bay-breasted are sufficiently large that this explanation of coexistence is quite reasonable.

The myrtle warbler is present in many habitats in the summer but is never abundant. It has a very large summer and winter range, feeds from the tree tops to the forest floor, and by rapid peering or by hawking. It makes frequent long flights and defends a large territory. Probably it can be considered a marginal species which, by being less specialized and thus more flexible in its requirements, manages to maintain a constant, low population (Figure 10).

The Cape May warbler is in a different category, at least in the region near the southern limit of its range. For here it apparently depends upon the occasional outbreaks of superabundant food (usually spruce budworms) for its continued existence. The bay-breasted warbler, to a lesser degree, does the same thing. During budworm outbreaks, probably because of their extra large clutches, they are able to increase more rapidly than the other species, obtaining a temporary advantage. During the years between outbreaks they suffer reductions in numbers and may even be eliminated locally. Lack's hypothesis, that the clutch is adjusted so as to produce the maximum number of surviving offspring, provides a suitable explanation of the decrease during normal years of these large-clutched species. It may be asked why, if Lack's hypothesis is correct, natural selection favored large clutches in Cape May and bay-breasted. Cheshire's (1954) censuses suggest a tentative answer. During his years of censusing, increases in the bay-breasted warbler population reached a figure of over 300% per year. This probably far exceeds the maximum possible in-

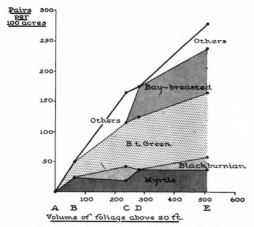

FIG. 10. Composition of the warbler population in plots A, B, C, D, and E. "Others" refers to other warbler species which feed at greater heights than 20 feet above the ground. The units of volume measurement are only proportional to the volume, but each unit roughly equals 1500 cubic feet per acre.

crease due to survival of nestlings raised in that place; probably immigration is the explanation. But if the species with large clutches search for areas in which food is superabundant and immigrate into these regions, then, for the species as a whole, the large clutch may be adapted to the maximum survival of offspring. Cape May and bay-breasted warblers may therefore be considered to be good examples of fugitive species (Hutchinson 1951).

Thus, of the five species, Cape May warblers and to a lesser degree bay-breasted warblers are dependent upon periods of superabundant food, while the remaining species maintain populations roughly proportional to the volume of foliage of the type in which they normally feed. There are differences of feeding position, behavior, and nesting date which reduce competition. These, combined with slight differences in habitat preference and perhaps a tendency for territoriality to have a stronger regulating effect upon the same species than upon others, permit the coexistence of the species.

ACKNOWLEDGMENTS

Prof. G. E. Hutchinson and Dr. S. D. Ripley have played indispensable roles in the development of this work, providing advice, encouragement, and support. The author had valuable discussion with Dr. A. F. Skutch, James Bond, and Paul Slud concerning birds on their wintering grounds. Dr. L. R. Holdridge provided invaluable help in Costa Rica, acting as naturalist and guide. Sincere thanks also go to the following persons. Miss Helen T. Mac Arthur prepared the illustrations. Drs. R. A. Paynter and Dean Amadon gave information about or provided access to the collections under their supervision. Drs. C. L. Remington, W. R. Henson, and P. B. Dowden provided entomological information. Dr. J. W. Mac Arthur helped with the observations. Finally, the author wishes to thank his wife who helped with observations, prepared the manuscript, and provided encouragement.

The work was supported by grants from the Peabody Museum of Natural History of Yale University and from the Chapman Memorial Fund of the American Museum of Natural History.

REFERENCES

Amadon, D. 1950. The Hawaiian honeycreepers (Aves, Drepaniidae). Bull. Amer. Mus. Nat. Hist. **95**: 157-262.

Andrewartha, H. G. and L. C. Birch. 1954. The distribution and abundance of animals. Chicago: Univ. Chicago Press.

Banerjee, S. K. and K. R. Nair. 1940. Tables of confidence intervals for the median in samples from any continuous population. Sankya **4**: 551-558.

Bent, A. C. 1953. Life histories of the North American wood warblers. U. S. Nat. Mus. Bull. **203**.

Betts, M. 1955. The food of titmice in an oak woodland. Jour. Anim. Ecol. **24**: 282-323.

Bond, J. 1937. The Cape May warbler in Maine. Auk **54**: 306-308.

———. 1957. North American wood warblers in the West Indies. Audubon Mag. **59**: 20-23.

Bourne, W. R. P. 1957. The breeding birds of Bermuda. Ibis **99**: 94-105.

Breckenridge, W. J. 1956. Measurements of the habitat niche of the least flycatcher. Wilson Bull. **68**: 47-51.

Cheshire, W. P. 1954. Bird populations and potential predation on the spruce budworm. Canada Dept. Agric. Sci. Serv., Annual Tech. Report, Green River Project 1953, Sect. **14**.

Cruickshank, A. D. 1956. Aud. Field Notes **10**: 431-432. (and earlier censuses of the same plot).

———. 1956a. Nesting heights of some woodland warblers in Maine. Wilson Bull. **68**: 157.

Darwin, C. R. 1859. The origin of species by means of natural selection or the preservation of favoured races in the struggle for life. London: Murray.

Dixon, W. J. and F. J. Massey. 1951. Introduction to statistical analysis. New York: McGraw-Hill.

Eaton, S. W. 1953. Wood warblers wintering in Cuba. Wilson Bull. **65**: 169-174.

Forbush, E. H. 1929. The birds of Massachusetts and other New England states. Vol. III. Boston: Mass. Dept. Agric.

Gardner, L. L. 1925. The adaptive modifications and the taxonomic value of the tongue in birds. Proc. U. S. Nat. Mus. 67, art. **19**: 1-49.

Gause, G. F. 1934. The struggle for existence. Baltimore: Williams and Wilkins.

Greenbank, D. O. 1956. The role of climate and dispersal in the initiation of outbreaks of the spruce budworm in New Brunswick. Can. Jour. Zool. **34**: 453-476.

Grinnell, J. 1914. Barriers to distribution as regards birds and mammals. Amer. Nat. **48**: 249-254.

——. 1921. The principle of rapid peering in birds. Univ. Calif. Chron. **23**: 392-396.

——. 1922. The trend of avian populations in California. Science **56**: 671-676.

—— and T. Storer. 1924. Animal life in the Yosemite. Berkeley: Univ. Calif. Press.

Griscom, L. 1938. The birds of Lake Umbagog region of Maine. Compiled from the diaries and journals of William Brewster. Bull. Mus. Comp. Zool. **66**: 525-620.

Grosenbaugh, L. R. 1952. Plotless timber estimates—new, fast, easy. Jour. Forestry **50**: 33-37.

Haldane, J. B. S. 1955. Review of Lack (1954). Ibis **97**: 375-377.

Hausman, L. A. 1927. On the winter food of the tree swallow (*Iridoprocne bicolor*) & the myrtle warbler (*Dendroica coronata*). Amer. Nat. **61**: 379-382.

Hensley, M. M. and J. B. Cope. 1951. Further data on removal and repopulation of the breeding birds in a spruce-fir forest community. Auk **68**: 483-493.

Hinde, R. A. 1956. The biological significance of territories of birds. Ibis **98**: 340-369.

Hoel, P. G. 1954. Introduction to mathematical statistics. New York: Wiley.

Hutchinson, G. E. 1951. Copepodology for the ornithologist. Ecology **32**: 571-577.

Huxley, J. 1942. Evolution, the modern synthesis. New York: Harper.

Kendeigh, S. C. 1945. Community selection birds on the Helderberg Plateau of New York. Auk **62**: 418-436.

——. 1947. Bird population studies in the coniferous forest biome during a spruce budworm outbreak. Biol. Bull. **1**, Ont. Dept. Lands and For.

Kluijver, H. N. and N. Tinbergen. 1953. Regulation of density in titmice. Arch. Ned. Zool. **10**: 265-289.

Knight, O. W. 1908. The birds of Maine. Bangor.

Lack, D. 1947. Darwin's finches. Cambridge: Cambridge Univ. Press.

——. 1954. The natural regulation of animal numbers. Oxford: Oxford Univ. Press.

——. 1955. The mortality factors affecting adult numbers. In Cragg, J. B. and N. W. Pirie 1955. The numbers of man and animals. London: Oliver and Boyd.

MacKenzie, J. M. D. 1946. Some factors influencing woodland birds. Quart. Jour. For. **40**: 82-88.

Mayr, E. 1950. Speciation in birds. Proc. Xth Int. Ornith. Cong. Uppsala.

McAtee, W. L. 1926. The relation of birds to woodlots in New York State. Roosevelt Wildlife Bull. **4**: 7-157.

——. 1932. Effectiveness in nature of the so-called protective adaptations in the animal kingdom, chiefly as illustrated by the food habits of Nearctic birds. Smithsonian Misc. Coll. **85**(7): 1-201.

Mitchell, R. T. 1952. Consumption of spruce budworms by birds in a Maine spruce-fir forest. Jour. For. **50**: 387-389.

Palmer, R. S. 1949. Maine birds. Bull. Mus. Comp. Zool. **102**: 1-656.

Pettingill, O. S. 1951. A guide to bird finding east of the Mississippi. New York: Oxford Univ. Press.

Salomonsen, F. 1954. Evolution and bird migration. Acta **XI** Cong. Int. Orn. Basil.

Skutch, A. F. 1954. Life histories of Central American birds. Pacific Coast Avifauna No. 31.

Smith, W. P. 1953. Aud. Field Notes **7**: 337 (and earlier censuses of the same plot).

Stewart, R. E. and J. W. Aldrich. 1951. Removal and repopulation of breeding birds in a spruce-fir forest community. Auk **68**: 471-482.

——. 1952. Ecological studies of breeding bird populations in northern Maine. Ecol. **33**: 226-238.

Street, P. B. 1956. Birds of the Pocono Mountains, Pennsylvania. Delaware Valley Ornith. Club. Philadelphia.

Swaine, J. M. and F. C. Craighead. 1924. Studies on the spruce budworm (*Cacoecia fumiferana Clem.*) Part I. Dom. of Canada Dept. Agric. Bull. **37**: 1-27.

Swed, F. S. and C. Eisenhart. 1943. Tables for testing randomness of grouping in a sequence of alternatives. Ann. Math. Stat. **14**: 66-87.

Volterra, V. 1926. Variazione e fluttuazione del numero d'individiu in specie animali conviventi. Mem. Accad. Lincei **2**: 31-113. (Translated in Chapman, R. N. 1931. Animal ecology. New York: McGraw-Hill.)

Wangersky, P. J. and W. J. Cunningham. 1956. On time lags in equations of growth. Proc. Nat. Acad. Sci. **42**: 699-702.

Williams, A. B. 1950. Aud. Field Notes **4**: 297-298 (and earlier censuses of the same plot).

Wilson, E. B. 1952. An introduction to scientific research. New York: McGraw-Hill.

Reprinted from *Ecol. Monographs,* 36(4), 328–329, 376–381, 383 (1966),
with permission of the publisher, Duke University Press, Durham, N. C.

MESOPSOCUS POPULATIONS ON LARCH IN ENGLAND — THE DISTRIBUTION AND DYNAMICS OF TWO CLOSELY-RELATED COEXISTING SPECIES OF PSOCOPTERA SHARING THE SAME FOOD RESOURCE

Edward Broadhead and Anthony J. Wapshere

[*Editors' Note:* In the original, material precedes this excerpt.]

GENERAL BIOLOGICAL FEATURES

The two *Mesopsocus* species possess features which render them particularly suitable as subjects for population study in the field. They have one generation a year. Their entire lives, from egg to adult, are passed in the same habitat—on the bark surface of the trees. Their egg batches are readily distinguishable and their nymphs are easily identified both to species and to instar. The females of both species are wingless and their populations on the larch trees at Harrogate are highly interspersed so that a single sampling procedure can give, for both species, comparable estimates of densities at all stages of the life cycle.

Mesopsocus immunis and *M. unipunctatus* are among the largest of the psocids occurring in Britain, adult females having a body length of 3-4 mm. *M. unipunctatus* is slightly larger than *immunis.* They are not structurally differentiated from each other to any appreciable extent. The males of both species are winged and these can be recognised to species only by examination of the genital armature. The females of *immunis* show very little variation in colour, but those of *unipunctatus* vary considerably, the palest specimens being almost indistinguishable from female *immunis.* The shape of the terminal lobe of the subgenital plate is the most reliable character for specific diagnosis. The eggs of both species are laid in clusters of 5-8, covered by hard black faecal material over which a dense silken web is spun in *immunis* but not in *unipunctatus.* The eggs are arranged within the egg batch in a characteristic way in each species so that egg batches of *immunis* from which the silk has eroded away over the winter are usually distinguishable from those of *unipunctatus.* Specific diagnosis of the nymphs at all stages, even before they hatch from the egg, is quickly and easily made by reference to the striking difference between the two species in the colour pattern of the clypeus. The successive instars are distinguished by head width and wing pad development and the sex of the

nymphs can be recognized after the 3rd instar by wing pad size (Broadhead & Wapshere, in press).

These two *Mesopsocus* species have almost identical phenologies and they are both attacked by the same two species of parasite, the mymarid *Alaptus fusculus* (Haliday *in* Walker) and a braconid, a *Leiophron* (= Euphorus) species near *similis* (Curt.). The *Mesopsocus* eggs hatch in early spring (late March to early April) and the nymphs pass through 6 instars of approximately equal duration. Adults appear towards the end of June and the oviposition period extends to early August by which time all individuals have died. Newly laid eggs develop rapidly and by September embryonic development is complete and the fully formed nymph can be seen through the chorion. These nymphs overwinter within the egg shell.

The mymarid, *Alaptus fusculus,* whose flight period of 3-4 weeks is synchronized with the oviposition period of its two hosts, oviposits in the newly laid eggs of both *Mesopsocus* species. The egg of the parasite develops rapidly and this is evidently necessary in order to arrest the development of the host egg at a very early stage. The stages of growth of the larva in the host egg are very similar to those described by Bakkendorf (1934) for *Alaptus minimus* Haliday *in* Walker. The eggs, which are of the pedunculate type characteristic of Mymaridae (Clausen 1940: 104), develop within a period of 4 weeks through a mandibulate-caudate type of 1st instar and a sac-like 2nd instar to a large sacculate larva which has ingested all the yolk and which now completely fills the host egg shell. By this time a white crescentic spot has developed in the posterior part of the larva, probably an excretory product since it is dissolved rapidly in KOH solution. Parasitized eggs in September are therefore characterized by their uniformly yellow colour with the white spot or crescent clearly visible through the chorion. They remain in this condition throughout the winter and early spring, and by mid-June the larva has pupated. The pupa, which is visible through the chorion of the host egg, remains transparent with practically all the yolk absorbed. About a week before adult emergence, an orange patch appears in the abdomen. The pupa changes to the adult form within the host egg, the eyes becoming red and the body cuticle darkening as it becomes sclerotized. Adults at this stage can be dissected out and they are fully developed **with** the wings and wing bristles fully expanded, so that the pupal skin has evidently been shed within the host egg. The adults emerge in July and appear on the twigs in largest numbers when most of the *Mesopsocus* eggs have just been laid. Only a single parasite larva is found in any one host egg. Among the thousands of parasitized eggs observed in September in the years 1958-1961, an egg containing two parasites has been recorded on only two or three occasions. A hundred parasitized eggs of *M. unipunctatus*, collected in July 1961 when the mymarid was ovipositing, each contained only one parasite larva in either the 1st or the 2nd instar, so that it seems likely that a female *Alaptus* lays only one egg in each host and that the *Mesopsocus* eggs once parasitized are then avoided by other female *Alaptus*.

The braconid parasite evidently oviposits in the 5th and 6th instars of both *Mesopsocus* species, since the parasite's flight period is at this time and since it is in these nymphs that 1st instar braconid larvae have been found by dissection. This 1st instar has a well developed head, bearing mandibles, and an elongate body with a tail—representing the mandibulate-caudate type characteristic of *Leiophron* (= *Euphorus*) (Clausen 1940: 49). It moults into an elongate worm-like 2nd instar and then passes through further instars growing to fill almost completely the abdomen of the 6th instar nymph of the host. A single host individual was always found to contain only one parasite larva. Although the muscles of the abdomen are damaged, the gut is never destroyed. Parasitized 6th instar nymphs of *Mesopsocus* do not moult into adults in the field but remain in this instar until the parasite emerges, by which time the unparasitized nymphs have changed into adults. The braconid larva, when fully grown, emerges through the abdominal wall of the host, which then soon dies. Larvae emerging from *Mesopsocus* nymphs in the laboratory fell to the floor of the container where they immediately spun silken cocoons in which they pupated. In the woodland they presumably pupate in the soil or litter beneath the trees. Some of the parasites which pupated in the laboratory emerged as adults several weeks later, probably a premature emergence. It seems probable that under field conditions the adult braconid emerges from the pupal case but remains within the cocoon throughout the winter to emerge the following June when the *Mesopsocus* are in the 5th instar. Cocoons, obtained in the laboratory, revealed, when opened in January, adult braconids with the wings fully expanded, alive but in a torpid condition.

Apart from the two species of *Mesopsocus*, four other species of psocid feed on *Pleurococcus* on these larch trees at Harrogate and are sufficiently abundant there to merit a quantitative examination of their life histories. These are *Philotarsus picicornis* (F.), *Elipsocus hyalinus* (Steph.), *Elipsocus westwoodi* McL. and *Amphigerontia bifasciata* (Latr.). The eggs of the first three of these species are indistinguishable from each other. From such eggs specimens of an

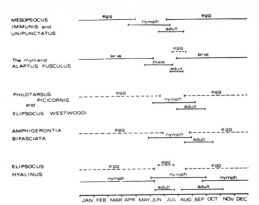

FIG. 1. The phenology of six psocid species and of one mymarid parasite on larch trees at Harrogate, 1958. (continuous line indicates occurrence observed, dotted line occurrence inferred).

Alaptus species have been bred out, and an *Alaptus* also parasitizes the eggs of *A. bifasciata*. A report on the taxonomy of all these forms of *Alaptus* will be published separately (Clark & Broadhead, in preparation). The nymphs of all these four species of psocids can be identified to species and to instar (Broadhead and Wapshere, 1961). *A. bifasciata* appears to be free from parasitization by braconids but braconid larvae have been dissected out from nymphs of *P. picicornis*, *E. hyalinus* and *E. westwoodi*. Parasites from the two latter hosts have been successfully reared through to the adults, identified as *Leiophron* (= *Euphorus*) *fulvipes* (Curt.). Since the parasites are in the same stage at the same time in all these hosts, the braconid living in *P. picicornis* is probably also of this same species. The 1st instar larvae of this *Leiophron* is found in the 1st, 2nd and 3rd instar nymphs of *P. picicornis* so that its larval development is prolonged compared with that of the species which attacks *Mesopsocus*. It does, also, however, emerge from the 6th instar nymph and drop to the ground where it spins a silken cocoon and pupates.

The phenologies of these six species of psocid feeding upon *Pleurococcus* and of the mymarid which parasitizes the two *Mesopsocus* species are set out in Fig. 1. This is based on an analysis of the weekly samples taken in the field from March to September in 1958. The precise limits of the egg stage in all, except *Mesopsocus* where they are directly known from the collections, are arbitrarily fixed as beginning just after the appearance of the first adults and ending just before the last occurrence of 1st instar nymphs. Three phenological types are recognizable. Firstly, the early summer species—*M. immunis* and *M. unipunctatus*. Both species hatch early in spring (April) and the population of each species feeds maximally, as judged by biomass, in the fortnight of late June to early July. Secondly, the late summer species—*P. picicornis*, *E. westwoodi* and *A. bifasciata*. These species have, like *Mesopsocus*, only one generation a year but they hatch later in spring (June) and their populations are taking most food in August and early September. In all stages of the life history, *A. bifasciata* is a week or more behind the other

TABLE 29. The numbers of animals and of old *Mesopsocus* egg batches occupying space on 16 sample twigs taken from each of 8 larch blocks, Harrogate, in the week of peak incidence of mature females of the two *Mesopsocus* species, 29 June 1959.

Larch block no.	*M. immunis* and *M. unipunctatus*	Other psocid species	Large stem aphis	Collembolan *E. nivalis*	Mite, *Angstis*	Hemerobiid larvae	Coccinellid larvae, pupae and adults	Psyllids	Other animals *	Total individuals	Previous year's egg batches of both *Mesopsocus* species still present
13	5	18	4	6	4	9	4	4	7	61	48
12	9	23	12	3	3	2	4	1	3	60	28
2	7	32	15	0	5	5	1	1	1	67	37
8	7	36	12	4	2	2	3	1	5	72	19
3	16	22	1	4	2	3	3	0	7	58	20
6	7	32	3	1	1	0	0	3	4	51	26
5	23	5	2	6	5	0	0	1	6	48	31
11	4	14	29	5	4	3	1	1	1	62	30
Total numbers on 128 twigs	78	182	78	29	26	24	16	12	34	479	239**

*Caterpillars, spiders, the Anthocorid *Tetraphleps*, parasitic Hymenoptera, syrphid larvae, etc.
**Containing 688 adult mymarids on the point of emergence.

two species. Thirdly, the species with two generations a year—*E. hyalinus*. In this parthenogenetic species, adult females appear in June and early July and again in August, September and October. There is one complete generation during the summer period and the animals overwinter as 2nd-4th instar nymphs and as eggs. Three other psocid species feeding on *Pleurococcus* also occur in this larch plantation at Harrogate but are present there in only very small numbers. They are *Graphopsocus cruciatus* (L.), *Stenopsocus immaculatus* (Steph.) and *Cuneopalpus cyanops* (Rost.). They also appear to have two generations a year but the numbers collected are insufficient to establish this with certainty.

[*Editors' Note:* Material has been omitted at this point.]

DISCUSSION

The Gause hypothesis that two species can coexist in an undiversified environment only if they occupy different ecological niches has stimulated over the past twenty years a great deal of ecological work, both on experimental laboratory populations and on natural populations of animals. There has been an interesting dichotomy of objectives in these two approaches. The laboratory approach has been concerned primarily with the study of the process and outcome of interaction when two related species are put together under conditions such that the one is bound to interfere greatly with the other. The work of Park, discussed in its more general aspects in the paper of Neyman, Park and Scott (1956), has shown the great potential of this laboratory approach in framing concepts which can serve as stimulating hypotheses for the field ecologist. On the other hand, field ecologists have for the most part proceeded by discovering situations where species, which have very similar ecological requirements and which therefore

may be competing for some limited resource, are nevertheless coexisting in an apparently stable way. More detailed examination of such cases has nearly always indicated some biological difference, often a a difference in food preference (Lack, 1944), which is presumed to be sufficiently great to minimise competition between the species and allow their coexistence. An adequate analysis of such natural situations will involve a careful comparison of all the ecological attributes of each species to find out just how extensive is the niche overlap. It will involve an enquiry as to whether each resource which is required by both species is limiting and it should also involve a study of the factors responsible for the equilibration of these populations. In the last resort, an adequate study of the coexistence of closely related species in nature involves a study of the mechanism of population regulation and an assessment of all the factors which influence the equilibrium level attained. Such a study is more likely to be successful, if it is concerned with relatively sedentary species exploiting food resources which are steadily renewed at any one place, than if concerned with species depending for their maintenance in any area upon adequate dispersal and the searching out of highly circumscribed ephemeral habitats.

The psocids occurring on woodland trees are particularly suitable animals for this type of study. They pass their whole life cycle in the one habitat and they are relatively sedentary. The two *Mesopsocus* species have two additional useful properties, namely that the adult females are wingless and that eggs and nymphs as well as adults are readily identifiable to species. Their food is self-renewing and widely dispersed throughout the habitat in which they move.

On larch trees at Harrogate and in many other places in northern England (Broadhead, 1958, Tables 5 and 12), 5-6 psocid species, as well as a Collembolan, live together in considerable numbers on the same trees and feed on the same food, i.e. the intimate mixture of *Pleurococcus* and fungal spores present on the bark of trunk and branches. In such circumstances, interspecific competition for food would be expected to occur at some time and in some places, competition being here used in the sense of a modification of mortality and/or natality rates of one species resulting from a food shortage induced by the demands made on this food by the other species. Partial spatial or temporal separation of the feeding stages of these species would reduce the intensity and the incidence of competition between them. Spatial separation on the larch tree of closely related common species which feed on the same food is well marked in the two *Amphigerontia* species and in the two *Elipsocus* species, *westwoodi* and *hyalinus* (Broadhead, 1958, Table 7), as also it is in the two much rarer *Peripsocus* species (Broadhead and Datta, 1960). In each of these cases, one species occurs for the most part on the living branches whereas its close relative occurs mainly on the dead ones. On the other hand, disregarding these taxonomic categories, of the six common *Pleurococcus*-feeding species occurring on the larch branches at Harrogate (Fig. 13), the three which are the most abundant are also those showing the least spatial separation on the tree. These three—the two *Mesopsocus* species and *Philotarsus picicornis*—each occur at approximately the same densities on living as on dead branches, whereas *E. westwoodi* occurs mainly on the living and *A. bifasciata* and *E. hyalinus* mainly on the dead branches (Broadhead, 1958, Table 7). The extent of temporal separation of these species is shown in Fig. 13. The two *Mesopsocus* populations feed early in the summer followed by the other four psocid species in the late summer, the *Mesopsocus* populations making the greatest as well as the earliest demand on the food (Fig. 14). Feeding by these late summer populations is unlikely to affect seriously the food available to the *Mesopsocus* populations the following spring since the winter allows a period of respite for the food organisms (Fig. 14) but it seems likely that in some years the *Mesopsocus* populations will reduce the food level sufficiently to induce a food shortage for the late summer species. This appears to have happened in 1960 in the Harrogate plantation (Fig. 13). Comparison of nymphal abundance in 1959 and 1960 shows an increase over this period for all species except *E. westwoodi*. The increase was particularly large in *P. picicornis*, relatively large numbers of adults being produced in the late summer of 1959 and these giving rise to the exceptionally large number of 1st instar nymphs in 1960. In the early spring of 1960 (Fig. 14) the food failed to regenerate to its normal level before psocid grazing started. The two *Mesopsocus* populations grazed this reduced supply first and, judging from the numbers of adults produced, did not suffer a noticeable food shortage. Very few adults of *P. picicornis*, however, were produced from the very large number of 1st instar nymphs, and correspondingly few nymphs emerged in the spring of 1961. Similar results are noted for *E. westwoodi*, *E. hyalinus* and *A. bifasciata*, in all of which fewer adults appeared in the late summer of 1960 with a reduction of the population in 1961. The relatively greater reduction of *P. picicornis*

may be attributed to the more sedentary habits of the nymphs, which spin loose silken tents and so are probably more susceptible to local food shortage. This interaction between the early summer feeding populations and the late summer ones appears in this instance to have been initiated by a sudden reduction in growth rate of the food due to causes unrelated to psocid grazing. It leaves unresolved the question of how frequently such populations would increase, in the absence of extraneous disturbing influences, to levels which would induce a food shortage. Neither partial spatial nor partial temporal separation of the various species populations, however, could itself eliminate competition entirely. Yet the coexistence of all these psocid species appears to be a permanent and fairly stable one.

The general ecological problem may be stated more precisely thus: with reference to one food resource, what conditions allow its exploitation by the greatest number of herbivores? This question will now be elaborated theoretically and then discussed with reference to what is considered to be the most critical material, namely the dynamics of coexistence of the two closely related *Mesopsocus* species in a number of replicate larch blocks in the Harrogate plantation.

There is now much evidence to show that, in a closed or self-contained homogeneous environment, two species cannot coexist if they require the same resource. In a purely theoretical way this was expressed in the Lotka-Volterra equations of interaction between two non-predators. It has been demonstrated empirically in the simpler ecological systems by many laboratory studies, particularly those of Park (1956) on mixed *Tribolium confusum* and *castaneum* populations, where the habitat was reasonably homogeneous and emigration and immigration excluded. It evidently does not apply to the more complex systems. The Gause hypothesis did represent a bold and stimulating extrapolation of such a simple model but was inadequate when applied to the much more complex situations in the field. The possibilities of reducing or obviating the results of competition between two such species requiring the same food resource by introducing environmental heterogeneity may be stated as follows: 1. *Coexistence of potentially competing species.* Two species requiring the same food may coexist in a stable way provided that the two populations are separately governed, each by its own density dependent factor, at a level below that which would result in food shortage. 2. *Coexistence of competing species.* Two species competing for food may coexist indefinitely provided that there exists a balance of advantages, which may be operative·(a) in time, (b) in space, or (c) through a common enemy. (a) As Park has shown, the competitive prowess of two species may be reversed by climatic change, so that an environment whose climate oscillates with an appropriate frequency around an appropriate mean value would allow prolonged coexistence of the two competing species whose fluctuations would be out of phase. (b) Coexistence is possible provided that one species is favoured, or has a breeding refuge, in one part of the habitat and the other species in another part. Movement from these refuges would maintain a permanent zone of coexistence. (c) A common enemy such as a parasite or predator could act as a stabilizer to the populations of two competing species provided that it could develop a preference for either of the host (or prey) species depending on their relative abundance, i.e. that it prefers whichever species happens to be the commoner at any particular time. Another model incorporating a balance of advantages, in this case with reference to two hypothetical plant species, has been described by Skellam (1951). 3. *Transient coexistence.* Many examples of coexistence in nature, especially in disturbed habitats, may well represent a transient and unstable condition, one species being in process of decline to extinction. Should such local extinctions in a series of semi-isolated populations be followed by local re-establishment due to immigration, the coexistence of these species would be greatly prolonged, especially if the competitive powers of the two species were sufficiently similar to result in alternative outcomes to the competition process in replicate populations as described by Park (1956). The more similar the responses of two species to shortage of any one resource, the longer will they coexist in competition for this resource. These ideas can be regarded as *a priori* models available for testing and for the interpretation of ecological situations, either those much simplified by design in the laboratory or those more complex and largely uncontrollable in the field.

The two *Mesopsocus* species at Harrogate provide critical material for the study of the dynamics of coexistence of competing or potentially competing species in nature since they are so very similar to each other in so many ways and since the plantation at Harrogate provides a number of replicate mixed species populations, genetically similar because the males are winged, but ecologically largely isolated because females are wingless. Similarities in the ecological attributes of these two species are many. They are almost identical in size and body form, in the number of nymphal instars (Fig. 11) and in the phenology of all stages in the life cycle (Fig. 10). The habitats of the two species are coextensive over eleven tree species in a mixed woodland (Table 2) and, within the larch and pine plantation at Harrogate, individuals of both species occur together in all the blocks (Appendix A) and in all parts of the blocks (Table 4). The dispersion patterns of the two species within blocks are very similar on pine and not very different from each other on larch (p. 334). The populations of both species are dispersed throughout the entire larch tree with very similar trends in the vertical distribution of eggs on the tree (Table 7). The difference between the two species in proportional distribution on dead and on living branches is only very slight (Table 5) and the density patterns of their nymphs along the larch branch are very similar (Fig. 8). The two species respond in the same way to climatic variations from year to year with regard to the hatching of the eggs and the onset of oviposition (Fig. 12a, b). They both act as hosts to the

same two parasite species. The population changes in 8 blocks of trees over a 6-year period recorded in Fig. 16a shows that these two species are responding to their total environment, both climatic and biotic, in a remarkably similar way. This is confirmed by comparison of the survivorship curves for the nymphal and adult periods in Fig. 18 and by the changes in numbers in 23 larch blocks over a 4-year period (Fig. 3). These similarities, together with the observation that both species take the same food in nature throughout the nymphal and adult period (Table 1) and respond in a similar way to shortage of their preferred food (Broadhead, 1958, Table 4), point to the following possibilities. They may at some time or other be competing with each other for food. If so a balance of advantages must be sought if the system appears to be a stabilized one (model 2 above). On the other hand, the system may have the appearance of stability but be fundamentally an unstable one, with competition proceeding towards extinction of one of the species in subpopulations but with subsequent reinforcement by migration (model 3). However, should competition be shown to be negligible in incidence and severity, then the clue to an understanding of their coexistence is likely to be found in the differences between the two species (model 1). These will now be examined in turn.

That the coexistence of these two species is stable is suggested both by the distribution studies and by the study of their population dynamics. The former studies all indicate that the two species are invariably present together in local areas and this must be considered in conjunction with the fact that females are wingless, movement from one population unit to another being thereby severely restricted. Thus both species occurred together on each of the 11 tree species at Fountains Abbey and they occurred together in each of the 23 larch blocks over the 4 years of this study (Appendix A). The latter studies relate to the highly correlated changes in the populations of the two species from year to year both in the 23 larch blocks over the 4-year period (Table 6) and in 8 of these blocks over a 6-year period (Fig. 16a).

This information has a bearing also on the problem whether or not interspecific competition occurs in these replicate mixed species populations in the larch blocks at Harrogate. The lack of any significant negative correlation between the numbers of the two species on living branches on the sides of all the 13 larch blocks and the 21 pine blocks indicates that, spatially within this plantation, areas of high *immunis* density are not associated with areas of low *unipunctatus* density and vice versa. This points, with due regard to the limitations specified on p. 332, to a lack of any significant competitive interaction. This tentative conclusion is reinforced by reference to the significant positive correlation between the numbers of the two species in time both in the 23 blocks (p. 335) and in the 8 blocks of larch (p. 356). In no larch block is the increase of one species associated with the decrease of the other over the 4-year period. The two species either increase together or decrease together from one year to the next. These considerations render very unlikely the possibility that the *Mesopsocus* populations in this plantation represent a mosaic of local extinctions, or reductions, of one or the other species which are the local results of an intense competitive interaction between them. Taken together with the results of the population dynamics study, which will be discussed below, they render unlikely the supposition that these two populations are simply undergoing a "random walk" (Park, 1956, Fig. 3) from which, sooner or later, they will depart with the then predictable elimination of one species. This sharply distinguishes the ecological situation involving these mixed species populations of *Mesopsocus* from that modelled by Park for the mixed species populations of *Tribolium*.

The coincident increases, and decreases, in the populations of the two *Mesopsocus* species have been shown to be, in part, due to changes in natality from year to year (Fig. 31). There is little doubt that if food were ever in short supply, then the first and major response of the *Mesopsocus* populations would be a reduction in natality rather than an increased mortality. The food available per unit psocid biomass is minimal when females are just emerging as adults and mobilizing protein for the production of their large yolky eggs (Fig. 14). The correspondence between natality changes in both species and changes in food level from year to year (Fig. 24) suggests the possibility of a causal connection. Although the two species show roughly the same degree of change in natality from year to year, further analysis (p. 369) indicates that the relationship between oviposition rate and *Pleurococcus* is significant for *immunis* but not for *unipunctatus*. This could be interpreted as follows. It is almost inconceivable that two species will respond to food shortage in exactly the same way. Should food shortage occur and should *unipunctatus* be more efficient in searching for food than *immunis* (the greater freedom of movement of *unipunctatus* within the larch blocks has already been commented upon on p. 335), then the oviposition rate of *immunis* would be expected to show a closer relationship with food level than would that of the more successful species, *unipunctatus*. It will be remembered that in the studies of distribution within the larch blocks (p. 334) a closer relation was noted between food level and nymphal densities in *immunis* than in *unipunctatus*. In parenthesis, it may be noted here that Harrogate at 500 feet altitude is in the foothills of the Pennines where, above 800 feet, *unipunctatus* exists but *immunis* cannot survive (Broadhead, 1958, p. 240). The greater winter loss of egg batches in *immunis* as compared with *unipunctatus* and the greater sensitivity of *immunis* to lower temperatures with regard to oviposition have already been suggested as factors contributing to the extinction of this species at places above 800 feet altitude. The competitive effect of *unipunctatus* upon *immunis* cannot be excluded as another possible contributory factor.

These comments on the likelihood of competition are speculative. Reasons have already been given for regarding the yearly changes in oviposition rate as genuine changes in natality but the possibility still remains that a migration component may be involved in these estimates (p. 368). Moreover the graphs in Fig. 14 refer only to food within the sample zone and to the *Mesopsocus* recorded in this zone. While these figures will reflect in a general way the changes in food level on the tree as a whole, they give no indication of absolute amounts of food available, and the individuals recorded in the samples no doubt spent part of their time further down the branches outside the sample zone where (Fig. 8c) *Pleurococcus* was more abundant. These results must be considered also in conjunction with the results of the multiple regression analyses relating numbers at the beginning and at the end of the two periods of adult life (Table 24). No significant relationship was found between changes in numbers of each of the *Mesopsocus* species and the numbers of the other species present at the same time for either of these two periods. This is negative evidence but the data on which it is based are adequate to indicate the significance of other factors, notably temperature. A strong competitive interaction at this stage of the life cycle over these 4 years does not, therefore, seem to occur. It must be remembered that all these regression analyses use the data for the 8 blocks of trees for the 4 years as a single series of 32 sets, so that the effects of interspecific competition occurring in perhaps only one of the four years may be, statistically, masked by inclusion in the whole series.

None of the evidence presented in this paper points clearly to the existence of a balance of advantages of any of the types described in model 2 above. There is indicated in Table 4 some differentiation of habitat within the larch blocks in this plantation, *immunis* being more abundant on the west side than on the east side of the block and *unipunctatus* more abundant on the east than on the west side. However this differentiation is not evident in the pine blocks where the two species also coexist. The slight differentiation of habitat on the larch tree shown in Table 5 suggests the possibility that *immunis* is favoured on the living branches and *unipunctatus* is favoured on the dead ones, movement of the animals from the one to the other part of this habitat maintaining a coextensive distribution of the two species over the entire tree (model 2b). Further assessment of this interesting possibility would require a knowledge of mortality rates of each species on dead, as well as on living, branches and an examination of the significance of the differences between the species in colour and in colour variation (p. 328). Repeated reversals of the outcome of competition dependent upon environmental fluctuations in time (model 2a) appear highly unlikely since they would not produce the highly positively correlated changes in the populations from year to year which are recorded for these two *Mesopsocus* species (pp. 335

and 356). This latter finding also renders unlikely the possibility that the mymarid parasite operated over these four years in the manner described in model 2c.

With regard, then, to the data already discussed, there is some suggestive evidence that the two species may have been competing for food at some time during these four years with a relatively greater adverse effect upon *immunis*. This evidence is not strong and other findings point to the lack of any competitive interaction of importance. In the absence of intense competition between the species it is unlikely that a balance of advantages would be readily discovered. On the other hand the coexistence of the two species appears to be a stable one so that a clue to an understanding of this stability must be sought for in the differences between the species. The outstanding differences are few. The extinction of *immunis* but not of *unipunctatus* above 800 feet altitude has already been discussed. The colour difference has also been mentioned. The egg batch of *immunis* is covered with a strong closely-woven silken web, which is completely absent in *unipunctatus*. The ecological consequence of this difference is that the mymarid can attack *unipunctatus* much more successfully than it can attack *immunis*, a situation which may in the past have determined the direction of selection pressure on the braconid, which can only survive in this habitat if it is more efficient in its attack on *immunis* than on *unipunctatus*. The last major difference between the species is in the pattern of density distribution of the eggs along the larch branch (Figs. 4, 5, 6). The ecological significance of this difference will now be discussed.

In contrast to the weakness of evidence for interspecific competition, the evidence for intra-specific competition in the adult stage in each species is strong and leads to the conclusion that this intraspecific competition is for oviposition sites and not for food resources. The regression equations relating numbers at the end to numbers at the beginning of each of the two periods of adult life (Table 24) describe a density governing process for each *Mesopsocus* species. In the preoviposition period, and probably also in the oviposition period, this process is a density induced movement of adult females out of the sample zone when densities there are high and into this zone when densities are low. The population outside the sample zone is acting as reservoir and regulator of the population inside this zone. Movement into the sample zone when numbers are low can be produced only if the sample zone is a region preferred by the adult females of both species because it provides some special resource. This resource must be different in the two species, since no relationship was found in these regression analyses between changes in number of one species and the number of the other species. This resource cannot be food, since both species require the same food and since food is present more abundantly outside the sample zone than within it (Fig. 5). The only resource, required

by adult females at this time and satisfying these requirements, is oviposition site. Both species lay their eggs at greatest densities within the sample zone (the terminal 75 cm lengths of living branches) and modal differences between the species exist both for the regions within the zone where egg density is greatest (Fig. 5) and for the actual sites preferred (Table 12). Such differences could operate to allow intraspecific competition and minimize interspecific competition only if they reflect innate differences between the two species in the behaviour of ovipositing females. That this is so is clearly indicated by the comparisons of the egg density distributions at Harrogate, where both species occur together, with those at Witton Fell and at Ottery St. Mary, where only one of the species occurs (Fig. 6, 7). It is appropriate in this context to indicate the greatest density of *Mesopsocus* egg batches ever recorded on any one sampling date during the four years of this study. The numbers of newly laid egg batches in the 128 twig sample unit taken on 7 August 1959 were 94 *immunis* (733 eggs) and 382 *unipunctatus* (2393 eggs), equivalent to 1 batch per 36 sq.cm and 1 per 9 sq.cm. of twig surface for the two species respectively. On this date the greatest number on any one 24 cm long twig was 15 (97 eggs) or 1 batch per 1.8 sq.cm. At these densities there has been an acute shortage of sites available with intense intraspecific competition. The actual number of oviposition sites available to these females is evidently very much less than the apparent number that superficial inspection of a twig may suggest. Table 29 records the numbers of animals and of the previous year's *Mesopsocus* egg batches present on the twigs on 29 June 1959, when the current year's eggs were being laid at their greatest rate. A total of 239 old egg batches still remained to occupy space on these twigs, thereby denying it to the ovipositing females, and the 479 animals (1 per 7 sq.cm) moving about on the twigs would not only occupy space but be a source of much disturbance to these females. They would be subject also to disturbance from the movement of the twigs by wind, from birds and from flies which were recorded in the samples. Clearly a larch twig is a very busy place, only a small proportion of its total area being available for oviposition, and fluctuations in numbers of all these other kinds of animals will produce changes in the number of available oviposition sites for the two *Mesopsocus* species from year to year.

In conclusion, the dynamics of the coexistence of these two *Mesopsocus* species appears to conform to model 1 described earlier in this discussion. Within this larch plantation over these four years, the populations of the two species were separately governed, each by intraspecific competition for its own particular oviposition site, at a level which induced little or no interspecific competition for food. Under these conditions, intraspecific competition being greater than interspecific competition, a stable coexistence is possible and appears to have existed. The two parasite species were also coexisting although they too exploit the same food resources, namely the two host species. In so far as an increase in one will lead to a decrease in the other (Table 28 and p. 375) they may be regarded as being in competition with each other. Although further information on the occurrence of the braconid in this *Mesopsocus* community at different places and over a longer time is desirable, their coexistence does appear to be stable (Fig. 25) and this can be ascribed to the difference in headquarters of the two species, the mymarid being more efficient in its attack on *unipunctatus* but the braconid being more efficient in its attack on *immunis*. These main conclusions are embodied in the quantitative model described in the last section of this paper (Table 26).

[*Editors' Note:* Material has been omitted at this point.]

REFERENCES

Bakkendorf, O. 1934. Biological investigations on some Danish Hymenopterous egg-parasites, especially in Homopterous and Heteropterous eggs, with taxonomic remarks and descriptions of new species. Ent. Meddel. 19: 1-135.

Betts, M. M. 1955. The food of titmice in oak woodland. J. Anim. Ecol. 24: 282-323.

Betts, M. M. 1956. A list of insects taken by titmice in the forest of Dean (Gloucestershire). Ent. mon. Mag. 92: 68-71.

Broadhead, E. 1958. Some records of animals preying upon psocids. Ent. mon. Mag. 94: 68-9.

———. 1958. The psocid fauna of larch trees in northern England—an ecological study of mixed species populations exploiting a common resource. J. Anim. Ecol. 27: 217-63.

——— & B. Datta. 1960. The taxonomy and ecology of British species of *Peripsocus* Hagen (Corrodentia, Pseudocaeciliidae). Trans. Soc. Brit. Ent. 14: 131-46.

——— & I. W. B. Thornton. 1955. An ecological study of three closely related psocid species. Oikos 6: 1-50.

——— & A. J. Wapshere. 1961. Notes on the eggs and nymphal instars of some psocid species. Ent. Mon. Mag. 96: 162-6.

——— & ———. 1966. Notes on the biology of British *Mesopsocus* species and on the weights of some psocid species. Ent. Mon. Mag. in press.

——— & ———. 1966. The life cycles of four species of Psocoptera on larch trees in Yorkshire. Trans. Soc. Brit. Ent. 17: 95-103.

Clark, M. L. & E. Broadhead. Taxonomic observations on some species of *Alaptus* (Hymenoptera, Mymaridae). in preparation.

Clausen, C. P. 1940. Entomophagous insects. New York.

Cole, L. C. 1960. Competitive exclusion. Science 132: 348-9.

Davies, O. L. 1961. Statistical methods in research and production. London.

Hincks, W. D. 1959. The British species of the genus *Alaptus* Haliday *in* Walker (Hym., Chalc., Mymaridae). Trans. Soc. Brit. Ent. 13: 137-48.

Kendeigh, S. C. 1961. Animal Ecology. London.

Lack, D. 1944. Ecological aspects of species-formation in passerine birds. Ibis 86: 260-86.

Nicholson, A. J. 1933. The balance of animal populations. J. Anim. Ecol. 2: 132-78.

——— & V. A. Bailey. 1935. The balance of animal populations. Part 1. Proc. Zool. Soc. Lond.: 551-98.

Park, T. *in* Neyman, J., T. Park & E. L. Scott. 1956. Struggle for Existence. The *Tribolium* model: biological and statistical aspects. Proc. 3rd Berkeley Symposium on Math. Statistics and Probability, 1954-55; 41-79.

Skellam, J. G. 1951. Random dispersal in theoretical populations. Biometrika 38: 196-218.

Varley, G. C. & G. R. Gradwell. 1960. Key factors in population studies. J. Anim. Ecol. 29: 399-401.

38

Reprinted from *Ecology*, **47**(3), 393–407 (1966), with permission of the publisher, Duke University Press, Durham, N. C.

SOME ASPECTS OF THE ECOLOGY OF MIGRANT SHOREBIRDS

Harry F. Recher[1]

Department of Biology, University of Pennsylvania, Philadelphia, Pennsylvania

Abstract. During migration, shorebirds form dense multispecific aggregations within relatively uniform and limited marine littoral habitats. The amount of available feeding space in the habitats frequented fluctuates widely with the daily and seasonal changes in the tidal rhythm. Shorebird species broadly overlap in their periods of peak abundance, inter- and intrahabitat distributions, and in the food organisms preyed upon. However, the totality of species differences and the transient character of migratory assemblages apparently minimizes interspecific interactions that might result in competitive exclusion. The staggering of peak population densities and differences in distribution is most pronounced among morphologically similar species. The number of individuals and species that occur in an area is apparently determined by the amount of available feeding space and the physical diversity of the habitat. Food appears to be generally abundant relative to the requirements of the birds in all the habitats studied. The environmental conditions encountered during migration and the interactions with other individuals have evidently been important factors in the evolution of morphological and behavioral differences among shorebird species. Individuals must be as able to survive during the nonbreeding season as they are during the breeding season.

Introduction

Whenever assemblages of organisms are studied it is found that the organisms comprising the assemblage utilize different but overlapping segments of the environment (Gibb 1954; MacArthur 1958; Kohn 1959; Dixon 1961; Hamilton 1962; MacIntosh 1963; and others). Coexisting animal species differ to varying degrees in food organisms preyed upon, in feeding behavior, and in the part of the habitat utilized for feeding and reproduction. Unfortunately, the majority of studies upon animal assemblages have been concerned with breeding bird populations among which the requirements of reproduction place a premium upon the acquisition of food. Among most terrestrial bird species, population density is minimal during this period. Studies of breeding bird populations investigate interspecific interactions at a period when such interactions may be subordinate to concurrent intraspecific interactions and when population densities are closely regulated through intraspecific territorial behavior. It is therefore of some interest to consider a population of organisms which are not actively engaged in reproduction, where the amount of food appears adequate relative to the demands of the individual, but where the amount of space may be considerably restricted through environmental variables and very high population densities. This is the situation prevailing during migration among many groups of birds in the North Temperate Zone.

Migrating individuals associate in aggregations which have greater population densities than those of the breeding season. Species also come together which would otherwise remain geographically or ecologically separated. The high population densities prevailing during migration increase the probability of interindividual contact and therefore create an ideal situation for studying ecological and behavioral interactions among species.

The importance of the migratory and nonbreeding phases of the avian life cycle has certainly been underestimated. Not only may there be considerable numbers of individuals which do not breed during a particular year, but it is typical of many migratory birds to spend more than three-fourths of the year in migration and on the wintering grounds. Therefore, it might reasonably be expected that environmental conditions encountered during migration play an important role in the evolution of migratory birds. Morphology and behavior must be as suitable for the conditions of migration as for those of the nesting season.

Migrant waders (Order Charadriiformes) were chosen for the present study because a large number of species with similar requirements come together during migration forming dense multispecific aggregations in relatively uniform, limited habitats. Furthermore, in marine habitats frequented by migrant shorebirds, the amount of available feeding space fluctuates with the tidal rhythm.

This paper is primarily concerned with ecological relationships among migrant waders and will mention behavioral interactions only briefly. A second paper will deal more extensively with behavioral interactions and the effect of such interactions upon the structure and organization of populations and communities.

I wish to thank Robert MacArthur for many helpful suggestions during the preparation of the

[1] Present address: Biology Department, Princeton University, Princeton, New Jersey.

manuscript. Paul Ehrlich and Richard Holm provided invaluable assistance during the course of this research at Stanford University and in the preparation of the dissertation from which this report is in part taken.

The research was supported by National Science Foundation predoctoral fellowships and United States Public Health Training Grants 5T1 GM-365-01 and -02 at Stanford University. The manuscript was prepared during my tenure as a Postdoctoral Fellow of the National Institutes of Mental Health at the University of Pennsylvania.

PROCEDURE

The ideas presented here are based upon 2 years of field observation along the central coast of California. Work was begun in August 1961, and, as described here, includes observations made through September 1963. An effort was made to investigate the problem from all possible aspects, whereas the work was essentially restricted to field observation. The environment was neither manipulated nor controlled, but its natural variation provided a continuous series of "experiments."

Areas of observation

Along the central coast of California the most extensive habitat utilized by waders is the marine littoral zone. Suitable fresh water and upland areas are very restricted. Observations were made on as wide a range of habitat types as possible within the coastal area bounded by Salmon Creek, Sonoma County in the north and the Carmel River, Monterey County in the south. Major areas of observation were Bodega Bay, Sonoma County; various portions of San Francisco Bay; and Elkhorn Slough on Monterey Bay, Monterey County. The Palo Alto tidelands on San Francisco Bay were chosen for intensive investigation. This area has extensive mudflats subject to tidal variation and supports dense aggregations of waders throughout migration and during the winter. Areas of tidal sandflat were used for observation at Bodega Bay and at Elkhorn Slough. The most extensive observations on areas of open sandy beach and of the rocky intertidal zone were made along Monterey Bay, Santa Cruz and Monterey Counties, and the San Mateo County Coast. (For a more detailed description of marine habitats along the California coast see Ricketts and Calvin 1939).

Observations were also made along the New Jersey shore between Tuckerton and Atlantic City during the spring and fall of 1964. These data are included where they supplement the more extensive California observations.

Census techniques

In each region of observation, an area for census was established within which all shorebirds present were counted. The Palo Alto area was censused an average of once every 10 days. Other areas were censused each time they were visited. Since census areas differed in size, the census data are not comparable quantitatively except on the basis of proportional species composition. However, within any area, numbers may be compared from census to census. It was found in the Palo Alto area that an accurate representation of the species composition within the area could only be obtained by censusing on the falling phase of a minus or near minus tide after at least 30%, but not more than 60%, of the tidal flats are exposed. (Storer 1951, followed a similar procedure in censusing the tideflats at Bay Farm Island, Alameda County). The number of shorebirds feeding on the Palo Alto tidal flats during a falling tide continues to increase until at least 30% of the tidal area is exposed. Beyond this point, the number of birds remains relatively constant. Difficulty in censusing, however, becomes evident as the falling tide exposes more than 60% of the mud flat, for the census area then becomes indistinguishable from surrounding regions.

The smaller areas involved at the other observation sites permitted censuses to be made under most tidal conditions. Whenever possible, the number of birds in an area was determined by more than one count, preferably involving both a falling and rising tide, with a check count made on the following day(s). However, in the Palo Alto area, it often was impossible to census all species on the same day within the accepted tidal limits because of the high population densities. In such cases, species not censused were counted the following day. Subdivisions were made within each census area in which different habitat or substrate types were present for the purpose of making intra-area comparisons.

Survey of food availability and utilization

In some instances, it was possible to determine the degree of feeding success and what prey organisms were taken by close observation of feeding birds. However, in order to obtain a more accurate picture of the feeding ecology of these birds, it was necessary to analyze stomach contents. An area at the mouth of San Francisquito Creek within the Palo Alto tideland area was selected for the collection of specimens. Collection was begun in September 1962, and continued through May 1963. Stomach analysis was supplemented by a limited survey of the invertebrates available as food to shorebirds within this area.

Comparative distribution of migrant shorebirds

The comparative species composition in different habitats can be determined from the census data. However, it was necessary to develop methods for investigating the intrahabitat distribution of species. In part, this can be determined by subdividing the census areas. In this way, intrahabitat differences in distribution (resulting from variations in substrate, exposure to wind, or the distribution and abundance of food organisms) were recorded as part of the census procedure.

However, this approach cannot answer questions concerning the effect of horizontal diversity, resulting from tidal movements, upon the distribution of individuals. This horizontal diversity is determined by the amount of water retained by the substrate above the water's edge, and by the depth of water beyond the water's edge. Therefore, the intrahabitat distribution of species was further investigated by dividing the habitat into a series of zones roughly parallel to the water's edge, and recording the number of birds occurring along the water's edge and within each zone.

On the Palo Alto mudflats, the area beyond the water's edge was arbitrarily divided into two zones. The first of these (zone A) extended from the water's edge to a parallel line approximately 1 ft from the water's edge. The second, zone B, included the area beyond the 1-ft parallel. Above the water's edge, the first zone was recognized as the area on which a surface film of water remained visible. The second zone above the water's edge consisted of those areas which lacked a surface film of water. Compensation was made for tidal pools or ridges by assigning them to one zone or another according to the depth of water or the amount of moisture retained by the substrate.

GENERAL ECOLOGY

Migration

Nonbreeding waders are found throughout the year along the coast of central California. How-

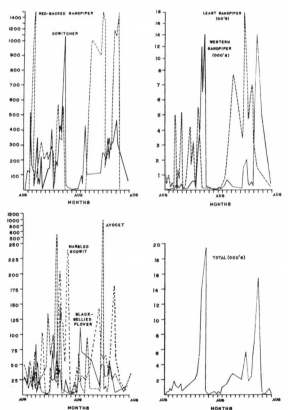

FIG. 1. Numbers of shorebirds at Palo Alto from August 1961 to August 1963.

ever, summering populations are small and very local in distribution. At Palo Alto, the spring migration begins during the latter part of January. Maximum population densities occur during the last 2 weeks of April after which there is a pronounced decrease in numbers. Northward-bound individuals are found throughout the month of May. The first fall migrants arrive early in July and the fall migration may be considered as continuing through November. The peak of the fall migration is reached in late October or early November, at Palo Alto. Total numbers decline through November, but remain relatively constant throughout December and the early part of January. Nonbreeding birds found during June are considered summering individuals. Birds present during December or January are considered to have reached the southernmost limit of their migrations and are therefore wintering. Individuals undoubtedly spend more than 2 months on their wintering grounds just as some individuals probably spend more than 1 month on a summering area.

When the census data for individual species are plotted, it appears that migration is not a gradual increase and decrease in the number of migrants (Fig. 1). Rather the impression is of successive groups of birds migrating into and out of an area. Similar observations have been made on migrating waders by Storer (1951) at Bay Farm Island, Alameda County on San Francisco Bay and by Urner and Storer (1949) along the New Jersey shore. The graphs show also that, at least during the spring, the peak population densities of some species tend to be staggered.

The wave-like character of migration is largely obscured when the total numbers of all species are plotted (Fig. 1), a good indication that species tend to migrate independently of each other.

In part, the fluctuations observed in population composition may represent local movements with individuals using a number of alternate feeding areas. It is also certain that individuals vary in the length of time they spend at any area during migration (Swinebroad 1964). At Palo Alto, occasional distinctive individuals were continually observed over a 1-month period. These individuals were remarkably constant as to the feeding and loafing sites frequented. Other equally distinctive birds would be seen once and then evidently leave the observation area.

It is known from banding studies done in Europe that breeding populations of some shorebirds frequent separate wintering areas (Salomonsen 1955). If these populations were to remain apart during migration, it could account for the different

waves of a species to pass through an area. There are also indications that among some species there is a partial segregation of sexes during migration. During the fall migration, the adults of many species migrate south in advance of the young and represent the first wave(s) to pass south (Bent 1927, 1929, Pitelka 1950).

It is generally accepted that the rate and duration of migratory movements are affected by prevailing climatic conditions. However, there are differences between the fall and spring migrations of waders which do not result directly from climatic conditions, and which can not be explained entirely on the basis of differences in "sexual drive" between spring and fall birds. On both coasts of North America, the rate at which the spring migration proceeds appears greater than during the fall. The graphs of spring migration (Fig. 1) show fewer peaks in population density than during the fall, but each peak tends to represent a much greater number of individuals. It seems that not only are individuals moving more rapidly during spring migration, but also the main period of migration is considerably compressed. Thus the population densities observed during the spring tend to be higher than those seen in the fall.

Available space rather than the amount of available food appears the environmental factor most important in limiting the size and density of migrant shorebird populations. Space is restricted on all high tides and during the first stages of the falling tide. However, the most significant restriction of space may occur during those tidal sequences in which only a limited area is exposed during low tide. The extent of daily tidal fluctuations (hence, the amount of feeding space available to waders) on the West Coast differs considerably between the spring and fall (Fig. 2). The difference in available feeding area may in part account for the differences observed in the rates of spring and fall migrations. During the

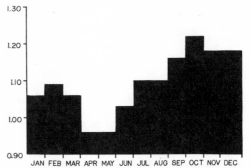

FIG. 2. Average monthly depth of low tide (in ft) above mean low-low datum (0.0 ft).

spring the average depth of the low tides above the mean low-low tends to be lower than during the autumn or winter. In addition, the tidal fluctuation is less during the spring. As a result, the tidal area exposed at low tide during the spring months tends to be greater than in the fall and to fluctuate less from one tidal sequence to the other.

The greater feeding area available during the spring may permit a larger number of individuals to aggregate without decreasing the feeding efficiency of the individuals present. It may also explain, in part, the "advantage" obtained among many species of waders among which the adults migrate south in advance of the young. One can presume the young of the year are less efficient feeders than the more experienced adult birds, and hence might be at a disadvantage if forced to compete with adults for food and space under conditions where the amount of space available for feeding was limited.

Migration as loosely integrated groups would tend to avoid population "pile-ups" in which the population density might increase to the point of significantly reducing the feeding efficiency of the individuals present. Because of their common environment, it can be assumed that the individuals comprising a group are at similar physiological and psychological levels and would, therefore, respond in a similar manner to environmental stimuli (Welch 1963). A communal response among the individuals of a group, rather than a haphazard response among aggregated individuals, would enable a more precise regulation of population density to some optimum level. As the number of individuals increases within a given area there is an increase in interindividual interactions followed by an increase in the level of activity among the bird species present (Recher MS). Such an increase of activity ultimately leads to dispersal within and migration out of the area concerned. Individuals comprising a group would tend to respond together to the influx of new "nongroup" individuals, and would therefore migrate as a group. Newly-arrived individuals, being at different pyhsiological and psychological levels, would not respond but would tend to remain within the area. Other stimuli probably are as effective and important in coordinating group movements as increasing population density. Climatic conditions, available food and the level of sexual excitement have all been mentioned as affecting the rate and form of migratory movements (Dorst 1962).

Morphologically similar species which utilize the same habitat may be able to coexist at times of low population densities, but come into competition at high population densities. Such species would be expected to migrate at different times. Among migrant waders along the coast of central California, there is only one example of a pair of morphologically similar, abundant, and yet, coexisting species, the Least and Western Sandpipers (*Erolia minutilla* and *Ereunetes mauri*, respectively). All other morphologically similar shorebirds tend to frequent different habitats. Behavioral interactions among individuals of these two species are more frequent than among any other pair of waders studied (Recher MS). The Western Sandpiper is slightly larger, and a more aggressive bird, and it appears to exclude the Least Sandpiper from many feeding sites. During spring migration, the peak population densities of these two species at Palo Alto are staggered (Fig. 1), the main migration of least sandpipers taking place 6 to 8 weeks before the April peak in numbers of the Western Sandpiper. During their peak, Western Sandpipers are widely distributed throughout the available feeding area and intraspecific aggression is more frequent than at any other time of the year. Presumably, if Least Sandpipers migrated at the same time, they would be subjected to an excessive degree of interspecific interindividual aggression. The coexistence of these two species becomes possible largely because the Least Sandpiper, by migrating early, is able to avoid the greatest population densities of the Western Sandpiper.

On the East Coast the similar Least and Semipalmated Sandpipers (*Ereunetes pusillus*) migrate at approximately the same time (Urner and Storer 1949; Stewart and Robbins 1958). However, on the East Coast there appears to be a greater degree of habitat separation between these two species than between the Least and Western Sandpipers. This is perhaps a result of the more extensive marsh areas which the Least Sandpiper seems to "prefer" along the East Coast as compared to the West Coast.

Because of their similar requirements, minimizing interindividual interactions is of greatest advantage among conspecific individuals. It may also be more advantageous among individuals of morphologically similar species which frequent the same habitats than among species which are ecologically more distinct. The spreading of individuals through a series of consecutive waves and the staggering of peak population densities among species segregate individuals through time. Alternatively, species may utilize different habitats or migration routes and thereby achieve spatial segregation.

Distribution

Most waders utilize a variety of different marine habitats—littoral areas that differ as to substrate composition and invertebrate fauna. During migration, 34 species of waders can be found along the central coast of California (Grinnell and Miller 1944). They are not all equally abundant, nor do all frequent the same habitats. However, they all do frequent habitats which are similar in several important respects.

Compared to forest communities, marine littoral habitats lack, from the point of view of this study, vertical diversity above the substrate surface. There is, however, vertical diversity within the substrate itself and horizontal diversity along the substrate surface. If parallel lines are drawn to the water's edge during low tide, diversity along a horizontal gradient can be expressed in terms of how long the area has been above water and how wet it has remained. Beyond the water's edge, diversity can be measured in terms of water depth. Horizontal diversity also results from the intermixture of different substrates. Vertical diversity below the substrate surface results from a

TABLE I. Per cent species composition of four census areas differing in substrate composition[a]

	Census areas and substrate composition			
Species	Bodega Bay	Palo Alto	Asilomar State Park	New Brighton Beach State Park
	Tidal sandflat	Tidal mudflat	Rocky inter-tidal zone	Open sandy beach
Black Oystercatcher	..	..	3.1	..
Semipalmated Plover	<0.5	<0.5	..	..
Snowy Plover	0.6	<0.5	..	..
Killdeer Plover	<0.5	<0.5	..	<0.5
Black-bellied Plover	1.8	1.0	0.7	<0.5
Surfbird	..	..	7.7	..
Ruddy Turnstone	<0.5	<0.5	..	<0.5
Black Turnstone	1.1	..	36.2	1.3
Long-billed Curlew	<0.5	<0.5	..	<0.5
Whimbrel	<0.5	<0.5	2.1	0.8
Wandering Tattler	..	..	3.9	..
Willet	2.5	8.4	2.5	3.9
Greater Yellowlegs	..	<0.5	..	..
Lesser Yellowlegs	..	<0.5	..	..
Knot	<0.5	<0.5	..	..
Least Sandpiper	9.8	6.6	..	..
Red-backed Sandpiper	22.0	12.1	..	..
Dowitcher Species	4.8	6.5	..	..
Western Sandpiper	23.6	60.4	..	..
Marbled Godwit	27.7	2.0	..	4.5
Sanderling	5.8	<0.5	44.2	89.4
Avocet	<0.5	2.3	..	..
Black-necked Stilt	..	<0.5	..	..
Total number	20,200	145,330	432	1,982
Number of censuses	7	70	18	16

[a]Figures based upon total number of birds censused between August 1961 and August 1963.

TABLE II. Per cent species composition of three different habitats within the Elkhorn Slough area, Monterey County. Figures based upon the total number of shorebirds censused between April 1962 and April 1963 (12 censuses)

Species	Open sandy beach	Marsh and marsh area	Sand and mudflat
Semipalmated Plover	..	1.0	<0.5
Snowy Plover	3.6	1.0	..
Killdeer Plover	..	0.5	0.5
Black-bellied Plover	<0.5	1.0	0.8
Ruddy Turnstone	<0.5	..	..
Long-billed Curlew	<0.5	1.0	0.7
Whimbrel	<0.5	..	..
Willet	59.6	56.5	30.2
Greater Yellowlegs	..	<0.5	<0.5
Knot	..	..	<0.5
Least Sandpiper	..	3.1	11.2
Red-backed Sandpiper	..	4.9	3.7
Dowitcher Species	..	5.7	<0.5
Western Sandpiper	..	<0.5	5.8
Marbled Godwit	4.6	15.0	20.4
Sanderling	31.2	2.1	26.5
Avocet	..	1.5	..
Black-necked Stilt	..	1.0	<0.5
Red Phalarope	..	..	<0.5
Wilson's Phalarope	..	0.5	..
Northern Phalarope	..	5.1	<0.5
Total number	1,605	1,617	2,327
Average species diversity	0.951	1.118	1.658
Number of equally common species	2.6	3.0	5.2

stratification of invertebrate food organisms. Additionally, tidal movements cause a temporal diversity which, not unlike the alternation between day and night, enables different assemblages of organisms to utilize the same area without coming into direct contact.

Interhabitat distribution

If comparisons are made, it is seen that species tend to be more abundant in one area than in another (Tables I, II). In part, the differences in species composition between two areas such as Bodega Bay and Elkhorn Slough may reflect species differences in migration routes. However, differences in population composition between adjacent areas seem to be closely correlated with substrate composition. Thus, the species found in the rocky intertidal zone differ from those found on adjacent sandy beaches. It appears that the effects of substrate composition accrue from differences among substrates in the composition of the invertebrate fauna and of the distribution of invertebrates within the substrate. Area differences in substrate composition affect the distribution and abundance of shorebird species in three ways: 1) Insofar as substrate composition determines the abundance and availability of food organisms, it also determines the absolute abundance of shorebirds. 2) The greater the diversity of the substrate(s) within an area, the greater the birds species diversity will be. 3) The more two areas differ in the substrates present, the greater the species replacement between the two localities.

MacArthur and MacArthur (1961) have shown for terrestrial communities that the bird species diversity of an area can be predicted from the vegetation profile. That is, there is a direct relationship between habitat diversity, as measured by physical characteristics of the habitat, and bird species diversity. One can predict that there would be a similar relationship between habitat diversity and bird species diversity among aggregations of migrant birds.

Unfortunately, it has proven difficult to arrive at a measure of habitat diversity in littoral areas which can be determined with the same precision as vegetation profiles. However, bird species diversities can be calculated for different habitats and these compared with a gross evaluation of habitat diversity.

Bird species diversity is calculated by the formula:

$$- \sum_i p_i \log_e p_i$$

in which p_i is the proportion of all the bird individuals belonging to the ith species (MacArthur and MacArthur 1961). The number of "equally common" species is determined by:

$$e^{(- \sum_i p_i \log_e p_i)}$$

In effect this is a measure of the "probability" that the second bird individual encountered will differ from the first, the third from the second and first, and so forth. Thus, the higher the bird species diversity (i.e., the greater the number of equally common species) the greater is the probability that the second individual encountered will differ from the first. In the rest of this report H is used to denote species diversity and e^H the number of equally common species.

When species diversities for different habitats are compared (Table III), it is found that diversity differs among habitats and changes throughout migration. In determining the "average" bird species diversity of an area, the average is taken of the diversities as calculated for the separate censuses. The change in population composition through migration is viewed in this case as a succession of different bird "communities." Therefore, to determine the bird species diversity by lumping these "communities" and taking the diversity of the total number of birds seen through-

TABLE III. Species diversities of five census areas representing different substrates or intermixtures of substrates

Month of census	Palo Alto (Mudflat)		Bodega Bay (Sandflat)		Elkhorn Slough (Open beach, mud- and sandflat, marsh)		Asilomar (Rocky intertidal zone)		New Brighton (Open sandy beach)	
	H	e^H	H	e^H	H	e^H	H	e^H	H	e^H
Jan.	1.388	4.0	1.220	4.5	2.048	7.7	1.358	3.9	0.365	1.4
	1.510	4.5	..	..						
Feb.	1.630	5.1			1.129	3.1	1.260	3.5	0.135	1.1
	2.641	14.0	..	..	..	..	0.830	2.3	..	..
	1.418	4.1	..	..			...	...		
	1.672	5.3	..	..	..	..	..	..		
Mar.	1.592	4.9	..	..	1.825	6.2	..	..	0.442	1.6
			..	..					0.000	1.0
Apr.	1.009	2.7	..	..	1.729	5.6	1.286	3.6	0.651	1.9
May	0.406	1.5	..	..	1.654	5.2	0.636	1.9	0.408	1.5
June	..	..								
July	1.146	3.1	1.516	4.5	1.376	3.9	0.677	2.0		
	..	..					0.981	2.7		
Aug.	1.502	4.5	1.848	6.3	1.296	3.6	0.530	1.7		
	2.073	7.9	1.557	4.7	..	..	..	..		
Sept.	1.825	6.2	1.610	5.0	1.966	7.1	1.401	4.1	0.355	1.4
			1.985	7.2						
Oct.	1.350	3.9	1.618	5.0	1.932	6.9	0.663	1.9	0.242	1.3
			..	..			0.467	1.6	1.170	3.2
Nov.	1.657	5.2			2.038	7.6	1.206	3.3	0.577	1.8
Dec.	1.792	6.0	..	..	1.914	6.8	0.310	1.4	0.232	1.3
	..	..	..	..	2.053	7.8	0.874	2.4	..	..
	..	..			..	..	0.969	2.6	..	..
Average	1.538	5.1	1.622	5.2	1.747	6.0	0.896	2.6	0.458	1.6

out migration would be the same as lumping a number of different habitats together and determining the diversity of the combined populations. In both cases, a greater diversity results, which is misleading when one wishes to deal with relatively homogeneous communities, or aggregations of individuals which are presumed to be relatively "stable units" arising through the processes of natural selection.

If our original prediction is correct, then differences among habitats in species diversity indicate corresponding differences in habitat diversity. Diversity in littoral habitats can exist in two planes: horizontally across the substrate surface, and vertically within the substrate (and its covering waters). If different littoral marine habitats are compared, it can be seen that they do differ in these parameters of diversity, and that the less diverse habitats (from the point of view of a shorebird) are also those with a lower shorebird species diversity.

Diversity is greater in areas with an interspersion of different substrate types. Elkhorn Slough, which consists of a number of different habitat or substrate types, has a relatively high species diversity throughout migration. However, areas of uniform substrate composition within the greater slough area have lower species diversities than the slough as a whole (Tables II, III). This effect can be duplicated in terrestrial habitats by increasing the area censused so as to include a greater number of habitat types (MacArthur, MacArthur and Preer 1962). Such diversity is determined by the variety of habitats selected for study but does illustrate the increase in environmental diversity with an intermixture of substrates.

Comparisons of bird species diversities (Table III) indicate that the rocky intertidal zone and open sandy beaches are less diverse than areas of tidal flat. Theoretically, horizontal diversity resulting from tidal movements will be constant within all tidal habitats. However, wave action along open sandy beaches and the rocky intertidal zone prevents shorebirds from utilizing any feeding areas found beyond the water's edge. Moreover, these two habitats lack diversity below the substrate surface—the substrate of the rocky intertidal zone is largely impenetrable, and the open sandy beach is characterized by an impoverished invertebrate fauna.

Tidal sand and mud flats can be expected to have much the same habitat diversity. This, as indicated by the bird species diversity, is evidently the case. Tidal flats normally have extensive horizontal diversity above and below the water's edge as well as vertical diversity within the sub-strate. The totality of this diversity is available to shorebirds and results in high bird species diversities.

An increase in the diversity of invertebrate food organisms would not necessarily in itself result in an increase in the bird species diversity. Of the habitats studied, the rocky intertidal zone undoubtedly has the greatest diversity of food organisms but these are largely unavailable to feeding shorebirds. (Many invertebrates found in the rocky intertidal zone are unsuitable as food for shorebirds, and many others frequent inaccessible places.) Consequently, the bird species diversity of this habitat is reduced relative to that of a tidal flat. One can also imagine that if a tidal flat supported only one kind of invertebrate or only a limited size range of food organisms, the number of different kinds of birds that could feed on this flat would be restricted. But one does not often encounter this situation on tidal flats. Therefore, the limiting factor to shore bird species diversity appears to be the diversity of the habitat and not the diversity of food organisms. However, the greater diversity of tidal flats as compared to habitats in which the substrate is impenetrable is, in part, a result of the stratification of food organisms within the substrate, to which the diversity of food organisms may be a contributing factor.

Intrahabitat distribution

Plotting the distribution of shorebirds over the tidal flats at Palo Alto shows a horizontal segregation of species (Fig. 3). Similar differences in intrahabitat distribution are seen along the New Jersey shore. However, the composition of the migrant shorebird populations in this area differs considerably from that of California. Fewer species are found together and population densities tend to be much lower. In particular, the "large" shorebirds either do not occur or occur in very small numbers along the New Jersey shore. The Short-billed Dowitcher (*Limnodromus griseus*) and the Knot (*Calidris canutus*) are the largest of the common shorebirds occurring along the central New Jersey coast during migration. The Western Sandpiper is replaced by its ecological counterpart, the Semipalmated Sandpiper. Greater Yellowlegs (*Totanus melanoleucus*) are more abundant than in California.

As along the California coast, species tend to differ in their intrahabitat distributions. It should be noted here that in the areas where these observations were made, the extent of exposed tidal mudflat above the water's edge is considerably foreshortened in the New Jersey areas relative to that found on San Francisco Bay at Palo Alto.

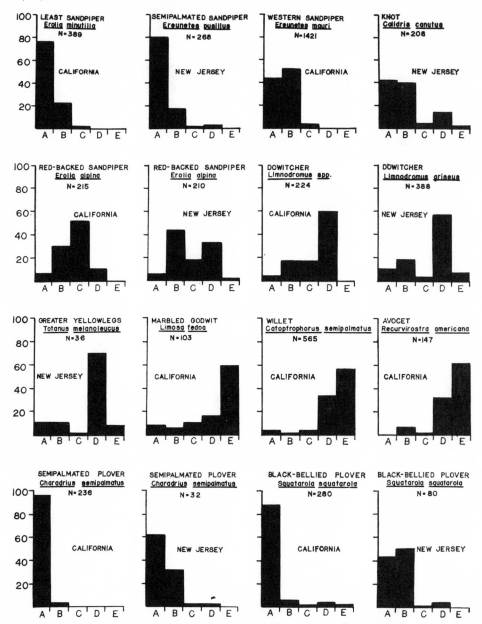

FIG. 3. Distribution (% occurrence) of migrant shorebird species on the Palo Alto census area and the New Jersey shore within a series of imaginary zones roughly parallel to the water's edge at low tide. The water's edge is represented by C. Above the water's edge, zone A = areas lacking a surface film of water; and zone B = areas retaining a surface film of water on the substrate. Beyond the water's edge, zone D = the area between the water's edge and a contour line 1 ft beyond and parallel to the water's edge; and zone E = the area beyond the 1-ft contour line.

This results in a more equal distribution of individuals between zones A and B (see legend of Fig. 3) on the New Jersey plots than was found with the same species in California, for example, the Black-bellied Plover (*Squatarola squatarola*) and the Semipalmated Plover (*Charadrius hiaticula*). With the exception of the Least Sandpiper, it appears that species occurring on both coasts have essentially the same distribution on tidal mudflats in relation to the water's edge.

In view of their morphological and behavioral similarity, it is appropriate to compare the distribution of Least, Western, and Semipalmated Sandpipers. The Least Sandpiper occurs with the Western Sandpiper along the California coast and with the Semipalmated Sandpiper along the New Jersey shore. In the areas of observation, the Western Sandpiper and the Semipalmated Sandpiper do not normally occur together. At Palo Alto, in California, the Least and Western Sandpipers were frequently observed feeding on the same tidal mudflats, though the Least Sandpiper also frequented marsh areas. On the tidal flats these two species had different but overlapping distributions (Fig. 3). In New Jersey, the Least Sandpiper was not observed to occur with any frequency on the tidal mudflats with the Semipalmated Sandpiper. Instead, it seemed to "prefer" areas of tidal marsh. It is therefore of interest to note the similarity of distribution on tidal mudflats of Semipalmated Sandpipers on the East Coast and of Least Sandpipers on the West Coast. Morphologically, the Least Sandpiper and the Semipalmated Sandpiper are the most similar of the three. An index based upon the ratio of bill length gives a figure of 1.1 for the Least and Semipalmated Sandpipers. Hutchinson (1959) states that for species to coexist, they must differ by a ratio (Brown and Wilson 1956) of at least 1.2 to 1.4. In view of their morphological similarity, it appears that the Least Sandpiper and the Semipalmated Sandpiper are not able to coexist in the same habitat (i.e., tidal mudflats). However, on the West Coast, the morphological differences between the Western and Least Sandpipers are sufficient (bill length ratio = 1.5) to permit a partial overlap in distribution.

The distribution and behavior of waders is essentially the same wherever there are extensive areas of tidal mud- and sandflats. Species segregation increases as the tide falls and is most pronounced when extensive tidal areas are uncovered on minus tides. The available feeding area utilized by shorebirds is limited when the tide first begins to fall. This results in a higher density of feeding individuals and a less distinct segregation of species than at a later time when a more extensive feeding area has been exposed.

As the tide falls, the birds follow and concentrate about the receding water's edge. At the maximum extent of low tide, individuals still tend to be concentrated immediately above and below the water's edge; species segregation is pronounced. On the low tide, birds begin to spread back over the exposed flats, the uppermost areas of which are almost completely deserted. With the rising tide, the birds fall back before the advancing water and the segregation of species once again becomes less distinct. As the tide rises more rapidly, individuals desert the flats for loafing sites or alternate feeding areas. This desertion becomes a mass exodus as the tide nears its upper limits.

On tidal flats, the increasing depth of water beyond the water's edge limits the distance in this direction that wading birds can feed. Waders with long legs and bills can feed in deeper water than those with shorter legs and bills. Phalaropes, by swimming, and stilts and yellowlegs, by feeding from the water rather than the substrate, are able to feed in deeper waters. On tidal flats, areas which do not have standing water are not utilized as frequently as wet areas by species which probe deeply into the substrate. Instead, they are frequented by visual and surface feeders such as willets, plovers, and smaller sandpipers.

Species which probe deeply into the substrate are presumably more efficient when the substrate is covered by water. It is probable that the distribution and behavior of invertebrate food organisms within the substrate changes as tidal areas are exposed by a falling tide. Clams, for example, may partially or completely withdraw their siphons. One can imagine that a clam with an extended siphon is easier to locate by virtue of its "larger" size and that birds feeding upon clams would tend to feed below and not above the water's edge. It is also probable that probing birds depend upon tactile stimuli received while only lightly probing or touching the substrate surface. Food organisms would, therefore, be more easily detected near the surface (as they are most apt to be when covered by water) than when the substrate is exposed.

Distribution is also affected by the clumping of conspecific individuals. During high population densities, a species is generally distributed throughout the area of observation. However, during low population densities, conspecific individuals tend to form flocks which remain as integrated units. As a result, during low population densities, a species will be absent from the greater part of the available habitat, but may be locally abundant. Flocks are formed during high population densities, but they appear to be less stable, birds continually entering and leaving. As the population density increases, there is a tendency for certain areas to be occupied before others. Once the density in such areas increases beyond a certain point, other less "preferred" areas are utilized.

A segregation of species somewhat similar to that found on tidal flats can be recognized among shorebirds feeding along open sandy beaches. At times, when water conditions permit, Willets (*Catoptrophorus semipalmatus*) follow the retreating waves further than either Sanderlings (*Crocethia alba*) or Marbled Godwits (*Limosa fedoa*). Marbled Godwits normally move only as far as the upper edge of the area wet by wave action, whereas Sanderlings follow close to the edge of the retreating wave. Snowy Plovers (*Charadrius alexandrinus*), preferring the dry areas above the high tide mark, seldom feed along the surf zone. Whimbrels (*Numenius phaeopus*) tend to feed in much the same area as Snowy Plovers. These five species are the only shorebirds that can be considered regular inhabitants of the open beach. Other species will be found along open beaches, but only sporadically or where there is an intermixture of sandy beach with other habitats.

Open coastal areas where there is an intermixture of the rocky intertidal zone with sandy beach often support large and diverse populations of migrant shorebirds. Such areas are normally sheltered from direct wave action and have a greater diversity of feeding sites and food organisms than either the rocky intertidal zone or open sandy beach by themselves. Within this mixed habitat, unlike the rocky intertidal zone itself, a certain amount of horizontal and vertical diversity exists which results in a partial segregation of species.

Despite what appears an extremely abundant food supply, the rocky intertidal zone seldom supports dense aggregations of migrant waders. Wave action, the limited exposure on most low tides, and the relative briefness of this exposure are undoubtedly contributing factors to the small numbers of migrant waders which utilize the rocky intertidal zone. The most abundant species within this habitat, the Sanderling and the Black Turnstone (*Arenaria melanocephala*), utilize and probably depend upon alternate feeding areas. The Sanderling frequents adjacent sandy beaches whereas the Black Turnstone utilizes food organisms found in the windrows of kelp above the high tide line. None of the other waders charac-

teristic of the rocky intertidal zone are particularly abundant. The Black Oystercatcher (*Haematopus bachmani*), the Surfbird (*Aphriza virgata*), and the Wandering Tattler (*Heteroscelus incanum*) are not normally found outside of this habitat, and their small numbers may reflect the spatial limitations of the rocky intertidal zone.

There are no intrahabitat differences in individual distribution. Within an area, the individuals of any particular species move over the entire range of the species distribution. At Palo Alto, Elkhorn Slough, and Bodega Bay, individuals were observed to feed along the edge of the marsh and then wander over the flats to the water's edge. Similarly, individuals would move from the water and begin feeding on the exposed flats. This type of behavior was observed among all the shorebird species encountered during this study.

Food and feeding

Waders are opportunists, feeding on whatever foods are available. Consequently, there are considerable differences in the diets of birds collected at different localities or at different times of the year (McAtee and Beal 1912; Bent 1927, 1929; Reeder 1951). Local concentrations of food attract and hold dense aggregations of migrant birds, and within such areas all species will have much the same diet. However, in areas where food organisms are more evenly distributed and where the variety of available food organisms is greater, dietary differences appear. In an area such as Palo Alto where the kinds of food are relatively

TABLE IV. Number of invertebrate food organisms per ft² of surface area on tidal mudflats at Palo Alto, Alameda County[a]

Invertebrate species	Average number invertebrates per ft²
Annelida	
Polychaeta	
Neanthes succinea	44.3
Mollusca	
Gastropoda	
Ilyanassa obsoleta ($<\frac{1}{4}$ in.)	0.5
Ilyanassa obsoleta ($>\frac{1}{4}$ in.)	0.5
Pelecypoda	
Modiolus demissus	+
Mya arenaria and *Macoma inconspicua*	7.5
Gemma Gemma	31.7
Arthropoda	
Crustacea	
Ostracod species	4.3
Amphipod species	52.0
Hemigrapsus oregonensis	+
Total no. invertebrates per ft²	140.8+

[a]Figures based upon 24 samples collected during April and August 1962.

TABLE V. Per cent composition of food items occurring in the gizzard contents of the shorebirds collected at Palo Alto

Invertebrates recovered	Bird species with no. gizzards analyzed									
	Semipalmated Plover (3)	Black-bellied Plover (3)	Avocet (9)	Dowitcher (27)	Least Sandpiper (38)	Western Sandpiper (78)	Red-backed Sandpiper (46)	Knot (3)	Marbled Godwit (9)	Willet (16)
Amphipod species	—	—	4.0	—	21.1	8.6	8.9	—	—	—
Gemma gemma	2.8	12.7	52.0	6.4	5.3	8.6	2.4	55.3	6.2	44.6
Neanthes succinea	94.5	16.4	16.0	71.4	5.3	8.6	70.0	44.0	76.0	4.6
Ostracod species	0.5	—	4.0	9.6	57.8	62.8	9.7	—	—	—
Ilyanassa obsoleta <¼ in	2.2	1.8	24.0	7.4	10.5	11.4	5.8	—	2.5	9.9
>¼ in	—	65.5	—	2.1	—	—	1.6	0.7	8.7	33.2
Modiolus demissus	—	—	—	—	—	—	—	—	—	0.8
Mya arenaria and Macoma inconspicua	—	3.6	—	3.1	—	—	1.6	—	6.6	3.8
Hemigrapsus oregonensis	—	—	—	—	—	—	—	—	—	3.1
Average no. items per gizzard	137	55	25	94	38	35	121	159	195	132

limited (Table IV), there is a considerable species overlap in food organisms utilized (Table V).

The question therefore arises as to whether such an overlap in food organisms utilized indicates the existence of interspecific competition. Certainly the pattern followed by feeding shorebirds in moving with the tides results in all species utilizing the same feeding sites. But whereas shorebirds must resort to the same horizontal plane in feeding, differences in distribution, morphology and feeding behavior are such that species utilize different spatial and temporal segments of their otherwise common environment.

The horizontal distribution of species results in a temporal succession of species from point to

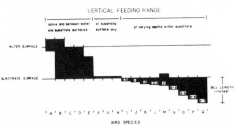

FIG. 4. Schematic diagram of the vertical feeding range of migrant shorebird species commonly observed at Palo Alto. Distribution is based upon observations of feeding behavior and bill length. Species: (A) Northern Phalarope, *Lobipes lobatus;* (B) Wilson's Phalarope, *Steganopus tricolor;* (C) Greater Yellowlegs, *Totanus melanoleucus;* (D) Black-necked Stilt, *Himantopus mexicanus;* (E) Avocet, *Recurvirostra americana;* (F) Killdeer, *Charadrius vociferus;* (G) Semipalmated Plover, *Charadrius semipalmatus;* (H) Black-bellied Plover, *Squatarola squatarola;* (I) Least Sandpiper, *Erolia minutilla;* (J) Western Sandpiper, *Ereunetes mauri;* (K) Red-backed Sandpiper, *Erolia alpina;* (L) Knot, *Calidris canutus;* (M) Willet, *Catoptrophorus semipalmatus;* (N) Dowitcher, *Limnodromus* sp.; (O) Whimbrel, *Numenius phaeopus;* (P) Marbled Godwit, *Limosa fedoa;* (Q) Long-billed Curlew, *Numenius americanus.*

point. Moreover, species tend to feed at different levels within or above the substrate (Fig. 4). Vertical feeding ranges broadly overlap, but species tend to take different parts of food populations which have a wide vertical range beneath the substrate surface. Coupled with the abundance of food in marine littoral habitats, the combination of temporal and horizontal segregation among shorebird species probably minimizes the degree of interspecific food competition. The staggering of peak population densities during migration further serves to reduce species competition and permits species to coexist which might otherwise interfere with one another.

A feeding shorebird may take every food organism as it is encountered, or it may select certain kinds or sizes of organisms. Dietary differences among species appear to reflect food preferences as well as variation in the composition of food organisms found within the different feeding ranges of shorebird species. The diversity of food organisms taken can be calculated in the same way as bird species diversity. It can be predicted that species feeding randomly will tend to have a higher food species diversity than species which feed selectively. A comparison of food species diversities for the shorebird species collected at Palo Alto leads to several interesting conclusions regarding the feeding behavior of different-sized birds.

An initial prediction, based upon observations of feeding behavior, that the Avocet (*Recurvirostra americana*), which feeds randomly by "filtering" the substrate surface, and the Semipalmated Plover, which feeds visually and is highly selective in the food organisms taken, will have high and low food species diversities, respectively, is verified when food species diversities are calculated (Table VI).

In general (the exceptions being the Semipal-

TABLE VI. Food diversity of 10 shorebird species collected at Palo Alto

Species	Food diversity (H)	Number of equally common species (e^H)
Semipalmated Plover	0.188	1.2
Black-bellied Plover	1.022	2.8
Avocet	1.234	3.4
Least Sandpiper	1.190	3.3
Western Sandpiper	1.172	3.2
Red-backed Sandpiper	1.060	2.9
Dowitcher	1.020	2.8
Marbled Godwit	0.861	2.4
Willet	1.326	3.9
Knot	0.722	2.1

mated Plover, the Willet, and the Avocet), there is an inverse relationship between the diversity of food organisms taken and the size of the shorebird. It therefore appears that larger shorebirds feed more selectively (on the Palo Alto mudflats) than smaller shorebird species. The Avocet, as a large bird feeding randomly, and the Semipalmated Plover, as a small bird feeding selectively, are exceptions. The Willet evidently is a much less selective feeder than other large shorebirds, tending to take a variety of large and small prey species. In many ways, the Willet has a wider range of behavior than other shorebirds, a fact which is reflected both in the diversity of food organisms taken and in the number of different habitats it frequents. At Palo Alto, the number of smaller-sized food species is greater than the number of larger-sized food species, and because the larger shorebirds tend to feed upon the larger prey orga-

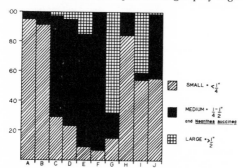

FIG. 5. Percentage of small, medium and large-sized food organisms in the diets of 10 shorebird species analyzed at Palo Alto during migration. Species: (A) Least Sandpiper, *Erolia minutilla;* (B) Western Sandpiper, *Ereunetes mauri;* (C) Red-backed Sandpiper, *Erolia alpina;* (D) Dowitcher, *Limnodromus* spp.; (E) Marbled Godwit, *Limosa fedoa;* (F) Semipalmated Plover, *Charadrius semipalmatus;* (G) Black-bellied Plover, *Squatarola squatarola;* (H) Avocet, *Recurvirostra americana;* (I) Willet, *Catoptrophorus semipalmatus;* (J) Knot, *Calidris canutus.*

nisms (Fig. 5), this results in a lower food diversity for the larger shorebirds. Whether or not large shorebirds feed as selectively in areas with a greater diversity of large food organisms than at Palo Alto is, therefore, open to question. The conclusion drawn from the Palo Alto data is that large shorebirds tend to feed selectively on the larger food species available, and although they do take smaller organisms, they do not appear to take each one as it is encountered.

The manner in which an animal feeds is determined by the energy expended balanced against that gained. In theory, the most efficient process would be to take each food organism as it is encountered. This, however, is dependent upon the rate of feeding possible. A large bird, attempting to feed on small food species, might not be able to feed rapidly enough regardless of the abundance of food if each item has to be taken singly. Consequently, most large animals feeding on very small prey tend to be filter feeders, gathering large numbers of prey with each feeding effort (e.g., the Avocet). A better strategy for large organisms in general is to ignore smaller prey organisms and use the time and energy thereby saved to search for larger food items. Conversely, small predators are probably most efficient when they take every prey species encountered.

We can predict from this that small predators would tend to be more numerous than large predators. Selection should tend to maximize the energy utilized by a species. Thus, for a species feeding upon a small-sized, but abundant prey organism, the most efficient strategy would be to be numerous but individually small, or rare and very large, as in the case of the filter feeders (Slobodkin 1964). Predators specializing on larger food species would tend to be less numerous but of an individually greater size. Besides the physical necessity of being larger in order to overcome or swallow larger prey species, a larger size permits the predator to search for longer periods of time without requiring food. (The assumption is made that larger prey organisms are rarer than smaller prey organisms). Needless to say the population sizes of both large and small predators are ultimately controlled by the absolute abundance of their prey species so that in some instances large predators may be more abundant than small predators.

DISCUSSION

During migration, shorebird species differ in their distributions through time and space, but these differences are statistical rather than absolute. In effect, species overlap broadly in their requirements. Because migrant shorebirds must resort to the same horizontal plane, spatial segregation among morphologically similar species is not possible. The high degree of morphological divergence among shorebird species may, therefore, be a means by which ecological segregation is achieved in a habitat where spatial segregation is not possible (MacArthur and Levins 1964). Morphologically similar species tend to frequent different habitats, to migrate along different routes, or to migrate at different times.

An essential difference between migratory and breeding shorebird populations is the dense multispecific aggregations found throughout migration. Among breeding shorebirds there is a pronounced intraspecific territorial behavior, and populations are relatively dispersed. The habitats frequented by migrant shorebirds are remarkable in the abundance and constancy of food. This is not true of the breeding grounds where the food available has a pronounced yearly cycle of abundance. It is logical to presume that juvenile birds are less efficient feeders than are adults. In the case of a fluctuating food supply, selection would therefore favor the evolution of a system in which the young of precocial species would be subject to a minimum of competition with adults. Territorial behavior, habitat selection, the early migration of adults, or of one sex from the nesting area, are all means to reduce inter- and intraspecific adult–young competition both on the breeding grounds and during migration.

As a result of tidal movements and cycles, the amount of available feeding space in the littoral zone habitats frequented by migrant shorebirds fluctuates throughout the day and from season to season. It seems likely that the greater population densities recorded during the spring migration are partly a result of the greater feeding area available during this season. However, this is not to say that the dense multispecific aggregations recorded correspond to stable community assemblages in which there is a minimal level of interspecific competition. Rather, the lower species diversities of wintering populations as compared to those recorded during migration (Table III) may indicate that the dense multispecific aggregations of migration are unstable. Competitive exclusion (Hardin 1960) may be avoided during migration only because of the transient character of migratory assemblages, the abundance of food, and the constancy of food along the path of migration. Species coexist because they migrate from point to point without ever fully exploiting the available food resources while minimizing inter-

individual interference through a partial degree of spatial and temporal segregation.

Morphological and behavioral differences among shorebird species seem to have evolved as means of exploiting environmental resources of food. Divergence probably has been reinforced by the pattern of migration in which large numbers of species form relatively dense aggregations within essentially two-dimensional habitats. Within such aggregations, feeding efficiency may be impaired whenever population density exceeds a certain threshold. As a result, there would be a tendency for intraspecific mechanisms to evolve which regulate population density at a level below which interference between feeding individuals is minimal. Similarly, there would be a tendency for species to segregate in such a way that interspecific interference is minimized. Segregation might initially be behavioral, but there would be selection against individuals which are less efficient in their utilization of environmental food resources. As a result, species would tend to occupy different spatial segments of the environment and to diverge morphologically so as to exploit these segments with greater efficiency.

LITERATURE CITED

Bent, A. C. 1927. Life histories of North American shore birds, Pt. 1. Bull. U. S. Nat. Mus. No. 142.
———. 1929. Life histories of North American shore birds, Pt. 2. Bull. U. S. Nat. Mus. No. 146.
Brown, W. L. and Wilson, E. O. 1956. Character displacement. Sys. Zool. 5: 49-64.
Dixon, K. L. 1961. Habitat distribution and niche relationships in North American species of Parus, p. 179 to 216. In F. Blair [ed.] Vertebrate speciation. Univ. Texas Press, Austin.
Dorst, J. 1962. The migrations of birds. Houghton Mifflin Company, Boston.
Gibb, J. 1954. Feeding ecology of tits, with notes on Treecreeper and Goldcrest. Ibis 96(4): 513-543.
Grinnell, J. and Miller, A. H. 1944. The distribution of the birds of California. Pacific Coast Avifauna, No. 27.
Hamilton, T. H. 1962. Species relationships and adaptations for sympatry in the avian genus Vireo. Condor 64(1): 40-68.
Hardin, G. 1960. The competitive exclusion principle. Science 131: 1292-1297.
Hutchinson, G. 1959. Homage to Santa Rosalia or why are there so many different kinds of animals. Amer. Naturalist 93: 145-159.
Kohn, A. J. 1959. The ecology of Conus in Hawaii. Ecol. Monogr. 29: 47-90.
MacArthur, R. H. 1958. Population ecology of some warblers of northeastern coniferous forests. Ecology 39: 599-619.
——— and Levins, R. 1964. Competition, habitat selection, and character displacement in a patchy environment. Proc. Nat. Acad. Sci. 51: 1207-1210.
——— and MacArthur, J. W. 1961. On bird species diversity. I. Ecology 42: 594-598.

————, MacArthur, J. W. and Preer, J. 1962. On bird species diversity. II. Amer. Naturalist 96: 167-174.

MacIntosh, R. P. 1963. Ecosystems, evolution and relational patterns of living organisms. Amer. Scientist 51: 246-267.

McAtee, W. L. and Beal, F. E. L. 1912. Some common game, aquatic and rapacious birds in relation to man. U. S. Dept. of Agric. Farm. Bull. No. 497.

Pitelka, F. 1950. Geographic variation and the species problem in the shorebird genus *Limnodromus*. Univ. Calif. Publ. Zool. 50: 1-108.

Recher, H. F. Behavioral interactions among migrant shorebirds. (*In manuscript.*)

Reeder, W. G. 1951. Stomach analysis of a group of shorebirds. Condor 53: 43-45.

Ricketts, E. F. and Calvin, J. 1939. Between Pacific tides. Stanford Univ. Press, Stanford, California.

Salomonsen, F. 1955. The evolutionary significance of bird migration. Dansk. biol. Medd. 22: 1-62.

Slobodkin, L. B. 1964. The strategy of evolution. Amer. Scientist 52: 342-357.

Stewart, R. E. and Robbins, C. S. 1958. Birds of Maryland and the District of Columbia. North Amer. Fauna, 62. U. S. Gov. Printing Office, Washington.

Storer, R. W. 1951. The seasonal occurrence of shorebirds on Bay Farm Island, Alameda County, California. Condor 53: 186-193.

Swinebroad, J. 1964. Nocturnal roosts of migrating shorebirds. Wilson Bull. 76: 155-159.

Urner, C. and Storer, R. W. 1949. The distribution and abundance of shorebirds on the North and Central New Jersey coast, 1928-1938. Auk 66: 177-194.

Welch, B. L. 1963. Psychophysiological response to the level of environmental stimulation. 16th Int'l. Congr. Zool. VI: 269.

Part VII
CONCLUSION

Editors' Comments
on Papers 39 Through 41

As the conclusion to this book we offer three papers that have sought to give general statements for the niche concept. The first of these is Hutchinson's article titled, appropriately for this section, "Concluding Remarks" (Paper 39). In this article, a classic for niche theory, Hutchinson sought first to formulate his concepts of the fundamental and realized niche. Considering variables within the biotope—both physical and biotic—each species has a range of possible occurrence for each variable. Its limit points along these axes define a multidimensional hypervolume that he terms the "fundamental niche." In the presence of competing species, only some fraction of the fundamental niche may be occupied by the species, and this smaller volume is then the species' "realized niche." The realized niche corresponds essentially to the niche hypervolume described in the introduction to this book and in the last article (Paper 41). Modifications of his concept are indicated on page 417. Point 1 observes that frequencies in the hypervolume will have a denser center—as expressed elsewhere in this book as the "population cloud." Point 2 notes the difficulty of constructing axes for many niche variables, and point 3 the fact that time of activity becomes a means of niche differentiation. Points 4 and 3 suggest that all other species may be parts of the coordinate system for a given species, and propose projection onto a hyperspace of less than *n* dimensions. Implicit in these statements is the concept of a community niche hyperspace that the first editor has made central to his interpretation of species diversity (Whittaker, 1965, 1967, 1969, 1972).

Hutchinson continues with the Volterra–Gause principle, as he terms the principle of competitive exclusion, and the ways its implications can be escaped by unstable populations or qualified for stable ones. Application of the niche concept to community species diversity is considered, by way of MacArthur's broken stick or random niche boundary hypothesis, on the importance values of species in communities. The application was a most significant development for later diversity theory, although the broken stick hypothesis is now part of a more complex view of importance–value distributions (Whittaker, 1965, 1969; Cohen, 1968; May, 1975). Hutchinson concludes with the concept of "empty niches" and man's population increase. The editors, as indicated in Paper 41, do not consider the notion of empty niches to be very useful. Man, in his harvests from and manipulations of communities, is not filling empty niches but preempting resources that would otherwise be used, and altering communities. An extrapolation suggests that man will use ever more resources, while releasing ever more toxins, until he uses all biological resources of a world without other living organisms. We doubt that history will work out that way, although the problem does not seem less urgent now than it did to Hutchinson in 1957.

Robert MacArthur, a most distinguished student of Hutchinson who contributed to the origin of the fundamental niche concept, developed his views of niche theory in a distinctive direction in association with Richard Levins. The concept of "grain" (p. 161), referred to in other selections in this volume, is an important part of MacArthur's interpretation. The idea is interesting and relates to the concept of intracommunity pattern; but it is not clear that grain itself has been the subject of effective research or that there can be as many species as there are grain types (p. 163). Levin (Paper 8) and the introduction to Part III have commented on the suggestion of direct correspondence of resource number and species number; see also Wilbur (1972). The reasoning regarding alpha matrices (p. 171 to the end) is ingenious in principle but is essentially steady-state in logic. As Wilbur, Ayala, and Vandermeer have shown, understanding of niche must go beyond Lotka–Volterra equations. It should be noted that MacArthur is talking also of experiments with populations to determine alpha values, and the Wilbur (1972) article illustrates such experiments. Paper 40 is MacArthur's full article, given here because of its real interest and to represent a statement of niche theory that, although also influenced by Hutchinson, has developed in a direction somewhat different from ours.

The final article, Paper 41, represents our own synthesis with R. B. Root, and there is little more comment we can make. It was written because of increasing confusion, in concept and research, of niche and habitat that threatened to make the term "niche" useless. We hoped, by distinguishing the notions of niche and habitat from one another and re-

385

lating them to each other, to give both enhanced effectiveness for research and ecological interpretation. If the ideas of this article are quite similar to those of our introduction, we hope the reader has encountered a good share of other views of interest in between.

REFERENCES

Cohen, J. E. 1968. Alternate derivations of a species-abundance relation. *Amer. Naturalist* **102**: 165–172.

May, R. 1975. Patterns of species' abundance and diversity, in *The Ecology and Evolution of Communities*, eds. M. L. Cody and J. M. Diamond. Harvard University Press, Cambridge, Mass. (in press).

Whittaker, R. H. 1965. Dominance and diversity in land plant communities. *Science* **147**: 250–260.

Whittaker, R. H. 1967. Gradient analysis of vegetation. *Biol. Rev.* **42**: 207–264.

Whittaker, R. H. 1969. Evolution of diversity in land plant communities. *Brookhaven Symp. Biol.* **22**: 178–196.

Whittaker, R. H. 1972. Evolution and measurement of species diversity. *Taxon* **21** (2/3): 213–251.

Wilbur, H. M. 1972. Competition, predation, and the structure of the *Ambystoma-Rana sylvatica* community. *Ecology* **53**: 3–21.

39

Reprinted from *Cold Spring Harbor Symp. Quant. Biol.*, **22**, 415–427 (1957)

Concluding Remarks

G. Evelyn Hutchinson

Yale University, New Haven, Connecticut

This concluding survey[1] of the problems considered in the Symposium naturally falls into three sections. In the first brief section certain of the areas in which there is considerable difference in outlook are discussed with a view to ascertaining the nature of the differences in the points of view of workers in different parts of the field; no aspect of the Symposium has been more important than the reduction of areas of dispute. In the second section a rather detailed analysis of one particular problem is given, partly because the question, namely, the nature of the ecological niche and the validity of the principle of niche specificity has raised and continues to raise difficulties, and partly because discussion of this problem gives an opportunity to refer to new work of potential importance not otherwise considered in the Symposium. The third section deals with possible directions for future research.

The Demographic Symposium as a Heterogeneous Unstable Population

In the majority of cases the time taken to establish the general form of the curve of growth of a population from initial small numbers to a period of stability or of decline is equivalent to a number of generations. If, as in the case of man, the demographer is himself a member of one such generation, his attitude regarding the nature of the growth is certain to be different from that of an investigator studying, for instance, bacteria, where the whole process may unfold in a few days, or insects, where a few months are required for several cycles of growth and decline. This difference is apparent when Hajnal's remarks about the uselessness of the logistic are compared with the almost universal practice of animal demographers to start thinking by making some suitable, if almost unconscious, modification of this much abused function.

[1] I wish to thank all the participants for their kindness in sending in advance manuscripts or information relative to their contributions. All this material has been of great value in preparing the following remarks, though not all authors are mentioned individually. Where a contributor's name is given without a date, the reference is to the contribution printed earlier in this volume. I am also very much indebted to the members (Dr. Jane Brower, Dr. Lincoln Brower, Dr. J. C. Foothills, Mr. Joseph Frankel, Dr. Alan Kohn, Dr. Peter Klopfer, Dr. Robert MacArthur, Dr. Gordon A. Riley, Mr. Peter Wangersky, and Miss Sally Wheatland) of the Seminar in Advanced Ecology, held in this department during the past year. Anything that is new in the present paper emerged from this seminar and is not to be regarded specifically as an original contribution of the writer.

The human demographer by virtue of his position as a slow breeding participant observer, and also because he is usually called on to predict for practical purposes what will happen in the immediate future, is inevitably interested in what may be called the microdemography of man. The significant quantities are mainly second and third derivatives, rates of change of natality and mortality and the rates of change of such rates. These latter to the animal demographer might appear as random fluctuations which he can hardly hope to analyse in his experiments. What the animal demographer is mainly concerned with is the macrodemographic problem of the integral curve and its first derivative. He is accustomed to dealing with innumerable cases where the latter is negative, a situation that is so rare in human populations that it seems to be definitely pathological to the human demographer. Only when anthropology and archaeology enter the field of human demography does something comparable to animal demography, with its broad, if sometimes insufficiently supported generalisations and its fascinating problems of purely intellectual interest, emerge. From this point of view the papers of Birdsell and Braidwood are likely to appeal most strongly to the zoologist, who may want to compare the rate of spread of man with that considered by Kurtén (1957) for the hyena.

It is quite likely that the difference that has just been pointed out is by no means trivial. The environmental variables that affect fast growing and slow growing populations are likely to be much the same, but their effect is qualitatively different. Famine and pestilence may reduce human populations greatly but they rarely decimate them in the strict sense of the word. Variations, due to climatic factors, of insect populations are no doubt often proportionately vastly greater. A long life and a long generation period confer a certain homeostatic property on the organisms that possess them, though they prove disadvantageous when a new and powerful predator appears. The elephant and the rhinoceros no longer provide models of human populations, but in the early Pleistocene both may have done so. The rapid evolution of all three groups in the face of a long generation time is at least suggestive.

It is evident that a difference in interest may underlie some of the arguments which have enlivened, or at times disgraced, discussions of this subject. Some of the most significant modern work has arisen from an interest in extending the concepts of the struggle for existence put forward as an evolutionary mechanism by Darwin practically a century ago. Such work, of which Lack's recent contributions provide a distinguished example, tends to concentrate on relatively stable interacting populations in as undisturbed communition as possible. Another fertile field of research has been provided by the sudden increases in numbers of destructive animals, often after introduction or disturbance of natural environments. Here more than one point of view has been apparent. Where emphasis has been on biological control, that is, a conscious rebuilding of a complex biological association, a view point not unlike that of the evolutionist has emerged—where emphasis has been placed on the actual events leading to a very striking increase or decrease in abundance, given the immediate ecological conditions, the latter have appeared to be the most significant variables. Laboratory workers have moreover tended to keep all but a few factors constant, and to vary these few systematically. Field workers have tended to emphasize the ever changing nature of the environment. It is abundantly clear that all these points of view are necessary to obtain a complete picture. It is also very likely that the differences in initial point of view are often responsible for the differences in the interpretation of the data.

The initial differences of point of view are not the only difficulty. In the following section an analysis of a rather formal kind of one of the concepts frequently used in animal ecology, namely that of the *niche*, is attempted. This analysis will appear to some as compounded of equal parts of the obvious and the obscure. Some people however may find when they have worked through it, provided that it is correct, that some removal of irrelevant difficulties has been achieved. It is not necessary in any empirical science to keep an elaborate logicomathematical system always apparent, any more than it is necessary to keep a vacuum cleaner conspicuously in the middle of a room at all times. When a lot of irrelevant litter has accumulated the machine must be brought out, used, and then put away. It might be useful for those who argue that the word environment should refer to the environment of a population, and those who consider it should been the environment of an organism, to use the word both ways for a couple of months, writing "environment" when a single individual is involved, "Environment" when reference is to a population. In what follows the term will as far as possible not be used, except in the non-committal adjectival form environmental, meaning any property outside the organisms under consideration.

The Formalisation of the Niche and the Volterra-Gause Principle

Niche space and biotop space

Consider two independent environmental variables x_1 and x_2 which can be measured along ordinary rectangular coordinates. Let the limiting values permitting a species S_1 to survive and reproduce be respectively x'_1, x''_1 for x_1 and x'_2, x''_2 for x_2. An area is thus defined, each point of which corresponds to a possible environmental state permitting the species to exist indefinitely. If the variables are independent in their action on the species we may regard this area as the rectangle $(x_1 = x'_1, x_1 = x''_1, x_2 = x'_2, x_2 = x''_2)$, but failing such independence the area will exist whatever the shape of its sides.

We may now introduce another variable x_3 and obtain a volume, and then further variables $x_4 \ldots x_n$ until all of the ecological factors relative to S_1 have been considered. In this way an n-dimensional hypervolume is defined, every point in which corresponds to a state of the environment which would permit the species S_1 to exist indefinitely. For any species S_1, this hypervolume N_1 will be called the *fundamental niche*[2] of S_1. Similarly for a second species S_2 the fundamental niche will be a similarly defined hypervolume N_2.

It will be apparent that if this procedure could be carried out, all X_n variables, both physical and biological, being considered, the fundamental niche of any species will completely define its ecological properties. The fundamental niche defined in this way is merely an abstract formalisation of what is usually meant by an ecological niche.

As so defined the fundamental niche may be regarded as a set of points in an abstract n-dimensional N space. If the ordinary physical space B of a given biotop be considered, it will be apparent that any point $p(N)$ in N can correspond to a number of points $p_i(B)$ in B, at each one of which the conditions specified by $p(N)$ are realised in B. Since the values of the environmental variables $x_1 x_2 \ldots x_n$ are likely to vary continuously, any subset of points in a small elementary volume ΔN is likely to correspond to a number of small elementary volumes scattered about in B. Any volume B' of the order of the dimensions of the mean free paths of any animals under consideration is likely to contain points corresponding to points in various fundamental niches in N.

Since B is a limited volume of physical space comprising the biotope of a definite collection of species $S_1, S_2 \cdots S_n$, there is no reason why a given point in N should correspond to any points in B. If, for any species S_1, there are no points in

---

[2] This term is due to MacArthur. The general concept here developed was first put forward very briefly in a footnote (Hutchinson, 1944).

B corresponding to any of the points in N_1, then B will be said to be *incomplete* relative to S_1. If some of the points in N_1 are represented in B then the latter is *partially incomplete* relative to S_1, if all the points in N_1 are represented in B the latter is *complete* relative to S_1.

Limitations of the set-theoretic mode of expression. The following restrictions are imposed by this mode of description of the niche.

1. It is supposed that all points in each fundamental niche imply equal probability of persistance of the species, all points outside each niche, zero probability of survival of the relevant species. Ordinarily there will however be an optimal part of the niche with markedly suboptimal conditions near the boundaries.

2. It is assumed that all environmental variables can be linearly ordered. In the present state of knowledge this is obviously not possible. The difficulty presented by linear ordering is analogous to the difficulty presented by the ordering of degrees of belief in non-frequency theories of probability.

3. The model refers to a single instant of time. A nocturnal and a diurnal species will appear in quite separate niches, even if they feed on the same food, have the same temperature ranges etc. Similarly, motile species moving from one part of the biotop to another in performance of different functions may appear to compete, for example, for food, while their overall fundamental niches are separated by strikingly different reproductive requirements. In such cases the niche of a species may perhaps consist of two or more discrete hypervolumes in N. MacArthur proposed to consider a more restricted niche describing only variables in relation to which competition actually occurs. This however does not abolish the difficulty. A formal method of avoiding the difficulty might be derived, involving projection onto a hyperspace of less than n-dimensions. For the purposes for which the model is devised, namely a clarification of niche-specificity, this objection is less serious than might at first be supposed.

4. Only a few species are to be considered at once, so that abstraction of these makes little difference to the whole community. Interaction of any of the considered species is regarded as competitive in sense 2 of Birch (1957), negative competition being permissible, though not considered here. All species other than those under consideration are regarded as part of the coordinate system.

Terminology of subsets. If N_1 and N_2 be two fundamental niches they may either have no points in common in which case they are said to be *separate*, or they have points in common and are said to *intersect*. (Fig. 1)

In the latter case:

$(N_1 - N_2)$ is the subset of N_1 of points not in N_2
$(N_2 - N_1)$ is the subset of N_2 of points not in N_1

$N_1 \cdot N_2$ is the subset of points common to N_1 and N_2, and is also referred to as the *intersection subset*.

Definition of niche specificity. Volterra (1926, see also Lotka 1932) demonstrated by elementary analytic methods that under constant conditions two species utilizing, and limited by, a common resource cannot coexist in a limited system.[3] Winsor (1934) by a simple but elegant formulation showed that such a conclusion is independent of any kind of finite variations in the limiting resource. Gause (1934, 1935) confirmed this general conclusion experimentally in the sense that if the two species are forced to compete in an undiversified environment one inevitably becomes extinct. If there is a diversification in the system so that some parts favor one species, other parts the other, the two species can coexist. These findings have been extended and generalised to the conclusion that two species, when they co-occur, must in some sense be occupying different niches. The present writer believes that properly stated as an empirical generalisation, which is true except in cases where there are good reasons not to expect it to be true,[4] the principle is of fundamental importance and may be properly called the Volterra-Gause Principle. Some of the confusion surrounding the principle has arisen from the concept of two species not being able to co-occur when they occupy identical niches. According to the formulation given above, identity of fundamental niche would imply $N_1 = N_2$, that is, every point of N_1 is a member of N_2 and every point of N_2 a member of N_1. If the two species S_1 and S_2 are indeed valid species distinguishable by a systematist and not freely interbreeding, this is so unlikely that the case is of no empirical interest. In terms of the set-theoretic presentation, what the Volterra-Gause principle meaningfully states is that for any small element of the intersection subset $N_1 \cdot N_2$, there do not exist in B corresponding small parts, some inhabited by S_1, others by S_2.

Omitting the quasi-tautotogical case of $N_1 = N_2$, the following cases can be distinguished.

(1) N_2 is a proper subset of N_1 (N_2 is "inside" N_1)

   (a) competition proceeds in favor of S_1 in all the elements of B corresponding to $N_1 \cdot N_2$; given adequate time only S_1 survives.

   (b) competition proceeds in favor of S_2 in all elements of B corresponding to some part of the intersection subset and both species survive.

(2) $N_1 \cdot N_2$ is a proper subset of both N_1 and N_2; S_1 survives in the parts of B space

[3] I regret that I am unable to appreciate Brian's contention (1956) that the Volterra model refers only to interference, and the Winsor model to exploitation.
[4] cf. Schrödinger's famous restatement of Newton's First Law of Motion, that a body perseveres at rest or in uniform motion in a right line, except when it doesn't.

corresponding to $(\mathbf{N}_1 - \mathbf{N}_2)$, S_2 in the parts corresponding to $(\mathbf{N}_2 - \mathbf{N}_1)$, the events in $\mathbf{N}_1 \cdot \mathbf{N}_2$ being as under I, with the proviso that no point in $\mathbf{N}_1 \cdot \mathbf{N}_2$ can correspond to the survival of both species.

In this case the two difference subsets $(\mathbf{N}_1 - \mathbf{N}_2)$ and $(\mathbf{N}_2 - \mathbf{N}_1)$ are, in Gause's terminology, refuges for S_1 and S_2 respectively.

If we define the realised niche $\mathbf{N}'_1$ of S_1 in the presence of S_2 as $(\mathbf{N}_1 - \mathbf{N}_2)$, if it exists, plus that part of $\mathbf{N}_1 \cdot \mathbf{N}_2$ as implies survival of S_1, and similarly the realised niche $\mathbf{N}'_2$ of S_2 as $(\mathbf{N}_2 - \mathbf{N}_1)$, if it exists, plus that part of $\mathbf{N}_1 \cdot \mathbf{N}_2$ corresponding to survival of S_2, then the Volterra-Gause principle is a statement of an empirical generalisation, which may be verified or falsified, that realised niches do not intersect. If the generalisation proved to be universally false, the falsification would presumably imply that in nature resources are never limiting.

Validity of the Gause-Volterra Principle. The set-theoretic approach outlined above permits certain refinements which, however obvious they may seem, apparently require to be stated formally in an unambiguous way to prevent further confusion. This approach however tells us nothing about the validity of the principle, but merely where we should look for its verification or falsification.

Two major ways of approaching the problem have been used, one experimental, the other observational. In the experimental approach, the method (*e.g.* Gause, 1934, 1935; Crombie, 1945, 1946, 1947) has been essentially to use animal populations as elements in analogue computers to solve competition equations. As analogue computers, competing populations leave much to be desired when compared with the more conventional electronic machines used for instance by Wangersky and Cunningham. At best the results of laboratory population experiments are qualitatively in line with theory when all the environmental variables are well controlled. In general such experiments indicate that where animals are forced by the partial incompleteness of the **B** space to live in competition under conditions corresponding to a small part of the intersection subset, only one species survives. They also demonstrate that the identity of the survivor is dependent on the environmental conditions, or in other words on which part of the intersection subset is considered, and that when deliberate niche diversification is brought about so that at least one non-intersection subset is represented in **B**, two species may co-occur indefinitely. It would of course be most disturbing if confirmatory models could not be made from actual populations when considerable trouble is taken to conform to the postulates of the deductive theory.

The second way in which confirmation has been sought, namely by field studies of communities consisting of a number of allied species also lead to

a confirmation of the theory, but one which may need some degree of qualification. Most work has dealt with pairs of species, but the detailed studies on *Drosophila* of Cooper and Dobzhansky (1956) and of Da Cunha, El-Tabey Shekata and de Olivera (1957), to name only two groups of investigators, the investigation of about 18 species of *Conus* on Hawaiian reef and littoral benches (Kohn, in press) and the detailed studies of the food of six co-occurring species of *Parus* (Betts, 1955) indicate remarkable cases among many co-occurring species of insects, mollusks and birds respectively. However much data is accumulated there will almost always be unresolved questions relating to particular species, though the presumption from this sort of work is that, in any large group of sympatric species belonging to a single genus or subfamily, careful work will always reveal ecological differences. The sceptic may reply in two ways, firstly pointing out that the quasi-tautological case of $\mathbf{N}_1 = \mathbf{N}_2$ has already been dismissed as too improbable to be of interest, and that when a great deal of work has to be done to establish the difference, we are getting as near to niche identity as is likely in a probabilistic world. Occasionally it may be possible to use indirect arguments to show that the differences are at least evolutionarily significant. Lack (1947b) for instance points out that in the Galapagos Islands, among the heavy billed species of *Geospiza*, where both *G. fortis* Gould, and *G. fuliginosa* Gould co-occur on an island, there is a significant separation in bill size, but where either species exists alone, as on Crossman Island and Daphne Island the bills are intermediate and presumably adapted to eating modal sized food. This is hard to explain unless the small average difference in food size believed to exist between sympatric *G. fortis* and *G. fuliginosa* is actually of profound ecological significance. The case is particularly interesting as most earlier authors have dismissed the significance of the small alleged differences in the size of food taken by the species. Few cases of specific ecological difference encountered outside *Geospiza* would appear at first sight so tenuous as this.

A more important objection to the Volterra-Gause principle may be derived from the extreme difficulty of identifying competition as a process actually occurring in nature. Large numbers of cases can of course be given in which there is very strong indirect evidence of competitive relationships between species actually determining their distribution. A few examples may be mentioned. In the British Isles (Hynes, 1954, 1955) the two most widespread species of *Gammarus* in freshwater are *Gammarus deubeni* Lillj: and *G. pulex* (L.). The latter is the common species in England and most of the mainland of Scotland, the former is found exclusively in Ireland, the Shetlands, Orkneys and most of the other Scottish Islands and in Cornwall. On northern mainland Scotland only

G. *lacustris* Sars is found. Both *deubeni* and *pulex* occur on the Isle of Man and in western Cornwall. Only in the Isle of Man have the two species been taken together. It is extremely probable that *pulex* is a recent introduction to that island. *G. deubeni* is well known in brackish water around the whole of northern Europe. It is reasonable to suppose that the fundamental niches of the two species overlap, but that within the overlap *pulex* is successful, while *deubeni* with a greater tolerance of salinity has a refuge in brackish water. Hynes moreover shows that *G. pulex* has a biotic (reproductive) potential two or three times that of *deubeni* so that in a limited system inhabitable by both species, under constant conditions *deubeni* is bound to be replaced by *pulex*. This case is as clear as one could want except that Hynes is unable to explain the absence of *G. deubeni* from various uninhabited favorable localities in the Isle of Man and elsewhere. Hynes also notes that Steusloff (1943) had similar experiences with the absence of *Gammarus pulex* in various apparently favorable German localities. Ueno (1934) moreover pointed out that *Gammarus pulex* (*sens. lat.*) occurs abundantly in Kashmir up to 1600 meters, and is an important element in the aquatic fauna of the Tibetan highlands to the east above 3800 miles, but is quite absent in the most favorable localities at intermediate altitudes. These disconcerting empty spaces in the distribution of *Gammarus* may raise doubts as to the completeness of the picture presented in Hynes' excellent investigations.

Another very well analysed case (Dumas, 1956) has been recently given for two sympatric species of *Plethodon*, *P. dunni* Bishop, and *P. vehiculum* (Cooper), in the Coastal Ranges of Oregon. Here experiments and field observations both indicate that *P. dunni* is slightly less tolerant of low humidity and high temperature than is *P. vehiculum*, but when both co-occur *dunni* can exclude *vehiculum* from the best sites. However under ordinary conditions in nature the number of unoccupied sites which appear entirely suitable is considerable, so that competition can not be limiting except in abnormally dry years.

In both these cases, which are two of the best analysed in the literature, the extreme proponent of the Volterra-Gause principle could argue that if the investigator was equipped with the sensory apparatus of *Gammarus* or *Plethodon* he would know that the supposedly suitable unoccupied sites were really quite unsuitable for any self respecting member of the genus in question. This however is pure supposition.

Even in the rather conspicuous case of the introduction of *Sciurus caroliniensis* Gmelin and its spread in Britain, the popular view that the bad bold invader has displaced the charming native *S. vulgaris leucourus* Kerr, is apparently mythological. Both species are persecuted by man; *S. caroliniensis* seems to stand this persecution bet-

ter than does the native red squirrel and therefore tends to spread into unoccupied area from which *S. vulgaris leucourus* has earlier retreated (Shorten, 1953, 1954).

Andrewartha (see also Andrewartha and Birch, 1954) has stressed the apparent fact that while most proponents of the competitive organisation of communities have emphasised competition for food, there is in fact normally more than enough food present. This appears, incidentally, most strikingly in some of Kohn's unpublished data on the genus *Conus*.

The only conclusion that one can draw at present from the observations is that although animal communities appear qualitatively to be constructed as if competition were regulating their structure, even in the best studied cases there are nearly always difficulties and unexplored possibilities. These difficulties suggest that if competition is determinative it either acts intermittently, as in abnormally dry seasons for *Plethodon*, or it is a more subtle process than has been supposed. Thus Lincoln Brower (*in press*) investigating a group of species of North American *Papilio* in which one eastern polyphagous species is replaced by three western oligophagous species, has been impressed by the lack of field evidence for any inadequacy in food resources. He points out however, that specific separation of food might lower the probability of local high density on a given plant, and so the risk of predation by a bird that only stopped to feed when food was abundant (*cf.* de Ruiter, 1952).

Unfortunately there is no end to the possible erection of hypothesis fitted to particular cases that will bring them within the rubric of increasingly subtle forms of competition. Some other method of investigation would clearly be desirable. Before drawing attention to one such possible method, the expected limitations of the Volterra-Gause principle must be examined.

Cases where the Volterra-Gause principle is unlikely to apply. (a) Skellam (1951; see also Brian, 1956b) has considered a model in which two species occur one of (S_1) much lower reproductive potential than the other (S_2). It is assumed that if S_1 and S_2 both arrive in an element of the biotops S_1 always displaces S_2, but that excess elements are always available at the time of breeding and dispersal so that some are never occupied S_1. In view of the higher reproductive potential, S_2 will reach some of these and survive. The model is primarily applicable to annual plants with a definite breeding season, random dispersal of seeds and complete seasonal mortality so all sites are cleared before the new generation starts growing, S_2 is in fact a limiting case of what Hutchinson (1951, 1953) called a fugitive species which could only be established in randomly vacated elements of a biotop. Skellam's model requires clearing of sites by high death rate, Hutchinson's qualitative statement a formation of transient sites by random small catastrophes in the biotop. Otherwise the two concepts developed independently are identical.

(b) When competition for resources becomes a contest rather than a scramble in Nicholson's admirable terminology, there is a theoretical possibility that the principle might not apply. If the breeding population be limited by the number of territories that can be set up in an area, and if a number of unmated individuals without breeding territory are present, food being in excess of the overall requirements, it is possible that territories could be set up by any species entirely independent of the other species, the territorial contests being completely intraspecific. Here a resource, namely area, is limiting but since it does not matter to one species if another is using the area, no interspecific competition need result. No case appears yet to be known, though less extreme modifications of the idea just put forward have apparently been held by several naturalists. Dr. Robert MacArthur has been studying a number of sympatric species of American warblers of the genus *Dendroica* which might be expected to be as likely as any organism to show the phenomenon. He finds however very striking niche specificity among species inhabiting the same trees.

(c) The various cases where circumstances change in the biotop reversing the direction of competition before the latter has run its course. Ideally we may consider two extreme cases with regard to the effect of changing weather and season on competition. In natural populations living for a time under conditions simulating those obtaining in laboratory cultures in a thermostat, if the competition time, that is, the time needed to permit replacement of one species by another, is very short compared with the periods of the significant environmental variables, then complete replacement will occur. This can only happen in very rapidly breeding organisms. Proctor (1957) has found that various green algae always replace *Haematococcus* is small bodies of water which never dry up, though if desiccation and refilling occur frequently enough the *Haematococcus* which is more drought resistant than its competitors will persist indefinitely. If on the contrary the competition time is long compared with the environmental periods, then the relevant environmental determinants of competition will tend to be mean climatic parameters, showing but secular trends in most cases, and competition will inevitably proceed to its end unless some quite exceptional event intervenes.[5]

[5] If there were really three species of giant tortoise (Rothschild, 1915) on Rodriguez, and even more on Mauritius, and if these were sympatric and due to multiple invasion (unlike the races on Albemarle in the Galapagos Islands) it is just conceivable that the population growth was so slow that mixed populations persisted for centuries and that the completion of competition had not occurred before man exterminated all the species involved.

Between the two extreme cases it is reasonable to suppose that there will exist numerous cases in which the direction of competition is never constant enough to allow elimination of one competitor. This seems likely to be the case in the autotrophic plankton of lakes, which inhabits a region in which the supply of nutrients is almost always markedly suboptimal, is subject to continual small changes in temperature and light intensity and in which a large number of species may (Hutchinson, 1941, 1944) coexist.

There is interesting evidence derived from the important work of Brian (1956a) on ants that the completion of competitive exclusion is less likely to occur in seral than in climax stages, which may provide comparable evidence of the effect of environmental changes in competition. Moreover whenever we find the type of situation described so persuasively by Andrewartha and Birch (1954) in which the major limitation on numbers is the length of time that meteorological and other conditions are operating favorably on a species, it is reasonable to suppose that interspecific competition is no more important than intraspecific competition. Much of the apparent extreme difference between the outlook of, for instance, these investigators, or for that matter Milne on the one hand, and a writer such as Lack (1954) on the other, is clearly due to the relationship of generation time to seasonal cycle which differs in the insects and in the birds. The future of animal ecology rests in a realisation not only that different animals have different autecologies, but also that different major groups tend to have fundamental similarities and differences particularly in their broad temporal relationships. The existence of the resemblances moreover may be quite unsuspected and must be determined empirically. In another place (Hutchinson, 1951) I have assembled such evidence as exists on the freshwater copepoda, which seem to be reminiscent of birds rather than of phytoplankton or of terrestrial insects in their competitive relationships.

It is also important to realize, as Cole has indicated in the introductory contribution to this Symposium, that the mere fact that the same species are usually common or rare over long periods of time and that where changes have been observed in well studied faunas such as the British birds or butterflies they can usually be attributed to definite environmental causes in itself indicates that the random action of weather on generation is almost never the whole story. Skellam's demonstration that such action must lead to final extinction must be born in mind. It is quite possible that the change in the phytoplankton of some of the least culturally influenced of the English Lakes, such as the disappearance of *Rhizosolenia* from Wastwater (Pearsall, 1932), may provide a case of random extinction under continually reversing competition. The general evidence of considerable stability under most conditions would suggest that competitive action of some sort is nearly always of significance.

Rarity and commonness of species and the non-intersection of realised niches. Several ways of approaching the problem of the rarity and commonness of species have been suggested (Fisher, Corbet and Williams, 1943; Preston, 1948; Brian, 1953; Shinozaki and Urata, 1953). In all these approaches relatively simple statistical distributions have been fitted to the data, without any attempt being made to elucidate the biological meaning of such distribution. Recently however MacArthur (1957) has advanced the subject by deducing the consequences of certain alternative hypotheses which can be developed in terms of a formal theory of niches.

It has been pointed out in a previous paragraph that the Volterra-Gause principle is equivalent to a statement that the realised niches of co-occurring species are non-intersecting. Consider a **B** space containing an equilibrium community of n species $S_1 S_2 \cdots S_n$, represented by numbers of individuals $N_1 N_2 \cdots N_2$. For any species S_K it will be possible to identify in **B** a number of elements, each of which corresponds to a whole or part of $\mathbf{N}'_K$ and to no other part of **N**. Suppose that at any given moment each of these elements is occupied by a single individual of S_K, the total volume of B which may be regarded as the specific biotop of S_K will be $N_K \Delta \mathbf{B} (S_K)$, $\Delta \mathbf{B}(S_K)$ being the mean volume of **B** occupied by one individual of S_K. Since the biotop is in equilibrium with respect to the n species present, all possible spaces will be filled so that

$$\mathbf{B} = \sum_{K=1}^{n} N_K \Delta \mathbf{B} \; (S_K)$$

We do not know anything *a priori* about the distribution of $N_1 \Delta \mathbf{B}(S_1), \; N_2 \Delta \mathbf{B}(S_2) \cdots N_n \Delta \mathbf{B}(S_n)$,

except that these different specific biotops are taken as volumes proportional to $N_1 N_2 \cdots N_n$, which is a justifiable first approximation if the species are of comparable size and physiology. In general some of the species will be rare and some common. The simplest hypothesis consistent with this, is that a random division of **B** between the species has taken place.

Consider a line of finite length. This may be broken at random into n parts by throwing $(n-1)$ random points upon it. It would also be possible to divide the line successively by throwing n random pairs of points upon it. In the first case the division is into non-overlapping sections, in the second the sections overlap. MacArthur, whose paper may be consulted for references to the mathematical procedures involved, has given the expected distributions for the division of a line by these alternative methods (Fig. 2). He has moreover shown that with certain restrictions the distribution (I) which corresponds to non-intersecting specific biotops and so to non-intersecting realised niches, fits certain multispecific biological associations extremely well. The form of this distribution is independent of the number of dimensions in **B**. The alternative distribution with overlapping specific biotops predicts fewer species of intermediate rarity and more of great rarity than is actually found; proceding from the linear case (II), to division of an area or a volume, accentuates this discrepancy. Two very striking cases in which distribution I fits biological multispecific populations are given in Figure 3 from MacArthur and in Figure 4 from the recent studies of Dr. Alan Kohn (in press).

The limitation which is imposed by the theory is that in all large subdivisions of **B** the ratio of total number of individuals $(m = \sum_{i=1}^{n} N_i)$ to

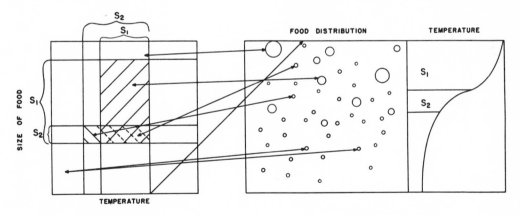

FIGURE 1. Two fundamental niches defined by a pair of variables in a two-dimensional niche space. Only one species is supposed to be able to persist in the intersection subset region. The lines joining equivalent points in the niche space and biotop space indicate the relationship of the two spaces. The distribution of the two species involved is shown on the right hand panel with a temperature depth curve of the kind usual in a lake in summer.

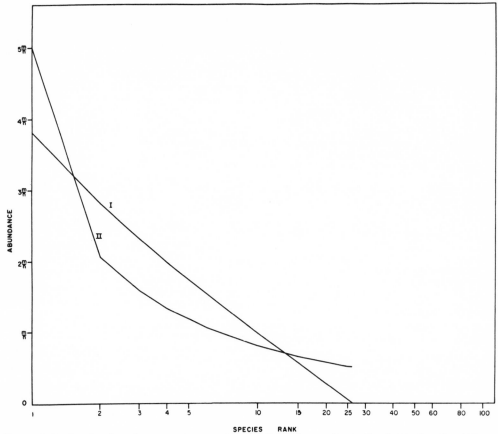

FIGURE 2. Rank order of species arranged per number of individuals according to the distributions I and II considered by MacArthur.

total number of species (n) must remain constant. This is likely to be the case in any biotop which is what may be termed *homogeneously diverse*, that is, in which the elements of the environmental mosaic (trees, stones, bushes, dead logs, etc.) are small compared with the mean free paths of the organisms under consideration. When a heterogeneously diverse area, comprising for instance stands of woodland separated by areas of pasture, is considered it is very unlikely that the ratio of total numbers of individuals to number of species will be identical in both woodland and pasture (if it occasionally were, the fact that both censuses could be added would not be of any biological interest). MacArthur finds that at least some bodies of published data which do not fit distribution I as a whole, can be broken down according to the type of environment into subcensuses which do fit the distribution. Data from moth traps and from populations of diatoms on slides submerged in rivers would not be expected to fit the distribu-

tion and in fact do not do so.[6] Such collection methods certainly sample very heterogeneously diverse areas.

The great merit of MacArthur's study is that it attempts to deduce operationally distinct differences between the results of two rival hypotheses, one of which corresponds essentially to the extreme density dependent view of interspecific interaction, the other to the opposite view. Although certain simplifying assumptions must be made in the theoretical treatment, the initial results suggest that in stable homogeneously diverse biotops the abundances of different species are arranged as if the realised niches were non-overlapping; this does not mean that populations may not exist under other conditions which would depart very widely from MacArthur's findings.

The problem of the saturation of the biotop. An important but quite inadequately studied aspect

[6] I am indebted to Dr. Ruth Patrick for the opportunity to test some of her diatometer censuses.

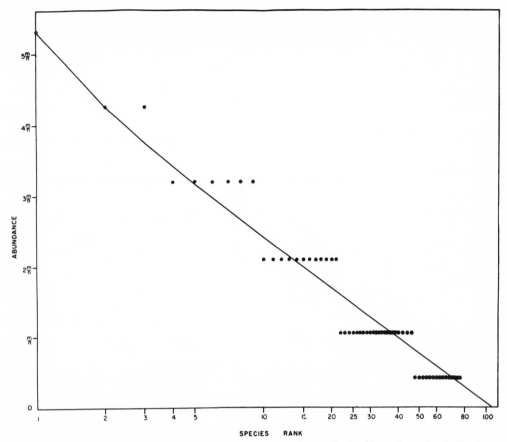

FIGURE 3. Rank order of species of birds in a tropical forest, closely following MacArthur's distribution I.

of niche specificity is that of the number of species that a given biotop can support. The nature of this problem can be best made clear by means of an example.

The aquatic bugs of the family *Corixidae* are of practically world wide distribution. Omitting a purely Australasian subfamily, they may be divided into the *Micronectinae* which are nearly always small, under 5 mm long and the *Corixinae* of which the great majority of species are over 5 mm long. Both subfamilies probably feed largely on organic detritus, though a few of the more primitive members of the *Corixinae* are definite predators. Some at least suck out the contents of algal cells, but unlike the other Heteroptera they can take particulate matter of some size unto their alimentary tracts. There is abundant evidence that the organic content of the bottom deposits of the shallow water in which these insects live is a major ecological factor regulating their occurrence. No *Micronectinae* occur in temperate North America and in the Old World this subfamily is far more abundant in the tropics while the *Corixinae* are far more abundant in the temperate regions (Lundblad, 1934; Jaczewski, 1937). Thus in Britain there are 30 species of *Corixinae* and three of *Micronectinae* (Macan, 1956), in peninsular Italy 20 or 21 species of *Corixinae* and five of *Micronectinae* (Stickel, 1955), in non-Palaeartic India about a dozen species of *Corixinae* and at least ten species of *Micronectinae* (Hutchinson, 1940) and in Indonesia (Lundblad, 1934) only three *Corixinae* and 14 *Micronectinae*. A reasonable explanation of this variation in the relative proportions of the two subfamilies is suggested by the findings of Macan (1938) and the more casual observations of other investigators that *Micronecta* prefers a low organic subtratum; in tropical localities the high rate of decomposition would reduce the organic content.

In certain isolated tropical areas at high altitudes, notably Ethiopia and the Nilghiri Hills of southern India the decline in the numbers of *Micronectinae* with increasing altitudes, and so

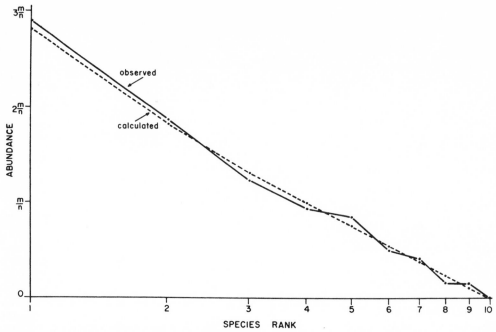

FIGURE 4. Rank order of species of *Conus* on a littoral bench in Hawaii (Kohn).

lower average water temperatures, is most noticable, but there is no increase in the number of *Corixinae*, presumably because the surrounding fauna is not rich enough to have permitted frequent invasion and speciation. Thus in the Nilghiri Hills between 2100 and 2300 m, intense collecting yielded three *Corixinae* of which two appear to be endemic, and one non-endemic species of *Micronecta*. Very casual collecting below 1000 m in south India has produced two species of *Corixinae* and five species of *Micronectinae*. The question raised by cases like this is whether the three Nilghiri *Corixinae* fill all the available niches which in Europe might support perhaps 15 or 20 species, or whether there are really empty niches. Intuitively one would suppose both alternatives might be partly true, but there is no information on which to form a real judgment. The rapid spread of introduced species often gives evidence of empty niches, but such rapid spread in many instances has taken place in disturbed areas. The problem clearly needs far more systematic study than it has been given. The addition and the replacement of species of fishes proceeding down a river, and the competitive situations involved, may provide some of the best material for this sort of study, but though much data exists, few attempts at systematic comparative interpretation have been made (*cf.* Hutchinson, 1939).

THE FUTURE OF COMPARATIVE DEMOGRAPHIC STUDIES

Perhaps the most interesting general aspect of the present Symposium is the strong emphasis placed on the changing nature of the populations with which almost all investigators deal. In certain cases, notably in the parthenogenetic crustacean *Daphnia* (Slobodkin, 1954), it is possible to work with clones that must be almost uniform genetically, but all the work on bisexual organisms is done under conditions in which evolution may take place. The emergence in Nicholson's experiments of strains of *Lucilia* in which adult females no longer need a protein meal before egg laying provides a dramatic example of evolution in the laboratory; the work reported by Dobzhansky, by Lewontin, and by Wallace, in discussion, shows how experimental evolution, for which subject the Carnegie Laboratory at Cold Spring Harbor was founded, has at last come into its own.

So far little attention has been paid to the problem of changes in the properties of populations of the greatest demographic interest in such experiments. A more systematic study of evolutionary change in fecundity, mean life span, age and duration of reproductive activity and length of post reproductive life is clearly needed. The most interesting models that might be devised would be those in which selection operated in favor of low fecundity, long pre-reproductive life and on any aspect of post-reproductive life.

There is in many groups, notably *Daphnia*, dependence of natality on food supply (Slobodkin, 1954) though the adjustment can never be instantaneous and so can lead to oscillations. In the case of birds the work of the Oxford school (Moreau, 1944; Lack, 1947a, 1954 and many papers quoted in the last named) indicates that in many birds natality is regulated by natural selection to correspond to the maximum number of young that can be reared in a clutch. In some circumstances the absolute survival of young is greater when the fecundity is low than when it is high. The peculiar nature of the subpopulations formed by groups of nestlings in nests makes this reasonable. Slobodkin (1953) has pointed out that in certain cases in which migration into numerous limited areas is possible, a high reproductive rate might have a lower selective advantage than a low rate. Actually in a very broad sense the bird's nest is a device to formalise the numerous limited areas, the existence of which permits such a type of selection. It should be possible with some insects to set up population cages in which access to a large number of very small amounts of larval food is fairly difficult for a fertile female. If the individual masses of larval food were such that there was an appreciable chance that many larvae on a single mass would die of starvation while a few larvae would survive, it is possible that selection for low fecundity might occur. This experiment would certainly imitate many situations in nature.

The evolutionary aspects of the problem raised by those cases where there is a delay of reproductive activity after adult morphology has been achieved is much harder to understand. Some birds though they attain full body size within a year (or in the case of most passerines in the nest) are apparently not able to breed until their third or later year. It is difficult to see why this should be so. In any given species there may be good endocrinological reasons for the delay, but they can hardly be evolutionarily inevitable. The situation has an obvious *prima facie* disadvantage, since most birds have a strikingly diagonal survivorship curve after the first year of life and this in itself indicates little capacity for learning to live. One would have supposed that in the birds, mainly but not exclusively large sea birds, which show the delay, any genetic change favoring early reproduction would have a great selective advantage. Any experimental model imitating this situation would be of great interest.

The problem of possible social effects of long post-reproductive life, which can hardly be subject to direct selection, provides another case in which any hints from changes in demographic parameters in experiments would be most helpful. The experimental study of the evolutionary aspects of demography is certain to yield surprises. While we have Nicholson's work, in which

the amplitude of the oscillation in *Lucilia* populations appear to be increased or at least not decreased as a result of the evolution he has observed, though the minima are less low and the variation less regular, we do not know if this sort of effect is likely to be general. Utida's elegant work on bean weevils appears to be consistent with some evolutionary damping of oscillations which would be theoretically a likely result.

The most curious case of a genetic change playing a regular part in a demographic process is certainly that in rodents described by Chitty. In view of the large number of simple ways which are now available to explain regular oscillations in a population, it is extremely important to heed Chitty's warning that the obvious explanation is not necessarily the true one. To the writer, this seems to be a particular danger in human demography, though the mysteries of variation of the human sex ratio, so clearly expounded by Colombo, should be a warning against over-simple hypotheses, for here no reasonable hypotheses have been suggested. Human demography relies too much on what psychologists call intervening variable theory. The reproducing organisms are taken for granted; when their properties change, either as the result of evolution or of changes in learned behaviour, the results are apt to be upsetting. The present "baby boom" is such an upset, and here a tendency to over-simplified thinking is also apparent. If, as appears clear at least for parts of North America, the present birth rate is positively correlated with economic position, it is easy to suppose that couples now have as many children as they can afford, just as most small birds appear to do. There is, however, a difference. If at any economic level a four child family was desired, but occasionally owing to the imperfections of birth control a five child family was actually achieved, we should not expect the fifth child to have a negligible expectation of life at birth, so that the total contribution to the population per family would be the same from a four and a five child family. Yet this is exactly what Lack and Arn (1947) found for the broods of the Alpine swift *Apus melba*. In man the criterion is never purely economic; it is not how large a brood can be reared, but how large a brood the parents think they can rear without undue economic sacrifice. Such a method of setting limits to natality is obviously extremely complicated. It involves an equilibrium between a series of desires, partly conscious, partly unconscious, and a series of estimates of present and future resources. There is absolutely no reason to suppose that the mean desired family size determined in such a way is a simple function of economics, uninfluenced by a vast number of other cultural factors. The assumption that a large family is *per se* a good thing is obviously involved; this may be accepted individually by most parents even though it is at present a very dubious assumption on general grounds of social well being. Part of the acceptance of such an assumption is certain to be due to unconscious factors. Susannah Coolidge in a remarkable, as yet unpublished, essay,[7] "Population *versus* People," suggests that for many women a new pregnancy is an occasion for a temporary shifting of some of the responsibility for the older children away from the mother, and so is welcomed. She also suspects that it may be an unconscious expression of disappointment over, or repudiation of, the older children and so be essentially a repeated neurotic symptom. Moreover, the present highly conspicuous fashion for maternity, certainly a healthy reaction from the seclusion of upper-class pregnant women a couple of generations ago, is also quite likely fostered by those business interests which seem to believe that an indefinitely expanding economy is possible on a non-expanding planet.

An adequate science of human demography must take into account mechanism of these kinds, just as animal demography has taken into account all the available information on the physiological ecology and behaviour of blow flies, *Daphnia* and bean weevils. Unhappily, human beings are far harder to investigate than are these admirable laboratory animals; unhappily also, the need becomes more urgent daily.

REFERENCES

ANDREWARTHA, H. G., and BIRCH, L. C., 1954, The Distribution and Abundance of Animals. Chicago, University of Chicago Press. xv, 782 pp.

BETTS, M. M., 1955, The food of titmice in oak woodland. J. Anim. Ecol. *24:* 282–323.

BIRCH, L. C., 1957, The meanings of competition. Amer. Nat. *91:* 5–18.

BRIAN, M. V., 1953, Species frequencies from random samples in animal populations. J. Anim. Ecol. *22:* 57–64.

1956a, Segregation of species of the ant genus *Myrmica*. J. Anim. Ecol. *25:* 319–337.

1956b, Exploitation and interference in interspecies competition. J. Anim. Ecol. *25:* 339–347.

COOPER, D. M., and DOBZHANSKY, TH., 1956, Studies on the ecology of *Drosophila* in the Yosemite region of California. I. The occurrence of species of *Drosophila* in different life zones and at different seasons. Ecology *37:* 526–533.

CROMBIE, A. C., 1945, On competition between different species of graminivorous insects. Proc. Roy. Soc. Lond. *132*B: 362–395.

1946, Further experiments on insect competition. Proc. Roy. Soc. Lond. *133*B: 76–109.

1947, Interspecific competition. J. Anim. Ecol. *16:* 44–73.

DA CUNHA, A. B., EL-TABEY SHEKATA, A. M., and DE OLIVIERA, W., 1957, A study of the diet and nutritional preferences of tropical of *Drosophila*. Ecology *38:* 98–106.

DUMAS, P. C., 1956, The ecological relations of sympatry in *Plethodon dunni* and *Plethodon vehiculum*. Ecology *37:* 484–495.

FISHER, R. A., CORBET, A. S., and WILLIAMS, C. B., 1943, The relation between the number of species and the number of individuals in a ransom sample of an animal population. J. Anim. Ecol. *12:* 42–58.

GAUSE, G. F., 1934, The struggle for existence. Baltimore, Williams & Wilkins. 163 pp.

1935, Vérifications expérimentales de la théorie mathématique de la lutte pour la vie. Actualités scientifiques *277*. Paris. 63 pp.

HUTCHINSON, G. E., 1939, Ecological observations on the fishes of Kashmir and Indian Tibet. Ecol. Monogr. *9:* 145–182.

1940, A revision of the Corixidae of India and adjacent regions. Trans. Conn. Acad. Arts Sci. *33:* 339–476.

1941, Ecological aspects of succession in natural populations. Amer. Nat. *75:* 406–418.

1944, Limnological studies in Connecticut. VII. A critical examination of the supposed relationship between phytoplankton periodicity and chemical changes in lake waters. Ecology *25:* 3–26.

1951, Copepodology for the ornithologist. Ecology *32:* 571–577.

1953, The concept of pattern in ecology. Proc. Acad. Nat. Sci. Phila. *105:* 1–12.

HYNES, H. B. N., 1954, The ecology of *Gammarus deubeni* Lilljeborg and its occurrence in fresh water in western Britain. J. Anim. Ecol. *23:* 38–84.

1955, The reproductive cycle of some British freshwater Gammaridae. J. Anim. Ecol. *24:* 352–387.

JACZEWSKI, S., 1937, Allgemeine Zügeder geographischen Verbreitung der Wasserhemiptera. Arch. Hydrobiol. *31:* 565–591.

KOHN, A. J., The ecology of *Conus* in Hawaii. (Yale Dissertation 1057, in press.)

KURTÉN, B., 1957, Mammal migrations, cenozoic stratigraphy, and the age of Peking man and the australopithecines. J. Paleontol. *31:* 215–227.

LACK, D., 1947a, The significance of clutch-size. Ibis *89:* 302–352.

1947b, Darwin's finches. Cambridge, England. x, 208 pp.

1954, The natural regulation of animal numbers. Oxford, The Clarendon Press, viii, 343.

LACK, D., and ARN, H., 1947, Die Bedeutung der Gelegegrösse beim Alpensegler. Ornith. Beobact. *44:* 188–210.

LOTKA, A. J., The growth of mixed populations, two species competing for a common food supply. J. Wash. Acad. Sci. *22:* 461–469.

LUNDBLAD, O., 1933, Zur Kenntnis der aquatilen und semi-aquatilen Hemipteren von Sumatra, Java, und Bali. Arch. Hydrobiol. Suppl. *12:* 1–195, 263–489.

MACAN, T. T., 1938, Evolution of aquatic habitats with special reference to the distribution of Corixidae. J. Anim. Ecol. *7:* 1–19.

1956, A revised key to the British water bugs (Hemiptera, Heteroptera) Freshwater Biol. Assoc. Sci. Publ. *16:* 73 pp.

MACARTHUR, R. H., 1957, On the relative abundance of bird species. Proc. Nat. Acad. Sci. Wash. *43:* 293–295.

MOREAU, R. E., 1944, Clutch-size: a comparative study, with special reference to African birds. Ibis *86:* 286–347.

PEARSALL, W. H., 1932, The phytoplankton in the English lakes II. The composition of the phytoplankton in relation to dissolved substances. J. Ecol. *20:* 241–262.

PRESTON, F. W., 1948, The commonness, and rarity, of species. Ecology *29:* 254–283.

PROCTOR, V. W., 1957, Some factors controlling the distribution of *Haematococcus pluvialis*. Ecology, *in press*.

ROTHSCHILD, LORD, 1915, On the gigantic land tortoises of the Seychelles and Aldabra-Madagascar group with some notes on certain forms of the Mascarene group. Novitat. Zool. *22:* 418–442.

RUITER, L. DE, 1952, Some experiments on the camouflage of stick caterpillars. Behaviour *4:* 222–233.

[7] I am greatly indebted to the author of this work for permission to refer to some of her conclusions.

SHINOZAKI, K., and URATA, N., 1953, Researches on population ecology II. Kyoto Univ. (not seen; ref. MacArthur, 1957).

SHORTEN, M., 1953, Notes on the distribution of the grey squirrel (*Sciurus carolinensis*) and the red squirrel (*Sciurus vulgaris leucourus*) in England and Wales from 1945 to 1952. J. Anim. Ecol. *22:* 134–140.

1954, Squirrels. (New Naturalist Monograph 12.) London 212 pp.

SKELLAM, J. G., 1951, Random dispersal in theoretical populations. Biometrika *38:* 196–218.

SLOBODKIN, L. B., 1953, An algebra of population growth. Ecology *34:* 513–517.

1954, Population dynamics in *Daphnia obtusa* Kurz. Ecol. Monogr. *24:* 69–88.

STEUSLOFF, V., 1943, Ein Beitrag zur Kenntniss der Verbreitung und der Lebensräume von *Gammarus*-Arten in Nordwest-Deutschland. Arch. Hydrobiol. *40:* 79–97.

STICKEL, W., 1955, Illustrierte Bestimmungstabellen der Wanzen. II. Europa. Hf. 2, 3. pp. 40–80 (Berlin, apparently published by author).

UENO, M., 1934, Yale North India Expedition. Report on the amphipod genus *Gammarus*. Mem. Conn. Acad. Arts Sci. *10:* 63–75.

VOLTERRA, V., 1926, Vartazioni e fluttuazioni del numero d'individui in specie animali conviventi. Mem. R. Accad. Lincei ser. 6, *2:* 1–36.

WINSOR, C. P., 1934, Mathematical analysis of the growth of mixed populations. Cold Spr. Harb. Symp. Quant. Biol. *2:* 181–187.

40

Reprinted from *Population Biology and Evolution,* R. C. Lweontin, ed.,
Syracuse University Press, Syracuse, N. Y., 1968, pp. 159–176

The Theory of the Niche

ROBERT MAC ARTHUR

Princeton University
Princeton, New Jersey

INTRODUCTION

A few words about the role of theory in ecology in general will be in
order before we turn to the theory of the niche in particular. Ecological
patterns, about which we construct theories, are only interesting if they
are repeated. They may be repeated in space or in time, and they may be
repeated from species to species. A pattern which has all of these kinds of
repetition is of special interest because of its generality, and yet these very
general events are only seen by ecologists with rather blurred vision. The
very sharp-sighted always find discrepancies and are able to say that there
is no generality, only a spectrum of special cases. This diversity in outlook,
has proved useful in every science, but it is nowhere more marked than in
ecology. Here, aside from a few physical, chemical, and ecological constraints, the patterns are shaped only by natural selection. The constraints set limits beyond which the system cannot wander; within these
limits there is an evolutionary optimum toward which organisms should
converge. But the uncertainties of the environment make the goal change
from place to place and time to time, and the organisms under the influence
of this selection only show a rough tendency to conform to one another and
to the optimum. Clearly, no pattern of such organisms can be expected to
coincide with numercial precision from situation to situation. Fortunately,
numerical precision is not the only aim of science; the new hypotheses and
simplification of education which come from qualitative theories can be
just as rewarding.

A theory must eventually be falsifiable to be useful to a scientist, but it
does not *in itself* have to be directly and easily verified by measurement.
More often it is the consequences of the theory that are verified or proved
false. In what follows we shall see some of the ecological constraints and
some of the evolutionary optima in a disprovable form.

STRENGTHS AND WEAKNESSES OF EARLY
DEFINITIONS OF NICHE

The term niche was almost simultaneously defined by Elton and Grinnell to mean two different things. To Elton, the niche was, somewhat vaguely, the animal's "role" in the community. This included its position in the food web as well as miscellaneous other habits. To Grinnell, the niche was a subdivision of the environment, a somewhat finer subdivision than the life zone. Some features of each were incorporated into the niche as Hutchinson (1958) has redefined it. In Hutchinson's niche, each measureable feature of the environment was given one coordinate in a infinite-dimensional space. The region in this space in which fitness of an individual was positive was called that individual's niche. This provided, for the first time, a definition precise enough that the term "niche" could enter into falsifiable statements. Now the statement "Two species cannot co-exist if their niches are identical" became true but trivial (since no two individuals, let alone two species, have identical niches). The statement "Two species can coexist if their niches do not overlap" becomes plausibly false, since by feeding in different places, two species would occupy non-intersecting niches even if they both depended on and competed for the same highly mobile food supply.

Hutchinson proposed to relate niche to competition by locating homogeneous patches of environment lying within the intersection of two species' niches. There, he conjectured, only one of the species would persist within an enclosure preventing migration. Although this conjecture suggests an interesting experiment which has, I believe, never been done, the experiment could not, for three reasons, falsify the hypothesis. To falsify the hypothesis we would have to know that the enclosure did indeed be within the intersection of the niches. But, as Hutchinson was among the first to realize:

1. The niche space is infinite dimensional, and we do not know how the addition of dimensions might shrink the niche intersection.

2. The niche coordinates are ambiguous, and we do not know how their alteration might change the niche intersection.

3. Where a variety of mobile resources (large and small, for example) are found in the same area, it is not clear whether that area lies in the niche intersection of a species which is efficient on large foods but able to eat small, and a second species which is efficient on small foods but able to eat large. If the area is in the intersection, then the conjecture appears false; if the area is not in the intersection, the conjecture appears untestable.

People who insist that all such terms be operational will reject "niche" just as they must reject "phenotype" and "genotype," as involving an infinite number of measurements; but some statements about *differences* between niches are perfectly testable, which is all that matters. (In fact, the parallel between "niche" and "phenotype" is more than formal. "Phenotype" embodies all the measurements which can be made on an individual during its lifetime, including those measurements which constitute its niche. Thus "phenotype" includes "niche." And to the extent that all relevant phenotype parameters affect fitness, "niche" almost includes "phenotype.") Hence, the term "niche" will mainly appear in comparative statements. We can compare niches of two similar species, or we can compare the niche of one species at two places or two times. Usually these differences will involve only one or two measurements. Levins has discussed the evolution of the niche in a fashion which I cannot improve, so I turn to other aspects.

COMPETITION AND THE STRUCTURE OF THE ENVIRONMENT

The environment, especially the terrestrial environment, has such a complex geometrical structure that it is no wonder a term like niche seems to be either ambiguous or inadequate. If animals were unable to use the structure, it would not affect the coexistence of species; but when the structure is large compared to the organisms, some analysis of the structure itself is necessary.

To help understand the structure of the environment, Levins and I introduced the concept of grain. We now call a patch of environment "fine-grained," relative to a species, if that species comes upon the resources and other components of that patch in the proportion in which they occur. Conversely, if the species can spend a disproportionate amount of its time on one resource or other component, then we called the patch coarse-grained. Thus, a warbler feeding among the tree tops in a forest possibly treats the different deciduous tree species as a fine-grained mixture and, if there is twice as much maple canopy as beech canopy, spends twice as much time in maple as in beech. Even more plausibly, the warbler comes upon the different defoliating insects in a small piece of canopy in the proportion in which they occur, so that within that piece the food supply is fine-grained. Almost certainly it sees everything within some small area reflecting the visual field of the bird. To the insects, however, especially the monophagous ones, the same canopy is obviously coarse-grained since each species may spend all of its time in one species of tree, and even in one part of the tree. A filter-feeding pelagic copepod may treat depth as a coarse-grained feature of its environment, since it selects one depth at which to concentrate its feeding, but within that depth it comes upon all particles in the proportion in which they occur. Although it only selects some of these to eat, its potential food supply is fine-grained. Finally, notice that a species may treat the environment as fine-grained even though each individual treats it as coarse. For instance, a plant species may distribute its seeds randomly over various patches of its environment, but each germinated seedling will spend its life in one patch. The sharp-sighted will, of course, object that no environment is ever perfectly fine-grained, but science is made up of such fictions. (Think how physics would be without its frictionless pulleys, conservative fields, ideal gases, and the like), and to the extent that we can approach a truly fine-grained situation the term can be useful. Of course, to make comparative statements we must have some measure of the degree of departure from fine-grainedness. (Here is one such measure: let $p_1, p_2, p_3 \ldots p_n$ be the proportions of components as they occur in nature and let $t_1, t_2, t_3 \ldots t_n$ be the corresponding numbers of independent observations of a given duration [say 5 seconds] of the organisms' time spent in these components. The

sum of the t_i, T, must be fixed. Then $\displaystyle\sum_{i=1}^{n} \frac{(t_i - Tp_i)^2}{Tp_i}$ has a χ^2 distribution with mean $n - 1$ and standard deviation $\sqrt{2(n - 1)}$ so that

$$D = \sum_{i=1}^{n} \frac{\dfrac{(t_i - Tp_i)^2}{Tp_i} - (n - 1)}{\sqrt{2(n - 1)}}$$

is a measure of departure which is nearly independent of n, and its comparison is independent of the fixed number of seconds of observation.)

I now return to the problems of understanding how species coexist. To accomplish this I make the following conjectures:

(a) In a patch which is either a fine-grained mixture or homogeneous and structureless, species only coexist by virtue of resource subdivision. (Downy and hairy woodpeckers may come upon the same food items, but the downy select the small and the hairy select the large, so that coexistence is possible.)

(b) A real environment can be assembled from bricks, or building blocks, of fine-grained patches.

These conjectures, which are the basis of what I will say, require a little discussion. That a fine-grained medium cannot support more species than it has resources seems to be the message of the bottle experiments of the last thirty years. It is also rendered likely by a theoretical argument (Mac Arthur and Levins, 1964). But the usual exceptions must be made; namely, if anything keeps the populations too low to exhaust the resources, that thing may not be acting in a fine-grained fashion and could allow extras to persist. Also, of course, the definition of resource is fuzzy. To be two resources, two items of food must be independently harvestable. For instance, bark and leaves and seeds of a tree may well be three (at least) resources; but if the bark is eaten to such a level that it interferes with leaf production, it is no longer a separate resource. No amount of harvesting of seeds will interfere with leaf production, at least in the short run, and so seeds are clearly a different resource. If the first conjecture proves valid, then we only need to look at resource coordinates of niches in fine-grained environments, which is a vast improvement.

The second conjecture is more subtle. It may turn out that some, but not all, environments can be assembled from fine-grained components. If so, we will certainly need to learn how to tell which environments can. For these environments it will be quite easy to understand coexistence. I shall do this in steps, summarized in Table I.

TABLE I

Number of resources	Environment	Maximum number of species
1	fine-grained	1
2	" "	2
n	" "	n
1	coarse-grained; m-grain types	m
1 or many	coarse-grained; continuous	many
1 or many	$\left\{\begin{array}{l}\text{(coarse-grained for some species)} \\ \text{(fine grained for others)}\end{array}\right\}$	?

The main new feature added by coarsening the grain is the possibility of habitat subdivision; barring other factors limiting the number of species, there can be as many species as there are grain types, even with but one resource in the system. There can also be fewer species than grain types, but sometimes this will be accomplished by one species occupying more than one grain type indiscriminately—in other words, it will treat these in a fine-grained fashion. It might feed in one or both grain types but still in a coarse-grained fashion. Here the time wasted hurrying through the poor patch types might more than compensate for feeding specializations within good patches, and a "patch generalist" would be in a position to out-compete two patch specialists. The details of when this would happen depend upon whether the specialists can travel a path which avoids poor patches, or whether the patches are more isolated. Thus some measure of patch connectedness will be called for. I have no useful measure to propose.

To make these statements testable, we return to the measure of departure from fine-grainedness. If we choose some standard departure (Say $D = K$), then there is a host of possible, and testable, hypotheses:

(a) If, for one species, $D < K$, then no other species will be present (unless they differ in resources).

(b) If, for one species, $D > K$, then other species will be present and, adding together the utilizations of all the species, $D < K$. These are just guesses, but they illustrate the machinery for testing grain statements. We might also predict that K would be less in stable environments.

Finally I have talked about coexistence within a fine-grained patch by virtue of resource subdivision and about coexistence by virtue of coarse grained utilization of the habitat. Both of these concepts can be made more numerical, with the accompanying chances of being wrong, as I now will try to show. I shall base my discussion on joint papers with Levins (Mac Arthur and Levins, 1964, 1967).

First I deal with coexistence by virtue of resource subdivision within a fine-grained patch. Resources are consumed and renewed, and the level of abundance which they reach depends upon the balance of consumption and renewal. A heavier consumption will normally lower the abundance of the resource. There will normally be some threshold level of the resource below which the consumer cannot feed rapidly enough to maintain its population. Thus, a bird may have to rear a certain number of young to compensate for its annual mortality and must gather a certain number of insects per hour in order to feed those young. This determines the threshold density of the insect resource. When the resources exceed this threshold, each consumer can gather more than enough food and its population can rise, which will cause the resource level to drop. Hence, the equilibrium level of the resource will normally coincide with the threshold level marking how few individuals the harvesters can use to perpetuate their population. (This all assumes a resource-limited consumer, of course.) This threshold level to which the consumer reduces its resources I call T. Hence, as a not-too-rough approximation, the population S_1 of a resource-limited consumer might grow according to the formula

$$\frac{ds_1}{dt} = rs_1 f(g(R_1, R_2 \ldots) - T)$$

where f is a monotonic function such that $f(o) = o$, and, where $R_1, R_2 \ldots$ etc. are the densities of the various resources in the fine-grained mixture. For pictorial purposes we consider a two-resource system, and we plot $\frac{ds_1}{dt} = o$ in the graph whose coordinates are R_1 and R_2 (see the right-hand

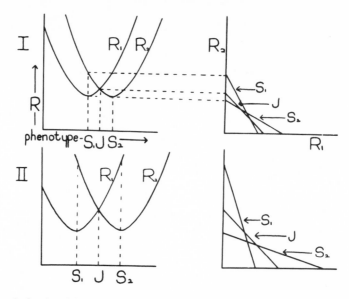

Figure 1. In the right-hand graphs, the levels to which species with phenotypes S_1, S_2, and J can reduce resources R_1 and R_2 are indicated by lines (isoclines). These lines would be curved if the resources are not alternatives to one another. In the top right-hand graph, J can invade a community containing S_1 and S_2, while in the lower right-hand graph the J line lies outside the intersection of the S_1 and S_2 lines showing J cannot invade. In the left-hand graphs, the intercepts of the isoclines of the right graphs are plotted. Here the R_1 and R_2 intercepts can be plotted as a continuous, but rather arbitrary, function of phenotype, the low point on the R_1 curve being the phenotype, S_1, which is specialized for this resource. The phenotype J is a jack-of-both-trades in the sense that it can reduce both resources equally. These graphs can be used to predict evolutionary convergence and divergence: when the resources are as different as in graphs II, J is inferior and phenotypes near J will diverge toward S_1 and S_2; when the resources are as similar as in graphs I, phenotypes S_1 and S_2 will converge toward J.

parts of Figure 1). From what we have said, $\dfrac{ds_1}{dt} = o$ if $g(R_1, R_2) = T$ which is the equation to plot on the graph.

In case $g(R_1, R_2) = A_1 R_1 + A_2 R_2$, then the resources are interchangeable and $\dfrac{ds_1}{dt} = o$ is a straight line as shown in the figures. If the resources provide different dietary requirements, the curve $g(R_1 R_2) = T$ bends inward toward the origin and may even be asymptotic to the coordinates. At any rate, the population of resource-limited species s_1 will increase until the resources are reduced to some value along the curve $g(R_1 R_2) = T$ at which time $\dfrac{ds_1}{dt} = o$ and there will be no further increase. We now ask whether a second and third species, with populations s_2 and J can invade. If the curve $\frac{ds_2}{dt} = o$ does not intersect $\frac{ds_1}{dt} = o$, they cannot co-exist in equilibrium; but if they do intersect, equilibrium may (but need not always) be possible. If they do coexist, it must be at the resource level corresponding to the intersection of the curves $\frac{ds_1}{dt} = o$ and $\frac{ds_2}{dt} = o$. In this case, J can invade if, and only if, the line $\frac{dJ}{dt} = o$ lies inside the intersection point; and again we must decide whether J will replace both or coexist with one. (It cannot coexist stably with both because there is no value of R_1 and R_2 lying on all three lines; even if the three lines intersected at one point, it can be shown that at least one would go extinct.) Whether one or two consumer species persist depends upon whether the resource species are able to replace themselves at the level indicated by the intersection of the curves. To answer this we must also examine the equations governing the resource population growth. Both consumers will persist if $\frac{dR_1}{dt} = o$ and $\frac{dR_2}{dt} = o$ have a positive solution when the R_1 and R_2 values at the intersec-tion are put in. The main point is that no more species than resources will persist and perhaps not as many consumers as resources. Since selec-tion of diet is not influenced by competitions, the species, to consume different diets, must have quite different phenotypes; so the species which do coexist within a fine-grained patch should differ in phenotype. These same graphs can be related to phenotype and used to predict character divergence and convergence (see the figure), but I shall instead turn to the situation in a coarse-grained environment.

In addition to resource subdivision, the coarse-grained environment offers opportunities for habitat separation, and it is this new aspect which I shall describe here, using the same kind of graph (Fig. 2a, 2b). Now R_1 will not be one kind of resource, but rather the quantity of the mixed resources in grains of type 1. As before our graphs show us that, by virtue of the grains of habitat, no more species than grain types can persist (and perhaps fewer). And, as before, if the resources in the grains of habitat are interchangeable, then the lines will be straight. In this case, knowledge of the intercepts is sufficient, and we inquire to what level a predator can reduce its resources in each kind of habitat grain. This will be closely related to the proportion of its time the predator spends in that kind of grain.

In the figures the upper limits to the amounts of R_1 and R_2 are marked by the stippled zone. That is, the resources are only capable of renewing within the stippled zone so that only these are feasible resource levels. In

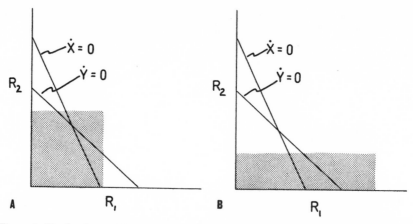

Figure 2. The levels at which the rescurces can renew themselves are superimposed with stippling onto graphs like the right-hand graphs of Figure 1. In 2a, both resources can reach the same level and the stippled zone includes the intersection of the isoclines for species X and Y. Hence X and Y will persist. In 2b, resource 2 reaches only a very low level, perhaps because it is confined to a small part of the environment. Hence, both X and Y cannot persist and X alone can always oust Y and prevent its reinvasion.

Figure 2a, two species can coexist (unless some other limitation on resources or consumers is acting), while in Figure 2b, the inequality of resource limits prevents more than one species from persisting.

Hence, by decomposing an environment into fine-grained building blocks, it seems possible to consider separately the resource coordinates of the Hutchinson niche and the other coordinates which determine the patch classification.

THE NICHE AND SPECIES DIVERSITY

My brother and I (Mac Arthur and Mac Arthur, 1961, and later) discovered that the number of bird species breeding in a small area of rather uniform aspect could be predicted in terms of the layers of vegetation and seemed independent of the number of plant species (see Figure 3). More specifically, we let p_1 be the proportion of the total vegetation lying in the herbaceous layer between the ground and about two feet, p_2 be the proportion of vegetation in the brush layer from about 2 to 25 feet, and p_3 be the proportion in the canopy about 25 feet. Then the logarithm of the

number of species was proportional to $-\sum_{i=1}^{3} p_i \ln p_i$ which we called the foli-

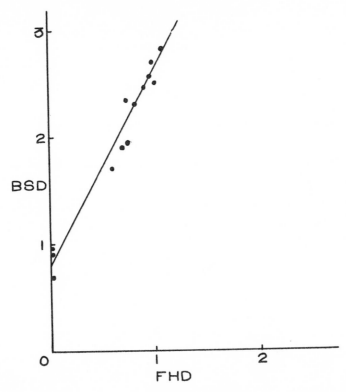

Figure 3. The empirical relation between bird species diversity, B.S.D., (roughly the logarithm of the number of species) and the foliage height diversity, F.H.D., calculated as described in the text. The bird censuses are from deciduous and grassland habitats in temperate North America.

age height diversity. In other words, the number of bird species could be predicted in terms of the structure of the habitat by means of a particular formula. (For comparisons involving different geographical areas, more than structural differences need to be considered [Mac Arthur, Recher, and Cody, 1966], and the climatic stability and environmental productivity and, perhaps, history are influential.) Here I shall show how the theory presented in the last section not only predicts the qualitative aspects of this relation but also the exact formula for foliage height diversity (although this last may be a coincidence).

We saw in Figure 2 that the number of species persisting depended upon whether the intersection of the isoclines lay in the stippled zone. And the dimensions of the stippled zone are determined by the proportion of the total area which belongs to the various patches. If patch 1 has ¾ of the area and patch 2 has ¼, then the resources in patches of type 1 should reach about three times the abundance of those in patches of type 2. In other words, the dimensions should be roughly proportional to ¾ and ¼. So the volume of the stippled area is ¾ × ¼ or, in general, a product of p_i's. Still more generally, there will be different resources in each patch type. If the patch type is absent ($p_i = o$), then the number of resources it contains must also be zero, and it seems reasonable to suppose that the number of resources is proportional to p_i. This is equivalent to a linear species-area curve, which is a first approximation to the truth. Hence the product of p_i's will contain p_1 about np_1 times, p_2 about np_2 times, and p_3 about np_3 times; that is, we will plot a graph with n times p_1 coordinates for resources in patches of type 1 and with $n \times p_2$ coordinates for resources in patches of type 2, etc. The product of all of these numbers is $p_1{}^{np_1}\ \ p_2{}^{np_2}\ \ p_3{}^{np_3}\ \ldots$ which should be related to the number of species so that the logarithm of the number of species should be determined by the logarithm of $p_1{}^{np_1}\ \ p_2{}^{np_2}$ ... which is $n \Sigma p_i\ ln\ p_i$, the formula which we have seen is actually useful. This explanation may be spurious, but for the present it serves to connect the theory of competition with the facts of species diversity. Of course, different organisms would not recognize layers of vegetation as the coarse grains of their environments, so the determination of which patches are relevant must be made empirically.

When other factors than competition limit species diversity, such as predation or history, then there is no reason to expect this theory to be relevant.

NICHE EXPANSION AND COMPETITIVE RELEASE

One of the most frequently documented bits of evidence for competition is the habitat expansion of species on islands where they are freed from their mainland competitors. For instance, Cameron (1958) pointed out that the arctic hare *(Lepus arcticus)* on Newfoundland formerly had expanded its habitat to include not only the tundra where it lives on the mainland but also the forest land throughout the island. The obvious cause was the release from competition with the snowshoe hare (*Lepus americanus*) which occupies the forests on the mainland but had not colonized Newfoundland. This example is specially interesting because the snowshoe hare was subsequently introduced and the arctic hare contracted back into its tundra stronghold. This case (like most of the others) was not as carefully studied as we might wish, but Crowell is now performing the same sorts of experiments with mice on islands off the Maine coast, and the process should soon be well-documented.

Pianka and I (Mac Arthur and Pianka, 1966), and independently Emlen (1966), gave a theoretical explanation for this kind of expansion which involves some new and testable consequences. We were concerned with optimal diet and optimal patch utilization based on economic grounds. That part of the argument which deals with niche expansion under release from competition runs as follows. The decision which patch to feed in is made before the food is located, while the decision whether to pursue and eat an item is made after the food item is located. Hence, the patch decision must be made on the grounds of expectations of yield while the diet decisions are made on the merits of the item already located. An item already located is worth pursuing if, in the time it takes to catch and eat it, no more rewarding item would be found. This decision is not based on the abundance of the item under consideration (although the rarity of other kinds of items will enter the decision), and so a reduction in the abundance of that item should not affect its acceptability in the diet. A reduction in the abundance of all items within a patch will greatly affect the expected harvest within that patch and should, therefore, change the decision of where to feed. That is, competitive reduction in food should alter the patches in which the species searches (often reducing the habitat), but should not very greatly alter the acceptable range of foods within a patch. (The diet itself, reflecting both acceptability and abundance of the foods may change.) Hence, we often expect a habitat contraction in the presence of competitors, but the range of acceptable items should be relatively unchanged. The habitat expansion is, as we have seen, frequently documented; the range of items in the diet is just a testable conjecture at this time.

The same theory leads to other testable predictions, and I include one here because optimal diet is so relevant to the study of the niche. Once the species finds an item of food, the decision of whether or not to try for it depends upon the likelihood of coming upon a more rewarding item during the time it would take to capture and eat the first one. If capture and eating take very little time compared to search (the species is a "searcher"), then the likelihood is negligible and all palatable items should be eaten. If capture and eating are relatively very time-consuming and location of other items is quick (the species is a pursuer), it will always pay to specialize. Finally, if the density of food is increased, it will cut search but not pursuit time so the species will become more of a pursuer and should specialize more. Recher (in press) has verified some of these conclusions in his heron studies.

I shall end by giving one final numerical example of how theory can suggest experiments. For this purpose I go to the Volterra (1926) equations for two consumers X_1 and X_2 and their resources R_1 and R_2, although more general equations could be used.

411

$$\left.\begin{array}{l} \dfrac{dx_1}{dt} = X_1[\alpha_{11}R_1 + \alpha_{12}R_2 - T_1] \\[2ex] \dfrac{dx_2}{dt} = X_2[\alpha_{21}R_1 + \alpha_{22}R_2 - T_2] \end{array}\right\} \quad \dot{X} = X[AR - T]$$

$$\left.\begin{array}{l} \dfrac{dR_1}{dt} = \dfrac{r_1 R_1}{S_1}[S_1 - \gamma_{11}X_1 - \gamma_{12}X_2 - R_1] \\[2ex] \dfrac{dR_2}{dt} = \dfrac{r_2 R_2}{S_2}[S_2 - \gamma_{12}X_1 - \gamma_{22}X_2 - R_2] \end{array}\right\} \quad \dot{R} = \dfrac{r}{S}R[S - R - GX]$$

(The right hand equations are the matrix equivalents with boldface letters being column vectors and A the matrix of the alphas, and G of the gammas.)

A couple of comments are in order here. First, α_{ij} is the probability that a given item of resource j is eaten by a given individual of species i in a unit of time, and the alphas should be multiplied by w_j, the weight of resource j, if the weights vary. γ_{ij} is roughly $\alpha_{ji}S_i/r_i$. Second, I have added one new feature into the Volterra equations: the self-limitation of the resources, by means of the R_1 and R_2 terms on the extreme right. These put an upper limit (S_1, S_2) on the population of R_1 or R_2 in the absence of consumers corresponding to the stippled regions of Figure 2. There are doubtless also limits on the consumers in the presence of superabundant resources, but I will not be concerned with that case. These equations should be viewed as a plausible linear approximation to the true equations and are thus useful only near the equilibrium. That is, they are useful in studying the statics but not the dynamics of the situation. This misunderstanding has led the overenthusiastic to use Volterra equations far from equilibrium and the overcritical to assert that these equations are useless or even harmful to ecologists. I believe we should view them pragmatically, and if they can suggest novel and useful experiments they will have proved their worth. I shall use them to suggest experiments, but whether these experiments prove useful remains to be seen.

First, I shall relate the alphas to the intensity of competition. I have already done this in one way in Figures 1 and 2, where X_1 clearly outcompetes X_2 if the alphas of X_2 are so low that its isocline lies outside the X_1 isocline. I can also relate these four equations to the two classical competition equations by solving the second two for R in terms of X, at equilibrium and substituting into the first:

At equilibrium $\dot{R} = 0$ so that $R = S - GX$.

Substituting this into the equation for X we get

$$\begin{aligned} \dot{X} &= X[AS - GX\} - T] = X[(AS - T) - AGX] \\ &= X[K - AGX] \quad \text{where} \quad K = AS - T. \end{aligned}$$

This is of the form of the competition equations where the competition coefficients are the terms in the symmetric matrix AG. Again we see that the alphas determine the intensity of competition. To use the equations in this form, we convert back to the long-hand version:

$$\dot{X}_1 = X_1[K_1 - \beta_{11}X_1 - \beta_{12}X_2]$$
$$\dot{X}_2 = X_2[K_2 - \beta_{21}X_1 - \beta_{22}X_2]$$

where $\beta_{11} = (\alpha_{11}\gamma_{11} + \alpha_{12}\gamma_{21})$, $\beta_{12} = (\alpha_{11}\gamma_{12} + \alpha_{12}\gamma_{22})$, etc.

As is well known, β_{12} / β_{11} and β_{21} / β_{22} cannot be too large or stable co-existence is impossible.

The main question is, "How do we measure the alphas or betas in a real situation?" Here are two ways, motivated by the theory: Let K_1 be altered and we measure the change in the equilibrium populations. To alter K_1 (we don't need to know by how much it is altered) we can either go to a different environment or can act as predators ourselves, removing X_1 individials at the rate pX_1; this will reduce K_1 to $K_1 - p$.

At the former equilibrium

$$X_1 = \frac{K_1\beta_{22} - K_2\beta_{12}}{\beta_{11}\beta_{22} - \beta_{21}\beta_{12}}$$

$$X_2 = \frac{K_2\beta_{11} - K_1\beta_{21}}{\beta_{11}\beta_{22} - \beta_{21}\beta_{12}}$$

and at the new equilibrium, with K_1 replaced by K'_1, we simply substitute K'_1 for K_1 in these equations. We subtract the old from the new to get the change in equilibrium populations:

$$\Delta X_1 = \frac{K'_1\beta_{22} - K_1\beta_{22}}{\beta_{11}\beta_{22} - \beta_{21}\beta_{12}}$$

$$\Delta X_2 = \frac{K_1\beta_{21} - K'_1\beta_{21}}{\beta_{11}\beta_{22} - \beta_{21}\beta_{12}}$$

whence

$$\frac{\Delta X_1}{\Delta X_2} = -\frac{\beta_{22}}{\beta_{21}}$$

This is just the quantity we need to predict the stability of the species co-existence. (By altering R_1 we could have measured the ratio of alphas if we wished that instead.) Hence the theory tells us that by acting as predators on one species and watching the change in equilibrium populations we can measure the intensity of competition. This gives the precise form of the experiment we must perform, and if the results are well correlated with our predictions (that the ratios of the β's cannot be too large), then our use of the equations will have been justified. We should not expect the observations to conform perfectly, and to the extent that they do not, we must repair the equations. This successive use and repair of the theory is, of course, they way science works.

The derivation of the competition equations has another virtue: it gives us a second recipe for calculating the coefficients from observations of the time and energy budget of the species. Thus

$$\frac{\beta_{12}}{\beta_{11}} = \frac{\alpha_{11}\gamma_{21} + \alpha_{12}\gamma_{22}}{\alpha_{11}\gamma_{11} + \alpha_{12}\gamma_{21}} = \frac{\sum_i \alpha_{1i}\alpha_{2i}\dfrac{w_i S_i}{r_i}}{\sum_i \alpha_{1i}^2 \dfrac{w_i S_i}{r_i}}$$

for instance, and each α_{ij} is a product of the proportion of time that consumer i spends where it would find resource j, times the probability that, during a unit of hunting time in this place, the consumer will actually locate, catch, and eat a particular unit of resource. This may seem pretty unmeasurable, but there is immediate hope along at least one line of attack. Closely related species which subdivide the coarse grains of the habitat and are morphologically similar need not have large differences in search and pursuit efficiency. It would seem more likely *a priori* that their efficiencies are only barely different—just enough to make each superior in its own patch type. In this case only the time budgets enter into the calculation of $\dfrac{\beta_{21}}{\beta_{11}}$, and the limiting similarity of coexisting species can be related to the time budget.

In this case, if r, S, and w terms are equal

$$\frac{\beta_{12}}{\beta_{11}} = \frac{t_{11}t_{21} + t_{12}t_{22}}{t_{11}^2 + t_{22}^2}$$

or more generally

$$\frac{\beta_{ij}}{\beta_{ii}} = \frac{\sum_k t_{ik}t_{jk}}{\sum_k (t_{ik})^2}$$

For example, the number of seconds the warblers feed in different parts of a spruce tree can be used to give preliminary equations (X_1 = myrtle, X_2 = Black-throated green, X_3 = blackburnian, X_4 = bay-breasted), with each equation divided by β_{ii}.

$$\frac{dx_1}{dt} = r_1 X_1 [6 \cdot 190 - X_1 - \cdot 490\, X_2 - \cdot 480\, X_3 - \cdot 420\, X_4]$$

$$\frac{dX_2}{dt} = r_2 X_2 [9 \cdot 082 - \cdot 519\, X_1 - X_2 - \cdot 959\, X_3 - \cdot 695\, X_4]$$

$$\frac{dX_3}{dt} = r_3 X_3 [6 \cdot 047 - \cdot 344\, x_1' - \cdot 654\, X_2 - X_3 - \cdot 363\, X_4]$$

$$\frac{dX_4}{dt} = r_4 X_4 [9 \cdot 014 - \cdot 545\, X_1 - \cdot 854\, X_2 - \cdot 654\, X_3 - X_4]$$

Here the coefficients of the X's are given by equation (1) using the data from Figure 4, and the constant terms in the brackets come from knowing the populations of the species. For instance, since there were 2 pairs of myrtles per 5 acres, 5 of black-throated green, 1 of blackburnian and 3 of bay breasted, we get $K_1 = 2 + 5 \times \cdot 490 + 1 \times \cdot 480 + 3 \times \cdot 420 = 6 \cdot 190$. From these equations we can calculate nearly anything we wish. Here I shall illustrate by finding the values of K_1, K_2, and K_3 which will prevent the bay-breasted warbler, X_4, from invading a community containing only the first three. For this purpose I abbreviate the matrix of X_1, X_2, X_3 coefficients by B:

$$B = \begin{bmatrix} 1 & \cdot 490 & \cdot 480 \\ \cdot 519 & 1 & \cdot 959 \\ \cdot 344 & \cdot 654 & 1 \end{bmatrix}$$

so that the equilibrium K values of the first three species are given by the column vector

$$\begin{bmatrix} K_1 \\ K_2 \\ K_3 \end{bmatrix} = B \begin{bmatrix} X_1 \\ X_2 \\ X_3 \end{bmatrix}$$

We know the bay-breasted can invade only when $K_4 - \Sigma \beta x > o$ for small X_4 and the values of X_1, X_2, and X_3 determined by equations (2). That is, it can invade when

$$K_4 > \begin{bmatrix} \cdot 545 & \cdot 854 & \cdot 654 \end{bmatrix} \begin{bmatrix} X_1 \\ X_2 \\ X_3 \end{bmatrix} = \begin{bmatrix} \cdot 545 & \cdot 854 & \cdot 654 \end{bmatrix} (B^{-1}) \begin{bmatrix} K_1 \\ K_2 \\ K_3 \end{bmatrix}$$

and, performing the inversion of the matrix B and premultiplying by the row vector, I get as the condition for invasion:

$$K_4 > \cdot 1392\, K_1 + 1 \cdot 0775\, K_2 - \cdot 4461\, K_3$$

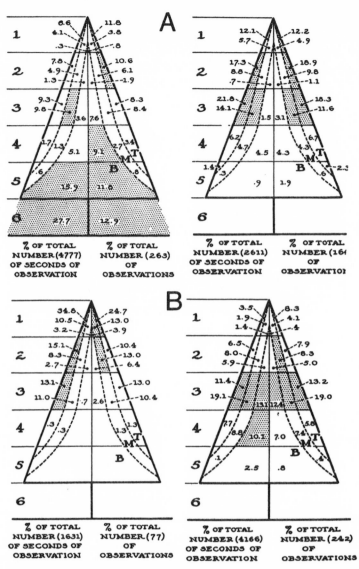

Figure 4. The figures are diagrammatic spruce trees subdivided into zones, with the time budgets of four species of warblers entered into the zones. From upper left to lower right, the species are myrtle, black-throated green, blackburnian, and bay breasted. The shaded zones indicate the most concentrated activity (from Mac Arthur, 1958).

When the distributions of foods are altered so that the K's no longer satisfy this inequality, we expect the bay-breasted warbler to be absent from the community. Notice that the negative term in the equation shows that an increase in blackburnian warblers should, by itself, make invasion easier for the bay-breasted. There are a great many testable predictions hidden in the equations. For instance, in the linear approximation on an island covered with similar forest but lacking all but the black-throated green, the density of this species should rise from 5 pairs per 5 acres to 9.082 pairs. The K's can be independently estimated from the time budget (from the formula $K = As - T$) so that $K_1 = \sum_j a_{ij}S_j - T_1$. By our assumptions, S_j may be roughly equal to the volume of the patch times the resource density within it and a_{ij} proportional to the proportion of its time the consumer spends within that kind of patch). These estimates can be directly compared with the other estimates from the censuses. A more accurate analysis of the warbler data would involve selecting feeding areas of equal S.

ACKNOWLEDGEMENTS

Most of this work is a result of studies carried out jointly with R. Levins. Drs. H. Horn and E. Leigh provided useful criticisms.

LITERATURE CITED

Cameron, W. Auston, 1958, "Mammals of the Islands in the Gulf of S. Lawrence," *Nat. Mus. Canada, Bull.*, 154.

Emlen, J. Merrit, 1966, "The Role of Time and Energy in Food Preference," *Amer. Natur.*, 100:611–617.

Hutchinson, G. E., 1958, "Concluding Remarks," Cold Spring Harbor Symp. Quant. Biol., 22:415–427.

Mac Arthur, R. H., 1958, "Population Ecology of Some Warblers of Northeastern Coniferous Forests," *Ecology*, 39:599–619.

Mac Arthur, R. H., and R. Levins, 1964, "Competition, Habitat Selection and Character Displacement in a Patchy Environment," *Proc. Nat. Acad. Sci.*, 51:1207–1210.

————., 1967, "The Limiting Similarity, Convergence, and Divergence of Coexisting Species." *Amer. Natur.*, 101:377–385.

Mac Arthur, R. H., and J. W. Mac Arthur, 1961, "On Bird Species Diversity," *Ecology* 42: 594–598.

Mac Arthur, R. H., H. Recher, and M. Cody, 1966, "On the Relation Between Habitat Selection and Species Diversity," *Amer. Natur.*, 100:319–332.

Mac Arthur, R., and E. Pianka, 1966, "On Optimal Use of a Patchy Environment," *Amer. Natur.*, 100:603–609.

Volterra, V., 1926, "Variazione e Fluttuazione del Numero d'Indvidui in Specie Animali Conviventi," *Mem. Accad. Naz. Lincei*, 2:31–113.

Reprinted from *Amer. Naturalist,* **107**(955), 321–338 (1973)

NICHE, HABITAT, AND ECOTOPE

R. H. WHITTAKER,* S. A. LEVIN,† AND R. B. ROOT‡

Cornell University, Ithaca, New York 14850

It is a maxim of the civil law that definitions are dangerous.
[SAMUEL JOHNSON, *The Rambler*]

It is regrettable that two of the most important terms in ecology, "niche" and "habitat," are now among the most confused in usage. To the extent that use of these is confused, use of other terms and concepts which depend on them will be confused as well. It is the purpose of this paper to sort out these concepts and to suggest a clear assignment of terms.

HISTORY OF NICHE CONCEPTS

The confusion comes from use of the same word, "niche," for different concepts. We shall distinguish three senses of the word as: (*a*) the niche as the position or role of a species within a given community—the functional concept of niche; (*b*) the niche as the distributional relation of a species to a range of environments and communities—the niche as habitat, or the place niche concept; and (*c*) the niche as an amalgam of both these ideas, and thus defined by both intracommunity and intercommunity factors. Of these (*b*) makes "niche" synonymous with "habitat"; for the present, we term (*c*) the "habitat + niche" concept.

Grinnell (1917, p. 433) first proposed "niche" to designate the place in an association occupied by a single species, emphasizing for the California thrasher not only food relations but dependence on cover, and adaptation thereto in physical structure and temperament. The thrasher's "ultimate associational niche . . . is one of the minor niches which with their occupants all together make the chaparral association." Grinnell (1924, p. 227) later called the ecologic niche the ultimate unit of habitat, and still later (1928, p. 435) defined it as the "ultimate distributional unit, within which each species is held by its structural and functional limitations." Grinnell's intention was probably closest to concept (*c*) of the habitat

---

* Section of Ecology and Systematics.

† Department of Theoretical and Applied Mechanics, Section of Ecology and Systematics, and Center for Applied Mathematics.

‡ Department of Entomology and Section of Ecology and Systematics.

+ niche, but from these statements stems use of "niche" in sense (*b*), as undistinguished from "habitat" (e.g., Hesse, Allee, and Schmidt 1937, p. 135; Allee et al. 1949, p. 234).

The functional concept of niche (*a*) was developed by Elton (1927) and others. For Elton (1927, p. 63) the niche is "the status of an animal in its community," its place in its biotic environment, particularly its relations to food and enemies. Elton's concept was adopted by Gause (1934, p. 19) as a basis for his development of the idea now termed the principle of Gause and Volterra or of competitive exclusion. "A niche indicates what place the given species occupies in a community, i.e. what are its habits, food, and mode of life." Similar statements were made in other books (Andrewartha and Birch 1954, p. 3; Bodenheimer 1958, p. 166; Kendeigh 1961, p. 16; Andrewartha 1961, p. 4; Whittaker 1970, p. 16; see also articles by Savage 1958; Udvardy 1959; DeBach 1966). Dice (1952, p. 227) summarized changing statements of the concept and defined niche as "the ecological position that a species occupies in a particular ecosystem." For Odum (1953, p. 15; 1959, p. 27), "the ecological niche, . . . is the position or status of an organism within its community and ecosystem resulting from the organism's structural adaptations, physiological responses, and specific behavior (inherited and/or learned)." Clarke (1954, p. 468) indicates that "niche" stresses "the function of the species in the community rather than its physical place in the habitat." "This 'functional niche' is more fundamental than the 'place niche,' but both concepts exist and should eventually be given different names." The uses by Dice (1952), Odum (1953), and Clarke (1954) might well have stabilized the term as the functional role in the community, but this stabilization did not occur.

The current difficulty derives primarily from different interpretations of a single, seminal paper by Hutchinson (1958). Hutchinson (1958, 1965, 1967) proposed that the environmental variables affecting a species be conceived as a set of *n* coordinates. For each of these coordinates limiting values exist, within which the species can survive and reproduce. The ranges of the coordinates within the limiting values define an *n*-dimensional hypervolume, at every point within which environmental conditions would permit the species to exist indefinitely. This hypervolume may be called the species' "fundamental niche." If both physical and biological variables (e.g., temperature difference with depth and food-size difference affecting a zooplankton community) are considered, the fundamental niche will completely define the species' ecological properties. The fundamental niche defined in this way is an abstract formalization of what is usually meant by ecological niche. If, within the ordinary physical space of a given biotope, there are points at which the conditions of this fundamental niche are fully realized, then the biotope is "complete" relative to that species. Because of competition and other interactions the species may be excluded from some parts of the fundamental niche. The reduced hypervolume in which a species then exists is termed its "realized niche." Hutchinson (1958) further observes: (1) Though the formulation suggests equal probability of survival of a species at all points up to the boundaries of its niche hypervolume, there will ordinarily be an optimum part of the niche and suboptimum conditions near the boundaries. (2) Linear ordering of all environmental variables is assumed, though this is not in practice possible. (3) The formulation refers to an instant in time, but time must also be considered a variable: A nocturnal and a diurnal species, with the same food and temperature ranges, etc., will occupy quite separate niches. Similarly, species that are motile in the biotope in different ways occupy different niches, even though they compete for food.

A number of other writers (including two authors of this paper) have considered that Hutchinson's fundamental niche formulated concept (c) of the habitat + niche. However, it should be emphasized that Hutchinson (1958) was concerned with the ecological requirements of individual species and competing species within biotopes, that is, within the environments of particular communities. The direction of his argument was toward interpretation of the principle of Gause and Volterra, and toward MacArthur's (1957) approach to niche space division by the birds of a "homogeneously diverse" biotope (that is, of a given community) and the question of the implication of niche specificity for the number of species that a given biotope can support. His formulation was of niche concept (a), with the niche defined purely *intensively*—that is, by variables as they apply within the biotope (Hutchinson 1967, p. 232). Factors of habitat, in contrast, have spatial extension. Hutchinson was not defining habitat; other authors (Ramenski 1924, 1930; Ellenberg 1950, 1952; Whittaker 1951, 1952, 1956; Bray and Curtis 1957) had used multidimensional treatments of habitat relations and had, moreover, sought to characterize species habitats through population measures. The multidimensional concept of habitat thus did not originate with Hutchinson and was not his concern in statement of the fundamental niche, although some other authors have approached species distributional or habitat relations through the multidimensional concept and termed the result "niche" analysis. That which was most original with Hutchinson and of profound significance was the application of the multidimensional concept of niche relationships to the coexistence of species in biotopes and the richness in species of communities.

Odum's (1971, p. 234) new edition has sought clarification, but we feel it has not succeeded. Odum accepts Grinnell's later (habitat + niche) position as the general concept of "ecological niche," and three aspects of this are distinguished as the "spatial or habitat niche," Hutchinson's "multidimensional or hypervolume niche," and for the Eltonian concept the "trophic niche." This division of the concept, however, leaves us with several difficulties. Elton, as an animal ecologist, indeed emphasized trophic relations in his discussions of the niche (cf. Weatherly 1963). However, the niche, as the species' place in the community, involves not only food, but also shelter and subtrate, vertical position, relation to patchiness or grain, diurnal and seasonal timing, control by predation, competition, allelochemic relations, etc. Also, the niche concept must apply to plants and saprobes as well as to animals, and the nutritional definition has little utility for discussing plants (Wuenscher 1969). Moreover, the critical dimensions of the niche may be not food resources but other limiting factors; it is the latter which, when a resource shared by two species is superabundant, may operate to permit their coexistence (Levin 1970). "Trophic niche" is not at all the same as *niche* in the functional or Eltonian sense (*a*), by which we mean the fully characterized position of the species in relation to intensive or intracommunity variables and other species within the community.

CONSEQUENCES OF CONFUSION OF CONCEPTS

Clearly the multiple usage of the term "niche" is confusing, and some narrowing of application for the sake of clarity is needed. If one were to discard the functional concept (*a*) in favor of either the niche as habitat (*b*) or as involving both habitat and niche factors (*c*), then one would have to accept certain undesirable consequences:

1. With either concept (*b*) or (*c*) application of the niche concept to community organization is obscured. The attributes of niche most relevant to a given community are those of the species' role in that community, not the species' distributional response to environmental gradients (*b*), or these plus different functional niche relationships in other biotopes (*c*).

2. Either of these concepts alters statements of the principle of Gause (1934). His conception was that two species could not persist in the same niche—that is, in direct competition in the same community. The central point that many niche differences are adaptive responses to competition in the community is altered if two species with "different niches" may either have different habitats (*b*) and hence not meet in competition in the same community, or differ in habitat + niche (*c*) but do not persist in the same community because their functional niches in that community are the same.

3. When the habitat + niche concept (*c*) is used, different (but equally important) evolutionary relationships are being compacted in the same terms. Niche difference within the community involves genetic characteristics evolved in relation to other species in the community. Habitat difference involves evolutionary response to gradients of environmental factors external to (though often modified by) the community. A product of niche differentiation among species is within-habitat or α diversity; the corresponding product of habitat differentiation is between-habitat or β diversity. Breadth of resources used or other coordinates occupied within the community is niche breadth and expresses intrapopulation genetic characteristics. Range of habitat gradients occupied is habitat amplitude or width, and this amplitude may express interpopulation differentiation based on mechanisms (ecotypic and subspecific differentiation, apomictic clone selection, etc.) different from those responsible for niche breadth.

4. Beyond this, if the use is not restricted to that of the functional concept (*a*), it may be unclear to which concept or to what combination of niche and habitat factors the word "niche" applies. For example, Levins

(1968) suggests as measurements of "niche breadth" both distribution of species over environments, and relation to such intracommunity variables as food source and season. Having measured niche breadth by concept (c), Levins (1968) goes on to discuss niche relationships within a community and thus applies, without indicating the change of perspective and measurements appropriate, concept (a). McNaughton and Wolf (1970) also sought a correlation between dominance (which may be considered an expression of resource use and hence of niche breadth) and "niche width" (for which they use a measure of habitat width). Green (1971) applied multiple discriminant analysis to bivalve mollusc species, and the physical and chemical characteristics of samples from lakes in which they occur, to ordinate the species and the lakes. The species were shown to differ in the ranges of environmental characteristics over which they occur, with the exception of one pair of *Pisidium* species with closely similar distributions in one lake; at least partial trophic separation was suggested for this pair. The study, though termed an analysis of "niche," is an analysis of habitat; and in relation to this the point of the *Pisidium* pair may be that species can have closely similar habitats if they differ in niche. No criticism of such studies as research is intended in the observation that their uses of the term "niche" do not facilitate communication of the significance of their research.

In summary, use of "niche" in sense (b) or (c) both leaves nameless the concept of most significance for community theory and fails to distinguish between intracommunity and intercommunity relationships. The crucial question for usage is whether clarity is served when "niche" and "habitat" are overlapping terms, with habitat attributes either identical with, or a subset of, niche attributes. For us, clarity is better served by a usage in which "niche" and "habitat" are complementary terms for different sets of attributes—respectively, intra- and intercommunity. However, if the term "niche" is assigned to the functional sense, a term is also needed for another concept, that of the habitat + niche.

SUGGESTED USAGE

On use of these terms we suggest:

1. The term "niche" should apply exclusively to the intracommunity role of the species.

2. When the broader concept entailing both intercommunity and intracommunity variables, hence habitat + niche, is used, a different term is appropriate; and for this we suggest "ecotope" (Schmithüsen 1968, p. 128; Troll 1968). The term is currently variously used as equivalent to habitat, biotope, microlandscape, or biogeocenose, but is not needed as a synonym of one of these. We suggest that henceforth it represent the species' relation to the full range of environmental and biotic variables affecting it. It should be pointed out that this concept is still useful when the distinction between inter- and intracommunity variables is not feasible or not desired; the concepts of "niche" and "habitat" in contrast require this distinction.

3. The terms "habitat" and "biotope" have been used almost interchangeably (the former more in English and the latter more in other European languages) for environment in its physical and chemical aspects. Two concepts are being covered by these terms, for they apply to the environment of the community at a given place, and to the environment of a species including the range of situations in which it occurs. No serious confusion seems to result from use in English of "habitat" for the environments of both communities and species. When distinction is desired, however, "biotope" should apply to the community's environment; "habitat," to the species' environment (Udvardy 1959). In the next section we shall define habitat by environmental gradients that may be considered in some sense external to (even though modified by) communities, but it is common informal practice to specify a kind of community as a habitat (for example, the habitat of the red crossbill in North Carolina is the spruce-fir forest). Such a statement is not inconsistent with our treatment, for it also implies specification of a species' habitat in terms of a range of environments occupied by a community type. James (1971) ordinates bird species by their relations to characteristics of plant communities; the result is properly termed an ordination by habitat, since it relates the distribution of the bird populations to intercommunity gradients of vegetation structure that express differences in environment of the vegetation.

FORMULATION

a) Interrelation of Concepts

While we think meaning of niche, habitat, and ecotope should normally be clear in practice, a formal statement of their relations to one another (and Hutchinson's "fundamental niche") is not simple. Three kinds of variables are to be considered in the definitions: (1) intercommunity or habitat variables, that is, environmental variables with an extensive spatial component (e.g., elevation, slope exposure, soil moisture from valley bottom to open south slope, soil fertility as affected by parent materials, etc., and community gradients consequent on these); (2) intracommunity or niche variables, that is, intensive or local environmental variables (height above ground, relation to intracommunity pattern, seasonal time, diurnal time, prey size, ratio of animal to plant food, etc.); and (3) population response variables (density, coverage, frequency of utilization, reproductive success, fitness, etc.). Though distinctions among these are important, they are not discontinuous with one another. When intercommunity and intracommunity variables are difficult to separate, they may be grouped as "environmental variables."

A habitat gradient may be treated for discussion as if external to or independent of species' population responses to it, but primary habitat gradients are to varying degrees modified or determined by the species and communities that occur along them. Thus a gradient of decreasing elevation and soil moisture in California mountains may be occupied by a spectrum of intergrading communities (pine forest, pine-oak woodland, chaparral, desert). Presence of the communities modifies or produces additional environmental factor gradients (of vegetation height and cover, exposure to sunlight, microclimatic humidity, soil organic-matter content, etc.) to which some species populations respond. Species populations are then distributed along a gradient we may conceive as a complex gradient (Whittaker 1967), or an assemblage of environmental factor gradients that change together through space along this elevation gradient (but may be quite differently related to one another along other primary habitat gradients). Each species is distributed along the complex gradient on the basis of its own environmental requirements and population dynamics, and the population response of a given species may have a significant effect on the environments of its own members (as needle litter of and transpiration of soil water by the pines affect the root environments of those pines). The response of a population to the gradient may affect other species, and consequently the niche and habitat relationships to these, as well.

For example, the niche of the California thrasher as described by Grinnell (1917) includes its nesting in dense masses of foliage 2–6 feet above the ground. Along the California mountain gradient this requirement is met only by the shrubs in that part of the community spectrum we term "chaparral." The habitat of the thrasher may consequently be determined by its population response to niche requirements, as these, in turn are affected by population responses of the shrub species that dominate the chaparral to the habitat gradient. Thus habitat, niche, and population variables are variously interlinked in the thrasher's total relationship to environment, which we term its "ecotope." This interlinkage of variables makes no less significant the distinction between the California thrasher's niche within the chaparral (in which it differs from other bird species occurring there), and its habitat in the middle elevations occupied by chaparral (in which it differs from the Crissal thrasher of the deserts at lower elevations). Further complications involve the migrations and movements in feeding of animal populations; the ecotope of a bird population may include population movement between two different habitats occupied (with different niche behaviors) at different times.

Troublesome as these interrelations may be for our definitions, the problems of definition are much like those encountered in other areas where terms are found to be interlinked in clusters, as observed by Dewey and Bentley (1949). We seek to order our cluster of terms, and the concepts they stand for, as parts of a conceptual system, a system including other concepts we shall relate to niche, habitat, and ecotope (see Whittaker 1967). The system may well be thought of as occupying a hyperspace defined by the three kinds of variables—of habitat, niche, and population measure—as axes. We shall deal with our conceptual axes two at a time—first habitat and population variables, then niche and population variables—before drawing them together in the concept of ecotope.

424

b) Habitat

Consider a landscape, locations or sites in which may be characterized by m environmental variables (extensive variables in the sense of Hutchinson [1967], such as, elevation, slope exposure, soil fertility, etc.). These variables may be conceived as axes of an m-dimensional coordinate system defining an environmental or *habitat hyperspace* (Goodall 1963; Whittaker 1967). Communities occur at locations in the landscape pattern, to which correspond points in the habitat hyperspace. (In this usage we understand by "community" an entity defined in the neighborhood of a point location in the landscape. Thus the landscape need not be viewed as dotted with a collection of discrete communities, but may often be conceived as a continuum of communities blending into one another.) The location of the community is its community habitat or *biotope*, characteristics of which are specified by a point or set of points in the hyperspace. (The combination of the variables bearing upon a particular organism or species within a biotope is referred to as an *environmental complex* [Billings 1952].) Each species in the landscape occurs over some range of the environmental variables, the limits of which for the species outline a habitat hypervolume as a fraction of the habitat hyperspace.

Habitat is usually conceived as the range of environments or communities over which a species occurs; the habitat hypervolume is an abstract formulation of this range in terms of extensive environmental variables and the species' limits in relation to them. The habitat hypervolume will not, however, be delimited simply as a region within which the species could persist indefinitely, as defined by physiological tolerances. Not only may interactions with other species exclude it from some environments in which it is physiologically able to exist, but the species may also occur in environments that are at times unfavorable to it. Organisms regularly confront conditions in their environments that are temporarily beyond their "limits of tolerance," however these limits are defined. If the environment did not change, a population of such organisms would become extinct unless sustained by immigration from more favorable regions. However, environments do change, and a population's survival in a changing environment may depend crucially on its responses to unfavorable periods, during which the best it can do is to "cut its losses," while making up for these during more favorable periods. We thus cannot include only favorable environmental conditions in our discussion of habitat. The hypervolume concept (Hutchinson 1958), when applied to habitat gradients, may not adequately define a species' habitat; and some measure of the species' distribution over its habitat is desirable.

Along a given habitat gradient the species shows response curves for various population measurements that may be applied to it. Population density curves are typically bell-shaped, apparently Gaussian in form (fig. 1), tapering on each side of the mode or population optimum toward ill-defined limits (the asymptotic tails of the curves). (The population

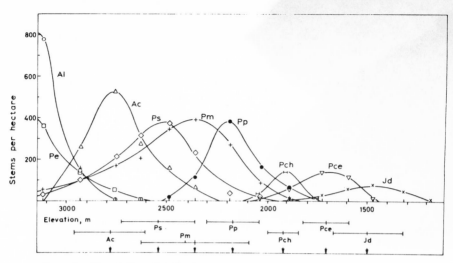

FIG. 1.—Population distributions on a habitat gradient. Coniferous tree species populations are plotted in relation to the elevation gradient on north-facing slopes in the Santa Catalina and Pinaleño Mountains, Arizona (data of Whittaker and Niering 1965). The tree species, competing as community dominants, form a replacement series along the gradient (comparable series formed by oak species along the same gradient are shown by Whittaker [1969]). Whereas the modes of species populations are apparently distributed randomly along the gradient when there are many competing species (Whittaker 1956), the distribution of modes of these competing dominants appears to be regular. Furthermore, in those cases in which the bars for habitat widths overlap, the species of an overlapping pair are differently distributed along the topographic moisture gradient. Species pairs of which the first occupies the more mesic and the second the more xeric topographic position at a given elevation are: Al and Pe, Ac and Ps, Pm and Ps, Pm and Pp, Pce and Jd. Data are numbers of tree stems over 1 cm dbh per hectare, based on sets of five 0.1-hectare samples grouped by elevation intervals and plotted at the mean elevations for the groups. The bars below the abscissa give means and habitat widths expressed as one standard deviation, except for Pm, in which the apparent mode has been used as its center rather than its mean (also indicated). Tree species are indicated by genus and species initials: *Picea engelmanni, Abies lasiocarpa, Abies concolor, Pinus strobiformis, Pseudotsuga menziesii, Pinus ponderosa, Pinus chihuahuana, Pinus cembroides,* and *Juniperus deppeana.* These species are dominants of a community gradient from subalpine forest (Pe and Al), through montane forests (Ac, Ps, Pm, and Pp) to submontane woodlands (Pch, Pce, and Jd).

optimum is not the same as the physiological optimum; moreover, some species have more than one local population optimum, for different eco-types.) Along any two axes of the hyperspace the population densities may form a Gaussian response surface (fig. 2; Whittaker 1956); in response to more than two axes of the hyperspace they form an "atmospheric" distribution (Bray and Curtis 1957). The species in its hypervolume forms not a sharply bounded distribution but a population cloud. Characterization through experiments of the species' habitat hypervolume is a principal goal

; Wuenscher 1969), but experiments in the
adequately describe the population cloud

...ttering of their population centers in the habitat
...ference in habitat (Whittaker 1956, 1967, 1969).
...ds of species normally, however, overlap broadly if their
...her; only exceptionally are species sharply bounded
...sion (Whittaker 1956, 1962; Terborgh 1970). In
many species populations consequently form
a continuum. Techniques of gradient analysis

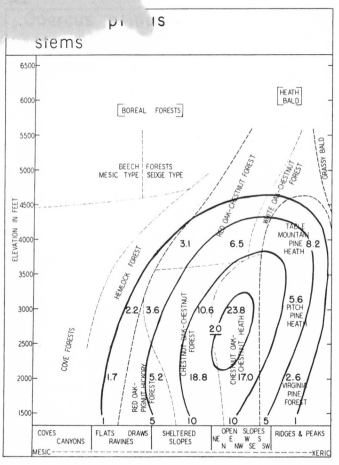

Fig. 2.—A Gaussian response surface or "binomial solid" for the habitat
of *Quercus prinus* (chestnut oak) in the Great Smoky Mountains, Tennessee
(Whittaker 1956). The population is centered in the chestnut oak heath, at the
middle elevations in moderately dry sites, and decreases in all directions away
from this, as indicated by the population contour lines. Data points are
percentages of stems over 1 cm dbh in composite samples of approximately 1,000
stems each.

seek understanding of this continuum and species habitat relations within it by: (1) reduction of the m-dimensional continuum to a coordinate system of a few major axes (either recognized complex gradients of many correlated environmental factors, e.g., elevation, or compositional axes derived from indirect ordination); major relationships of communities to biotopes in the landscape may then be understood as a pattern in this coordinate system; (2) representation of the distributions of the species populations in this coordinate system (fig. 2), so that relations of species to one another and communities may be observed; (3) ordination of the species by locating in the coordinate system the centers of their hypervolumes or population clouds (fig. 3 and Whittaker 1967; for applications to animal species see Pennak 1951; Whittaker 1952; Bond 1957; Whittaker and Fairbanks 1958; Beals 1960; Fager and McGowan 1963; Terborgh 1970; Green 1971; James 1971).

c) Niche

Consider a community in the landscape; the intracommunity variables to which species respond may be represented by n "niche variables," which

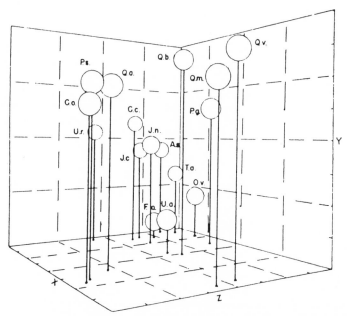

Fig. 3.—A habitat ordination of tree species by the centers of their distributions in a three-dimensional hyperspace (Bray and Curtis 1957). Tree species are indicated by genus and species initials: *Acer saccharum, Carya cordiformis, C. ovata, Fraxinus americana, Juglans cinerea, J. nigra, Ostrya virginiana, Populus grandidentata, Prunus serotina, Quercus alba, Q. borealis, Q. macrocarpa, Q. velutina, Tilia americana, Ulmus americana,* and *U. rubra.* Successional relations toward increasing mesophytism are significant along the X axis, soil drainage conditions along the Y axis, and disturbance effects along the Z axis.

will in general include axes representing other member species of the community as well as more general niche variables such as height above ground, prey size, etc. These variables as axes define a multidimensional niche hyperspace interrelating the species of the community. Each species in the community utilizes, or occurs in, or is affected by, some range of these axes, the limits of which outline its niche hypervolume, or realized niche in the sense of Hutchinson (1958, 1967).

The limits of this hypervolume do not adequately characterize the species' niche relationships as relative success or concentration of the species population (as expressed in density, fitness, frequency of resource utilization, or other measures) changes along the axes. To complete the definition of the niche, one superimposes on the hypervolume a measure of the population response of the species at each point. These response variables emphasize the functional relationships of the species in the community, and thereby link Hutchinson's realized niche and the functional niche concept of Elton. The *niche* then becomes the species' position in the hyperspace as represented by a response surface or cloudlike population measure within its niche hypervolume. MacArthur (1958; MacArthur, MacArthur, and Preer 1962) has done most to characterize the niche clouds of animal species populations in a community. Figure 4 represents in this form two dimensions of the niche of the blue gray gnatcatcher as studied by Root (1967).

It is no contradiction of the principle of Gause that species clouds may overlap broadly in the niche hyperspace. Species evolve toward scattering of their centers in the hyperspace—toward niche difference. Species may be

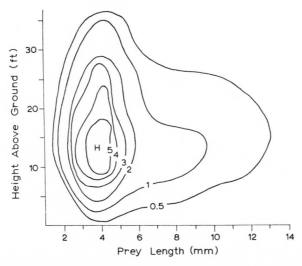

FIG. 4.—A niche response surface, representing capture of prey of different sizes taken at different heights above the ground by the blue gray gnatcatcher (*Polioptila caerulea*). The contour lines map the feeding frequencies (in terms of percentage of total diet) to these two niche axes for adult gnatcatchers during the incubation period in July and August, in oak woodlands in California (data of Root [1967]).

added to the community; along a given niche or resource gradient the additions may be seen as increased species packing (MacArthur 1970), implying increase in α diversity. It is because species that occur in the same community differ in niche that discontinuities at which species exclude one another are few (and some of these few are recent contacts) among the many broadly overlapping distributions of species that are partial competitors (Whittaker 1965). When major niche relationships in a community can be reduced to a few niche axes, species can be ordinated in a manner analogous to habitat ordination (MacArthur, MacArthur, and Preer 1962; Cody 1968).

A niche is an evolved, multidimensional attribute of a particular species population (see Colwell and Futuyma 1971). The range of an external niche axis for the species (e.g., prey size, or light intensity), and the species' response in that range (pursuit and consumption, plant height and photosynthetic adaptation) are coupled or complementary (Wuenscher 1969). Since a niche is a set of relationships for a particular species population in a given community, there is no need to postulate the existence of empty niches. Much debate on the nature of the niche, its measurement, formulation, and demonstration independent of the species can be resolved by recognizing the nature of the concept. It is a construct, one of a class of concepts not subject to direct observation but postulated to explain a range of observations. A niche is as much an attribute of a species as a personality (as expressed in interaction with others) is of a person, but this fact does not deprive of their interest efforts at quantitative formulation, grouping of species' niches into guilds (Root 1967; Price 1971), and comparison of niche relationships in different communities (Cody 1968).

A niche hyperspace is an evolved, multidimensional attribute of a particular community. The niche hyperspace also is a construct, postulated as a basis for interpreting community organization, species packing, and α diversity. The "structure" of the hyperspace is inseparable from the community of which this structure is an abstraction. There is no vacant niche awaiting the arrival of spider monkeys on the pampas; the pampas have not evolved a niche hyperspace including a potential hypervolume for the monkeys. There was no niche awaiting deer in New Zealand. The deer, when added to a community, laid claim to resources that had been otherwise used (by saprobe species, if not by animals) and altered the niche hyperspace of the community; the altered hyperspace includes a hypervolume occupied by the deer.

d) Ecotope

Consider, finally, the landscape of communities. A species in the landscape has its "place" in relation to $m + n'$ variables of habitat and niche. (The n' is used to indicate that the n niche axes as formulated above for a particular community have been extended to apply to the full range of communities in the landscape. The transformed niche hypervolume thereby defined we shall not distinguish further from the hypervolume discussed in the preceding section.) Niche and habitat variables may be combined to form an $(m + n')$-dimensional compound hyperspace, representing the full range of external circumstances to which species in the landscape are adapted (Whittaker 1969). In accordance with suggested usage (2), above, the compound hyperspace may be termed an "ecotope hyperspace." Each species in the landscape is adapted to some range of the environmental factors that are axes of the hyperspace. The limits of that range for the species outline its ecotope hypervolume. When a population measure is superimposed on the $m + n'$ hypervolume, the resulting cloud describes the species' relation to both habitat and niche; this we call the ecotope. The ecotope hypervolume provides a more formal statement of the habitat + niche concept.

It is critical to note that the ecotope as herein developed describes the species' response to the full range of environmental variables to which it is exposed. As such, the ecotope definition can be made independent of the notions of niche and habitat, and is unaffected if the notions of intercommunity and intracommunity variables are indistinct. The ecotope indeed is the ultimate evolutionary context of a species, even if the niche is the proximate one. Species' distributions over ranges of habitats, and migrations between communities as evolutionary responses, are to be understood in terms of the ecotope. The niche may moreover be regarded as the restriction of the ecotope to a particular community, however that community is defined.

Species of a landscape (or area or island) evolve toward different positions in the ecotope hyperspace; they may evolve simultaneously toward niche and habitat difference from one another (Whittaker 1969). From different aspects of this evolution result niche packing, α diversity, and dominance-diversity structure within the community; and habitat packing, β diversity, and community distribution along habitat gradients. For the landscape as a whole these processes in evolutionary time are expressed in biotic richness, area, or gamma diversity.

SUMMATION

The confusion affecting use of "niche" and related terms can be resolved as follows:

1. The m variables of physical and chemical environment that form spatial gradients in a landscape or area define as axes a habitat hyperspace. The part of this hyperspace a given species occupies is its habitat hypervolume. The species' population response to habitat variables within this hypervolume, as expressed in a population measure, describes its *habitat*. The environment of a particular community in the landscape is a community habitat or biotope.

2. The n variables by which species in a given community are adaptively related define as axes a niche hyperspace. The part of this hyperspace in which a species exists is its niche hypervolume, or realized niche in the sense of Hutchinson. The species' population response within its niche hypervolume describes its *niche*.

3. The variables of habitats and niches may be combined to define as axes an $(m + n')$-dimensional ecotope hyperspace. The part of this hyperspace to which a given species is adapted is its ecotope hypervolume. When a population measure is superimposed on this hypervolume, the *ecotope* of the species is described.

4. Niche and niche hyperspace are complementary constructs important for the interpretation of community organization. Clarity in the use of these and other terms is served by applying "niche" to the role of the species within the community, "habitat" to its distributional response to inter-community environmental factors, and "ecotope" to its full range of adaptations to external factors of both niche and habitat.

Figure 5 is an abstract representation of the relations of these concepts in a conceptual hyperspace with habitat, niche, and population variables as axes. Habitat variables are described along an "axis" perpendicular to the page. In general there will, of course, be more than one habitat variable, but for idealized representation the m-dimensional *habitat hyperspace* is reduced to a single axis. Similarly, niche variables are described along an axis parallel to this line of print, though in truth this axis stands for an n'-dimensional *niche hyperspace* (for each set of values of the habitat variables). One can then visualize these n'-dimensional niche hyperspaces, built or fibered upon the m-dimensional habitat hyperspace as a base space, to form an $(m + n')$-dimensional *ecotope hyperspace* (reduced for illustration to a plane defined by habitat and niche variables). Finally, the population measures superimposed on the $(m + n')$-dimensional hyperspace unite to form the $(m + n' + k)$-dimensional surface pictured, the species *ecotope*. (We have for this representation reduced to one the various possible population measures.)

The vertical projection of this ecotope onto the ecotope hyperspace "floor" of the figure defines the *ecotope hypervolume*. Projection of the ecotope hypervolume onto habitat hyperspace defines the *habitat hypervolume;*

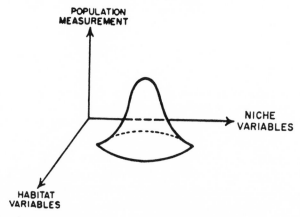

FIG. 5.—A diagrammatic representation of the relations to one another of a system of concepts including niche, habitat, and ecotope. See text.

projection of the ecotope hypervolume onto the niche hyperspace defines the *niche hypervolume*. Horizontal projection of the ecotope onto the "wall" of the habitat and population variables describes the species' range of habitat response. A cross section through the ecotope with an $(n' + k)$-dimensional hyperplane parallel to the page (i.e., for fixed values of the habitat variables) describes the niche of the species (for those habitat variables).

ACKNOWLEDGMENTS

For comments on the manuscript we thank P. F. Brussard, W. L. Brown, Jr., L. C. Cole, H. G. Gauch, Jr., D. C. Lewin, P. L. Marks, R. K. Peet, D. Pimentel, O. Sexton, and, especially, G. E. Hutchinson and the late R. H. MacArthur. The contributions of the first two authors were supported by National Science Foundation grants and that of the third author by Hatch funds.

LITERATURE CITED

Allee, W. C., O. Park, A. E. Emerson, T. Park, and K. P. Schmidt. 1949. Principles of animal ecology. Saunders, Philadelphia. 837 p.

Andrewartha, H. G. 1961. Introduction to the study of animal populations. Methuen, London. 281 p.

Andrewartha, H. G., and L. C. Birch. 1954. The distribution and abundance of animals. Univ. Chicago Press, Chicago. 782 p.

Beals, E. W. 1960. Forest bird communities in the Apostle Islands of Wisconsin. Wilson Bull. 72:156–181.

Billings, W. D. 1952. The environmental complex in relation to plant growth and distribution. Quart. Rev. Biol. 27:251–265.

Bodenheimer, F. S. 1958. Animal ecology today. Junk, Den Haag. 276 p.

Bond, R. R. 1957. Ecological distribution of breeding birds in the upland forests of southern Wisconsin. Ecol. Monogr. 27:351–384.

Bray, J. R., and J. T. Curtis. 1957. An ordination of the upland forest communities of southern Wisconsin. Ecol. Monogr. 27:325–349.

Clarke, G. L. 1954. Elements of ecology. Wiley, New York. 534 p.

Cody, M. L. 1968. On the methods of resource division in grassland bird communities. Amer. Natur. 102:107–147.

Colwell, R. K., and D. J. Futuyma. 1971. On the measurement of niche breadth and overlap. Ecology 52:567–576.

DeBach, P. 1966. The competitive displacement and coexistence principles. Annu. Rev. Entomol. 11:183–212.

Dewey, J., and A. F. Bentley. 1949. Knowing and the known. Beacon, Boston. 334 p.

Dice, L. R. 1952. Natural communities. Univ. Michigan, Ann Arbor. 547 p.

Ellenberg, H. 1950. Landwirtschaftliche Pflanzensoziologie. I. Unkrautgemeinschaften als Zeiger für Klima und Boden. Ulmer, Stuttgart. 141 p.

———. 1952. Landwirtschaftliche Pflanzensoziologie. II. Wiesen und Weiden und ihre standörtliche Bewertung. Ulmer, Stuttgart. 143 p.

Elton, C. 1927. Animal ecology. Sidgwick & Jackson, London. 209 p.

Fager, E. W., and J. A. McGowan. 1963. Zooplankton species groups in the North Pacific. Science 140:453–460.

Gause, G. F. 1934. The struggle for existence. Reprint ed., Hafner, New York, 1964. 163 p.

Goodall, D. W. 1963. The continuum and the individualistic association [French summary]. Vegetatio 11:297–316.

Green, R. H. 1971. A multivariate statistical approach to the Hutchinsonian niche: bivalve molluscs of central Canada. Ecology 52:225–229.

Grinnell, J. 1917. The niche-relationships of the California thrasher. Auk 34:427–433.

———. 1924. Geography and evolution. Ecology 5:225–229.

———. 1928. Presence and absence of animals. Univ. California Chron. 30:429–450.

Hesse, R., W. C. Allee, and K. P. Schmidt. 1937. Ecological animal geography. Wiley, New York. 597 p.

Hutchinson, G. E. 1958. Concluding remarks. Cold Spring Harbor Symp. Quant. Biol. 22:415–427.

———. 1965. The ecological theater and the evolutionary play. Yale Univ. Press, New Haven, Conn. 139 p.

———. 1967. A treatise on limnology. Vol. 2. Introduction to lake biology and the limnoplankton. Wiley, New York. 1,115 p.

James, F. C. 1971. Ordinations of habitat relationships among breeding birds. Wilson Bull. 83:215–236.

Kendeigh, S. C. 1961. Animal ecology. Prentice-Hall, Englewood Cliffs, N.J. 468 p.

Levin, S. A. 1970. Community equilibria and stability, and an extension of the competitive exclusion principle. Amer. Natur. 104:413–423.

Levins, R. 1968. Evolution in changing environments. Princeton Univ. Press, Princeton, N.J. 120 p.

MacArthur, R. H. 1957. On the relative abundance of bird species. Nat. Acad. Sci., Proc. 45:293–295.

———. 1958. Population ecology of some warblers of northeastern coniferous forests. Ecology 39:599–619.

———. 1970. Species packing and competitive equilibrium for many species. Theoret. Pop. Biol. 1:1–11.

MacArthur, R. H., J. W. MacArthur, and J. Preer. 1962. On bird species diversity. II. Prediction of bird census from habitat measurements. Amer. Natur. 96:167–174.

McNaughton, S. J., and L. L. Wolf. 1970. Dominance and the niche in ecological systems. Science 167:131–139.

Odum, E. P. 1953. Fundamentals of ecology. Saunders, Philadelphia. 384 p. 2d ed., 1959, 546 p. 3d ed., 1971, 574 p.

Pennak, R. W. 1951. Comparative ecology of the interstitial fauna of fresh-water and marine beaches. Année Biol., ser. 3, 27:449–480.

Price, P. W. 1971. Niche breadth and dominance of parasitic insects sharing the same host species. Ecology 52:587–596.

Ramensky, L. G. 1924. Die Grundgesetsmässigkeiten im Aufbau der Vegetationsdecke [in Russian]. Vêstnik Opytnogo Dêla, Voronezh, p. 37–73. (Abstr. in Bot. Centralblatt, n.s. 7:453–455, 1926.)

———. 1930. Zur Methodik der vergleichenden Bearbeitung und Ordnung von Pflanzenlisten und anderen Objecken, die durch mehrere, verschiedenartig wirkende Faktoren bestimmt werden. Beitr. Biol. Pflanzen 18:269–304.

Root, R. B. 1967. The niche expoitation pattern of the blue-gray gnatcatcher. Ecol. Monogr. 37:317–350.

Savage, J. M. 1958. The concept of ecologic niche, with reference to the theory of natural coexistence. Evolution 12:111–112.

Schmithüsen, J. 1968. Allgemeine Vegetationsgeographie. In E. Obst and J. Schmithüsen [ed.], Lehrbuch der allgemeinen Geographie, pt. 4. 3d ed. De Gruyter, Berlin. 463 p.

Terborgh, J. 1970. Distribution on environmental gradients: theory and a preliminary interpretation of distributional patterns in the avifauna of the Cordillera Vilcabamba, Peru. Ecology 52:22–40.

Troll, C. 1968. Landschaftsökologie [English summary]. In R. Tüxen [ed.], Pflanzensoziologie und Landschaftsökologie. Ber. Symp. Int. Vergl. Vegetationskunde, Stolzenau/Weser 1963, 7:1–21.

Udvardy, M. F. D. 1959. Notes on the ecological concepts of habitat, biotope and niche. Ecology 40:725–728.

Weatherley, A. H. 1963. Notions of niche and competition among animals with special reference to freshwater fish. Nature 197:14–17.

Whittaker, R. H. 1951. A criticism of the plant association and climatic climax concepts. Northwest Sci. 25:17–31.

———. 1952. A study of summer foliage insect communities in the Great Smoky Mountains. Ecol. Monogr. 22:1–44.

———. 1956. Vegetation of the Great Smoky Mountains. Ecol. Monogr. 26:1–80.

———. 1962. Classification of natural communities. Bot. Rev. 28:1–239.

———. 1965. Dominance and diversity in land plant communities. Science 147:250–260.

———. 1967. Gradient analysis of vegetation. Biol. Rev. 42:207–264.

———. 1969. Evolution of diversity in plant communities. Brookhaven Symp. Biol. 22:178–196.

———. 1970. Communities and ecosystems. Macmillan, New York. 162 p.

Whittaker, R. H., and C. W. Fairbanks. 1958. A study of plankton copepod communities in the Columbia Basin, southeastern Washington. Ecology 39:46–65.

Whittaker, R. H., and W. A. Niering. 1965. Vegetation of the Santa Catalina Mountains, Arizona. II. A gradient analysis of the south slope. Ecology 45:429–452.

Wuenscher, J. E. 1969. Niche specification and competition modeling. J. Theoret. Biol. 25:436–443.

435

AUTHOR CITATION INDEX

SUBJECT INDEX

BENCHMARK BOOKS PUBLISHING PROGRAM

The BENCHMARK BOOKS publishing program presently includes twenty Series of volumes of "Benchmark Papers" in the pure and applied sciences. Each Series will contain from twelve to forty or more volumes of classic and recent papers representing the landmark developments within the particular subject area of the series. The papers contained in each volume are selected by experts for contemporary impact, historical significance, and scientific elegance.

NICHE: Theory and Application

Edited by
ROBERT H. WHITTAKER and
SIMON A. LEVIN
Cornell University

Since Grinnell's 1917 classic article, "The Niche-Relationships of the California Thrasher," ecologists have come to recognize the complex concept of niche as a vital key to understanding natural communities and species evolution. Now two recognized authorities, Robert H. Whittaker and Simon A. Levin, have organized and introduced 46 landmark papers on the way species relate to other species in the same environment and community. They provide an effective introduction and a major reference work for scientists involved with this burgeoning field.

The editors' general introduction provides a summation of the state of the art as well as a guide to the subject's literature. Additional commentaries introduce and relate the primary writings in each of the book's seven units. The first section, on the origin of the concept, includes the classic writings of Grinnel, Elton, and Gause. Subsequent sections cover the competitive exclusion principle, niche axes, niche dimensions and their relation to habitat dimensions, and variations in space and time. Case histories are given in the sixth section, while the final section brings together three major theoretical statements by Hutchinson, MacArthur, and Whittaker, Levin and Root.

(continued on back flap)